Algorithmic geometry

Algorithmic Geometry

Jean-Daniel Boissonnat
Mariette Yvinec
INRIA Sophia-Antipolis, France

Translated by Hervé Brönnimann
INRIA Sophia-Antipolis, France

CAMBRIDGE
UNIVERSITY PRESS

PUBLISHED BY THE PRESS SYNDICATE OF THE UNIVERSITY OF CAMBRIDGE
The Pitt Building, Trumpington Street, Cambridge CB2 1RP, United Kingdom

CAMBRIDGE UNIVERSITY PRESS
The Edinburgh Building, Cambridge CB2 2RU, United Kingdom
40 West 20th Street, New York, NY 10011-4211, USA
10 Stamford Road, Oakleigh, Melbourne 3166, Australia

First published in French by Edisciences, 1995 as Géométrie Algorithmique
ⓒ 1995 Edisciences International, Paris

English edition ⓒ Cambridge University Press 1998

First published in English 1998

A catalogue record for this book is available from the British Library

ISBN 0 521 56322 4 hardback
ISBN 0 521 56529 4 paperback

Transferred to digital printing 2005

To Bertrand,
    Martine,
    Cécile,
    Clément,
    Alexis,
    Marion,
    Quentin,
    Romain,
    Eve,
    and the others. . .

Table of contents

Preface

A new field

Many disciplines require a knowledge of how to efficiently deal with and build
geometric objects. Among many examples, one could quote robotics, computer
vision, computer graphics, medical imaging, virtual reality, or computer aided
design. The first geometric results with a constructive flavor date back to Euclid
and remarkable developments occurred during the nineteenth century. However,
only very recently did the design and analysis of geometric algorithms find a
systematic treatment: this is the topic of computational geometry which as a field
truly emerged in the mid 1970s. Since then, the field has undergone considerable
growth, and is now a full-fledged scientific discipline, of which this text presents
the foundations.

Contents and layout of this book

The design of efficient geometric algorithms and their analysis are largely based on
geometric structures, algorithmic data structuring techniques, and combinatorial
results.

A major contribution of computational geometry is to exemplify the central
role played by a small number of *fundamental geometric structures* and their
relation to many geometric problems.

Geometric data structures and their systematic analysis guided the layout of
this text. We have dedicated a part to each of the fundamental geometric structures: convex hulls, triangulations, arrangements, and Voronoi diagrams.

In order to control the complexity of an algorithm, one must know the complexity of the objects that it generates. For example, it is essential to have a
sharp bound on the number of facets of a polytope as a function of the number
of its vertices: this is the celebrated upper-bound theorem proved by McMullen
in 1970. *Combinatorial geometry* plays an essential role in this book and the
first chapters of each part lay the mathematical grounds and prove the basic
combinatorial properties satisfied by the corresponding geometric structures.

At the same time as geometric data structures of general interest were being studied, new *algorithmic techniques* were devised. To general algorithmic paradigms, computational geometry added its own geometric techniques. The first purely geometric paradigm in the history of the field, the sweep method, was originally used by Bentley and Ottmann in an algorithm that computes the intersection of a set of line segments in the plane. Subsequent developments of general techniques soon encountered important theoretical difficulties which led to quite sophisticated variants and theoretical constructions without truly affecting the practice of the field. As a reaction against this tendency, a few authors decided it was more desirable to look for simple algorithms which were efficient on the average, rather than algorithms whose good behavior in the worst case did not guarantee good behavior in practical instances of the problem.

The recent body of work on *randomization* gave the most significant answer in this direction. An algorithm is said to be randomized if, after making random choices during its execution, it gives the solution to a purely deterministic problem. No probabilistic assumptions are made about the input objects, and randomness is used only to choose the path that the algorithm will follow to the solution. Randomized algorithms are often simple to conceive and to program, and their average complexity (over all the random choices made during the execution) is usually very good, often even optimal. Randomization leads to general methods for the design and analysis of algorithms, and allows efficient computation of geometric structures, both in theory and practice. For these reasons, randomization holds a central position in this book. The first three chapters in the first part contain all the generic material related to randomization, and instances of randomized algorithms are presented throughout the subsequent chapters.

Goals and limits of this book

The goal of this book is twofold. In the first place, it aims at giving a coherent exposition of the field rather than a collection of results, and at presenting only methods that possess a certain degree of generality. The algorithms presented in this book have been selected to work in all dimensions whenever this was possible: the case of dimension 2 only receives special treatment when particular methods lead to significant improvement, which happens surprisingly seldom.

In the second place, this book aims at presenting solutions which, while theoretically efficient and relevant, remain relatively simple and applicable in practical situations. Most of these algorithms have been implemented by their authors and their practical behavior has turned out to agree with the analyses developed in this book.

Nevertheless, this book does not claim to be a comprehensive treatment of the whole field of computational geometry. In particular, the reader will find mention of geometric data structures for queries and multidimensional searching only in the exercises. Also, this book is mostly concerned with Euclidean geometry, so the reader should not expect a full treatment of problems dealing with curves and algebraic surfaces. However, possible extensions to curves and surfaces are systematically pointed out, and some exercises as well as the bibliographical notes indicate how to extend the results in these directions.

From algorithms to programs

Rather than focusing on details of the implementation, this book emphasizes the principles underlying the algorithms. There is thus an important step to be taken from the descriptions given here to the actual implementation.

Unfortunately, however, there is currently no satisfactory treatment of problems related to the implementation of geometric algorithms. This topic is gaining more momentum and other books will undoubtedly fill in this gap in the near future.

The first problem raised by the implementation of geometric algorithms has to do with the finite precision used by the computer. More often than not, and as is done in this book, algorithms are designed and analyzed in a model of computation where computers can deal with arbitrarily long real numbers, and all operations give an exact result. Of course, this is not the case in practice, and a naive implementation that uses a standard floating point representation of real numbers can lead to fatal errors during the execution.

The other major problem has to do with so-called degenerate cases (points on a single line, for instance). In this book, the objects are usually assumed to be in general position, which excludes degenerate cases. This allows a more synthetic treatment and focuses better on the underlying principles, rather than on the particular details allowing correct performance in degenerate situations. Clarity is thus preferred to a comprehensive treatment which would require much longer discussions.

How to read this book

This book assumes no particular knowledge from the reader and should be accessible to any enthusiastic geometer. Its contents have been taught in several graduate courses both in mathematics and in computer science. It is aimed both at mathematicians interested in a constructive approach to geometry, and at computer scientists in need of an accurate treatment of computational geometry. Students, researchers, and engineers in more practical fields will find here a useful methodology and practical algorithms.

There is more than one way to read this book. The authors have tried to respect the *unalienable rights of the reader*, as defined by Daniel Pennac.[1] In particular, one may read parts, jump ahead, or search here and there for the needed piece of information. Chapters 1 and 2 are mere reviews that the informed reader may skip. Chapters 4, 5, and 6 introduce the formal framework for randomized algorithms. They hold a central position in this book. Nevertheless, the reader who is mostly interested in the applications may skip them in a first reading and just get familiar with the results. Or this reader may read chapter 5 to get a further glimpse of the algorithms without reading the full text. The subsequent parts are dedicated to, respectively, convex hulls, triangulations, arrangements, and Voronoi diagrams. They are essentially independent. As a sampler for an introductory course, the teacher may choose to present chapters 8 or 9, 12, then the section in chapter 14 about dimension 2, then chapters 15 and 17, and chapter 19 if time permits.

Exercises are included to complement the text, rather than to solidify the understanding. They offer extensions as well as applications of the results. The solutions to the most difficult ones are sketched in the hints.

Bibliographical notes at the end of each chapter give pointers to the literature available on the topics of the chapters as well as on the exercises. These references are by no means comprehensive and only introductory references are given for related topics.

---

[1]In his essay *Comme un Roman*, the French writer Daniel Pennac describes the *unalienable rights of the reader* as:
1. The right not to read.
2. The right to jump ahead.
3. The right not to finish a book.
4. The right to read again.
5. The right to read anything.
6. The right to Bovarysm (textually transmissible disease).
7. The right to read here and there.
8. The right to thesaurize.
9. The right to read aloud.
10. The right not to say anything.

Translator's Preface

The original text was written in French. The translator's task was constrained by the fact that most of the French words used in the original text were originally coined by their authors in English publications, or have a commonly accepted translation into English. The problem was thus one of *reverse engineering*! Fortunately, there are now many textbooks in computational geometry which helped to resolve conflicts in terminology. Whenever possible, the translation conformed to the standard terminology or, for the more specialized vocabulary, to the terminology set up in the original papers.

For graphs, however, the use of the word *edge* overlapped with that of 1-faces for common geometric structures. Similarly, the word *vertex* is also used for polytopes in a different meaning than for graphs. The situation is somewhat complicated by the fact that sometimes graphs are introduced whose nodes are edges of a polygon. We have followed the French text in systematically using the words *node* and *arc* for the set underlying a graph and the symbolic links between the elements of this set. The terminology related to graphs is recalled in subsection 2.2.1.

We have departed from the French text for the word *saillant* (meaning salient) to follow the usage with *convex vertices/edges*, as opposed to *reflex*. Although a vertex or an edge is always convex in the original meaning of convexity, here it means (as most people would understand it) that the internal angle around the vertex or around the edge is smaller than π. Luckily, this definition is never used for higher-dimensional faces, and therefore should not create confusion.

Vertical decompositions as they are introduced in this book have also been called by various names, such as *trapezoidal maps*, *vertical partitions*, and *vertical visibility maps*. As with other authors, we have preferred the phrase *vertical decomposition* or even *decomposition* for short, in order to emphasize the relation with other geometric decomposition schemes, for example decompositions of arrangements, polygons, or polyhedra into simplices (also called triangulations). We should properly speak of the decomposition of the plane induced by a set of segments. The reader will forgive us for using the phrase *decomposition of (a set of) line segments*.

A translation is an excellent opportunity to enhance the text with added references, a broader index, more exercises, more explanatory figures, and sometimes more concise proofs. Examples of these are found everywhere in this book, especially in the exercises about data structures (exercise 2.6) or in the proof of the upper-bound theorem (see also exercises 7.8 and 7.10). The translator wishes to thank the authors for their guidance, their willingness to answer his questions, and for bringing him back to orthodoxy when his mood was getting whimsical.

Hervé Brönnimann.

Acknowledgments

This book benefited from the joint work of researchers of the PRISME project at INRIA and is inspired by much common work with Panagiotis Alevizos, André Cérézo, Olivier Devillers, Katrin Dobrindt, Franco Preparata, Micha Sharir, Boaz Tagansky, and Monique Teillaud. To proofread the manuscript, Jean Berstel and Franck Nielsen provided their help with an unfailing friendship. The translation has been carried out by Hervé Brönnimann who not only translated but also corrected the original manuscript in many places. Many thanks to all of them! A book about geometry could not exist without drawings, and a book about computational geometry could not exist without computer generated drawings. The JPdraw software provided the ruler and compass, and together with its designer Jean-Pierre Merlet it played an essential role in the conception of this book.

Part I

Algorithmic tools

The first part of this book introduces the most popular *tools* in computational geometry. These tools will be put to use throughout the rest of the book.

The first chapter gives a framework for the analysis of algorithms. The concept of complexity of an algorithm is reviewed. The underlying model of computation is made clear and unambiguous.

The second chapter reviews the fundamentals of data structures: lists, heaps, queues, dictionaries, and priority queues. These structures are mostly implemented as balanced trees. To serve as an example, red–black trees are fully described and their performances are evaluated.

The third chapter illustrates the main algorithmic techniques used to solve geometric problems: the incremental method, the divide-and-conquer method, the sweep method, and the decomposition method which subdivides a complex object into elementary geometric objects.

Finally, chapters 4, 5, and 6 introduce the randomization methods which have recently made a distinguished appearance on the stage of computational geometry. Only the incremental randomized method is introduced and used in this book, as opposed to the randomized divide-and-conquer method.

Chapter 1

Notions of complexity

Computational geometry aims at designing the most efficient algorithms to solve geometric problems. For this, one must clearly agree on the criteria to estimate or measure the efficiency of an algorithm or to compare two different algorithms. This chapter recalls a few basic notions related to the analysis of algorithms. These notions are fundamental to understanding the subsequent analyses given throughout this book. The first section recalls the definition of algorithmic complexity and the underlying model of computation used in the rest of this book. The second part introduces the notion of a lower bound for the complexity of an algorithm, and optimality.

1.1 The complexity of algorithms

1.1.1 The model of computation

From a practical standpoint, the performances of an algorithm can be evaluated by how much time and memory is required by a program that encodes this algorithm to run on a given machine. The running time and space both depend on the particular machine or on the programming language used, or even on the skills of the programmer. It is therefore impossible to consider them relevant measures of efficiency that could serve to compare different algorithms or implementations of the same algorithm. In order to compare, one is forced to define a standard model of a computer on which to evaluate the algorithms, called the *model of computation*. Thus, to define a model of computation is essentially to define the units of time and space. The unit of space specifies what types of variables a memory cell can hold; these are called the *elementary variables* (or *elementary types*). The model specifies what *elementary operations* can be realized in one time unit. The *running time complexity* is therefore the number of elementary operations that have to be performed in order to realize the operations as described in the algorithm. Likewise, the *spatial complexity* describes how many memory units are needed in order to store all the data required for the execution of the corresponding program.

The model of computation underlying all the algorithms given in this book is the so-called *real RAM model*. In this model, each memory unit can hold the representation of a real number, and accessing a memory location takes constant time, that is, time independent of the particular location to be accessed. The machine can work on real numbers of arbitrary precision for the same cost. The elementary operations are:

1. the comparison of two real numbers,

2. the four arithmetic operations,

3. all the usual mathematical functions, such as logarithm, exponential, trigonometric functions, etc.

4. the integer part computation.

The assumption that all numbers can be represented exactly allows us to ignore all the problems related to numerical accuracy, as they occur in the real world. In particular, the otherwise very relevant problems of robustness of these algorithms in relation to rounding and numerical inaccuracies are not mentioned in this book.

1.1.2   Notions of complexity

Worst-case and average-case complexity

Each instance of a problem (be it geometric or not) is specified by a set of data called the *input* to the problem. The *size* of the input is the number of memory units needed to represent this input. When all the input data are elementary, that is, can be represented in a bounded number of memory cells, this input size is simply proportional to the number of input data.

When an algorithm is run on a given set of data, one expects the number of elementary operations executed to depend primarily on the size of the input. However, the running time also depends on the input itself. The *worst-case complexity* of an algorithm, or complexity *in the worst case*, is a function $f(n)$ that gives an upper bound on the number of elementary operations run by the algorithm when the input size is n.

This worst-case complexity is a pessimistic estimator of the running time of an algorithm. For many algorithms, the upper bound on the number of operations is reached, or even approached, only for very peculiar inputs which occur marginally if at all. Sometimes the worst case can easily be avoided by an appropriate preprocessing. Therefore a better choice to evaluate the efficiency of an algorithm is often the *average-case complexity*, or complexity *on the average*. This is a function $g(n)$ that gives the average number of operations (in the statistical sense) if a probability measure is given on all inputs of size n. The average-case complexity is generally harder to estimate than its worst-case counterpart. Moreover, it also depends on the probability measure over the space of all inputs of size n. It is only useful when this probability measure accurately models the real distribution of the input to the algorithm.

In the same way, we can define the *worst-case* and the *average-case space complexities* of an algorithm.

In this book, the complexities will only be given for the worst case, and the word *complexity* will be used as a shorthand for the worst-case complexity (in time or space). Occasionally, we say that an algorithm *runs in time* $f(n)$ when its worst-case time complexity is $f(n)$. Likewise, we say that an algorithm *requires space* (or *storage*) $g(n)$ when its worst-case space complexity is $g(n)$.

Output-sensitive complexity

An algorithm that solves a given problem usually builds, for a given input, a result called the *output*, which embodies the solution to the problem. The *size* of the output equals the number of memory units needed to store this result. Obviously, the size of the output depends on the size of the input, but also on the input itself.

For a given problem, the *worst-case output size*, or *output size in the worst case*, is the function $s(n)$ that upper bounds the output size for all inputs of size n. The algorithm under consideration needs to at least write the output, therefore the size of the output in the worst case is an elementary lower bound on the running time complexity in the worst case.

For a given problem and a given input size, however, the output size can sometimes change a lot depending on the actual input given to the algorithm. For instance, consider the problem of computing all the intersecting pairs of a set of line segments in the plane. For a set of n segments, the input consists of $4n$ real numbers, two for each endpoint. There might be as few as no intersections, and as many as $\frac{n(n-1)}{2}$. In this case it is interesting to have at hand *adaptive* algorithms whose time complexity is a *function of the output size*. The number of elementary operations executed by such an algorithm depends on the size of the output for the instance of the problem, and not on the size of the output in the worst case. For instance, in the problem of reporting all pairs of intersecting line segments, the number of elementary operations carried out by the algorithm should be a function of the number of intersecting pairs, which is not true of the naive algorithm that tests all the pairs for intersection.

An adaptive algorithm can be analyzed in terms of both variables n and s, the respective sizes of the input and the output. The worst-case complexity of an adaptive algorithm is the function $f(n, s)$ that upper bounds the number of elementary operations needed for solving all the instances of the problem with input size n and output size s. Likewise, the average-case complexity of such an algorithm is the function $g(n, s)$ that upper bounds the number of operations carried out by the algorithm, averaged over all the instances of the problem with input size n and output size s.

The complexity of randomized algorithms

In this book, the reader will find many *randomized* algorithms, that is, algorithms whose execution is to some extent random. Such an algorithm will make random choices during its execution, and these choices will influence its subsequent behavior. In all cases, the algorithm will output the correct answer to the given problem, but the number of elementary operations needed for this will greatly depend on the random choices. The efficiency of a randomized algorithm is then evaluated as an average over all possible random choices. The analysis is then called a *randomized analysis*. However, such an analysis by no means involves any statistical hypothesis on the data itself. Rather, the complexity is averaged over all possible executions of the algorithm in the worst case for the input.

Preprocessing, queries, amortized analysis

It happens frequently that we have to answer many different questions of the same kind about a given set of data. For example, given a set of lines in the plane, the questions might ask for some kind of localization. Each query consists of a point in the plane, and the question asks for the enclosing cell in the subdivision of the plane induced by the lines. In cases such as this, it often pays off to compute a data structure during a preprocessing phase, which in turn will be queried repeatedly for all the different requests. The analysis therefore concerns both the complexity of the preprocessing phase and that of answering the requests. In some cases, the data structure is semi-dynamic, which means that it is possible to add more data on-line; it may also be fully dynamic, meaning that deletions as well as insertions are allowed. Each type of operation (insertion, deletion, query) has its own associated cost. Sometimes, the cost of a single operation is hard to evaluate, but one may estimate the compounded cost of a number of these operations. The complexity of such a sequence divided by the number of operations gives the *amortized complexity* of one operation. Such an analysis is then called an *amortized analysis*.

1.1.3 Asymptotic behavior, notation

The choice of an algorithm for solving a given problem is guided by the associated complexity, in time or in space, in the worst case or on the average. This choice is not crucial if the input set remains small. The complexity analysis really matters when the input size becomes big enough. As a consequence, we are mostly interested in the growth of the complexity as a function of the input size n, that is the asymptotic behavior of this function when the variable n approaches infinity. To analyze an algorithm is thus to determine or at least to upper bound the dominating term in the time or space complexity. Most of the time, the order of magnitude will suffice, and we will neglect the numerical constants. We will then speak of the order of magnitude of the asymptotic behavior of the complexity.

The usual functions $1, \log n, n, n \log n, n^2, n^3, \ldots, 2^n$, whose orders of magnitude form an increasing sequence, give a natural scale for comparing or evaluating the complexity of the algorithms.[1] Of course, this scale can be refined to arbitrary precision by factoring in other slow-growing functions. Such functions encountered later in this book are the iterated logarithm $\log^{(i)} n$, the very slow-growing $\log^* n$ function, or the inverse Ackermann $\alpha(n)$ function. These functions are defined as follows. The i-th iterated logarithm $\log^{(i)} n$ of a number n is the number $\log \log \cdots \log n$ obtained by composing the logarithmic function with itself i times and evaluating it on n. The function $\log^* n$ of n stands for the number of successive iterations of the logarithm function needed to yield a number smaller than or equal to 1, starting from a value n. In other words,

$$\log^* n = i \iff \begin{cases} \log^{(i-1)} n & > \quad 1 \\ \log^{(i)} n & \leq \quad 1 \end{cases}$$

The value of the $\log^*$ function remains smaller than 5 for all numbers from 1 up to $2^{65,535}$.

The *Ackermann function* is obtained by expanding the following recurrence:

$$\begin{aligned} A_1(n) &= 2n \\ A_k(n) &= A_{k-1}^{(n)}(1), \end{aligned}$$

where $A_k^{(n)}$ is the function obtained by composing the function A_k with itself n times. Henceforth, we will write $A(n)$ for $A_n(n)$. The Ackermann function is increasing, and its rate of growth is very fast. Here are the first values of this function: $A(1) = 2$, $A(2) = 4$, $A(3) = 16$, $A(4)$ is a tower of 65,536 powers of 2. The functional inverse of this function, defined by $\alpha(n) = \min\{p \geq 1 : A(p) \geq n\}$,

---

[1]The notation log stands for the logarithm in base 2, which in this book will be assumed as the base for all logarithm functions unless otherwise stated.

is thus an extremely slow-growing function. In fact, $\alpha(n)$ is at most 4 for all practical purposes.

In order to compare the growth of different functions, the following notation is extremely useful. Let f and g be two positive real-valued functions of the integer-valued variable n.

- By $f(n) = O(g(n))$, to be read "$f(n)$ is a big-oh of $g(n)$," we express the fact that there is an integer n_0 and a real-valued constant c such that

$$\forall n \geq n_0, \ f(n) \leq cg(n).$$

- By $f(n) = \Omega(g(n))$, to be read "$f(n)$ is a big-omega of $g(n)$," we express the fact that there is an integer n_0 and a real-valued constant c such that

$$\forall n \geq n_0, \ f(n) \geq cg(n).$$

- By $f(n) = \Theta(g(n))$, to be read "$f(n)$ is a big-theta of $g(n)$," we express the fact that there is an integer n_0 and two real-valued constants c_1 and c_2 such that
$$\forall n \geq n_0, \ c_1 g(n) \leq f(n) \leq c_2 g(n).$$

In particular, a function $f(n)$ is $O(1)$ if and only if it is bounded above by a constant. A function $f(n)$ is said to be *linear* if $f(n) = \Theta(n)$ and *quadratic* if $f(n) = \Theta(n^2)$.

Let A be an algorithm and $f(n)$ its complexity, for instance in the running time and in the worst case. We consider the complexity of the algorithm to be known if we can specify another function $g(n)$ such that $f(n) = \Theta(g(n))$. It will only be bounded from above if $f(n) = O(g(n))$ and from below if $f(n) = \Omega(g(n))$.

A last piece of notation is sometimes useful. We will write $f(n) = o(g(n))$, to be read "$f(n)$ is a little-oh of $g(n)$," if and only if

$$\lim_{n \to \infty} \frac{f(n)}{g(n)} = 0.$$

Note that this necessarily implies that $f(n) = O(g(n))$.

An algorithm is *a priori* all the more interesting if its complexity is of small order of magnitude. Indeed, all resources being equal, such an algorithm will work for a greater input size. Nevertheless, one must remain aware of the shortcomings of such a limited view. The asymptotic analysis predicts the existence of certain constants but does not give any information about the values of these constants. Consider, for example, two algorithms A and B solving the same problem. Suppose further that the complexity $f(n)$ of algorithm A is dominated by the complexity $g(n)$ of algorithm B, that is $f(n) = O(g(n))$. The latter asymptotic statement implies that, from a certain input size on, algorithm A will beat its competitor B in terms of running time. Nothing is said, however, about the threshold beyond which this is the case (the value of this threshold depends on the constants n_0 and c concealed by the big-oh notation). One must therefore refrain from choosing, for a particular practical situation, the algorithm whose asymptotic analysis yields a complexity with the smallest order of magnitude. The elegance and simplicity of an algorithm are both likely to lower the order of magnitude of the concealed constants, and should be taken into consideration if appropriate. For these reasons, this book usually presents several algorithms for solving the same problem.

1.2 Optimality, lower bounds

1.2.1 The complexity of a problem

Given a model of computation, the *complexity of a problem* is the minimum number of elementary operations needed by any algorithm, known or unknown, to solve that problem. To put it differently, it is a lower bound on the complexity of all possible algorithms solving that problem, its running time being evaluated in the worst case.

An algorithm A that solves a problem P is called *optimal* if its running time complexity has the same order of magnitude as the complexity of the problem, when the size of the input approaches infinity. By definition, any algorithm A that solves a problem P whose complexity is $g(n)$ has a complexity $f(n)$ such that $f(n) = \Omega(g(n))$. It is therefore optimal if moreover $f(n)$ satisfies $f(n) = O(g(n))$.

It is of the highest importance to determine the complexity of a problem for which one seeks a solution, since this complexity bounds the complexity of any algorithm that solves the problem. The size n of the input and the size $s(n)$ of the output are natural lower bounds on the complexity $g(n)$ of a problem P:

$$g(n) = \Omega(\max(n, s(n))).$$

By definition, the complexity $f(n)$ of any algorithm A that solves problem P is an upper bound on the complexity of P:

$$g(n) = O(f(n)).$$

In particular, if there is an algorithm A with complexity $f(n) = O(\max(n, s(n)))$ that solves problem P, then this algorithm is optimal and the complexity of the problem is $g(n) = \Theta(\max(n, s(n)))$. In other cases, the complexity of a problem is much more difficult to establish, and there is no general method for this purpose. The next two subsections give two short examples of methods that might be used to determine the complexity of a problem.

1.2.2   The example of sorting: decision trees

Sorting n numbers according to the natural (increasing) order is one of the rare problems whose complexity can be found by direct reasoning. Given a finite sequence of n numbers, $\mathcal{X} = (x_1, \ldots, x_n)$, all in some totally ordered set (for instance, $\mathbb{N}$ or $\mathbb{R}$), to sort them is to determine a permutation σ of $\{1, \ldots, n\}$ such that the sequence

$$\mathcal{Y} = (y_1, \ldots, y_n), \quad y_j = x_{\sigma(j)}$$

is totally ordered, that is,

$$y_1 \leq y_2 \leq \cdots \leq y_n.$$

If no particular property of the elements of X is known, the only operation at hand to sort $\mathcal{X}$ is the comparison of two elements. One speaks of sorting by comparison only. The following theorem shows that the complexity of the sorting problem is $\Omega(n \log n)$, in the comparison model of computation. In the same model, there are algorithms that solve this problem in the corresponding $O(n \log n)$ time: for instance, such an algorithm is given in chapter 3. In this model, the complexity of the problem is thus $\Theta(n \log n)$.

Theorem 1.2.1 *Sorting n numbers using only comparisons requires at least $\Omega(n \log n)$ comparisons.*

Proof. The proof of this theorem is based on the idea of a *decision tree*. One can always assume that the sequence under consideration does not contain the same element twice; thus all the numbers in the sequence are distinct. For lack of other information on the input data, the algorithm can only perform comparisons, then branch accordingly, depending on the result of this comparison. Branching is a binary process since there can be only two results to the comparison. The execution of such an algorithm can be represented by a binary tree, the *decision tree*. Each leaf represents a possible output from the algorithm; in our case, an output is one of the $n!$ possible permutations of the set $\{1, \ldots, n\}$. Each internal node represents some state of the algorithm, at which the algorithm will perform a comparison. Depending on the result of this comparison and on its current state, the algorithm will then branch to its right or left descendant in the tree, and subsequently perform the comparison stored at that node, or output the corresponding permutation if it reaches a leaf. All computations begin at the root of the tree, and each execution therefore corresponds to a path from the root to a leaf of the tree. The number of comparisons performed by the algorithm in the worst case is thus the height of the decision tree. A possible decision tree sorting three elements a, b, c is shown in figure 1.1. For our sorting problem, the

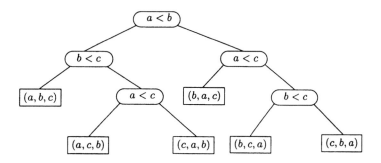

Figure 1.1. A decision tree sorting three elements a, b, c.

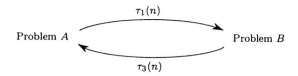

Figure 1.2. Transforming one problem into another.

decision tree has at least $n!$ leaves, and its height h is thus at least $\log(n!)$ which, according to Stirling's approximation formula,[2] is $\Omega(n \log n)$.   □

1.2.3 Lower bounds by transforming one problem into another

The *reduction* method is undoubtedly the method used the most frequently to determine the complexity of a problem. It consists of transforming the instance of the problem into an instance of another problem, the complexity of which is well known, or conversely of transforming the instance of another problem into an instance of the problem under consideration. To make the method more explicit, let A and B be two problems. We say A is transformed into B in time $\tau(n)$ if:

1. the input to problem A can be converted into an input suitable for problem B, using $\tau_1(n)$ elementary operations,

2. it is possible to convert the solution to problem B on the latter input into a solution to problem A on the former input, using $\tau_3(n)$ elementary operations, and

3. $\tau_1(n) + \tau_3(n) = \tau(n)$.

---

[2]Stirling's approximation formula states that $n! = \sqrt{2\pi n} \left(\frac{n}{e}\right)^n \left(1 + \frac{1}{12n} + o(\frac{1}{n^2})\right)$, where e stands for the base of natural logarithms.

Theorem 1.2.2 *If a problem A, whose complexity is $f(n)$, can be transformed in time $\tau(n)$ into a problem B whose complexity is $g(n)$, then*

$$f(n) = O(g(n) + \tau(n)),$$

$$g(n) = \Omega(f(n) - \tau(n)).$$

Proof. If the complexity of B is $g(n)$, then there is an algorithm that solves problem B in $g(n)$ elementary operations, and the transformation allows problem A to be solved using $g(n) + \tau(n)$ elementary operations. Conversely, if $f(n)$ is the complexity of problem A, there is no algorithm that solves B using less than $f(n) - \tau(n)$ operations.   □

Hence, the complexity of B gives an upper bound on the complexity of A, and that of A gives a lower bound for the complexity of B. We will show below that numerous geometric problems contain a sorting problem, for instance computing the convex hull or the Voronoi diagram of n points in the plane. The lower bound $\Omega(n \log n)$ holds for these problems in a suitable model of computation.

1.3   Bibliographical notes

The exposition of the concepts of complexity and optimality given in this chapter is purposely kept to a strict minimum. The reader seeking a more detailed discussion on all these notions is referred to one of the classical textbooks on the analysis of algorithms such as those by Aho, Hopcroft, and Ullman [6], Knuth [142], Sedgewick [200], Cormen, Leiserson, and Rivest [72], and Froidevaux, Gaudel and Soria [108] (in French).

Chapter 2

Basic data structures

Data structures are the keystone on which all algorithmic techniques rely. The definition of basic yet high-level data structures, with precise features and a well-studied implementation, allows the designer of an algorithm to concentrate on the core issues of the problem. For the programmer, it saves the tedious task of creating and administrating each pointer.

Throughout this book, we describe data structures especially designed for representing geometric objects and dealing with them. But computational geometers also make extensive use of data structures that represent subsets or sequences of objects. These structures can be used directly by the algorithms, or modified and augmented for geometric use. The first part of this chapter recalls the terminology and features of each basic data structure used in this book. It is useful to know how these structures can be implemented and what their performances are. The most delicate problem is undoubtedly the one addressed by dictionaries and priority queues, which treat finite subsets of a totally ordered set (the universe). To achieve better efficiency, these structures are usually encoded as balanced binary trees. For instance, the second part of this chapter describes red–black trees, a class of balanced trees that can be used to implement dictionaries and priority queues. Finally, when the universe is finite, dictionaries and priority queues can be even more efficiently implemented by other more sophisticated techniques, the characteristics of which are given without proof in the third part of this chapter.

The sole purpose of this chapter is to present, as far as data structures are concerned, the information necessary for a thorough understanding of the forthcoming algorithms. In particular, the authors by no means claim to present a comprehensive account of this topic, and the interested reader is urged to refer to the references given in the bibliographical notes.

2.1   Terminology and features of the basic data structures

2.1.1   Lists, heaps, and queues

Lists are the basic data structures used to represent a sequence of elements of a set.

Let $\mathcal{X} = \{X_1, X_2, \ldots, X_n\}$ be such a sequence. Any structure that wishes to represent this sequence should, at the very least, allow sequential access to these elements. The basic operation that achieves this is the *successor* operation which gives a pointer to the element X_{i+1} following the current element X_i. In some situations, both directions may be needed, and the data structure should also allow the *predecessor* operation which gives a pointer to the element X_{i-1} immediately preceding the current element X_i.

A list must also handle insertions of new elements and deletions of any of its elements. Therefore the list should allow an *insert* operation, for inserting a new element after a given position, and a *delete* operation, for deleting a given element. Insertion is often a must, if only for building the data structure, and deletion is often required as well.

Finally, we sometimes need two other operations on lists: *concatenation* which appends one list at the end of the other, and a converse *split* operation which breaks up a list into two parts at a given position.

There are implementation variants that efficiently realize these operations. Most of the time, a singly or doubly linked list suffices (see figure 2.1). Each element of the list is put into some memory location called a *record*, which includes two fields: one contains the value of that element, and the other a pointer to the record of the next element in the order of the list. When the list is doubly linked, a pointer to the record of the previous element in the order of the list is also supplied.

The data structure also contains a pointer to the first element of the list. Sometimes a pointer to the last element is also useful.

Such a data structure occupies a space that is proportional to the number of elements in the list. If this number is n, the space needed for the structure is $O(n)$. It allows the following operations to be performed in constant time: successor, predecessor (for a doubly linked list), insertion, and deletion. Such a list can therefore be built and enumerated in $O(n)$ time if n is the length of the sequence it represents. Also, pointers to the first and last elements allow concatenation and partition to be performed in constant time.

Stacks and *queues* are particular implementations of lists when the insertions and deletions only occur in special positions.

In the case of a *stack*, all insertions and deletions happen after the last element

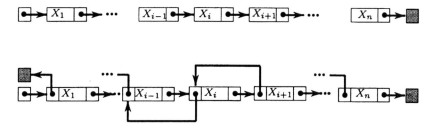

Figure 2.1. Singly and doubly linked lists.

in the list, which is called the top of the stack: to *stack* an element means to insert it as the last element of the list, and to *pop* consists of deleting the element that was stacked the most recently. Stacks are therefore particularly suited to process the elements of a set in the LIFO order, which stands for "last in, first out."

In the case of a *queue*, all insertions occur at the end of the list, whereas all deletions take place at the beginning of the list. Queues are therefore well suited to process elements in the FIFO order, which stands for "first in, first out." This is the normal order for a waiting line, or queue, hence the name given to this data structure.

Stacks or queues can always be implemented as general lists. There are more specific methods to implement these data structures but we will not expand on them in this book.

2.1.2 Dictionaries and priority queues

A data structure that represents a set S, subset of a universe U, must at least allow the following operations:

- *query*: given an element x of U, find out whether x belongs to S,

- *addition*: add an element x of U into S,

- *deletion*: delete an element x from the set S.

When the universe is totally ordered, the data structure is generally based on the order of the elements of S. Additions are then preferably called *insertions* and one may wish to perform *locations* (to be defined below) rather than queries. Elements of a totally ordered universe are called *keys* and the order on U is denoted by $\leq$. The words *smaller*, *greater*, *minimum*, and *maximum* refer to the total order on U. If S is a subset of the totally ordered universe U, then the data structure may be required to handle, in addition to the three previous operations, some of the following operations:

- *location:* given an element x of $\mathcal{U}$, find the smallest element y of $\mathcal{S}$ such that $x \leq y$,

- *minimum:* find the smallest element in $\mathcal{S}$,

- *maximum:* find the greatest element in $\mathcal{S}$,

- *predecessor:* find the element of $\mathcal{S}$ immediately preceding a given element x of $\mathcal{S}$,

- *successor:* find the element of $\mathcal{S}$ immediately following a given element x of $\mathcal{S}$.

We call a *dictionary* any data structure that can perform queries, insertions and deletions. If it supports searching for the minimum as well, we call it a *priority queue*. If it supports all the operations detailed above, we call it an *augmented dictionary*.

Priority queues and dictionaries can be implemented using lists or arrays. When the universe is totally ordered, it is often more efficient to use balanced data structures such as red–black trees, described below.

2.2   Balanced search trees

2.2.1   Graphs, trees, balanced trees

A *graph* is a pair $(\mathcal{X}, \mathcal{E})$, where $\mathcal{X}$ is a set of elements called *nodes* of the graph, and $\mathcal{E}$ is a set of pairs of nodes of $\mathcal{X}$, these pairs being called *arcs*. The graph is *directed* if the arcs are considered as ordered pairs. A *path* in the graph is an ordered sequence of nodes such that any two consecutive nodes are joined by an arc. The graph is *connected* if any two nodes can be joined by a path, and *acyclic* if no non-empty path can start and end at the same vertex without passing some other vertex of the graph at least twice. A *tree* is a directed, connected, and acyclic graph. As a consequence, one of the nodes stands out as having no arc coming into it; this node is commonly referred to as the *root*. Conversely, nodes having no arc coming out of them are called the *leaves*. Graphs, trees and their specific vocabulary are extensively described in the reference works cited in the bibliographical notes. In these references, nodes are sometimes also called *vertices* and arcs are commonly called *edges*. In this book, we stick to the words *nodes* and *arcs* for graphs, and restrict the use of the words *vertices* and *edges* to geometric objects. We invite the reader interested in further investigation to refer to these references if he or she should feel the need for it. Here, we content ourselves with reviewing how *balanced search trees* can be used to efficiently implement dictionaries and priority queues.

A *branch* of the tree is a path that stretches from the root to a leaf of the tree. A tree is considered to be *balanced* if all its branches have approximately the same length. This property, to be made precise below, ensures the efficiency of the data structure but complicates the insertion and deletion operations. Indeed, after each such operation, the structure must be rebalanced. There are several kinds of balanced trees, such as AVL trees, 2–3 or 2–3–4 trees, or even red–black trees. There are also many ways in which these variants can be used to implement dictionaries and priority queues. All the performances of these solutions are equivalent and optimal: if the set S stored in the data structure has n elements, the data structure occupies $O(n)$ space and any insertion, deletion, or query takes $O(\log n)$ time. For instance, the next section describes how to achieve these performances using red–black trees, and analyzes the corresponding cost of these operations.

2.2.2 Red–black trees

A *red–black tree* is a complete binary tree, that is, each node has either two children or none. The arcs, colored either red or black, satisfy the following constraints:

1. The paths from the root to all the leaves have the same number of black arcs.

2. All the leaves are related to their parent by a black arc.

3. There cannot be two consecutive red arcs along a path from the root to a leaf.

All the nodes have a *level*, which is the number of arcs on the path from the root to that node, and a *black level*, which is the number of black arcs on that path. The number of black arcs on a path from the root to a leaf is called the *black height of the tree*, since it does not depend on the particular leaf.

It is easy to see that a red–black tree is approximately balanced: the longest branch cannot have more than twice as many arcs as the shortest.

We propose to show how such a data structure can be used to implement a dictionary on a finite set S drawn from a totally ordered universe $\mathcal{U}$. The red–black tree is used as a searching data structure: to each node corresponds a key and two pointers towards its children. The keys attached to the leaves serve to represent the elements of S. The keys attached to the internal nodes serve as a guide for the searching operations. The key attached to an internal node must be greater than or equal to all the keys stored in its left subtree—the subtree rooted at its left child, and smaller than all the keys stored in its right subtree.

For instance, the key attached to an internal node can be systematically set to the greatest of the keys stored in its left subtree. A *left-hand depth-first traversal* of the tree visits all the nodes of the tree in the following order: the root first, then recursively the nodes in the left subtree, and finally the nodes in the right subtree. Such a traversal visits the leaves of the tree in the order of the elements of S.

Along with the key and the pointers to its children, the information stored at a node contains a special field to mark the color, either red or black, of the arc linking this node to its parent. To simplify the exposition, the color of an arc is often transferred to the node as well, and so we call a node black if it is linked to its parent by a black arc, and red if it is linked to its parent by a red arc. By convention, the root of the tree is always colored black. From now on, we denote by the same letter $N, O, P, Q, R, S, \ldots$ both the node and the key stored at that node.

Storage

Let S be a set of n elements, subset of the totally ordered universe. A red–black tree representing S has n leaves, and therefore has $n - 1$ internal nodes and $2(n - 1)$ arcs. The space required to store such a structure is thus $O(n)$.

Queries

To find out whether or not an element S of the universe $\mathcal{U}$ belongs to the set S, we need only follow a branch of the tree. At each internal node N, the next node in the branch is identified using a comparison between N and the key S that we are searching for. If $S \le N$, the search goes through the left child of N; if $S > N$ the search goes instead through the right child of N. The search always ends up at a leaf S' of the tree: the answer is that S is present if $S' = S$, and that S is missing from S if $S' \ne S$. In the latter case, S' is the element of S that immediately precedes or follows S according to the order on $\mathcal{U}$. The following theorem shows that, if there are n elements in S, such a search visits only $\Theta(\log n)$ nodes of the tree, and therefore runs in $\Theta(\log n)$ time.

Lemma 2.2.1 *If a red–black tree has n leaves, any path from the root to a leaf has at least $\frac{1}{2} \log n$ and at most $2 \log n$ arcs.*

Proof. The easiest proof of this result is to refer to a different kind of tree, the 2–3–4 tree. A *2–3–4 tree* is a tree whose nodes have either 2, 3, or 4 descendants, and all the paths from the root to the leaves have the same length, which is the *height* of the 2–3–4 tree. From a red–black tree, it is easy to make a 2–3–4 tree by merging all the nodes that are linked through red arcs (see figure 2.2). The height

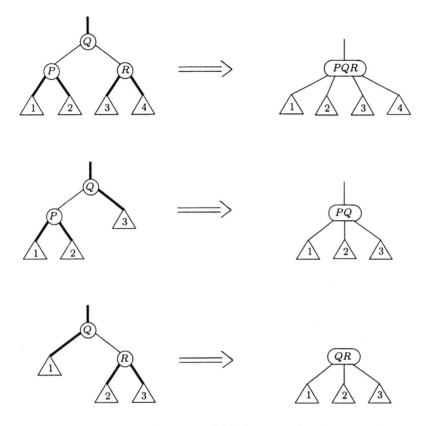

Figure 2.2. The correspondence between red–black trees and 2–3–4 trees.
In this and the subsequent pictures, the black arcs of red–black trees are represented in bold, the circles stand for internal nodes, rectangles for the leaves, and triangles stand for arbitrary subtrees.

h of this 2–3–4 tree is exactly the same as the black height of the corresponding red–black tree.

The red–black tree and its associated 2–3–4 tree have the same number of leaves, n, and the height h of the 2–3–4 tree satisfies

$$2^h \leq n \leq 4^h.$$

From this, it follows that the number h of black arcs on any branch is at least $\frac{1}{2} \log n$ and at most $\log n$. The total number of arcs on such a branch cannot be less than h, nor can it be more than $2h$. $\qquad\square$

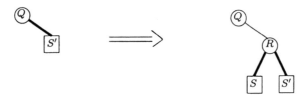

Figure 2.3. Red–black trees: insertion.

The picture assumes that $S < S'$. Just invert the two leaves S and S' if the converse is true.

Insertions

The insertion of an element S into the set $\mathcal{S}$, represented by a red–black tree, can be carried out in three stages.

First stage. We first find the location at which the new element must be inserted into the structure. For this, we follow a branch of the tree, in the same fashion as in the query explained above. The nodes along this branch are stored on a stack as we go along, the node encountered last being on the top of the stack. This stack will be used in the subsequent stages for rebalancing and recoloring the tree in a permissible manner. By assumption, the query gives a negative answer and ends on a leaf, the key S' of which is different from the key S of the element to be inserted.

Second stage. This is the phase where the actual insertion is performed. We replace the leaf S by an internal node R, with two newly created black leaves as its children, whose keys are S and S' in the appropriate order. The node R is linked by a red arc to the node Q that is stored on top of the stack. (This was the last node to be visited, and was the parent of S prior to the insertion, see figure 2.3.) The key of R is set to the smaller of the two keys of S and S'. The node R is stacked above Q. This way, the first two structural constraints of red–black trees are preserved; however, the last one is not satisfied if Q itself is a red node.

Third stage. This stage is a rebalancing stage whose purpose is to enforce the three structural constraints on red–black trees. During this stage, the algorithm maintains the following invariant: the top of the stack stores a red node R, and the two elements stored immediately before R in the stack are the parent Q and the grandparent P of R; Q and R are the only nodes that don't comply with the third constraint, therefore P itself is black, as well as the other child of Q and the two children of R. The node P itself may have both red children, or only Q is red and the other is black. The current step goes as follows:

1. Should P have both a black and a red child, then the third rule can be enforced by one of the following transformations (as in figure 2.4): simple

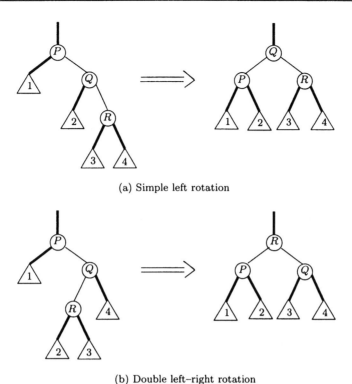

(a) Simple left rotation

(b) Double left–right rotation

Figure 2.4. Red–black trees: rotations.

left (resp. right) rotation if Q and R are both right (resp. left) children; double right–left (resp. left–right) rotation if Q is a right child and R a left child (resp. Q is a left child and R is a right child). Figure 2.4 shows only a simple left rotation and a double right–left rotation. We leave it to the reader to represent the symmetric rotations.

2. Should P have two red children, then the algorithm colors both children black and colors P red instead (see figure 2.5), unless P is the root of the tree in which case it is left black and nothing else is done. If the parent of P is black or at the root of the tree, then the third constraint has been restored, and the whole rebalancing task is over. If the parent of P is red, then the default in the third rule has been carried up two levels towards the root of the tree. Nodes R and Q are popped from the stack and the next step takes over with node P, its parent, and its grandparent.

The analysis of an insertion operation is almost immediate. The first stage requires $\Theta(\log n)$ operations if the set S to be searched has n elements. The actual insertion in the second stage can be carried out in constant time. As to

Figure 2.5. Red–black trees: changing the colors.

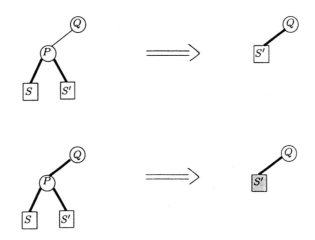

Figure 2.6. Red–black trees: deletions.

the third stage, only $O(\log n)$ nodes may need to be recolored, and only as many (simple or double) rotations may need to be performed. Red–black trees therefore allow a new element to be inserted in time $O(\log n)$.

Deletions

As with insertions, deletions can be performed in three stages.

First stage. A search for the key S to be deleted leads to the leaf that needs to be deleted, just as explained for queries or insertions. If x is not found to belong to S, then nothing else is done, otherwise the algorithm performs the second and third stages below.

Second stage. This is where actual deletion is performed: the leaf S that was located in the first stage, and its parent P, are removed from the tree; the sibling S' of S (the other child of P) is linked directly to the parent Q of P (see figure 2.6). If the arc linking P to Q is black, then the path leading from the root to S' lacks one black arc after the removal, and the first structural constraint no longer holds.

Third stage. We rebalance the tree obtained by the removal in the second stage. This operation is carried out in steps. At the current step, the tree contains one and only one *short* node: this is a node X such that the black height of the subtree rooted at X is one arc smaller than that of other subtrees rooted at the same black level in the tree. In the first step, the only short node is S'. Let X be the current short node, Q its parent, and R the other child of Q. Node X being the only short node, R cannot be a leaf of the tree.

1. Should R be black with two red children, then rebalancing can be obtained by performing the rotation depicted in figure 2.7, case 1.

2. Should R be black with both a black and a red child, rebalancing can be obtained by the double rotation depicted in figure 2.7, case 2.

3. Should R be black with two black children, two cases may arise. If node Q, the parent of X and R, is red, then the tree can be rebalanced by changing the colors as shown in figure 2.7, case 3a: Q is recolored in black and R in red. If Q is black, the tree cannot be rebalanced in a single step. Changing the colors as shown in figure 2.7, case 3b, makes the parent of Q become the short node, and the next step takes over with this node as the short node.

4. Finally, should R be a red node, the transformation explained in figure 2.7, case 4, will yield a tree whose short node has a black sibling and therefore can be taken care of by one of the transformations 1, 2, or 3a.

To summarize, the structural properties of a red–black tree can be restored with at most $O(\log n)$ transformations of type 3b, one transformation of type 4 followed by one of type 1, 2, or 3a, summing up as $O(\log n)$ elementary operations. The first stage requires only $O(\log n)$ operations and the removal can be performed with only a constant number of elementary operations. As a consequence, red–black trees can be used to represent a set S of n elements from a totally ordered universe $\mathcal{U}$, allowing us to perform deletions in time $O(\log n)$.

As described above, red–black trees can be used to implement a dictionary. To obtain a priority queue, it suffices to maintain a pointer towards the leaf storing the smallest (or the biggest) element of S. To have an augmented dictionary, one can add two pointers to each leaf, pointing to the previous and the next element of the set. Maintaining these additional pointers does not modify the complexity of the insertion and deletion operations, and allows the predecessor or the successor to be found in constant time.

The following theorem summarizes the performances of red–black trees.

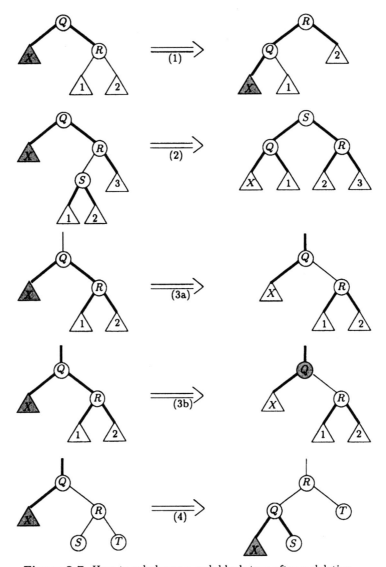

Figure 2.7. How to rebalance a red–black tree after a deletion.

Theorem 2.2.2 (Red–black trees) *Let S be a set of n elements drawn from a totally ordered universe $\mathcal{U}$. A red–black tree can be built in space $O(n)$ and time $O(n \log n)$ to represent this set. The tree allows each of the following operations to be performed in time $O(\log n)$: queries, insertions, deletions, minimum, maximum, and locations. Operations predecessor and successor take constant time.*

As an example of how to use such a data structure, we recall the problem of sorting n real numbers. The solution we give uses red–black trees to store the set of these n numbers. The tree can be built in $O(n \log n)$ time, uses $O(n)$ space, and the elements can be enumerated in order by performing a left-hand depth-first traversal of the tree. The only operation on the numbers used in this algorithm is comparison. Taking into account theorem 1.2.1, we have proved the following:

Theorem 2.2.3 (The complexity of sorting) *The problem of sorting n real numbers has complexity $\Theta(n \log n)$ in the comparison model.*

2.3   Dictionary on a finite universe

When the totally ordered universe is finite, it is possible to implement dictionaries and priority queues more efficiently by using more sophisticated structures. Henceforth, we will not use these techniques except in chapter 6, and only to show how to optimally implement dynamic randomized algorithms. Therefore, in this section, we will only recall the performances of these structures, without proof or further explanations. The reader may want to skip this section in a first reading.

In this entire section, we assume that the set S we want to maintain is a subset of a totally ordered universe $\mathcal{U}$, which is finite and of cardinality u. If u is not too big, the simplest way to implement a dictionary is to use an array of size u which allows queries, insertions, and deletions to be performed in constant time. To implement an augmented dictionary, one may have recourse to a data structure commonly referred to as a *stratified tree* (see also exercise 2.3) whose performances are given in the following theorem.

Theorem 2.3.1 (Stratified tree) *Let S be a subset of a totally ordered universe of finite size u. The set S may be stored in a stratified tree, which uses $O(u \log \log u)$ space and can be built in time $O(u \log \log u)$. Each of the insertion, deletion, location, minimum, predecessor, and successor operations can then be performed in time $O(\log \log u)$.*

Sometimes, the size of the underlying universe is just too big for this method to be practical. Hashing methods can then be used as a replacement.

Perfect dynamic hashing is a method that stores the dictionary over a finite, albeit huge, universe. In this method, random choices are made by the algorithm during the execution of the insertion and deletion operations. Such algorithms are called *randomized* below. The cost of these operations (insertions, deletions) depends on the random choices made by the algorithm and can only be evaluated on the average over all possible choices. Such an analysis is also said to be randomized. Moreover, it is impossible to bound the cost of a single operation.

	storage	insertion deletion location	minimum maximum	predecessor successor
red–black tree	n	$\log n$	$\log n$	1
stratified tree	$u \log \log u$	$\log \log u$	$\log \log u$	1
perfect dynamic hashing	n	1		
stratified tree with perfect dynamic hashing	n	$\log \log n$	1	1

Table 2.1. Implementations of dictionaries and priority queues.
The last three solutions assume that the universe is finite and of size u.

However, the cumulative cost of a sequence of m operations can be analyzed: in this case, the analysis is said to be *amortized*. The cost of a sequence of operations divided by the number of operations then stands as the *amortized cost* of an operation. The performances of perfect dynamic hashing are given in the following theorem (see also exercises 2.5 and 2.6).

Theorem 2.3.2 (Perfect dynamic hashing) *Let S be a subset with n elements of a totally ordered and finite universe with, say, u elements. Perfect dynamic hashing can be used to implement a dictionary for S using $O(n)$ space. Each query on the dictionary takes $O(1)$ time, and the amortized cost of an insertion or deletion is also $O(1)$.*

Finally, by combining both stratified trees and perfect dynamic hashing, one may build a data structure that performs well on all the operations of an augmented dictionary. Henceforth, this combination of data structures, a data structure in its own right, will be referred to as an *augmented dictionary on a finite universe*. The theorem below summarizes its characteristics.

Theorem 2.3.3 (Augmented dictionary on a finite universe) *Let S be a subset with n elements of a totally ordered and finite universe with u elements. An augmented dictionary for S may be built using $O(n)$ storage with the following performances: The minimum, predecessor, and successor operations can be performed in time $O(1)$, and location in time $O(\log \log n)$. Insertions and deletions run in amortized time $O(\log \log n)$ on the average.*

Table 2.1 summarizes further the performances of the different data structures discussed here that may be used to implement a dictionary or a priority queue.

2.4   Exercises

Exercise 2.1 (Segment trees) Segment trees were created to deal with a collection of intervals on the one-dimensional real line. Intervals may be created or deleted, provided that the endpoints belong to a set known in advance. The endpoints are sorted, and thought of as the integers $\{1, \ldots, n\}$ via a one-to-one correspondence that preserves the order. The associated segment tree is a balanced binary tree, each leaf of which represents an elementary interval of the form $[i, i+1]$. Each node of the tree therefore corresponds to an interval which is the union of all the elementary intervals associated with the leaves of the subtree rooted at that node. Intervals of this kind will be called *standard* intervals, and we will speak of a node instead of its associated standard interval.

The intervals of the collection are stored at the nodes of the tree. An interval I is stored in the structure at a few nodes of the tree: a node V stores I only if its associated standard interval is contained in I, but the standard interval of the parent of V is not.

1. Let l, resp. r, be the left, resp. right, endpoint of I. Let V_l be the standard elementary interval whose left endpoint is l, and let V_r be the standard elementary interval whose right endpoint is r. Let V_f be the smallest standard interval containing both V_l and V_r. The node V_f is the nearest common ancestor to both V_l and V_r, and it is called the *fork* of I. Show that the nodes which are marked as storing I are precisely the right children of the nodes on the path joining V_f to V_l in the tree, together with the left children of the nodes on the path joining V_f to V_r in the tree. Deduce from this that the nodes that store I correspond to a partition of I into $O(\log n)$ standard disjoint intervals, with at most two intervals at each level of the tree.

2. At each node, a secondary data structure accounts for the set of intervals stored by that node. According to the application, the data structure may list the intervals or simply maintain in a counter the number of these intervals. To add an interval to the segment tree simply consists of adding it to each of the secondary data structures of the nodes storing this interval, or incrementing the counter at these nodes. Deletions are handled similarly. Assume that only a counter is maintained. Show that an insertion or deletion can be performed in time $O(\log n)$. Show that the segment tree can be used to count the number of intervals containing a given real number x, in time $O(\log n)$.

Exercise 2.2 (Range trees) Given a set of n points S in $\mathbb{E}^d$, we wish to build a data structure to efficiently answer queries of the following kind: count the number of points inside an axis-oriented hyper-rectangle, or report them. One solution consists of building a range tree, a data structure particularly suited to this kind of query, which we describe now.

- The first level of the structure is a segment tree T_1 (see exercise 2.1) built on the first coordinates of the points in S, that is on the set $\{x_1(P) : P \in S\}$. For each node V of T_1, we denote by $S_d(V)$ the set of those points P of S whose first coordinate $x_1(P)$ belongs to the standard interval of V. The set $S_{d-1}(V)$ is the projection of $S_d(V)$ onto $\mathbb{E}^{d-1}$ parallel to the x_1-axis.

- If $d > 2$, every node V of T_1 has a pointer towards a range tree for the set of points $S_{d-1}(V)$ in $\mathbb{E}^{d-1}$.

1. We first assume that the queries ask for the number of points in S inside a given hyper-rectangle R_d (the counting problem). Let $q(S, R_d)$ be the time it takes to answer a query on the hyper-rectangle R_d. Let $\mathcal{V}_1$ stand for the collection of all the nodes storing the projection of R_d onto the x_1-axis, and R_{d-1} be the projection of R_d parallel to the

x_1-axis. Show that

$$q(\mathcal{S}, R_d) = O(\log n) + \sum_{V \in \mathcal{V}_1} q(\mathcal{S}_{d-1}(V), R_{d-1}).$$

From this, show that the maximum amount of time that a query can take on a set of n points in d dimensions is

$$q(n, d) = O(\log n)q(n, d - 1) = O\left((\log n)^d\right).$$

Show that a query in the reporting case can be answered in $O\left((\log n)^d + k\right)$ time if k is the number of points to be reported.

2. Show that the preprocessing space requirement and time are both $O\left(n(\log n)^d\right)$.

Exercise 2.3 (Stratified tree) Let $\mathcal{S}$ be a subset of a finite, totally ordered universe $\mathcal{U}$. Let u be the number of elements of $\mathcal{U}$, and without loss of generality assume that $u = 2^k$. For convenience, we identify the set of possible keys with $\{0, 1, \ldots, 2^k - 1\}$. A stratified tree $ST(\mathcal{U}, \mathcal{S})$ that implements an augmented dictionary on $\mathcal{S}$ is made up of:

- a doubly linked list which contains the elements of $\mathcal{U}$. Each record in this list has three pointers *sub*, *super*, and *rep*, and a boolean flag *marker* to identify the elements of $\mathcal{S}$.

- a representative R with two pointers to the maximal and minimal elements in $\mathcal{S}$, and a boolean *flag* to detect whether $\mathcal{S}$ is empty or not.

- $2^{\lceil k/2 \rceil}$ stratified trees $ST(\mathcal{U}_i, \mathcal{S}_i)$ for the sets $\mathcal{S}_i = \mathcal{S} \cap \mathcal{U}_i$ and the universes $\mathcal{U}_i = i\, 2^{\lfloor k/2 \rfloor} + \{0, 1, \ldots, 2^{\lfloor k/2 \rfloor} - 1\}$, with i ranging from 0 to $2^{\lceil k/2 \rceil - 1}$. Depending on the parity of k, each sub-universe $\mathcal{U}_i$ contains $\sqrt{u}$ or $\sqrt{2u}$ elements of $\mathcal{U}$.

- A stratified tree $ST(\mathcal{U}', \mathcal{R})$ for the set of representatives of $ST(\mathcal{U}_i, \mathcal{S}_i)$. The representative R_i of $ST(\mathcal{U}_i, \mathcal{S}_i)$ is the element whose key equals i in the set $\mathcal{U}' = \{0, 1, \ldots, 2^{\lceil k/2 \rceil - 1}\}$. Depending on the parity of k, the size of U' is $\sqrt{2u}$ or $\sqrt{u}$.

The trees $ST(\mathcal{U}_i, \mathcal{S}_i)$ and the tree $ST(\mathcal{U}', \mathcal{R})$ are called the sub-structures of $ST(\mathcal{U}, \mathcal{S})$. In turn, $ST(\mathcal{U}, \mathcal{S})$ is called a *super-structure* of those trees. Pointers *sub* and *super* keep a link between a record in the list and the corresponding record in the list of the sub-structure (resp. super-structure). The pointer *rep* points toward the representative of the structure.

1. Show that the stratified tree $ST(\mathcal{U}, \mathcal{S})$ can be stored in space $O(u \log \log u)$, and can be built for an empty set $\mathcal{S}$ in time $O(u \log \log u)$.

2. Show that each operation: insertion, deletion, location, minimum, predecessor, successor, can be performed in time $O(\log \log u)$.

Exercise 2.4 (Stratified trees and segment trees) Let $\mathcal{U}$ be a totally ordered, finite universe with $u = 2^k$ elements. Consider a complete and balanced binary tree $\mathcal{CBT}$ whose leaves are associated with the elements of $\mathcal{U}$. Consider further the set $\{0, 1 \ldots, k\}$ of levels of that tree, and build a segment tree $\mathcal{T}$ on this set. Show that you can do it in such a way so that each sub-structure of the stratified tree built on the universe $\mathcal{U}$ corresponds to a standard interval on $\mathcal{T}$.

Exercise 2.5 (Perfect dynamic hashing) Let $S = \{x_1, \ldots, x_n\}$ be a subset of a finite universe $\mathcal{U}$. Let u be the number of elements of $\mathcal{U}$, and without loss of generality assume that $u = 2^k$. A *hash function* h is any function from $\mathcal{U}$ to a set of keys $\mathcal{T}$ of size $t = 2^l$. We say that a class of hash functions is *universal* if, for any elements i, j in $\mathcal{U}$, the probability that $h(i) = h(j)$ for a random h in $\mathcal{H}$ equals $1/t$. The class is *almost universal* if the latter probability is $O(1/t)$.

1. Let $\mathcal{U}$ be the vector space $\mathbb{F}_2{}^k$ of dimension k over the field $\mathbb{F}_2$ with two elements, and $\mathcal{T}$ be the vector space $\mathbb{F}_2{}^l$. Let $\mathcal{H}$ be the class of all injective linear maps h from $\mathcal{U}$ to $\mathcal{T}$, that is, such that $h(a) = h(b)$ only if $a = b$. Show that $\mathcal{H}$ is a class of universal hash functions.

2. Let $\mathcal{H}$ be the class of all functions $h(x) = (kx \bmod p) \bmod t$, for a fixed prime number $p > u$ and all elements $k < p$. Show that $\mathcal{H}$ is an almost universal class of hash functions.

Exercise 2.6 (Perfect dynamic hashing) Let $S = \{x_1, \ldots, x_n\}$ be a subset of a finite universe $\mathcal{U}$. Hash functions and universal classes of hash functions are defined in the previous exercise. For convenience, we identify the universe with $\{0, 1, \ldots, u - 1\}$ and the set of keys with $\{0, 1, \ldots, t - 1\}$. Given a hash function h, we say that two elements a and b in $\mathcal{U}$ *collide at* j if $h(a) = h(b) = j$. Given a set S of n objects in $\mathcal{U}$, the problem is to find a hash function that gives as few collisions as possible over S.

1. Given a random hash function h in a class $\mathcal{H}$ of hash functions, let S_j be the set of elements in S that are mapped onto j, and let n_j be the size of S_j. Let N be the expected value, for a random element h of $\mathcal{H}$, of $\sum_{j=1}^{t} \binom{n_j}{2}$. Show that $N \leq \frac{n^2}{2t}$ if $\mathcal{H}$ is universal, and that $N = O(n^2/t)$ if $\mathcal{H}$ is almost universal.

From now on we assume that $\mathcal{H}$ is a universal class of hash functions.

2. Note that two elements collide at j if and only if $n_j > 1$. If $t = n^2$, show that the probability of having no collision over S for a random h in $\mathcal{H}$ is at least $1/2$. If $t = n$, show that $\sum_{j=1}^{t} n_j^2 \leq n$ with probability at least $1/2$ for a random function h in $\mathcal{H}$.

3. We now describe a two-level hashing scheme that has no collision with high probability. For any j in $\mathcal{T}$ such that $n_j > 1$, we set $t_j = n_j^2$ and pick a random hash function h_j in a universal class of hash function $\mathcal{H}_j$ onto a set $\mathcal{T}_j$ of t_j elements. The two-level hashing scheme first maps an element x to $j = h(x)$, and if $n_j > 1$, maps x onto $h_j(x)$ in $\mathcal{T}_j$. Assume that the sets $\mathcal{T}$ and $\mathcal{T}_j$ are all disjoint. Show that for a given S, the two-level hashing scheme has no collision, uses space $O(n)$, and has a query time $O(1)$ with probability at least $1/2$ over the choice of h and of the h_j's.

4. In order to make this scheme dynamic we use the standard doubling trick: instead of taking $t = n$, we take $t = 2n$ so that the first table can accommodate twice as many elements; similarly, for each j such that $n_j > 1$, we take $t_j = 4n_j^2$. When a collision occurs during the insertion of an element into a subtable, this subtable is rehashed. Also, when the number of elements effectively present in a table or a subtable exceeds twice (or falls below half) the number of elements present in the table during its last rehash, the size of this table is doubled (or halved) and this table is rehashed. Show that the expected number of times the table or subtables are rehashed is $O(1)$ if at most n operations are performed in a table that stores n elements. Conclude that this dynamic two-level hashing scheme uses space $O(n)$, has a query time $O(1)$, and allows insertions and deletions in time $O(1)$.

Exercise 2.7 (Persistent dictionary) A *persistent dictionary* is a data structure that maintains a subset S of a totally ordered universe while supporting insertions, deletions, and *locations in the past*. The data structure thus obtained depends on the *chronology* of the insertions and deletions: each insertion or deletion is given, as a *date*, its rank in the sequence of insertions and deletions. A location in the past has two parameters: one is the element x to search for in the universe and the other is the date at which the data structure is to be queried. For such a query, the data structure answers the smallest element greater than x that belongs to the set S at the date i, that is just after the i-th insertion or deletion operation.

To implement a persistent dictionary, it is convenient to use a red–black tree in which each node has $k + 2$ pointers, k being some non-negative integer constant. When during a rotation the node must change one or both of its children, the pointers to the children are not destroyed. Instead, the new child is stored in a new pointer among the $k + 2$. If all these pointers are in use, then a copy of the node is created with two pointers, one for each of the current children of the node. The parent of that node must also keep a pointer to the new node, and the same mechanism is used: if all the pointers in the parent node are used, a copy is made, and so forth. Each of these pointers has a time stamp that remembers the date of its creation. When the root itself is copied, a new entry is created in a dictionary of roots which is ordered chronologically. To perform a location in the past, the algorithm begins by looking up the dictionary for the most recent root prior to the requested date. Then location is performed in the normal fashion, taking care to follow the pointer to the children that is the most recent prior to the requested date.

Show that the space requirement of such a structure is $O(n)$ if n insertions or deletions are performed, starting with an empty set. Show that the cost of a location-in-the-past query is $O(\log n)$.

Hint: To estimate the storage requirements, an amortized analysis is useful. A node of the structure is said to be *active* if it can be reached from the most current root, and *dead* otherwise. The *potential function* of the data structure is defined as the number of active nodes minus a fraction $\frac{1}{k-1}$ of the number of free pointers available in the active nodes. The amortized cost of an operation (insertion or deletion) is simply the number of created nodes minus the change in the potential function, as far as the storage is concerned. Note that the potential is always positive, or zero for an empty structure. The total number of nodes created is bounded above by the total amortized cost. The amortized cost of copying is null, and to add a pointer to a node has an amortized cost of $\frac{1}{k-1}$. Therefore, the amortized cost (in storage) of an insertion or deletion is $O(1)$.

2.5 Bibliographical notes

Basic data structures are treated in any book on algorithms and data structure. The reader is invited to refer to Aho, Hopcroft, and Ullman [6], Knuth [142], Sedgewick [200], Cormen, Leiserson, and Rivest [72], or Froidevaux, Gaudel, and Soria [108] (in French).

The stratified tree (see exercise 2.3) is due to van Emde Boas, Kaas, and Zijlstra [216]. Our exposition is taken from Mehlhorn [163] and Mehlhorn and Näher [166]. Segment and range trees (exercises 2.1 and 2.2) are discussed in the book by Preparata and

Shamos [192]. Persistent data structures as described in exercise 2.7 are due to Sarnak and Tarjan [196]. Several geometric applications of persistent trees will be given in the exercises of chapter 3.

The perfect dynamic hashing method (see exercise 2.6) was developed by Dietzfelbinger, Karlin, Mehlhorn, auf der Heide, Rohnert, and Tarjan [84] and the augmented dictionary on a finite universe is due to Mehlhorn and Näher [165]. See also the book by Mehlhorn [163] for an extended discussion on hashing.

Chapter 3

Deterministic methods used in geometry

The goal of this and subsequent chapters is to introduce the algorithmic methods that are used most frequently to solve geometric problems. Generally speaking, computational geometry has recourse to all of the classical algorithmic techniques. Readers examining all the algorithms described in this book from a methodological point of view will distinguish essentially three methods: the incremental method, the divide-and-conquer method, and the sweep method.

The *incremental method* is perhaps the method which is the most largely emphasized in the book. It is also the most natural method, since it consists of processing the input to the problem one item at a time. The algorithm initiates the process by solving the problem for a small subset of the input, then maintains the solution to the problem as the remaining data are inserted one by one. In some cases, the algorithm may initially sort the input, in order to take advantage of the fact that the data are sorted. In other cases, the order in which the data are processed is indifferent, sometimes even deliberately random. In the latter case, we are dealing with the *randomized* incremental method, which will be stated and analyzed at length in chapter 5. We therefore will not expand further on the incremental method in this chapter.

The *divide-and-conquer* method is one of the oldest methods for the design of algorithms, and its use goes well beyond geometry. In computational geometry, this method leads to very efficient algorithms for certain problems. In this book for instance, such algorithms are developed to compute the convex hull of a set of n points in 2 or 3 dimensions (chapter 8), the lower envelope of a set of functions (chapter 16), a cell in an arrangement of segments in the plane (exercise 15.9), or even the Voronoi diagram of n points in the plane (exercise 19.1). In this chapter, the principles underlying the method are outlined in section 3.1, and the method is illustrated by an algorithm that has nothing to do with geometry: sorting a sequence of real numbers using merging (the so-called merge-sort algorithm).

The *sweep method*, in contrast to the divide-and-conquer method, is deeply linked with the geometric nature of the problems discussed thereafter. In 2 dimensions, numerous problems can be solved by sweeping the plane with a line. Computing the intersection points of a set of n line segments in the plane is a famous example of this kind of problem. In higher dimensions (3 and more), sweeping the space with a hyperplane often reduces a d-dimensional problem to a sequence of $(d-1)$-dimensional problems. The sweep method is described in section 3.2, and is exemplified by the problem of computing the intersection points of a set of n line segments in the plane. Uses of the sweep method can be found in chapter 12 as well, with the triangulation of a simple polygon, in chapter 15 for computing a single cell in an arrangement of line segments (exercises 15.8 and 15.9), and in chapter 19 for computing the Voronoi diagram of a set of points, or of a set of line segments.

Of course, there are other methods which are intimately related to the geometric nature of the problem at hand, and some are discussed and used in this book as well. A common instance is the technique of geometrically transforming a problem using some kind of *polarity* or *duality*, which converts a problem into a *dual* problem. In chapter 7, we show using this method that computing the intersection of a set of half-spaces is equivalent to computing the convex hull of a set of points. In chapter 17, a geometric transform is shown that reduces the computation of a Voronoi diagram to that of the intersection of half-spaces. Another characteristic of geometric algorithms is that they sometimes depend on decomposing the complex geometric objects that they process into *elementary* objects. These objects can be stored using a constant number of memory units and are handled more easily. Decompositions into simplices, or *triangulations*, are discussed in chapters 11, 12, and 13. Section 3.3 describes the *vertical decomposition* of a set of line segments in the plane. This decomposition refines the subdivision of the plane induced by the segments, by decomposing each cell into elementary trapezoids. Computing this decomposition serves as a running example throughout chapter 5 to exemplify the design of randomized algorithms. Such a decomposition can be seen as the prototype in a series of analog structures, also called *vertical decompositions*, which are of use in order to decompose various shapes of the d-dimensional Euclidean space $\mathbb{E}^d$ into elementary regions. For instance, vertical decompositions of polygons, of polyhedra, and of arrangements of various surfaces (hyperplanes, simplices, curves, or even two-dimensional surfaces) are presented throughout this book.

3.1 The divide-and-conquer method

3.1.1 Overview

The divide-and-conquer paradigm is a modern application of the old political saying: "divide your enemies to conquer them all". Solving a problem by the divide-and-conquer method involves recursive applications of the following scheme:

Dividing. Divide the problem into *simpler* subproblems. Such problems have a smaller input size, that is, if the input data are elementary, the input to these problems is made up of some but not all of the input data.

Solving. Separately solve all the subproblems. Usually, the subproblems are solved by applying the same algorithm recursively.

Merging. Merge the subproblem solutions to form the solution to the original problem.

The performance of the method depends on the complexities of the divide and merge steps, as well as on the size and number of the subproblems. Assume that each problem of size n is divided into p subproblems of size n/q, where p and q are some integer constants and n is a power of q. If the divide and merge steps perform $O(f(n))$ elementary operations altogether in the worst case, then the time complexity $t(n)$ of the whole algorithm satisfies the recurrence

$$t(n) = p\, t\left(\frac{n}{q}\right) + f(n).$$

Usually, the recursion stops when the problem size is small enough, for instance smaller than some constant n_0. Then $k = \lceil \log_q(n/n_0) \rceil$ is the depth of the recursive calls ($\log_q$ stands for the logarithm in base q), and the recurrence solves to

$$t(n) = O\left(p^k + \sum_{j=0}^{k-1} p^j f\left(\frac{n}{q^j}\right)\right).$$

In this expression, the first term corresponds to the time needed to solve all the elementary problems generated by the algorithm. The second term reflects the time complexity of all the merge and divide steps taken together. If f is a multiplicative function, *i.e.* such that $f(xy) = f(x)f(y)$ (which in particular is true when $f(n) = n^\alpha$ for some constant α), then $t(n)$ satisfies

$$t(n) = O\left(p^k + f(n) \sum_{j=0}^{k-1} \frac{p^j}{f(q)^j}\right),$$

or even, noting that $n = q^{\log n/\log q}$,

$$t(n) = \Theta\left(n^{\log p/\log q} + n^{\log f(q)/\log q} \sum_{j=0}^{k-1} \frac{p^j}{f(q)^j}\right).$$

If n is no longer assumed to be a power of q, it can nevertheless lie between two consecutive powers of q. The above analysis applies to the powers of q that bracket n and hence shows that $t(n)$ obeys the same asymptotic behavior. Summarizing, if f is a multiplicative function:

- If $p > f(q)$, then $t(n) = O\left(n^{\log p / \log q}\right)$.

- If $p = f(q)$, then $t(n) = O\left(n^{\log p / \log q} \log n\right)$, and further if $f(n) = n^\alpha$, then $t(n) = O\left(n^\alpha \log n\right)$.

- If $p < f(q)$, then $t(n) = O\left(n^{\log f(q) / \log q}\right)$, and further if $f(n) = n^\alpha$, then $t(n) = O\left(n^\alpha\right)$.

3.1.2 An example: sorting n numbers using merge-sort

Recall that the problem of sorting a finite sequence $\mathcal{X} = (x_1, \ldots, x_n)$ of n real numbers represented as, say, a list, consists of building a list $\mathcal{Y}$ formed of all the elements of $\mathcal{X}$ sorted in ascending order. The merge-sort algorithm involves the following stages:

Dividing. The list representing the sequence $\mathcal{X}$ is divided into two sublists that represent the sequences $\mathcal{X}_1 = (x_1, \ldots, x_m)$ and $\mathcal{X}_2 = (x_{m+1}, \ldots, x_n)$, with $m = \lfloor n/2 \rfloor$.

Solving. Each of the sequences $\mathcal{X}_1$ and $\mathcal{X}_2$ is sorted recursively using the same method. Let $\mathcal{Y}_1$ and $\mathcal{Y}_2$ stand for the two sequences resulting from the recursive sorting of the subsequences $\mathcal{X}_1$ and $\mathcal{X}_2$ respectively.

Merging. The sequence $\mathcal{Y}$ that solves the sorting problem on $\mathcal{X}$ can be obtained by merging the sequences $\mathcal{Y}_1$ and $\mathcal{Y}_2$ in the following manner. We simultaneously go through the corresponding lists, maintaining a pointer to the current element of each list. To start with, the current element of a list is its first element. As we skip through, we compare the current elements of both lists, and append the smaller at the end of the list representing $\mathcal{Y}$. The corresponding pointer is advanced to the next element in that list, which becomes the new current element of that list.

We can perform the dividing stage in linear time, and the merging stage also since each step involves making only one comparison and advances one element forward in one of the two lists for a maximum total of $2n$ comparisons. Therefore $f(n) = \Theta(n)$ and $p = q = 2$ since the original list is divided into two sublists of approximately equal size. From the preceding subsection, we can conclude that the complexity of sorting n real numbers using merge-sort is $\Theta(n \log n)$. Theorem 1.2.1 proves that this complexity cannot be improved by more than a constant factor. The following theorem summarizes this result:

Theorem 3.1.1 *The merge-sort algorithm sorts a sequence of n numbers in optimal $\Theta(n \log n)$ time.*

3.2 The sweep method

3.2.1 Overview

A sweep algorithm solves a two-dimensional problem by simulating a sweep of the plane with a line. Let us agree on a fixed direction in the plane, say that of the y-axis, which we will call the *vertical* direction. A line Δ parallel to this direction sweeps the plane when it moves continuously from left to right, from its initial position $x = -\infty$ to its final position $x = +\infty$.

Algorithms that proceed by sweeping the plane can in fact be very dissimilar. Their main common feature is the use of two data structures: one structure $\mathcal{Y}$ called the *state of the sweep* and another $\mathcal{X}$ called the *event queue*. Though the information stored in $\mathcal{Y}$ can vary from one algorithm to another, the following characteristics are always true:

1. the information stored in $\mathcal{Y}$ is related to the position of the sweep line, and changes when this line moves,

2. the structure $\mathcal{Y}$ must be modified only at a finite number of discrete positions, called *events*,

3. the maintenance of this structure yields enough information to build the solution to the original problem.

The event queue $\mathcal{X}$ stores the sequence of events yet to be processed. This sequence can be entirely known at the beginning of the algorithm, or discovered *on line, i.e.* as the algorithm processes the events. The sweep algorithm initializes the structure $\mathcal{Y}$ for the leftmost position $x = -\infty$ of the sweep line, and the sequence $\mathcal{X}$ with whatever events are known from the start (in increasing order of their abscissae). Each event is processed in turn, and $\mathcal{Y}$ is updated. Occasionally, new events will be detected and inserted in the queue $\mathcal{X}$, or, on the contrary, some events present in the queue $\mathcal{X}$ will no longer have to be processed and will be removed. When the event is processed, the queue $\mathcal{X}$ gives access to the next event to be processed.

When all the events are known at the start of the algorithm, the queue $\mathcal{X}$ may be implemented with a mere simply linked list. However, when some events are to be known only on line, the event queue must handle not only the minimum operation, but also queries, insertions, and sometimes even deletions: it is a priority queue (see chapter 2).

The choice of the data structure $\mathcal{Y}$ depends on the nature of the problem and may be handled through multiple components. More often than not, each of these components must handle a totally ordered set of objects, and the corresponding operations: query, insertion, deletion, sometimes even predecessor or successor. The appropriate choice is that of a dictionary, or an augmented dictionary (see chapter 2).

The sweep method can sometimes be useful in three or more dimensions. The generalization consists of sweeping the space $\mathbb{E}^d$ by a hyperplane perpendicular to the x_d-axis. The state of the sweep is stored in a data structure $\mathcal{Y}$ associated with the sweep hyperplane, and the set of events is the set of positions of the sweep hyperplane at which the state of the sweep $\mathcal{Y}$ changes. The data structure $\mathcal{Y}$ often maintains a representation of a $(d-1)$-dimensional object contained in the sweep hyperplane. The sweep method in higher dimensions, therefore, often consists of replacing a d-dimensional problem by a sequence of $(d-1)$-dimensional problems.

3.2.2 An example: computing the intersections of line segments

Let $\mathcal{S}$ be a set of line segments in the plane, the intersecting pairs of which we are interested in computing, together with the coordinates of each intersection point. The naive solution to the problem is to test all the $n(n-1)/2$ possible pairs. The resulting algorithm is therefore quadratic, *i.e.* runs in $\Theta(n^2)$ time. This is optimal in the worst case, since the number of intersecting pairs of a set of n line segments can be as high as $\Omega(n^2)$. However, the algorithm performs $\Omega(n^2)$ computations regardless of what the actual set of line segments is, whereas the number of intersection points is commonly much less than n^2. In those cases, an *output-sensitive* algorithm is more desirable (see section 1.1.2). The algorithm we present next possesses such a property: its running time complexity is $O\left((n+a)\log n\right)$ where a is the number of intersecting pairs in the set of line segments.

To simplify the description of the algorithm, let us assume that the line segments are in general position, which in this case amounts to saying that no three segments have a common intersection. Moreover, let us assume that all the endpoints have distinct abscissae. In particular, there can be no vertical line segment in the set $\mathcal{S}$. Should these assumptions be violated, it would be easy but tiresome to take special care of all the exceptions, and we leave the technicalities to the careful reader.

The algorithm is based upon the following remark: if two segments S and S' intersect, any vertical line Δ whose abscissa is close enough to that of $S \cap S'$ intersects both S and S', and these segments are consecutive in the vertically ordered sequence of all the intersections of a segment in $\mathcal{S}$ and Δ (see figure 3.1).

The sweep algorithm stores in the data structure $\mathcal{Y}$ the set of segments of $\mathcal{S}$

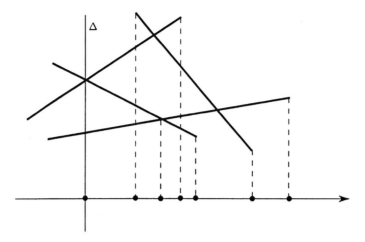

Figure 3.1. Computing the intersections of a set of line segments using the sweep method.

which intersect the vertical sweep line Δ. Such segments are said to be *active* at the current position of the sweep line. The structure $\mathcal{Y}$ stores the active segments in the order of the ordinates of their intersection point with the line Δ. The order of the sequence, or the sequence itself, is modified only when the line sweeps over the endpoint of a segment or over an intersection point.

1. If Δ sweeps over the left endpoint of a line segment S (that is to say, the endpoint with the smaller abscissa), this segment S is added to the structure $\mathcal{Y}$.

2. If Δ sweeps over the right endpoint of a line segment S (that is to say, the endpoint with the greater abscissa), this segment S is removed from the structure $\mathcal{Y}$.

3. If Δ sweeps over the intersection of two segments S and S', these segments S and S' switch their order in the sequence stored in $\mathcal{Y}$.

The set of events therefore includes the sweep line passing over the endpoints of the segments of $\mathcal{S}$, and over the intersections. The abscissae of the endpoints are known as part of the input, and we wish to compute the abscissae of the intersection points. A prospective intersection point I is known when two active segments become consecutive in the sequence stored in $\mathcal{Y}$. The corresponding event is then stored in the event queue $\mathcal{X}$. The state of the event queue is shown for a particular position of Δ on figure 3.1: each event is marked by a point on the x-axis.

At the beginning of the algorithm, the queue $\mathcal{X}$ stores the sequence of endpoints of the segments in $\mathcal{S}$ ordered by their abscissae. The data structure $\mathcal{Y}$ is empty.

As long as there is an available event in the queue $\mathcal{X}$, the algorithm extracts the event with the smallest abscissa, and processes it as follows.

Case 1. the event is associated with the left endpoint of a segment S. This segment is then inserted into $\mathcal{Y}$. Let $pred(S)$ and $succ(S)$ be the active segments which respectively precede and follow S in $\mathcal{Y}$. If $pred(S)$ and S (resp. S and $succ(S)$) intersect, their intersection point is inserted into $\mathcal{X}$.

Case 2. the event is associated with the right endpoint of a segment S. This segment is therefore queried and removed in the structure $\mathcal{Y}$. Let $pred(S)$ and $succ(S)$ be the active segments which respectively preceded and followed S in $\mathcal{Y}$. If $pred(S)$ and $succ(S)$ intersect in a point beyond the current position of the sweep line, this intersection point is queried in the structure $\mathcal{X}$ and the corresponding event is inserted there if it was not found.

Case 3. the event is associated with an intersection point of two segments S and S'. This intersection point is reported, and the segments S and S' are exchanged in $\mathcal{Y}$. Assuming S is the predecessor of S' after the exchange, S and its predecessor $pred(S)$ are tested for intersection. In the case of a positive answer, if the abscissa of their intersection is greater than the current position of the sweep line, this point is queried in the structure $\mathcal{X}$ and the corresponding event is inserted there if it was not found. The same operation is performed for S' and its successor $succ(S')$.

To prove the correctness of this algorithm, it suffices to notice that every intersecting pair becomes a pair of active consecutive segments in $\mathcal{Y}$, when the abscissa of the sweep line immediately precedes that of their intersection point. This pair is always tested for intersection at this point, if not before, therefore the corresponding intersection point is always detected and inserted into $\mathcal{X}$, to be reported later.

It remains to see how to implement the structures $\mathcal{Y}$ and $\mathcal{X}$. The structure $\mathcal{Y}$ contains at most n segments at any time, and must handle queries, insertions, deletions, and predecessor and successor queries: it is an augmented dictionary (see section 2.1). If this dictionary is implemented by a balanced tree, each query, insertion, and deletion can be performed in time $O(\log n)$, and finding predecessors and successors takes constant time.

The event queue $\mathcal{X}$ will contain at most $O(n + a)$ events, if a stands for the number of intersecting pairs among the segments in S. This structure must handle queries, insertions, deletions, and finding the minimum: it is a priority queue (see section 2.1). Again, a balanced binary tree will perform each of these operations in $O(\log(n + a)) = O(\log n)$ time.

The global analysis of the algorithm is now immediate. The initial step that sorts all the $2n$ endpoints according to their abscissae takes time $O(n \log n)$. The structure $\mathcal{X}$ is initialized and built within the same time bound. Next, each of the $2n + a$ events is processed in turn. Each event requires only a constant number of operations to be performed on the data structures $\mathcal{X}$ and $\mathcal{Y}$ and is therefore handled in time $O(\log n)$. Overall, the algorithm has a running time complexity of $O((n + a) \log n)$ and requires storage $O(n + a)$.

The algorithm can be slightly modified to avoid using more than $O(n)$ storage. It suffices, while processing any of cases 1 to 3, to remove from the event queue any event associated with two active but non-consecutive segments. In this way, the queue $\mathcal{X}$ contains only $O(n)$ events at any time, and yet the event immediately following the current position of the sweep line is always present in $\mathcal{X}$. Indeed, this event is associated either with an endpoint of a segment in $\mathcal{S}$, or with two intersecting segments which therefore must be consecutive in $\mathcal{X}$. Some events can be inserted into and deleted from $\mathcal{X}$ several times before they are processed, but this does not change the running time complexity of the algorithm, as the above scheme can be carried out using only a constant number of operations in the data structures $\mathcal{X}$ and $\mathcal{Y}$ at each step.

Theorem 3.2.1 *The intersection points of a set of segments in the plane can be computed using the sweep method. If the set of n segments in general position has a intersecting pairs, the resulting algorithm runs in $O((n+a) \log n)$ time and $O(n)$ space.*

3.3   Vertical decompositions

In this section we describe the vertical decomposition of a set of (possibly intersecting) line segments in the plane. A set $\mathcal{S}$ of segments induces a subdivision of the plane into *regions*, or *cells*, which are the connected components of $\mathbb{E}^2 \setminus \mathcal{S}$. The complexity of each cell, that is, the number of segment portions appearing on its boundary, is unbounded. The *vertical decomposition* of a set of segments is obtained by subdividing each cell into elementary trapezoidal regions. This decomposition of the plane can be considered as the prototype of a whole class of analogous geometric decompositions, similarly called *vertical decompositions*, presented in the remainder of this book.

3.3.1   Vertical decompositions of line segments

The *vertical decomposition* of a set of line segments in the plane is a structure which depends upon the choice of a particular direction. Here we assume this direction is that of the y-axis, which we call the *vertical* direction. When we want to refer to this direction explicitly, we speak of a *y-decomposition*.

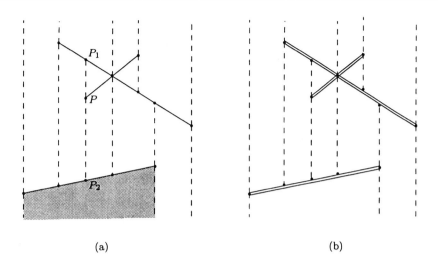

Figure 3.2. (a)The vertical decomposition $\mathcal{D}ec(\mathcal{S})$ of a set of line segments $\mathcal{S}$ in the plane. (b) Its simplified decomposition $\mathcal{D}ec_s(\mathcal{S})$.

Let $\mathcal{S}$ be a set of n segments in the plane. As previously, we suppose that the segments in $\mathcal{S}$ are in general position (meaning that no three segments have a common intersection) and that the abscissae of their endpoints are all distinct. In particular, this implies that no segment of $\mathcal{S}$ is vertical.

From each point P in the plane, we can trace two vertical half-lines both upward and downward, $\Delta_1 P$ and $\Delta_2 P$. Let P_i $(i = 1, 2)$ be the first point of $\Delta_i P$ distinct from P where this half-line meets a segment of $\mathcal{S}$. Should no such point exist, we make the convention that P_i is the point at infinity on the line $\Delta_i P$. Segments PP_1 and PP_2 are the *walls* stemming from point P. Hence, the walls stemming from a point P are the maximal vertical segments that have P as an endpoint and whose relative interiors do not intersect segments of $\mathcal{S}$ (see figure 3.2a).

We call *vertical decomposition* of the set $\mathcal{S}$ of segments, and we denote by $\mathcal{D}ec_y(\mathcal{S})$, or more simply by $\mathcal{D}ec(\mathcal{S})$ when the vertical direction y is clearly understood, the planar subdivision induced by the segments and the vertical walls stemming from the endpoints and from the intersection points of the segments in $\mathcal{S}$ (see figure 3.2a). The vertical decomposition of $\mathcal{S}$ can be described as a planar map whose vertices, edges, and regions subdivide the plane. The vertices of this map are the endpoints and intersection points of the segments of $\mathcal{S}$, and the endpoints of the walls. Their number is $O(n + a)$, if $\mathcal{S}$ has n segments with a intersecting pairs. Edges are either walls, or pieces of segments between consecutive vertices. Their number is therefore also $O(n + a)$. Euler's theorem for planar maps (see exercise 11.4) shows that the number of regions is also $O(n + a)$. Each region in the map has the shape of a trapezoid, the two parallel sides of which are vertical. Some degenerate ones are triangular (with only one vertical side), or semi-infinite (bounded at top or bottom by a segment portion with two semi-infinite walls on both sides), or doubly infinite (a slab bounded by two vertical lines on either side), or even a half-plane (bounded by only one vertical line).

It is easy to modify the above algorithm to compute not only the intersection points, but also the vertical decomposition of the given set of line segments.

Theorem 3.3.1 *A sweep algorithm builds the vertical decomposition of a set of n segments in general position with a intersecting pairs in time $O((n + a) \log n)$ and space $O(n + a)$.*

3.3.2 Vertical decompositions and simplified decompositions

Each region of a vertical decomposition is thus a trapezoid, or a degenerate one, and its boundary has at most four *sides*[1], two of which are vertical. Each vertical side of a trapezoid consists of one or two walls stemming from the same point. The non-vertical sides of a trapezoid are respectively called the *floor* and *ceiling* of the trapezoid. The floor or ceiling of a trapezoid is always included in some segment of S and its endpoints are vertices of $\mathcal{D}ec(S)$. Neither the floor nor the ceiling need be edges of the vertical decomposition $\mathcal{D}ec(S)$, however: they can be made up of several (up to $\Omega(n)$) edges of the planar map $\mathcal{D}ec(S)$. Indeed, several walls exterior to a trapezoid can butt against its floor or its ceiling, as is the case for the shadowed cell in figure 3.2a. With this understanding, the boundary of a region can be made up of $\Omega(n + a)$ edges of the planar map $\mathcal{D}ec(S)$, and thus have a non-bounded complexity.

From a slightly different point of view, we can describe a simplified decomposition scheme for which each trapezoidal region has a bounded complexity. The trick is to consider each segment as an infinitely thin rectangle. The boundary of such a rectangle is made up of two vertical sides of infinitely small length and two sides which are copies of the segment. The latter are called the *sides* of the segment (see figure 3.2b). Each side of the segment completely ignores how the walls abut against the opposite side, in other words the two sides of a segment are not connected. The floor and ceiling of a trapezoid are included in sides of two distinct segments. The simplified vertical decomposition of the set S can still be viewed as a vertical decomposition of the plane. Simply, in addition to the trapezoidal regions, there are (empty) regions included in the rectangles between the two sides of a segment. The trapezoidal regions are unchanged, except that they now have at most six edges on their boundaries: the floor and ceiling are each made up of a single edge, and there can be up to two walls per vertical side.

---

[1]Here, as in the previous subsection, the word *side* is used in its usual geometric meaning: a quadrangle is a geometric figure with four sides.

3.4 Exercises

Exercise 3.1 (Union, intersection of polygonal regions) By a *polygonal region*, we mean a connected area of the plane bounded by one or more disjoint polygons (a polygonal region may not always be simply connected, and may have holes). Show how to build the union or intersection of k polygonal regions using a sweep algorithm. Show that if the total complexity of the regions (the number of sides of all the polygons that bound it) is n, and the number of intersecting pairs between all the sides of all the polygonal regions is a, the algorithm will run in $O((n+a)\log n)$ time.

Exercise 3.2 (Detecting intersection) Show that to test whether any two segments in a set S intersect requires at least time $\Omega(n\log n)$. Show that the sweep algorithm can be modified to perform this test in time $O(n\log n)$.

Exercise 3.3 (Computing the intersection of curved arcs) Modify the sweep algorithm described in subsection 3.2.2 so as to report all the intersection points in a family of curved arcs. The arcs may or may not be finite. We further assume that any two arcs have only a bounded number of intersection points, which may be computed in constant time.

Hint: Do not forget to handle the events where the arcs have a vertical tangent.

Exercise 3.4 (Arbitrary sets of segments) Sketch the changes to be made to the sweep algorithm so that it still works on arbitrary sets of segments, getting rid of the assumptions about general position. The algorithm should run in time $O((n+a)\log n)$ where a is the number of intersecting pairs.

Exercise 3.5 (Location in a planar map) A planar map of size n is a planar subdivision of the plane $\mathbb{E}^2$ induced by a set of n segments which may intersect only at their endpoints. To locate a point in the planar map is to report the region of the subdivision that this point lies in. Show that a data structure may be built in time $O(n\log n)$ and space $O(n)$ to support location queries in time $O(\log n)$.

Hint: The vertical lines passing through the endpoints of the segment divide the plane into vertical strips ordered by increasing abscissae. The segments that intersect a strip form a totally ordered sequence inside this strip, and two sequences corresponding to two consecutive strips differ only in a constant number of positions. A sweep algorithm may use persistent structures (see exercise 2.7) to build the sequence of such lists.

Exercise 3.6 (Union using divide-and-conquer) Consider the n polygonal regions interior to n polygons. If any two such polygons intersect in at most two points, it can be shown that the union of these polygonal regions has complexity $O(n)$. Show that, in this case, it can be computed in $O(n\log^2 n)$ time.

Hint: The algorithm proceeds by using the divide-and-conquer method. Each merge step computes the union of two polygonal regions and can be performed using the sweep method. Each intersection between the edges of these regions is a vertex of their union, therefore there can be at most a linear number of such intersections.

Exercise 3.7 (Selecting the k-th element) Let S be a set of n elements, all belonging to a totally ordered universe. A k-th element of S is any element S of S such that there are at most $k-1$ elements in S strictly smaller than S and at least k elements smaller than or equal to S. Show that it is possible to avoid sorting S yet still compute a k-th element in time $O(n)$.

Hint: The algorithm is as follows:

> **select**(k, S)
> **if** $|S| < 50$, sort S and return any k-th element
> **otherwise**

1. Divide S into $\left\lfloor \frac{|S|}{5} \right\rfloor$ subsets of 5 elements each, with at most 4 elements in an additional set.

2. Compute the median of each subset, and the median M of the set $\mathcal{M}$ of all medians of these subsets by recursively calling $Select\left(\left\lfloor \frac{|\mathcal{M}|}{2} \right\rfloor, \mathcal{M}\right)$.

3. Let S_1, S_2, S_3 consist of those elements of S respectively smaller than, equal to, or greater than M.
   if $|S_1| \geq k$ return $Select(k, S_1)$.
   if $|S_1| + |S_2| \geq k$ return $Select(k - |S_1| - |S_2|, S_3)$.
   otherwise return M.

To analyze the complexity of such an algorithm, observe that $|S_1| \leq \frac{3}{4}n$ and that $|S_2| \leq \frac{3}{4}n$. Then, if $n \geq 50$, show that the time complexity of the algorithm satisfies the recurrence

$$t(n) \geq t(n/5) + t(3n/4) + cn,$$

where c is a constant. Show that this recurrence solves to $t(n) = O(n)$.

Exercise 3.8 (Union of parallel rectangles) Consider a set of axis-parallel rectangles in the plane $\mathbb{E}^2$. Propose an algorithm to compute the area, the perimeter, or even the boundary of the union of these rectangles.

Hint: One may use a sweep algorithm that maintains the intersection of the union of the rectangles with the sweep line, using a segment tree (see exercise 2.1). The perimeter or the area can be obtained in time $O(n \log n)$, and the complete description of the boundary in time $O((n + k) \log n)$ if this boundary has k edges.

3.5   Bibliographical notes

The sweep algorithm that computes the intersecting pairs of a set of n segments in the plane is due to Bentley and Ottmann [23]. This output-sensitive algorithm is not optimal, as the problem has complexity $\Theta(n + a)$ if, between the n segments, a pairs intersect. Indeed, the output has size $\Omega(n + a)$ and the result of exercise 3.2 shows that $\Omega(n \log n)$ is also a lower bound on the complexity of the problem. Chazelle and Edelsbrunner [49] give an optimal algorithm that computes, given a set of n segments with a intersecting pairs, the induced vertical decomposition in optimal $O(n \log n + a)$ time. In degenerate cases, the number b of intersection points can be much lower than a, and Burnikel, Mehlhorn and Schirra [40] have shown that it still is possible to compute the vertical decomposition in $O(n \log n + b)$ time. In chapter 5, we describe a randomized algorithm (that is, an algorithm which makes random choices during its execution) which runs in time $O(n \log n + a)$ on the average over all possible random choices it can make.

Persistent data structures and the idea of using them for locating a point in a planar map (as in exercise 3.5) are due to Sarnak and Tarjan [196]. Segment trees (see exercise 2.1) are especially suitable for solving many problems on rectangles. The solution to exercise 3.8 can be found in the book by Preparata and Shamos [192].

Chapter 4

Random sampling

The *randomization* method has proved useful in computational geometry. This usefulness can be ascribed in large part to a few probabilistic theorems which rely on combinatorial properties of certain geometric problems. The probabilities involved in those theorems concern random samples from the set of data, and do not involve statistical assumptions about the distribution of these data.

The goal of this chapter is to present those probabilistic theorems on which the analysis of the randomized incremental method is based. This method is described in chapters 5 and 6. We express these theorems in a framework general enough to be adaptable to different geometric settings. All the randomized algorithms presented below fit into the same framework as the one that we define here.

The first part of this chapter recalls the necessary definitions and notation. The second part proves the basic two theorems: the sampling theorem, and the moment theorem. These theorems provide the main tools to analyze the average performance of randomized algorithms.

4.1 Definitions

4.1.1 Objects, regions, and conflicts

In the framework presented here, any geometric problem can be formulated in terms of objects, regions, and conflicts between these objects and regions.

Objects are elements of a universe $\mathcal{O}$, usually infinite. The input to some problem will be a set $\mathcal{S}$ of objects of $\mathcal{O}$. The objects under consideration are typically subsets of the Euclidean space $\mathbb{E}^d$ such as points, line segments, lines, half-planes, hyperplanes, half-spaces, etc.

A *region* is a member of a set $\mathcal{F}$ of regions. Each region is associated with two sets of objects: those that *determine* it, and those that *conflict* with it.

The set of objects that determine a region is a finite subset of $\mathcal{O}$, of cardinality bounded by some constant b. The constant b depends on the nature of the problem, but not on the actual instance nor on its size. This restriction is required for all the probabilistic theorems to be expressed within the framework.

The set of objects that conflict with a given region is usually infinite and is called the *domain of influence* of the region.

Let $\mathcal{S}$ be a set of objects. A region F of $\mathcal{F}$ is *defined over* $\mathcal{S}$ if the set of objects that determines it is contained in $\mathcal{S}$. A region F is said to be *without conflict over* $\mathcal{S}$ if its domain of influence contains no member of $\mathcal{S}$, and otherwise is said to *have j conflicts over* $\mathcal{S}$ if its domain of influence contains j objects of $\mathcal{S}$.

For each geometric application, the notions of objects, regions, and conflicts are defined in such a way that the problem is equivalent to finding all the regions defined and without conflict over $\mathcal{S}$.

Let us immediately discuss a concrete example. Let $\mathcal{S}$ be a set of n points in the d-dimensional Euclidean space $\mathbb{E}^d$. The convex hull of $\mathcal{S}$ is the smallest convex set containing $\mathcal{S}$; suppose we wish to compute it. Assume the points are in general position[1]. The convex hull $conv(\mathcal{S})$ is a polytope whose special properties will be studied further in chapter 7. For now, it suffices to notice that, in order to compute the convex hull, we have to find all the subsets of d points in $\mathcal{S}$ such that one of the half-spaces bounded by the hyperplane passing through these d points contains no other point that belong to $\mathcal{S}$ (see figure 4.1). In this example, the objects are points, and the regions are open half-spaces in $\mathbb{E}^d$. Every set of d points determines two regions: the open half-spaces whose boundaries are the hyperplane passing through these points. A point is in conflict with a half-space if it lies inside it. To find the convex hull, one must find all the regions determined by points of $\mathcal{S}$ and without conflict over $\mathcal{S}$.

The preceding definitions call for a few comments.

Remark 1. A region is determined by a finite and bounded number of objects and this restriction is the only fundamental condition that objects, regions, and conflicts must satisfy. Nevertheless, we do not demand that all the regions be determined by exactly the same number of objects. In the case of the convex hull of n points in $\mathbb{E}^d$, all the regions are determined by exactly d points. One may envision other settings (as in the case of the vertical decomposition of a set a line segments in the plane, discussed in subsection 5.2.2), where the regions can be determined by a variable number i of objects, provided that $1 \leq i \leq b$ for some constant b.

Remark 2. A region does not conflict with the objects that determine it. This

---

[1]A set of points is in general position if every subset of $k + 1 \leq d + 1$ points is affinely independent, or in other words if it generates an affine subspace of dimension k.

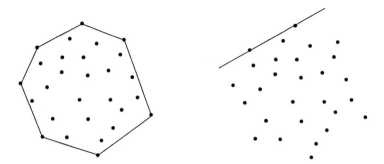

Figure 4.1. Convex hull and empty half-spaces.

simple convention greatly simplifies the statements and proofs of the theorems below, and does not modify their meaning. In the case of the convex hull, this can be easily achieved by defining the domain of influence of a region as an *open* half-space.

Remark 3. A region is characterized by two sets of objects: the set of objects that determine it, and the set of objects that conflict with it. Regions determined by different objects will be considered as different, even if they share the same domain of influence. In this context, a set S of objects is in general position precisely if any two regions determined by different subsets of S have distinct domains of influence.

Remark 4. A set of b or fewer objects may determine one, or more, or zero regions. Usually, the number of regions determined by a given set of (less than b) objects is bounded by a constant. For instance, in the case of convex hulls, every subset of d points determines exactly two regions. In this case, the total number of regions defined over a set of cardinality n is $O(n^b)$.

If S is a finite set of objects, say with n elements, we denote by $\mathcal{F}(S)$ the set of regions defined over S and, for each integer j in $[0, n]$, we denote by $\mathcal{F}_j(S)$ the set of all regions defined over S that have j conflicts over S. In particular, $\mathcal{F}_0(S)$ is the set of those regions that are defined over S and without conflict over S. Furthermore, we denote by $\mathcal{F}_{\leq k}(S)$ the subset of regions defined over S that have at most k conflicts over S.

When the regions are determined by a variable number i of objects ($1 \leq i \leq b$), the preceding notation may be refined to denote by $\mathcal{F}_j^i(S)$, $\mathcal{F}_{\leq k}^i(S)$, $\mathcal{F}_{\geq k}^i(S)$, the subsets of those regions defined by exactly i objects of S, with (respectively) exactly, at most, at least, k conflicts with the objects of S.

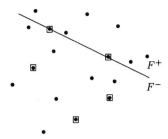

Figure 4.2. Instances of regions.

4.1.2 Random sampling

Let $\mathcal{R}$ be a subset of $\mathcal{S}$ with cardinality r. This subset $\mathcal{R}$ is a random sample of $\mathcal{S}$ if its elements are randomly chosen among all those of $\mathcal{S}$, such that each subset is equally likely to be chosen with probability $1/\binom{n}{r}$. In what follows, we shall call such a subset a *random r-sample* of the set $\mathcal{S}$.

The notation defined in the previous subsection is valid over any subset $\mathcal{R}$ of $\mathcal{S}$. In particular, $\mathcal{F}(\mathcal{R})$ is the set of regions defined over $\mathcal{R}$, $\mathcal{F}_j(\mathcal{R})$ is the set of regions defined over $\mathcal{R}$ that have j conflicts over $\mathcal{R}$, and $\mathcal{F}_j^i(\mathcal{R})$ is the set of of regions defined by exactly i objects of $\mathcal{R}$ that have j conflicts over $\mathcal{R}$. Since we may also be interested in the conflicts over $\mathcal{S}$ of a region defined over $\mathcal{R}$, or the converse, we will avoid any ambiguities by setting up a special terminology. Henceforth, by a *region defined and without conflict over* $\mathcal{R}$, we shall mean a region defined over $\mathcal{R}$ and without conflict over $\mathcal{R}$; these are the regions of $\mathcal{F}_0(\mathcal{R})$. Likewise, a *region defined and with j conflicts over* $\mathcal{R}$ is a region defined over $\mathcal{R}$ and that has j conflicts over $\mathcal{R}$; these are the regions of $\mathcal{F}_j(\mathcal{R})$.

In figure 4.2, the points of the subset $\mathcal{R}$ are enclosed by squares, the half-space F^+ belongs to $\mathcal{F}_6(\mathcal{S})$ and to $\mathcal{F}_0(\mathcal{R})$, while F^- belongs to $\mathcal{F}_{10}(\mathcal{S})$ and to $\mathcal{F}_3(\mathcal{R})$.

From now on, we are primarily interested in the regions defined over a random sample $\mathcal{R}$ from $\mathcal{S}$. Generally speaking, if $g(\mathcal{R})$ is a function of the sample $\mathcal{R}$, we denote by $g(r, \mathcal{S})$ the expected value of $g(\mathcal{R})$ for a random r-sample of $\mathcal{S}$. In particular, the following functions are defined: We denote by $f_j(\mathcal{R})$ the number of regions defined and with j conflicts over a subset $\mathcal{R}$ of $\mathcal{S}$ (in mathematical notation, $f_j(\mathcal{R}) = |\mathcal{F}_j(\mathcal{R})|$). Following our convention, $f_j(r, \mathcal{S})$ denotes the expected number of regions defined and with j conflicts over a random r-sample of $\mathcal{S}$. Likewise, $f_j^i(\mathcal{R})$ stands for the number of regions defined by i objects of $\mathcal{R}$ and with j conflicts over $\mathcal{R}$ (in mathematical notation, $f_j^i(\mathcal{R}) = |\mathcal{F}_j^i(\mathcal{R})|$). Then $f_j^i(r, \mathcal{S})$ is the expected number of such regions for a random r-sample of $\mathcal{S}$.

4.2   Probabilistic theorems

In this section, we prove two probabilistic theorems, the sampling theorem and
the moment theorem. These two theorems lay the foundations for our analysis
of randomized algorithms as described in chapters 5 and 6. The reader mostly
interested in the algorithmic applications of these theorems may skip this section
in a first reading. In order to understand the results, it would be enough to
memorize the definition of a moment, to look up lemma 4.2.5, and to admit
corollary 4.2.7.

The probabilistic theorems below are based on certain combinatorial properties
of the geometric objects. The probabilities involved concern mainly random
samples from the input data. In particular, these theorems do not make any
assumptions on the statistical distribution of the input data. The theorems are
stated in the formal framework introduced in the preceding section. Nevertheless,
to shape the intuition of the reader, we start by stating them explicitly for the
specific problem of computing the convex hull of a set of points in the plane.

Let S be a set of n points in the plane, assumed to be in general position,
let k be an integer smaller than n and let $\mathcal{R}$ be a random sample of S of size
$r = \lfloor n/k \rfloor$. The sampling theorem links the number of half-spaces defined over
S and containing at most k points of S, with the expected number of half-spaces
defined and without conflict over $\mathcal{R}$, which is precisely the number of edges of
the convex hull $conv(\mathcal{R})$. Let A and B be points of S. Segment AB is an edge of
the convex hull $conv(\mathcal{R})$ if and only if A and B are points of $\mathcal{R}$ and also one the
half-planes H_{AB}^{+} and H_{AB}^{-} bounded by the line AB does not contain any points
of $\mathcal{R}$. The *sampling theorem* relies on the fact that the segment AB joining two
points of S is an edge of the convex hull $conv(\mathcal{R})$ with a probability that increases
as the smallest number of points in either H_{AB}^{+} or H_{AB}^{-} decreases.

The *moment theorem* concerns the number of points in S and in its sample $\mathcal{R}$
that belong to some half-plane. If the size of $\mathcal{R}$ is large enough, the sample is
representative of the whole set, and the number of points of $\mathcal{R}$ in a half-plane is
roughly the number of points of S in this half-plane scaled by the appropriate
factor r/n.

In fact, the moment theorem is a little more restrictive and concerns only
those half-planes defined and without conflict over the sample. Any edge E of
$conv(\mathcal{R})$ corresponds to a region defined and without conflict over $\mathcal{R}$: the half-
plane $H^{-}(E)$ bounded by the line supporting E that contains no point of $\mathcal{R}$.
The first moment of $\mathcal{R}$ relative to S, or moment of order 1, is defined to be the
sum, over all edges E of the convex hull $conv(\mathcal{R})$, of the number of points of S
lying inside $H^{-}(E)$. In other words, the moment of order 1 of $\mathcal{R}$ with respect to
S counts each point of $S \setminus \mathcal{R}$ with a multiplicity equal to the number of edges
of $conv(\mathcal{R})$ whose supporting lines separate it from $conv(\mathcal{R})$ itself. Figure 4.3

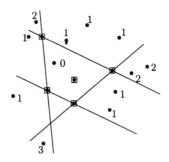

Figure 4.3. Moment of order 1.

indicates the multiplicity of each point, and the first-order moment of the sample is 16.

The moment theorem shows that, if the size of the sample is big enough, the expected moment of order 1 is at most $n - r$.

4.2.1 The sampling theorem

The sampling theorem yields an upper bound on the number of regions defined and with at most k conflicts over a set S of n elements. This bound depends on the expected number of regions defined and without conflict over a random $\lfloor n/k \rfloor$-sample of S. The proof of this theorem relies on the simple idea that, the fewer objects in conflict with a region, the more likely this region is to have no conflict with a random sample $\mathcal{R}$ of S. The proof uses the two fundamental lemmas below.

Lemma 4.2.1 *Let S be a set of n objects and F a region in conflict with j objects of S and determined by i objects of S. If $\mathcal{R}$ is a r-sample of S, the probability $p^i_{j,k}(r)$ that F be a region defined and with k conflicts over $\mathcal{R}$ is*

$$p^i_{j,k}(r) = \frac{\dbinom{j}{k} \dbinom{n-i-j}{r-i-k}}{\dbinom{n}{r}}.$$

Proof. Let $\mathcal{R}$ be a random r-sample of S. The region F of $\mathcal{F}^i_j(S)$ belongs to $\mathcal{F}^i_k(\mathcal{R})$ if it is determined by i objects in $\mathcal{R}$ and conflicts with k objects in $\mathcal{R}$. For this to be the case, the i objects determining F must be part of $\mathcal{R}$. The k objects of $\mathcal{R}$ conflicting with F must be chosen among the j objects of S that conflict with F. Finally, the $r - i - k$ remaining objects in $\mathcal{R}$ not in conflict with F must be chosen among the $n - i - j$ objects in $\mathcal{S}$ that do not determine or conflict with F. $\qquad\qquad\qquad\qquad\qquad\qquad\qquad\qquad\qquad\qquad\qquad\qquad\qquad\qquad\quad$ $\square$

We denote by $p_j^i(r)$ the probability $p_{j,0}^i(r)$ that a region F of $\mathcal{F}_j^i(\mathcal{S})$ be defined and without conflict over a random r-sample of $\mathcal{S}$:

$$p_j^i(r) = \frac{\dbinom{n-i-j}{r-i}}{\dbinom{n}{r}}.$$

Lemma 4.2.2 *Let $\mathcal{S}$ be a set of n objects and $\mathcal{R}$ a random r-sample of $\mathcal{S}$. The expected number $f_k^i(r, \mathcal{S})$ of regions determined by i objects of $\mathcal{R}$ that conflict with k objects of $\mathcal{R}$ is given by the formula*

$$f_k^i(r, \mathcal{S}) = \sum_{j=0}^{n-i} |\mathcal{F}_j^i(\mathcal{S})| \frac{\dbinom{j}{k}\dbinom{n-i-j}{r-i-k}}{\dbinom{n}{r}}.$$

Proof. The expected number of regions in the set $\mathcal{F}_k^i(\mathcal{R})$ is the sum, over all the regions determined by i objects of $\mathcal{S}$, of the probability that this region belongs to the set $\mathcal{F}_k^i(\mathcal{R})$. This probability is given by the lemma 4.2.1 above. $\qquad$ $\square$

Theorem 4.2.3 (Sampling theorem) *Let $\mathcal{S}$ be a set of n objects and k an integer such that $2 \leq k \leq \frac{n}{b+1}$. Then*

$$|\mathcal{F}_{\leq k}(\mathcal{S})| \leq 4(b+1)^b k^b f_0(\lfloor n/k \rfloor, \mathcal{S}).$$

where b is an upper bound on the number of objects that determine a region, $|\mathcal{F}_{\leq k}(\mathcal{S})|$ is the number of regions defined and with at most k conflicts over $\mathcal{S}$, and $f_0(\lfloor n/k \rfloor, \mathcal{S})$ is the expected number of regions defined and without conflict over a random $\lfloor n/k \rfloor$-sample of $\mathcal{S}$.

Proof. For each i, $1 \leq i \leq b$, we shall prove the following inequality bounding the number of regions determined by i objects:

$$|\mathcal{F}_{\leq k}^i(\mathcal{S})| \leq 4(b+1)^i k^i f_0^i(\lfloor n/k \rfloor, \mathcal{S}).$$

Then the theorem can be easily proved by summing over all the values of i between 1 and b.

Let k be an integer such that $2 \le k \le \frac{n}{b+1}$ and $\mathcal{R}$ a random sample of $\mathcal{S}$, of size $r = \lfloor n/k \rfloor$. From lemma 4.2.2, we recall that the expected number $f_0^i(r, \mathcal{S})$ of regions defined and without conflict over $\mathcal{R}$ is

$$f_0^i(r, \mathcal{S}) = \sum_{j=0}^{n-i} |\mathcal{F}_j^i(\mathcal{S})| \frac{\binom{n-i-j}{r-i}}{\binom{n}{r}} \ge |\mathcal{F}_{\le k}^i(\mathcal{S})| \frac{\binom{n-i-k}{r-i}}{\binom{n}{r}}.$$

The remainder of this proof is a mere computation on factorials, which shows that for each k such that $2 \le k \le \frac{n}{b+1}$, and $r = \lfloor n/k \rfloor$,

$$\frac{\binom{n-i-k}{r-i}}{\binom{n}{r}} \ge \frac{1}{4(b+1)^i k^i}.$$

Indeed,

$$\frac{\binom{n-i-k}{r-i}}{\binom{n}{r}} = \frac{r!}{(r-i)!} \frac{(n-i)!}{n!} \frac{(n-r)!}{(n-r-k)!} \frac{(n-i-k)!}{(n-i)!}.$$

We compute

$$\begin{aligned}
\frac{(n-r)!}{(n-r-k)!} \frac{(n-i-k)!}{(n-i)!} &\ge \left(\frac{n-r-k+1}{n-i-k+1} \right)^k \\
&\ge \left(\frac{n-n/k-k+1}{n-k} \right)^k \\
&\ge (1-1/k)^k \\
&\ge 1/4 \quad (\text{if } 2 \le k),
\end{aligned}$$

and

$$\begin{aligned}
\frac{r!}{(r-i)!} \frac{(n-i)!}{n!} &= \prod_{l=0}^{i-1} \frac{r-l}{n-l} \ge \prod_{l=1}^{i} \frac{r+1-l}{n} \\
&\ge \prod_{l=1}^{i} \frac{n/k - l}{n} \ge \frac{1}{k^i} \left(1 - \frac{bk}{n} \right)^i \\
&\ge \frac{1}{k^i (b+1)^i} \quad \left(\text{if } k \le \frac{n}{b+1} \right),
\end{aligned}$$

proving the inequality stated by the theorem. $\qquad\square$

Remark 1. The sampling theorem deals with the numbers $|\mathcal{F}_{\leq k}(\mathcal{S})|$ of regions with at most k conflicts, for values of k between 2 and $\frac{n}{b+1}$.

For the case of regions without or with at most one conflict, however, it is possible to prove the following bound

$$|\mathcal{F}_0(\mathcal{S})| \leq |\mathcal{F}_{\leq 1}(\mathcal{S})| \leq |\mathcal{F}_{\leq 2}(\mathcal{S})| \leq 4(b+1)^b 2^b f_0(\lfloor n/2 \rfloor, \mathcal{S}),$$

valid whenever $n \geq 2(b+1)$.

Moreover, for values of k close to n, there is always the trivial bound

$$|\mathcal{F}_{\leq k}(\mathcal{S})| \leq |\mathcal{F}(\mathcal{S})| = O(n^b)$$

if, as in remark 4 of subsection 4.1.1, we suppose that each subset of size at most b determines at most q regions, for a constant number q that depends on the interpretation of objects and regions.

Remark 2. The sampling theorem yields a deterministic combinatorial result when an upper bound on $f_0(\lfloor n/k \rfloor, \mathcal{S})$ can be derived. For instance, in chapter 14, we will use an upper bound on the number of faces of a d-dimensional polytope to yield, via the sampling theorem, an upper bound on the number of faces at level at most k in an arrangement of hyperplanes.

The following corollary is very useful for analyzing the average performance of randomized algorithms. It shows that the expected number of regions defined and with one or two conflicts over a random r-sample of a set $\mathcal{S}$ is of the same order of magnitude as the expected number of regions defined and without conflict over such a sample.

Corollary 4.2.4 *Let $\mathcal{S}$ be a set of n objects, with $n \geq 2(b+1)$. For each integer r such that $n \geq r \geq 2(b+1)$, we have*

$$\begin{aligned} f_1(r, \mathcal{S}) &\leq \beta f_0(\lfloor r/2 \rfloor, \mathcal{S}) \\ f_2(r, \mathcal{S}) &\leq \beta f_0(\lfloor r/2 \rfloor, \mathcal{S}) \end{aligned}$$

where $f_j(r, \mathcal{S})$ is the expected number of regions defined and with j conflicts over a random r-sample of $\mathcal{S}$, and β is the real constant

$$\beta = 4(b+1)^b 2^b.$$

Proof. Let $\mathcal{R}$ be a subset of $\mathcal{S}$ of size r, such that $2(b+1) \leq r$. Applied to $\mathcal{R}$, remark 1 following theorem 4.2.3 yields

$$|\mathcal{F}_1(\mathcal{R})| \leq 4(b+1)^b 2^b f_0(\lfloor r/2 \rfloor, \mathcal{R}).$$

The first inequality is obtained by taking expectations on the two members of this equation. Indeed, $f_0(\lfloor r/2 \rfloor, \mathcal{R})$ is the expected number of regions defined and without conflict over a random $\lfloor r/2 \rfloor$-sample of $\mathcal{R}$, and the expectation of this expected number when $\mathcal{R}$ itself is a random r-sample of $\mathcal{S}$ is simply $f_0(\lfloor r/2 \rfloor, \mathcal{S})$. The second inequality can be proved in much the same way. $\square$

4.2.2 The moment theorem

Let $\mathcal{S}$ be a set of n objects and $\mathcal{R}$ be a subset of $\mathcal{S}$. The moment theorem bounds the total number of conflicts between the objects of $\mathcal{S}$ and the regions defined and without conflict over $\mathcal{R}$.

Let k be a integer less than or equal to n. The *moment of order k* of $\mathcal{R}$ with respect to $\mathcal{S}$, denoted by $m_k(\mathcal{R}, \mathcal{S})$, is the sum

$$m_k(\mathcal{R}, \mathcal{S}) = \sum_{F \in \mathcal{F}_0(\mathcal{R})} \binom{|\mathcal{S}(F)|}{k},$$

where $\mathcal{F}_0(\mathcal{R})$ stands for the set of regions defined and without conflict over $\mathcal{R}$, and $|\mathcal{S}(F)|$ is the cardinality of the set $\mathcal{S}(F)$ of objects in $\mathcal{S}$ that conflict with a region F.

The moment of order 0, $m_0(\mathcal{R}, \mathcal{S})$, is simply the number of regions defined and without conflict over $\mathcal{R}$:

$$m_0(\mathcal{R}, \mathcal{S}) = |\mathcal{F}_0(\mathcal{R})|.$$

The moment of order 1, $m_1(\mathcal{R}, \mathcal{S})$, is the total number of conflicts between the elements of $\mathcal{S}$ and the regions defined and without conflict over $\mathcal{R}$:

$$m_1(\mathcal{R}, \mathcal{S}) = \sum_{F \in \mathcal{F}_0(\mathcal{R})} |\mathcal{S}(F)|.$$

The expectation of $m_k(\mathcal{R}, \mathcal{S})$ for a random r-sample $\mathcal{R}$ of $\mathcal{S}$ is denoted by $m_k(r, \mathcal{S})$. In particular, $m_0(r, \mathcal{S}) = f_0(r, \mathcal{S})$.

Lemma 4.2.5

$$m_k(r, \mathcal{S}) = \sum_{i=1}^{b} \sum_{j=0}^{n-i} |\mathcal{F}_j^i(\mathcal{S})| \binom{j}{k} p_j^i(r).$$

Proof. Recall that $p_j^i(r)$ stands for the probability that a given region F of $\mathcal{F}_j^i(\mathcal{S})$ be defined and without conflict over a random r-sample of $\mathcal{S}$, whence

$$m_k(r, \mathcal{S}) = \sum_{i=1}^{b} \sum_{j=0}^{n-i} \sum_{F \in \mathcal{F}_i^j(\mathcal{S})} \binom{j}{k} p_j^i(r). \qquad \square$$

Theorem 4.2.6 (Moment theorem) *Let $\mathcal{S}$ be a set of n objects. The expectation $m_k(r, \mathcal{S})$ of the moment of order k of a random r-sample of $\mathcal{S}$ is related to the expected number $f_k(r, \mathcal{S})$ of regions defined and with k conflicts over a random r-sample of $\mathcal{S}$ by the relation*

$$m_k(r, \mathcal{S}) \le f_k(r, \mathcal{S}) \frac{(n - r + k)! \, (r - b - k)!}{(n - r)! \, (r - b)!},$$

where each region is determined by at most b objects.

Proof. According to the previous lemma 4.2.5, and to lemma 4.2.1 which gives the expression for the probability $p_j^i(r)$, we have

$$m_k(r, \mathcal{S}) \; = \; \sum_{i=1}^{b} \sum_{j=0}^{n-i} |\mathcal{F}_j^i(\mathcal{S})| \binom{j}{k} \frac{\binom{n-i-j}{r-i}}{\binom{n}{r}}$$

$$= \; \sum_{i=1}^{b} \sum_{j=0}^{n-i} |\mathcal{F}_j^i(\mathcal{S})| \frac{\binom{j}{k} \binom{n-i-j}{r-i-k}}{\binom{n}{r}} \frac{(n-j-r+k)!}{(n-j-r)!} \frac{(r-i-k)!}{(r-i)!}$$

$$\leq \; \frac{(n-r+k)!}{(n-r)!} \frac{(r-b-k)!}{(r-b)!} \sum_{i=1}^{b} \sum_{j=0}^{n-i} |\mathcal{F}_j^i(\mathcal{S})| \frac{\binom{j}{k} \binom{n-i-j}{r-i-k}}{\binom{n}{r}}.$$

As proved by the same lemma 4.2.1, however, the factor

$$\frac{\binom{j}{k} \binom{n-i-j}{r-i-k}}{\binom{n}{r}}$$

is nothing else but the probability $p_{j,k}^i(r)$ that a region F of $\mathcal{F}_j^i(\mathcal{S})$ belong to $\mathcal{F}_k^i(\mathcal{R})$, whence

$$m_k(r, \mathcal{S}) \leq f_k(r, \mathcal{S}) \frac{(n-r+k)!}{(n-r)!} \frac{(r-b-k)!}{(r-b)!}. \qquad \square$$

Corollary 4.2.7 *Let $\mathcal{S}$ be a set of n objects. There exists a real constant γ and an integer r_0, both independent of n, such that for each $n \geq r \geq r_0$,*

$$m_1(r, \mathcal{S}) \; \leq \; \gamma \frac{n-r}{r} f_0(\lfloor r/2 \rfloor, \mathcal{S})$$

$$m_2(r, \mathcal{S}) \; \leq \; \gamma \frac{(n-r)^2}{r^2} f_0(\lfloor r/2 \rfloor, \mathcal{S}),$$

where $m_k(r, \mathcal{S})$ is the expected number of the k-th moment of a random r-sample of $\mathcal{S}$, and $f_0(r, \mathcal{S})$ is the expected number of regions defined and without conflict over a random r-sample of $\mathcal{S}$.

Proof. For $k = 1$, the moment theorem yields

$$m_1(r, \mathcal{S}) \leq f_1(r, \mathcal{S}) \frac{n-r+1}{r-b}$$

and the upper bound is a consequence of corollary 4.2.4. The second inequality can be proved very much the same way. $\qquad\square$

4.3 Exercises

Exercise 4.1 (Backward analysis) In this exercise, regions are determined by at most b objects of a set $\mathcal{S}$. Let $f_j(r, \mathcal{S})$ be the expected number of regions defined and without conflict over a random r-sample of $\mathcal{S}$. Corollary 4.2.4 to the sampling theorem proves that $f_1(r, \mathcal{S}) = O(f_0(r, \mathcal{S}))$. *Backward analysis* can be used to prove this without invoking the sampling theorem.

Let $\mathcal{R}$ be a subset of $\mathcal{S}$ of cardinality r, and $f_0(r-1, \mathcal{R})$ the expected number of regions defined and without conflict over a random sample of $\mathcal{R}$ of size $r - 1$. Show that

$$f_0(r - 1, \mathcal{R}) \;\leq\; \frac{1}{r}|\mathcal{F}_1(\mathcal{R})| + \frac{r-1}{r}|\mathcal{F}_0(\mathcal{R})| \tag{4.1}$$

$$f_0(r - 1, \mathcal{R}) \;\geq\; \frac{1}{r}|\mathcal{F}_1(\mathcal{R})| + \frac{r-b}{r}|\mathcal{F}_0(\mathcal{R})|. \tag{4.2}$$

From this, show that $f_1(r, \mathcal{S}) = O(f_0(r, \mathcal{S}))$. Similarly, show that $f_2(r, \mathcal{S}) = O(f_0(r, \mathcal{S}))$.

Hint: Backward analysis consists in observing that a random $(r-1)$-sample $\mathcal{R}'$ of $\mathcal{R}$ can be obtained by removing one random object from $\mathcal{R}$. Any region in $\mathcal{F}_0(\mathcal{R}')$ is defined over $\mathcal{R}$ and belongs either to $\mathcal{F}_0(\mathcal{R})$ or to $\mathcal{F}_1(\mathcal{R})$. A region F that belongs to $\mathcal{F}_0(\mathcal{R})$ determined by i objects is a region of $\mathcal{F}_0(\mathcal{R}')$ if the removed object is not one of the i objects that determine F; this happens with probability $\frac{r-i}{r}$. A region F that belongs to $\mathcal{F}_1(\mathcal{R})$ is a region of $\mathcal{F}_0(\mathcal{R}')$ if the removed object is precisely the one that was removed from $\mathcal{R}$, which happens with probability $\frac{1}{r}$. To show that $f_1(r, \mathcal{S}) = O(f_0(r, \mathcal{S}))$, it suffices to take expectations in equation 4.2 over all r-samples of $\mathcal{S}$ and to assume that $f_0(r, \mathcal{S})$ is a non-decreasing function of r.

Exercise 4.2 (The moment theorem, using backward analysis) Let $\mathcal{R}$ be a random r-sample of a set $\mathcal{S}$ of n objects, and O a random object of $\mathcal{S} \setminus \mathcal{R}$. Show that the expected number of regions defined and without conflict over $\mathcal{R}$ but conflicting with O is $O(\frac{1}{r+1} f_1(r+1, \mathcal{S}))$. From this, show that the expected value $m_1(r, \mathcal{S})$ of the moment of order 1 with respect to $\mathcal{S}$ of a random r-sample is $O(\frac{n-r}{r+1} f_1(r+1, \mathcal{S}))$. From this, deduce an alternative proof of the moment theorem by using the result of the previous exercise or corollary 4.2.4 to the sampling theorem.

Hint: Note that $\mathcal{R} \cup \{O\}$ is a random $(r+1)$-sample of $\mathcal{S}$ and that a region of $\mathcal{F}_0(\mathcal{R})$ that conflicts with O is a region of $\mathcal{F}_1(\mathcal{R} \cup \{O\})$ that conflicts with O.

Exercise 4.3 (An extension of the moment theorem) A function w is called *convex* if it satisfies, for all x, y in $\mathbb{R}$ and all α in $[0, 1]$,

$$w(\alpha x + (1 - \alpha)y) \geq \alpha w(x) + (1 - \alpha)w(y).$$

We are interested in regions determined by at most b objects of a set S of n objects. For each subset $\mathcal{R}$ of S and any convex function w, we define:

$$w_k(\mathcal{R}) = \sum_{F \in \mathcal{F}_0(\mathcal{R})} \binom{w(|S(F)|)}{k},$$

where $\mathcal{F}_0(\mathcal{R})$ is the set of regions defined and without conflict over $\mathcal{R}$ and $|S(F)|$ is the number of objects in S that conflict with F. Let $w_k(r, S)$ stand for the expected value of $w_k(\mathcal{R})$ for a random r-sample of S. Show that

$$w_k(r, S) \leq f_0(r, S)\, w\left(\frac{(n-r-k)!}{(n-r)!} \frac{(r-b-k)!}{(r-b)!} \frac{f_k(r, S)}{f_0(r, S)} \right).$$

Exercise 4.4 (Non-local subset of regions) We still work with the framework of objects, regions, and conflicts, each region being determined by at most b objects. In this exercise, we are mostly interested, for a subset $\mathcal{R}$ of objects in S, in a subset $\mathcal{G}_0(\mathcal{R})$ of regions defined and without conflict over $\mathcal{R}$. The definition of $\mathcal{G}_0(\mathcal{R})$ is not necessarily local, however: a region F of $\mathcal{F}_0(\mathcal{R})$ belongs to $\mathcal{G}_0(\mathcal{R})$ depending on *all* the elements of $\mathcal{R}$, not only those in conflict with F or that determine F. Nevertheless, suppose that the subsets of the form $\mathcal{G}_0(\mathcal{R})$ satisfy the following property: *If F is a region of $\mathcal{G}_0(\mathcal{R})$, $\mathcal{R}'$ a subset of $\mathcal{R}$, and if $\mathcal{R}'$ contains the elements that determine F, then F is a region of $\mathcal{G}_0(\mathcal{R}')$.*

Let $w_k(r, S)$ be the expected value of the sum

$$\sum_{F \in \mathcal{G}_0(\mathcal{R})} |S(F)|^k$$

where $|S(F)|$ is the number of objects of S in conflict with F. We are interested in showing the moment theorem for the regions in $\mathcal{G}_0(\mathcal{R})$, in other words that

$$w_k(r, S) = O\left(\frac{n^k}{r^k} g_0(r, S) \right),$$

where $g_0(r, S)$ is the expected number of regions in $\mathcal{G}_0(\mathcal{R})$ for a random r-sample of S.

Hint: 1. Let $p(r, F)$ be the probability that F be a region of $\mathcal{G}_0(\mathcal{R})$ for a random r-sample $\mathcal{R}$ of S. Show that, for all $t < r \leq n$,

$$p(r, F) \leq \frac{r!}{(r-b)!} \frac{(t-b)!}{t!} p(t, F).$$

2. Let us propose an incremental algorithm to compute $\mathcal{G}_0(S)$. The probability that a region F appear in $\mathcal{G}_0(\mathcal{R})$ precisely at step r is

$$\frac{b}{r} p(r, F).$$

The probability that it disappear from $\mathcal{G}_0(\mathcal{R})$ at the next step $r + 1$ is at least

$$\frac{|S(F)|}{n-r} p(r, F).$$

Show that, for all $r_1 < r_2$,

$$p(r_1, F) + \sum_{r_1+1}^{r_2} \frac{b}{r} p(r, F) \geq \sum_{r_1}^{r_2} \frac{|\mathcal{S}(F)|}{n-r} p(r, F).$$

3. Using the previous inequality, show by induction on k that

$$\forall r, r/2 \leq r_0 \leq r \quad \sum_{t=r_0}^{r} w_k(t, \mathcal{S}) \leq \gamma_k' r \frac{n^k}{r^k} g_0(r, \mathcal{S})$$

and that

$$w_k(r, \mathcal{S}) \leq \gamma_k \frac{n^k}{r^k} g_0(r, \mathcal{S}),$$

where γ_k' and γ_k are constants depending only on k.

Exercise 4.5 (Tail estimates) Let b be the maximum number of objects that determine a single region. Suppose again that a set of at most b objects determine at most q regions, q being a constant, or that the number of regions determined by a set $\mathcal{S}$ of n objects is $O(n^b)$.

1. Let $\mathcal{S}$ be a set of n objects and $\mathcal{R}$ a random r-sample of $\mathcal{S}$. Let α be a real constant in $]0, 1[$. Let $\pi_0(\alpha, r)$ denote the probability over all samples $\mathcal{R}$ that some region defined and without conflict over $\mathcal{R}$ have at least $\lceil \alpha n \rceil$ conflicts with $\mathcal{S}$. Show that, for r big enough,

$$\pi_0(\alpha, r) = O\left(r^b(1-\alpha)^r\right).$$

2. Show that for any constant $\lambda > b$, the probability $\pi_0(\lambda \log r / r, r)$ that some region F, defined and without conflict over $\mathcal{R}$, have at least $\lambda n \log r / r$ conflicts with $\mathcal{S}$ decreases to 0 as r increases.

Exercise 4.6 (Extension of the previous tail estimates) We propose to generalize the tail estimates given in exercise 4.5. Again, let b be the maximum number of objects that determine a single region, and suppose that the number of regions determined by a set $\mathcal{S}$ of n objects is $O(n^b)$.

Let $\mathcal{S}$ be a set of n objects, $\mathcal{R}$ a random r-sample of $\mathcal{S}$, α a real constant in $]0, 1[$, and m a positive integer. Denote by $\pi_m^-(\alpha, r)$ the probability that there exists a region F defined over $\mathcal{R}$ with at most m conflicts over $\mathcal{R}$, and at least $\lceil \alpha n \rceil$ conflicts over $\mathcal{S}$. Likewise, denote by $\pi_m^+(\alpha, r)$ the probability that there exists a region F defined over $\mathcal{R}$ with at least m conflicts over $\mathcal{R}$, and at most $\lceil \alpha n \rceil$ conflicts over $\mathcal{S}$.

Show that if the size r of the sample is big enough while still smaller than $\sqrt{n}/2$, then

$$\text{if } m \leq \alpha(r - b), \qquad \pi_m^-(\alpha, r) = \frac{O(r^b)}{(1-\alpha)^b} \left[\sum_{j=0}^{m} \binom{r}{j} \alpha^j (1-\alpha)^{r-j} \right],$$

$$\text{if } m \geq \alpha(r - b), \qquad \pi_m^+(\alpha, r) = \frac{O(r^b)}{(1-\alpha)^b} \left[\sum_{j \geq m} \binom{r}{j} \alpha^j (1-\alpha)^{r-j} \right].$$

Then show that, if $\alpha(r) = \lambda \log r / r$ and $m(r) = \log r / \log \log r$,

$$\lim_{r \to \infty} \pi^-_{m(r)} (\alpha(r), r) = 0.$$

Exercise 4.7 (An upper bound on $f_0(S)$) Consider the set $\mathcal{F}(S)$ of regions defined over a set S, each region being determined by at most b objects. Let $f_j(S)$ be the number of regions defined and having j conflicts with S, and $f_0(n)$ be the maximum of $f_0(S)$ over all sets S of n objects. Suppose that there is a relation between the number of regions defined and without conflict over S on one hand, and the number of regions defined over S and conflicting with one element of S on the other. Suppose further that this relation is of the type

$$c f_0(S) \le f_1(S) + d(n) \tag{4.3}$$

where c is an integer constant and $d(n)$ a known function of n. Let $t = b - c$. Show then that

$$f_0(n) = O\left(n^t \left(1 + \sum_{j=t+1}^{n} \frac{d(j)}{j^{t+1}} \right) \right).$$

In particular,

$$
\begin{array}{rcll}
f_0(n) & = & O(n^t) & \text{if } d(n) = O(n^{t'}) \text{ for } t' < t, \\
f_0(n) & = & O(n^t \log n) & \text{if } d(n) = O(n^t), \\
f_0(n) & = & O(n^{t'}) & \text{if } d(n) = O(n^{t'}) \text{ for } t' > t.
\end{array}
$$

Hint: Combining equation 4.2, written for a random $(n-1)$-sample of S, and equation 4.3 yields

$$
\begin{array}{rcl}
\dfrac{n-b+c}{n} f_0(S) & = & \dfrac{n-b}{n} f_0(S) + \dfrac{c}{n} f_0(S) \\[2mm]
& \le & \dfrac{n-b}{n} f_0(S) + \dfrac{1}{n}(f_1(S) + d(n)) \\[2mm]
& \le & f_0(n-1, S) + \dfrac{1}{n} d(n).
\end{array}
$$

Exercise 4.8 (Union of parallel hypercubes) Consider a set of *parallel* hypercubes in $\mathbb{E}^d$, that is, hypercubes whose sides are parallel to the axes.

Show that the union of n hypercubes has at most $O(n^{\lceil d/2 \rceil})$ faces for each $d \ge 1$.

Furthermore, show that the complexity of the union of n hypercubes of equal size is $O(n^{\lfloor d/2 \rfloor})$ when $d \ge 2$ and remains $O(n)$ in dimension 1.

Hint: Each vertex of the union belongs to a bounded number of faces of the union. Hence it suffices to bound the number of vertices of the union to bound the total complexity.

The proof works by induction on d. The proof is trivial in dimension 1, and easy in dimension 2.

In dimension d, each cube has $2d$ pairwise parallel facets. Let us denote by $F_j^+(C)$ the facet of the cube C that is perpendicular to the x_j-axis with maximal j-coordinate, and by $F_j^-(C)$ the facet of the cube C that is perpendicular to the x_j-axis with minimal

j-coordinate. Let C be a set of axis-parallel cubes in $\mathbb{E}^d$, and denote by $\mathcal{U}(C)$ the union of these cubes and $\mathcal{A}(C)$ their arrangement, that is, the decomposition of $\mathbb{E}^d$ induced by the cubes (see part IV for an introduction to arrangements). Each vertex of $\mathcal{U}(C)$ or of $\mathcal{A}(C)$ is at the intersection of d facets of cubes, one perpendicular to each axis direction. Such a vertex P is denoted by $(C_1^{\epsilon_1}, C_2^{\epsilon_2}, \ldots, C_d^{\epsilon_d})$ if at the intersection of facets $F_j^{\epsilon_j}(C_j)$, for $j = 1, \ldots, d$ and $\epsilon_j = +$ or $-$. The vertex P is called *outer* if it belongs to a $(d-2)$-face of one of the cubes (then not all the cubes C_j are distinct). It is called an *inner* vertex if it is at the intersection of d facets of pairwise distinct cubes. A vertex of $\mathcal{A}(C)$ is at *level* k if it belongs to the interior of k cubes of C. The vertices of the union are precisely the vertices at level 0 in the arrangement $\mathcal{A}(C)$. Let $w_k(C)$ be the number of inner vertices of $\mathcal{A}(C)$ at level k, and $v_k(C)$ be the number of outer vertices at level k, and $v_k(n, d)$ (resp. $w_k(n, d)$) the maximum of $v_k(C)$ (resp. of $w_k(C)$) over all possible sets C of n axis-parallel hypercubes in $\mathbb{E}^d$.

1. The maximum number $v_0(n, d)$ of outer vertices of the union is $O(n^{\lceil d/2 \rceil})$ (and $O(n^{\lfloor d/2 \rfloor})$ when the cubes have same size). Indeed, any outer vertex of $\mathcal{U}(C)$ belongs to a $(d-2)$-face H of one of the cubes in C and is a vertex (either outer or inner) of the union of all $(d-2)$-cubes $C \cap aff(H)$, where $aff(H)$ is the affine hull of H. Consequently,

$$v_0(n, d) \leq 2nd(d-1)\big(\hat{v}_0(n-1, d-2) + \hat{w}_0(n-1, d-2)\big),$$

where $\hat{v}_0(n-1, d-2)$ and $\hat{w}_0(n-1, d-2)$ respectively stand for the maximum numbers of outer or inner vertices in the union of $n-1$ cubes in a $(d-2)$-dimensional space lying inside a given $(d-2)$-cube.

2. Applying the sampling theorem (theorem 4.2.3) and its corollary 4.2.4, we derive a similar bound on the maximum number $v_1(n, d)$ of outer vertices at level 1.

3. To count the number of inner vertices, we use the following charging scheme. For each vertex $P = (C_1^{\epsilon_1}, C_2^{\epsilon_2}, \ldots, C_d^{\epsilon_d})$ of $\mathcal{U}(C)$, and each direction $j = 1, \ldots d$, slide along the edge of $\mathcal{A}(C)$ that lies inside the cube C_j (this edge is $\bigcap_{i \neq j} F_i^{\epsilon_i}(C_i)$) until the other vertex P' of this edge is reached.

If P' belongs to the facet $F_j^{-\epsilon_j}(C_j)$ of cube C_j, we do not charge anything. This case cannot happen unless the cubes have different side lengths and C_j is the smallest of the cubes intersecting at P.

If P' belongs to a $(d-2)$-face of one of the cubes C_i $(i \neq j)$ intersecting at P, P' is an outer vertex at level 1, and is charged one unit for P. Note that P' cannot be charged more than twice for this situation.

If P' belongs to another cube C' distinct from all the C_i intersecting at P, then P' is an inner vertex at level 1, and is charged one unit for P. Any inner vertex P' of this type may be charged up to d times for this situation. However, when it is charged more than once, say m times, we may redistribute the extra $m-1$ charges on the outer vertices at level 0 or 1, and these vertices will only be charged once in this fashion.

In the case of cubes with different sizes, the induction is

$$(d-1)w_0(C) \leq w_1(C) + 3v_1(C) + v_0(C).$$

In the case of cubes with identical sizes, we obtain

$$dw_0(C) \leq w_1(C) + 3v_1(C) + v_0(C).$$

It suffices to apply exercise 4.7 to conclude.

4.4   Bibliographical notes

Randomized methods revolutionized computational geometry. Most of the material in this chapter is taken from the ground-breaking work of Clarkson and Shor [71]. The randomized incremental algorithms in the next two chapters are concrete applications of the formalism developed in this chapter, and we invite the reader to consult the bibliographical notes of these chapters for more references. Clarkson and Shor proved the tail estimates and their extension as stated in exercises 4.5 and 4.6, which are the corner stone on which all analyses of randomized divide-and-conquer algorithms rely. In their article, they also prove the extension to the moment theorem proposed in exercise 4.3. This extension will prove useful in exercise 5.8 for the analysis of an algorithm that triangulates a simple polygon, due to Clarkson, Cole, and Tarjan [69].

The extension of the moment theorem to a non-local set of regions defined and without conflict over a random sample (exercise 4.4) is due to de Berg, Dobrindt, and Schwarzkopf [76]. The result stated in exercise 4.4 will be used in chapter 15 to analyze the randomized incremental algorithm that builds a single cell in an arrangement of line segments.

The *backward analysis* method proposed in exercises 4.1 and 4.2 was used by Chew [59] to analyze an algorithm that builds the Voronoi diagram of a convex polygon (see exercise 19.4). The method was used later in a systematic fashion by Seidel [203] and Devillers [80].

The method used in exercise 4.7 to obtain an upper bound on the expected number of regions defined and without conflict over a set of objects is due to Tagansky [212]. The analysis of the complexity of the union of parallel hypercubes in d dimensions (see exercise 4.8) given by Boissonnat, Sharir, Tagansky, and Yvinec [34] illustrates the power of this method.

Chapter 5

Randomized algorithms

A *randomized algorithm* is an algorithm that makes random choices during its execution. A randomized algorithm solves a deterministic problem and, whatever the random choices are, always runs in a finite time and outputs the correct solution to the problem. Therefore, only the path that the algorithm chooses to follow to reach the solution is random: the solution is always the same. Most randomized methods lead to conceptually simple algorithms, which often yield a better performance than their deterministic counterparts. This explains the success encountered by these methods and the important position they are granted in this book. The time and space used when running a randomized algorithm depend both on the input set and on the random choices. The performances of such an algorithm are thus analyzed on the average over all possible random choices made by the algorithm, yet in the worst case for the input. Randomization becomes interesting when this average complexity is smaller than the worst-case complexity of deterministic algorithms that solve the same problem.

The randomized algorithms described in this chapter, and more generally encountered in this book, use the randomized incremental method. The incremental resolution of a problem consists of two stages: first, the solution for a small subset of the data is computed, then the remaining input objects are inserted while the current solution is maintained. An incremental algorithm is said to be randomized if the data are inserted in a deliberately random order. For instance, sorting by insertion can be considered as a randomized incremental method: the current element, randomly chosen among the remaining ones, is inserted into the already sorted list of previously inserted elements.

Some incremental algorithms can work *on line*: they do not require prior knowledge of the whole set of data. Rather, these algorithms maintain the solution to the problem as the input data are successively inserted, without looking ahead at the objects that remain to be inserted. We refer to such algorithms as *on-line* or *semi-dynamic* algorithms. The order in which the data are inserted is imposed on the algorithm, and such algorithms cannot properly be called randomized, as their behavior is purely deterministic. Nevertheless, we may be interested in the behavior of such an algorithm when the insertion order is assumed to be random. We may then speak of the *randomized analysis* of an on-line algorithm.

In the first section of this chapter, the randomized incremental method is sketched in the framework introduced in the previous chapter, with objects, regions, and conflicts. The underlying probabilistic model is made clear. At any step, a randomized incremental algorithm must detect the conflicts between the current object and the regions created previously. One way of detecting these conflicts is to use a *conflict graph*. Algorithms using a conflict graph must have a global knowledge of the input and are thus *off-line*. Another way is to use an *influence graph*. This structure leads to semi-dynamic algorithms and allows the objects to be inserted on-line. The conflict graph is described in section 5.2 and the influence graph in section 5.3. In both cases, the method is illustrated by an algorithm that builds the *vertical decomposition* of a set of line segments in the plane. This planar map was introduced in section 3.3, and in particular one can deduce from it all the intersecting pairs of segments. Both methods lead to a randomized algorithm that runs in time $O(n \log n + a)$ on the average, where a is the number of intersecting pairs of segments, and this is optimal. Finally, section 5.4 shows how both methods may be combined and, sometimes, lead to *accelerated* algorithms. For instance, we show how to decompose the set of line segments forming the boundary of a simple polygon in time $O(n \log^* n)$ on the average (provided that the order of the edges along the boundary of the polygon is also part of the input).

We give several randomized incremental algorithms, for instance to compute convex hulls (chapter 8), to solve linear programs (chapter 10), to compute the lower envelope of a set of segments in the plane (chapter 15) or of triangles in three-dimensional space (chapter 16), or even to compute the k-level of an arrangement of hyperplanes (chapter 14) or a cell in an arrangement of segments (chapter 15) or of triangles (chapter 16).

5.1   The randomized incremental method

The problem to be solved is formulated in terms of objects, regions, and conflicts in the general framework described in the previous chapter. The problem now becomes that of finding the set $\mathcal{F}_0(\mathcal{S})$ of regions defined and without conflict over a finite set $\mathcal{S}$ of objects. The notation used in this chapter is that defined in subsections 4.1.1 and 4.1.2.

The initial step in the incremental method builds the set $\mathcal{F}_0(\mathcal{R}_0)$ of regions that are defined and without conflict over a small subset $\mathcal{R}_0$ of $\mathcal{S}$. Each subsequent step consists of processing an object of $\mathcal{S} \setminus \mathcal{R}_0$. Let $\mathcal{R}$ be the set of already processed objects and let us call *step r* the step during which we process the r-th object.

Let O be the object processed in step r. From the already computed set of regions defined and without conflict over $\mathcal{R}$, we compute in step r the set of regions defined and without conflict over $\mathcal{R} \cup \{O\}$.

- The regions of $\mathcal{F}_0(\mathcal{R})$ that do not belong to $\mathcal{F}_0(\mathcal{R} \cup \{O\})$ are exactly those regions in $\mathcal{F}_0(\mathcal{R})$ that conflict with O. These regions are said to be *killed* by O, and O is their *killer*.

- The regions of $\mathcal{F}_0(\mathcal{R} \cup \{O\})$ that do not belong to $\mathcal{F}_0(\mathcal{R})$ are exactly those regions $\mathcal{F}_0(\mathcal{R} \cup \{O\})$ that are determined by a subset of $\mathcal{R} \cup \{O\}$ that contains O. These regions are said to be *created* by O.

A region created by O during step r is said to be *created at step r*. The set of regions created by an incremental algorithm consists of all the regions created during the initial step or at one of the subsequent insertion steps.

The *chronological sequence* is the sequence of the objects of $\mathcal{S}$ in the order in which they are processed. The probabilistic assumption on which the randomized analysis of incremental algorithms relies is that the chronological sequence is any of the $n!$ possible permutations of the objects of $\mathcal{S}$ with uniform probability. As a consequence, the subset of objects $\mathcal{R}$ already processed at step r is a random r-sample of $\mathcal{S}$, and any subset of $\mathcal{S}$ is equally likely. The object O processed during step r is a random element of $\mathcal{S} \setminus \mathcal{R}$. Equivalently, it is any element of $\mathcal{R} \cup \{O\}$ with a uniform probability.

5.2 Off-line algorithms

5.2.1 The conflict graph

The first task of each step in an incremental algorithm is to detect the regions of $\mathcal{F}_0(\mathcal{R})$ that conflict with the object O to be processed at this step. These are the regions killed by O. To check all the regions in $\mathcal{F}_0(\mathcal{R})$ does not lead to efficient algorithms. Instead, in addition to the set $\mathcal{F}_0(\mathcal{R})$, we maintain a graph called the *conflict graph*. The conflict graph is a bipartite graph between the objects of $\mathcal{S} \setminus \mathcal{R}$ and the regions of $\mathcal{F}_0(\mathcal{R})$. Arcs belong to $\mathcal{F}_0(\mathcal{R}) \times (\mathcal{S} \setminus \mathcal{R})$, and are called the *conflict arcs*. There is a conflict arc (F, S) exactly for each region F of $\mathcal{F}_0(\mathcal{R})$ and each object S of $\mathcal{S} \setminus \mathcal{R}$ that conflicts with F.

The conflict graph allows us to find all the regions killed by an object O in time proportional to the number of those regions. Each step of the incremental algorithm must then update the conflict graph. The conflict arcs incident to the regions killed by O are removed and the new conflict arcs incident to the regions created by O are found. The complexity of each incremental step is thus at least bounded from below by the number of regions that are killed or created during this step, and by the number of conflict arcs that are removed or added during this step.

Update condition 5.2.1 (for conflict graphs) *A randomized incremental algorithm that uses a conflict graph satisfies the* update condition *if, at each incremental step:*

1. *updating the set of regions defined and without conflict over the current subset can be carried out in time proportional to the number of regions killed or created during this step, and*

2. *updating the conflict graph can be carried out in time proportional to the number of conflict arcs added or removed during this step.*

Lemma 5.2.2 *Let S be a set of n objects and F a region determined by i objects of S that has j conflicts with the objects in S.*

1. *The probability p'^i_j that F be one of the regions created by a randomized incremental algorithm processing S is*

$$p'^i_j = \frac{i!\,j!}{(i+j)!}.$$

2. *The probability $p'^i_j(r)$ that F be one of the regions created by the algorithm during step r is*

$$p'^i_j(r) = \frac{i}{r} p^i_j(r).$$

In these expressions, $p^i_j(r)$ stands for the probability that a region F of $\mathcal{F}^i_j(S)$ be defined and without conflict over a random r-sample of S, as given in subsection 4.2.1.

If we replace $p^i_j(r)$ by its expression in lemma 4.2.1, we obtain (see also exercise 5.1) that the probabilities p'^i_j and $p'^i_j(r)$ satisfy the relation

$$p'^i_j = \sum_{r=1}^{n} p'^i_j(r).$$

Proof. A region F of $\mathcal{F}^i_j(S)$ is created by a randomized incremental algorithm if and only if, during some step of the algorithm, this region is defined and without conflict over the current subset. Such a situation occurs when the i objects that determine F are processed before any of the j objects of S that conflict with F. Since all permutations of these objects are equally likely, this case happens with probability

$$\frac{i!j!}{(i+j)!},$$

proving the first part of the lemma. Let $\mathcal{R}$ be the set of objects processed in the steps preceding and including step r. For a region F to be created during step r, we first require that F be defined and without conflict over $\mathcal{R}$, which happens precisely with probability $p_j^i(r)$. If so, F is created at step r precisely if the object O processed during step r is one of the i objects of $\mathcal{R}$ that determine F. This happens with conditional probability i/r. $\square$

Theorem 5.2.3 (Conflict graph) *Let S be a set of n objects, and consider a randomized incremental algorithm that uses a conflict graph to process S.*

1. *The expected total number of regions created by the algorithm is*

$$O\left(\sum_{r=1}^{n} \frac{f_0(r,S)}{r}\right).$$

2. *The expected total number of conflict arcs added to the conflict graph by the algorithm is*

$$O\left(n\sum_{r=1}^{n} \frac{f_0(r,S)}{r^2}\right).$$

3. *If the algorithm satisfies the update condition, then its complexity (both in time and in space) is, on the average,*

$$O\left(n\sum_{r=1}^{n} \frac{f_0(r,S)}{r^2}\right).$$

In these expressions, $f_0(r,S)$ denotes the expected number of regions defined and without conflict over a random r-sample of S.

Thus, if $f_0(r,S)$ behaves linearly with respect to r ($f_0(r,S) = O(r)$), the total number of created regions is $O(n)$ on the average, the total number of conflict arcs is $O(n \log n)$ on the average, and the complexity of the algorithm is $O(n \log n)$ on the average. If the growth of $f_0(r,S)$ is super-linear with respect to r ($f_0(r,S) = O(r^\alpha)$ for some $\alpha > 1$), then the total number of created regions is $O(n^\alpha)$ on the average, the total number of conflict arc is $O(n^\alpha)$ on the average, and the complexity of the algorithm is $O(n^\alpha)$ on the average.

Proof.

1. We obtain the expectation $v(\mathcal{S})$ of the total number of regions created by the algorithm by summing, over all regions F defined over $\mathcal{S}$, the probability that this region F be created by the algorithm:

$$v(\mathcal{S}) = \sum_{i=1}^{b} \sum_{j=0}^{n-i} |\mathcal{F}_j^i(\mathcal{S})| \, p_j'^i = \sum_{i=1}^{b} \sum_{j=0}^{n-i} \sum_{r=1}^{n} |\mathcal{F}_j^i(\mathcal{S})| \, \frac{i}{r} \, p_j^i(r).$$

In this expression, we recognize the expected number of regions defined and without conflict over a random r-sample of $\mathcal{S}$ (lemma 4.2.2), so we get

$$v(\mathcal{S}) = O\left(\sum_{r=1}^{n} \frac{f_0(r, \mathcal{S})}{r} \right).$$

2. Let $e(\mathcal{S})$ be the expected total number of arcs added to the conflict graph. To estimate $e(\mathcal{S})$, we note that if a region F in conflict with j objects of $\mathcal{S}$ is a region created by the algorithm, then it is adjacent to j conflict arcs in the graph. Therefore,

$$e(\mathcal{S}) = \sum_{i=1}^{b} \sum_{j=0}^{n-i} |\mathcal{F}_j^i(\mathcal{S})| \, j \, p_j'^i, = \sum_{i=1}^{b} \sum_{j=0}^{n-i} \sum_{r=1}^{n} |\mathcal{F}_j^i(\mathcal{S})| \, \frac{ij}{r} \, p_j^i(r).$$

Apart from the factor i/r, we recognize in this expression the moment of order 1 of a random r-sample (lemma 4.2.5). Applying corollary 4.2.7 to the moment theorem, we get

$$\begin{aligned}
e(\mathcal{S}) \;&\leq\; b \sum_{r=1}^{n} \frac{m_1(r, \mathcal{S})}{r} \\
&=\; O\left(\sum_{r=1}^{n} \frac{n}{r^2} f_0(\lfloor r/2 \rfloor, \mathcal{S}) \right) = O\left(n \sum_{r=1}^{n} \frac{f_0(r, \mathcal{S})}{r^2} \right).
\end{aligned}$$

3. A given region is killed or created at most once during the course of the algorithm and, likewise, a given conflict arc is added or removed at most once. A randomized incremental algorithm which satisfies the update condition thus has an average complexity of at most

$$v(\mathcal{S}) + e(\mathcal{S}) = O\left(n \sum_{r=1}^{n} \frac{f_0(r, \mathcal{S})}{r^2} \right). \qquad \square$$

Exercise 5.2 presents a non-amortized analysis of each step of an algorithm that uses a conflict graph.

5.2.2   An example: vertical decomposition of line segments

Let us discuss again the problem of finding all the a intersecting pairs of a set of n line segments in the plane. The space sweep algorithm presented in chapter 3 solves this problem in time $O((n+a)\log n)$, which is suboptimal because a could be as large as $\Omega(n^2)$. The complexity of the problem is $O(n\log n + a)$, and is matched by a deterministic algorithm (see section 3.5). The algorithm we present here is randomized incremental, and its expected complexity also matches the optimal bound. It is much simpler than the deterministic algorithm, and generalizes easily to the case of curved segments (see exercise 5.5).

This algorithm builds in fact the vertical decomposition of the set of segments S, or more precisely its simplified decomposition $\mathcal{Dec}_s(S)$. This structure is discussed in detail in section 3.3. Let us simply recall that the vertical decomposition of S is induced by the segments of S and the vertical walls stemming from their endpoints and intersection points (see figure 3.2 or figure 5.4 below). The decomposition of the set S of segments with a intersecting pairs is of size $O(n+a)$, and can be used to report all the intersecting pairs of S in the same time bound $O(n+a)$.

To use the preceding framework, we define the set of objects $\mathcal{O}$ to be the set of all segments in the plane, the universe of regions $\mathcal{F}$ being the set of all possible trapezoids with parallel vertical sides, occasionally in some degenerate state such as triangles, semi-infinite trapezoids, vertical slabs, or half-planes bounded by a vertical line. A trapezoid F of $\mathcal{F}$ conflicts with a segment S if this segment intersects the interior of F. A trapezoid F is determined by a minimal set of segments $\mathcal{X}$, the decomposition of which includes F. Each region in the simplified decomposition $\mathcal{Dec}_s(S)$ is described using at most four segments of S. Indeed, the floor and ceiling are each supported by a single segment in S, and each vertical side consists of one or two walls stemming either from an endpoint of a segment, or from an intersection of two segments. In the former case, the vertical side consists either of one vertical wall stemming upward from an endpoint of the floor, or stemming downward from an endpoint of the ceiling, or of two vertical walls stemming in both directions from the endpoint of another segment. In the latter case, one of the intersecting segments is necessarily supporting the floor or ceiling of the trapezoid. In all cases, we see that four segments suffice to determine the trapezoid. Three may suffice if the trapezoid is a degenerate triangle. Partially infinite triangles may be determined by three, two, or even one segment. The cardinality of the set $\mathcal{X}$ that determines a region is thus at most four, and the constant b that upper-bounds the number of objects that determine a region is thus set to $b = 4$.

Therefore, the set of trapezoids in the vertical decomposition of the set S of segments is exactly the set of those segments defined and without conflict over S. According to our notation, this set is denoted by $\mathcal{F}_0(S)$.

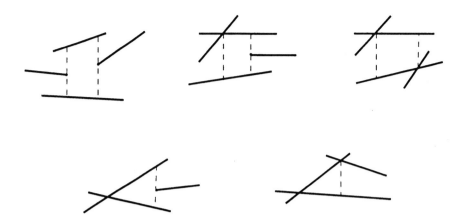

Figure 5.1. Instances of trapezoids in the simplified vertical decomposition.

The algorithm

According to the randomized incremental scheme, the algorithm processes the segments of S one by one in a random order. Again, we denote by $\mathcal{R}$ the set of the already processed segments at a given step r. The algorithm maintains the simplified vertical decomposition $\mathcal{D}ec_s(\mathcal{R})$ of this set of segments, as well as the corresponding conflict graph.

The decomposition $\mathcal{D}ec_s(\mathcal{R})$ is maintained in a data structure that encodes its simplified description. This structure includes the list of all trapezoids of $\mathcal{D}ec_s(\mathcal{R})$. Each trapezoid is described by the set of at most four segments that determine it, and by the at most six edges in the simplified decomposition (floor, ceiling, and at most four vertical walls) which bound it. The data structure also describes the vertical adjacency relationships in the decomposition, that is the pairs of trapezoids whose boundaries share a vertical wall. Such an adjacency is stored in a bidirectional pointer linking both trapezoids. Note that each trapezoid is vertically adjacent to at most four trapezoids.

The conflict graph has a conflict arc for each pair (F, S) of a trapezoid F of $\mathcal{D}ec_s(\mathcal{R})$ and a segment S of $S \setminus \mathcal{R}$ that intersects the interior of F. The conflict graph is implemented by a system of interconnected lists.

- For each segment S of $S \setminus \mathcal{R}$, the data structure stores a list $\mathcal{L}(S)$ representing the set of trapezoids of $\mathcal{D}ec_s(\mathcal{R})$ intersected by S. The list $\mathcal{L}(S)$ is ordered according to which trapezoids are encountered as we slide along S from left to right.

- For each trapezoid F of $\mathcal{D}ec_s(\mathcal{R})$, the algorithm maintains the list $\mathcal{L}'(F)$ of the segments in $S \setminus \mathcal{R}$ that conflict with F.

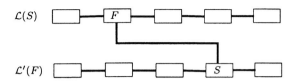

Figure 5.2. Representation of the conflict graph used to build the vertical decomposition of a set of line segments.

- The node that stores a segment S in the list $\mathcal{L}'(F)$ of a trapezoid F is interconnected via a bidirectional pointer with the node that stores the trapezoid F in the list $\mathcal{L}(S)$ of the segment S (see figure 5.2).

In the *initial step*, the algorithm builds the decomposition $Dec_s(\mathcal{R})$ for a subset $\mathcal{R}$ of S that contains only a single segment. This decomposition consists of four trapezoids. It also initializes the lists that represent the conflict graph. The initial decomposition is built in constant time, and the initial conflict graph in linear time.

The *current step* processes a new segment S of $S \setminus \mathcal{R}$: it updates the decomposition and the conflict graph accordingly.

Updating the decomposition. The conflict graph gives the list $\mathcal{L}(S)$ of all the trapezoids of $Dec_s(\mathcal{R})$ that are intersected by S. Each trapezoid is split into at most four *subregions* by the segment S, the walls stemming from the endpoints of S, and the walls stemming from the intersection points between S and the other segments in $\mathcal{R}$ (see figure 5.3).

These subregions are not necessarily trapezoids of $Dec_s(\mathcal{R} \cup \{S\})$. Indeed, S intersects some vertical walls of $Dec_s(\mathcal{R})$, and any such wall must be shortened: the portion of this wall that contains no endpoint or intersection point must be removed from the decomposition, and the two subregions that share this portion of the wall must be joined into a new trapezoid of $Dec_s(\mathcal{R} \cup \{S\})$ (see figure 5.4). Thus, any trapezoid of $Dec_s(\mathcal{R} \cup \{S\})$ created by S is either a subregion, or the union of a maximal subset of subregions that can be ordered so that two consecutive subregions share a portion of a wall to be removed. The vertical adjacency relationships in the decomposition that concern trapezoids created by S can be inferred from the vertical adjacency relationships between the subregions and from those between the trapezoids of $Dec_s(\mathcal{R})$ that conflict with S.

Updating the data structure that represents the decomposition $Dec_s(\mathcal{R})$ can therefore be carried out in time linear in the number of trapezoids conflicting with S.

Updating the conflict graph. When a trapezoid F is split into subregions F_i ($i \leq 4$), the list $\mathcal{L}'(F)$ of segments that conflict with F is traversed linearly, and a conflict list $\mathcal{L}'(F_i)$ is set up for each of the subregions F_i. During this

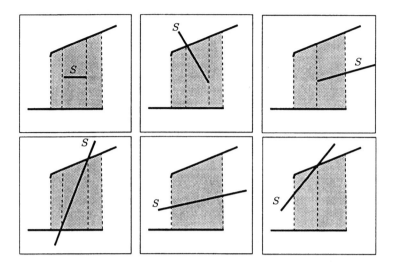

Figure 5.3. Decomposing a set of segments: splitting a trapezoid into at most four new trapezoids.

traversal, the list $\mathcal{L}(S')$ of each segment S' in $\mathcal{L}'(F)$ is updated as follows: each node pointing to F in such a list is replaced by the sequence of those subregions F_i that intersect S', in the left-to-right order along S'.

Consider now a sequence $F_1, F_2, \ldots, F_k$ of subregions that have to be joined to yield a trapezoid F' of $\mathcal{D}ec_s(\mathcal{R} \cup \{S\})$ created by S. We assume that the subregions are encountered in this order along S. To build the list $\mathcal{L}'(F')$, we must merge the lists $\mathcal{L}'(F_i)$ while at the same time removing redundant elements. To do this, we traverse successively each of the lists $\mathcal{L}'(F_i)$. For each segment S' that we encounter in $\mathcal{L}'(F_i)$, we obtain the entry corresponding to F_i in the list $\mathcal{L}(S')$ by following the bidirectional pointer in the entry corresponding to S' in the list $\mathcal{L}'(F_i)$. The subregions that conflict with S' and that have to be joined are consecutive in the list $\mathcal{L}(S')$. The nodes that correspond to these regions are removed from the list $\mathcal{L}(S')$, and for each entry F_j removed from the list $\mathcal{L}(S')$, the corresponding entry for S' in $\mathcal{L}'(F_j)$ is also removed. (This process is illustrated in figure 5.5.) In this fashion, we merge the conflict lists of a set of adjacent subregions while visiting each node of the conflict lists of these subregions once and only once. Similarly, the corresponding nodes in the conflict lists of the segments are visited once and only once. This ensures that the time taken to update the conflict graph is linear in the number of arcs of the graph that have to be removed.

Analysis of the algorithm

The preceding discussion shows that the algorithm which computes the vertical decomposition of a set of line segments using a conflict graph obeys the update

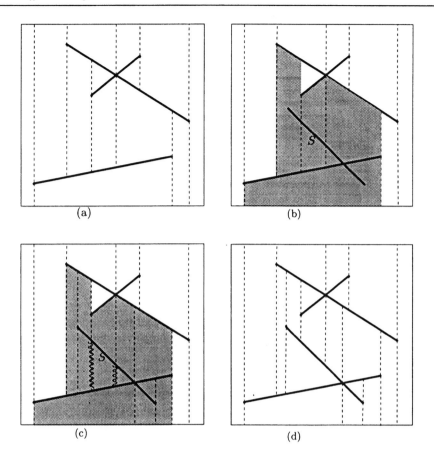

Figure 5.4. Decomposing a set of segments: the incremental construction
(a) The decomposition before inserting segment S.
(b) Shaded, the trapezoids that conflict with segment S.
(c) Splitting those trapezoids. Wavy lines indicate the portions of walls to be removed.
(d) The decomposition after inserting segment S.

condition 5.2.1. We may therefore quote theorem 5.2.3 to show that the average running time of the algorithm, given a set of n segments with a intersection points, is

$$O\left(n\sum_{1\leq r\leq n}\frac{f_0(r,\mathcal{S})}{r^2}\right).$$

Here, $f_0(r,\mathcal{S})$ is the expected number of trapezoids in the vertical decomposition of a random r-sample of $\mathcal{S}$. The following lemma estimates this number.

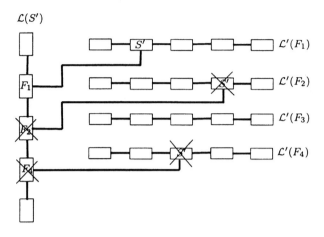

Figure 5.5. Decomposing a set of line segments: merging the conflict lists.

Lemma 5.2.4 *Let S be a set of n segments, a pairs of which have a non-empty intersection. The expected number $f_0(r, S)$ of trapezoids in the vertical decomposition of a random r-sample of S is $O(r + ar^2/n^2)$.*

Proof. Let $\mathcal{R}$ be a subset of r segments in S, and denote by $a(\mathcal{R})$ the number of intersecting pairs of segments in $\mathcal{R}$. The number of regions in the vertical decomposition of $\mathcal{R}$ is therefore $O(r + a(\mathcal{R}))$. If $\mathcal{R}$ is a random r-sample of S, however, the expected number $a(\mathcal{R})$ of intersections is $\frac{ar(r-1)}{n(n-1)}$. Indeed, an intersection point P between two segments of S is an intersection point of two segments of $\mathcal{R}$ if the two segments of S that intersect at P belong to $\mathcal{R}$, which happens with probability

$$\binom{n-2}{r-2} \bigg/ \binom{n}{r}.$$

$\square$

Therefore, we immediately derive the following theorem.

Theorem 5.2.5 *The vertical decomposition of a set S of n line segments in the plane with a intersecting pairs can be obtained by a randomized incremental algorithm that uses a conflict graph, in expected time $O(n \log n + a)$.*

Remark 1. The algorithm builds only the *simplified* representation $\mathcal{D}ec_s(S)$ of the decomposition. From this representation, however, we can derive the complete representation $\mathcal{D}ec(S)$ of the decomposition in $O(n + a)$ time. Exercise 5.3 shows that the complete representation can be directly computed by a variant of the preceding algorithm, while still using no more than $O(n \log n + a)$ running time.

Remark 2. The expected storage of the algorithm is $O(n \log n + a)$. In the variant mentioned in the above remark, it is possible to simplify the conflict graph: for each segment, we retain only a single conflict arc, for instance the conflict with the trapezoid which contains the left endpoint of the segment. We can still update the conflict graph in linear time, therefore the average running time is unchanged and still $O(n \log n + a)$, but the expected storage is lowered to $O(n + a)$ (see exercise 5.4).

5.3 On-line algorithms

Algorithms that use a conflict graph are incremental but static, that is, they require initial knowledge of all the segments to be inserted. In contrast, on-line (or semi-dynamic) algorithms maintain the solution to the problem as the input objects are inserted, with no preliminary knowledge of the input data. A possible way to transform an algorithm that uses a conflict graph into an on-line algorithm is to replace the conflict graph by a different kind of structure that can detect conflicts between *any* object and the regions defined and without conflict over the current set of objects. The *influence graph* is such a structure.

5.3.1 The influence graph

The influence graph is a structure that stores the history of the incremental construction and depends on the order in which the objects have been processed by the algorithm. This graph represents the regions created by the algorithm during the incremental construction, and can be used to detect the conflicts between these regions and a new object. When the algorithm uses a conflict graph, the set of data is known in advance, and the algorithm may then compute the objects in S that conflict with a given region. However, an on-line algorithm does not assume any knowledge of the objects to be processed. Thus it must be able to describe the entire domain of influence of a region which, as we recall, is the subset of all the objects in the universe that conflict with this region.

The influence graph is a directed, acyclic, and connected graph. It possesses a single root, and its nodes correspond to the regions created by the algorithm during its entire history. Therefore, a node corresponds to a region that was defined and without conflict over the current set of objects at some point during the execution of the algorithm. Properly speaking, this graph is not a tree: a node might have several parents. Nevertheless, the terminology of trees will be quite useful for describing it. In particular, a *leaf* is a node that has no children. The influence graph must possess two essential properties.

Property 5.3.1   *1. At each step of the algorithm, a region defined and without conflict over the current subset is associated with a leaf of the influence graph.*

   2. The domain of influence of a region associated with a node of the influence graph is contained in the union of the domains of influence of the regions associated with the parents of that node.

To shorten and simplify the terminology, we attach qualifiers normally used for a node to its corresponding region and vice versa. This allows us to speak of the domain of influence of a node instead of the domain of influence of its associated region. Likewise, a region has children or parents. This slight abuse in the terminology should not create any confusion.

The algorithm

The algorithm is incremental and maintains the set $\mathcal{F}_0(\mathcal{R})$ of regions defined and without conflict over the current set $\mathcal{R}$, together with the influence graph corresponding to the chronological sequence of objects in $\mathcal{R}$.

   The *initial step* processes a small set of r_0 objects. For instance, r_0 can be the minimal number of objects needed to determine a region. The algorithm computes the regions defined and without conflict over the set $\mathcal{R}_0$ of these r_0 objects. The influence graph is initialized by creating a root node, corresponding to a fictitious region whose influence domain is the universe $\mathcal{O}$ of objects in its entirety. A node whose parent is the root is created for each of the regions of $\mathcal{F}(\mathcal{R}_0)$.

   In the *current step*, the object O is added to $\mathcal{R}$. The work can be divided into two phases: we first locate O and then update the data structures.

   Locating. In this phase, we must find all the regions in $\mathcal{F}_0(\mathcal{R})$ that conflict with the new object O. Starting from the root of the influence graph, we recursively visit all the nodes that conflict with O, and their children. The regions of $\mathcal{F}_0(\mathcal{R})$ that conflict with O are said to be *killed* by O.

   Updating. We now have to update the data structure that represents the set of those regions defined and without conflict over the current subset of objects ($\mathcal{F}_0(\mathcal{R})$ becomes $\mathcal{F}_0(\mathcal{R}\cup\{O\})$). We also have to update the influence graph. A leaf of the influence graph is created for each of the regions in $\mathcal{F}_0(\mathcal{R} \cup \{O\}) \setminus \mathcal{F}_0(\mathcal{R})$. These are the regions *created* by O. Each of these leaves is linked to enough parents to satisfy property 2 of the influence graph. We never remove a node from the graph.

   The details of the implementation of these steps naturally depend on the problem. Typically, the set of regions created by O can be derived from the set of regions killed by O, and the parents of the leaves corresponding to created regions may be chosen among the nodes corresponding to regions killed by O. In this case, the information needed to carry out the update is gathered in the locating phase. In some cases, we may need to know some extra information, such as adjacency relationships between regions. The influence graph is then augmented with the required information.

Randomized analysis of the influence graph

A randomized on-line algorithm is not a randomized algorithm properly speaking. Indeed, the order in which the data are processed is imposed on the algorithm, and the algorithm makes no random choices whatsoever. The algorithm is therefore perfectly deterministic. Nevertheless, we can analyze the average performance of such an algorithm (in running time or storage) under the assumption that all inputs are equally likely or, more precisely, that any permutation of the input data is equally likely. The performances of the algorithm are valid for any input, and in particular no assumption is made on the distribution of the input. In this case, the analysis of the algorithm is said to be a *randomized analysis*. The randomized analysis of an on-line algorithm which has currently processed a set S of n objects assumes that the chronological sequence Σ, made up from the objects of S in the processing order, is a random sequence. This means that all $n!$ permutations of S are equally likely to occur in Σ. At any step, the current subset of objects already processed by the algorithm is thus a random sample of S.

Each node of the influence graph corresponds to a region created at some step of the algorithm. Such a region was, at this step, defined and without conflict over the current subset of objects. The set of those regions created by an incremental algorithm depends only on the order in which the objects are inserted. In particular, it does not depend on whether the incremental algorithm uses a conflict graph or an influence graph. The following theorem is thus an immediate consequence of theorem 5.2.3.

Theorem 5.3.2 *Let an on-line algorithm use an influence graph to process a set S of n objects. The expected number $v(S)$ of nodes in this influence graph is*

$$O\left(\sum_{r=1}^{n} \frac{f_0(r, S)}{r}\right).$$

In this expression, $f_0(r, S)$ denotes the expected number of regions defined and without conflict over a random r-sample of S.

To carry the analysis further, we must also be able to bound the number of arcs in the influence graph, since this number gives the time and storage taken to update the set of regions without conflict and the influence graph itself, as is done in the second phase of each insertion step of the algorithm. We also need a special assumption to control the complexity of testing whether there is a conflict between an object and a region.[1] The triple update condition stated below is actually satisfied by a large class of practical problems.

Update condition 5.3.3 (for influence graphs) *An on-line algorithm that uses an influence graph satisfies the* update condition *if:*

1. *the existence of a conflict between a given region and a given object can be tested in constant time.*

2. *the number of children of each node of the influence graph is bounded by a constant, and*

3. *the parents of a node created by an object O are nodes that are killed by O, and updating the influence graph takes time linear in the number of nodes killed or created at each step.*

Theorem 5.3.4 (Influence graph) *Consider an on-line algorithm that uses an influence graph to process a set S of n objects. If the algorithm satisfies the update condition 5.3.3, then:*

1. *The expected storage used by the algorithm to process the n objects is*

$$O\left(\sum_{r=1}^{n}\frac{f_0(r,S)}{r}\right).$$

2. *The expected time complexity of the algorithm is*

$$O\left(n\sum_{r=1}^{n}\frac{f_0(r,S)}{r^2}\right).$$

3. *The expected time complexity of the locating phase at step k is*

$$O\left(\sum_{r=1}^{k-1}\frac{f_0(r,S)}{r^2}\right).$$

4. *The expected time complexity of the updating phase at step k is*

$$O\left(\frac{f_0(k,S)}{k}+\frac{f_0(\lfloor(k-1)/2\rfloor,S)}{k-1}\right).$$

---

[1]Note that such an assumption is implicitly contained in the update condition 5.2.1 when the algorithm uses a conflict graph.

As always, $f_0(r, S)$ denotes the expected number of regions defined and without conflict over a random r-sample of S.

Thus, the expected time complexity of an on-line algorithm that uses an influence graph is identical to that of a similar incremental algorithm that uses a conflict graph, as long as the respective update conditions are satisfied.

If $f_0(r, S)$ behaves linearly with respect to r ($f_0(r, S) = O(r)$), the complexity of the algorithm is $O(n \log n)$ on the average, and the expected storage is $O(n)$. Introducing the n-th object takes time $O(\log n)$ for the locating phase, and constant time for updating the data structure and the influence graph.

If the growth of $f_0(r, S)$ is super-linear with respect to r ($f_0(r, S) = O(r^\alpha)$ for some $\alpha > 1$), the expected storage is $O(n^\alpha)$. Introducing the n-th object takes time $O(n^{\alpha-1})$ for the locating and updating phases.

Proof.

1. The upper bound on the expected storage is a direct consequence of theorem 5.3.2, which bounds the number of nodes in the influence graph, and of the second clause in the update condition, which bounds the number of children of each node.

2. The contribution to the running time complexity of the updating phases is proportional to the number of regions created, because of the third clause of the update condition. From theorem 5.2.3, we know that this number is

$$O\left(\sum_{r=1}^n \frac{f_0(r, S)}{r}\right).$$

We still must evaluate the cost of the locating phases. From the first clause of the update condition, we derive that the complexity of locating an object O is proportional to the number of nodes visited to locate O. If every child has a constant number of descendants (second clause in the update condition), however, the number of nodes visited during the locating phase of O is at most proportional to the nodes of the influence graph that conflict with O. The overall cost of the locating phases is therefore proportional to the total number of conflicts detected during the algorithm.

Let F be a region of $\mathcal{F}_j^i(S)$. If this region is created at some step of the algorithm, the corresponding node in the influence graph will be visited j times in the subsequent steps, and this happens each time we insert one of the j objects that conflict with F. For a given permutation of the input, an algorithm that uses an influence graph will not only create the same regions as another that uses a conflict graph, but will also detect a conflict with a given region as many times as there are conflict arcs incident to this region in the conflict graph.

As a consequence, the total expected complexity of the locating phases is proportional to the expected number of conflict arcs created in the conflict graph, and is given by theorem 5.2.3.

3. Finally, we can give a non-amortized analysis of each incremental step of an on-line algorithm. At step k, the locating phase takes time proportional to the number of nodes in the influence graph that conflict with O_k, the object introduced in this step. The average complexity of this locating phase during step k is thus proportional to $w(k, S)$, the expected number of nodes in the influence graph that conflict with O_k. From lemma 5.2.2, we know that a region F in $\mathcal{F}_j^i(S)$ is created at step r with a probability $p_j'^i(r) = \frac{i}{r} p_j^i(r)$. The conditional probability that this region conflict with O_k, knowing that F is created prior to step k, is $j/(n-r)$. Consequently,

$$w(k, S) \;=\; \sum_{i=1}^{b} \sum_{j=0}^{n-i} |\mathcal{F}_j^i(S)| \sum_{r=1}^{k-1} \frac{i}{r} \, p_j^i(r) \, \frac{j}{n-r}.$$

If we recognize the expression for the first order moment of a random r-sample of S given in lemma 4.2.5, and bound the sum above by using corollary 4.2.7 to the moment theorem, we obtain

$$
\begin{aligned}
w(k, S) \;&=\; \sum_{i=1}^{b} \sum_{r=1}^{k-1} \frac{i}{r(n-r)} \, m_1(r, S) \\
&=\; O\!\left(\sum_{r=1}^{k-1} \frac{f_0(\lfloor r/2 \rfloor, S)}{r^2} \right) = O\!\left(\sum_{r=1}^{k-1} \frac{f_0(r, S)}{r^2} \right).
\end{aligned}
$$

4. Now, updating the data structure and the influence graph at step k takes time proportional to the number of nodes created or killed by O_k. Let $v(k, S)$ be the expected number of regions created at step k. From lemma 5.2.2, we derive

$$
\begin{aligned}
v(k, S) \;&=\; \sum_{i=1}^{b} \sum_{j=0}^{n-i} |\mathcal{F}_j^i(S)| \, \frac{i}{k} \, p_j^i(k) \\
&=\; \frac{f_0(k, S)}{k}.
\end{aligned}
$$

Let now $v'(k, S)$ be the expected number of regions killed at step k. We denote by S_{k-1} the current subset immediately prior to step k. A region F in $\mathcal{F}_j^i(S)$ is a region killed at step k if it is a region of $\mathcal{F}_0(S_{k-1})$ that conflicts with O_k, which happens with probability

$$p_j^i(k-1) \frac{j}{n-k+1}.$$

Again, using lemma 4.2.5 and corollary 4.2.7, we get

$$
\begin{aligned}
v'(k,\mathcal{S}) &= \sum_{i=1}^{b}\sum_{j=0}^{n-i} |\mathcal{F}_j^i(\mathcal{S})|\, p_j^i(k-1)\, \frac{j}{n-k+1} \\
&= \frac{m_1(k-1,\mathcal{S})}{k-1} \\
&= O\left(\frac{f_0(\lfloor (k-1)/2 \rfloor,\mathcal{S})}{k-1}\right).
\end{aligned}
$$

This completes the proof of theorem 5.3.4. $\qquad\square$

Some remarks on the update condition

The update condition 5.3.3 is not mandatory and it is often possible to analyze an on-line algorithm that does not satisfy all of its clauses.

1. For instance, if the first clause is not satisfied, the cost of testing the conflicts may be added to the analysis. If this cost can be bounded, this bound appears as a multiplicative factor in the cost of the locating phases.

2. The analyses of on-line algorithms developed above and in the remainder of this section are still valid for less restrictive statements of the third clause. We may assume only that the cost of the update phase is proportional to the number of regions created or killed. We have preferred, however, to assume that the parents of nodes created by some object are killed by the same object. This assumption is satisfied by most of the algorithms given in this book, and it greatly simplifies the analysis of dynamic on-line algorithms given in the next chapter.

3. Lastly, the second clause can also be relaxed. Indeed, in order to bound the space needed to store the influence graph, it suffices to bound the total number of arcs in the entire graph and not necessarily the out-degree of each node. We may then generalize the analysis of the locating phase by using the notion of a biregion (see exercise 5.7). In particular, such an analysis applies to the case when the number of parents of a node is bounded, but not the number of children. We illustrate this situation in the case of the on-line computation of convex hulls (see exercise 8.5).

5.3.2 An example: vertical decomposition of line segments

Again, we discuss the problem of building the vertical decomposition of a set of line segments in the plane, and this time we show how to compute it on-line, using an influence graph. Each time a segment is inserted, the algorithm updates the decomposition of the current set of segments, called the *current*

decomposition for short. The notions of objects, regions, and conflicts are defined
as in subsection 5.2.2.

The trapezoids in the current decomposition are the regions defined and without conflict over the current set of segments and are linked to the corresponding
nodes in the influence graph. An internal node of this graph is associated with
a trapezoid which was in the current decomposition at some previous step of the
algorithm. In addition to the set of pointers that take care of the parent–child
relationships between the nodes, each node contains the following information:

- A description of the corresponding trapezoid and a list of the (at most four)
  segments that determine it.

- At most four pointers for the adjacency relationships through the vertical
  walls. As long as the node is a leaf of the influence graph, the corresponding
  trapezoid F belongs to the current decomposition and is adjacent to at most
  four leaves in the graph, each of which shares a vertical wall with F. When
  the node corresponding to F becomes an internal node, these pointers are
  not modified any more.

Therefore, a description of the simplified decomposition can be extracted from
the information stored at the leaves of the graph.

At each step, the new segment S is located in the influence graph, yielding the
list $\mathcal{L}(S)$ of trapezoids that it intersects. Each of these trapezoids is subdivided
into at most four subregions by S, by the walls stemming from the endpoints of
S, and by the walls stemming from the intersection points between S and some
previously inserted segment. In the influence graph, a temporary node is created
for these subregions. For each trapezoid F in the current decomposition that is
intersected by S, we create links to the temporary nodes for the subregions F_i inside F, and they become the children of F in the influence graph (see figure 5.3).
The list $\mathcal{L}(S)$ is not sorted, yet the vertical adjacency pointers allow a traversal
of the decomposition in the left-to-right order along S. This allows us to set
the vertical adjacencies of the subregions, and to identify which walls have to be
removed and which subregions have to be joined to obtain the simplified decom-
position $\mathcal{D}ec_s(\mathcal{R} \cup \{S\})$ after the insertion of S. We replace all the temporary
nodes that correspond to subregions to be joined by a single node that inherits
all the parents of the subregions. In this fashion, a leaf of the graph is created
for each trapezoid F created by S, and is linked to all the trapezoids F' in $\mathcal{L}(S)$
which intersect the interior of F (see figures 5.6, 5.7, and 5.8). Properties 1 and 2
of the influence graph are thus maintained from step to step. Vertical adjacency
relationships between trapezoids created by S can be derived from those of the
subregions. This completes the description of the update phase in an incremental
step.

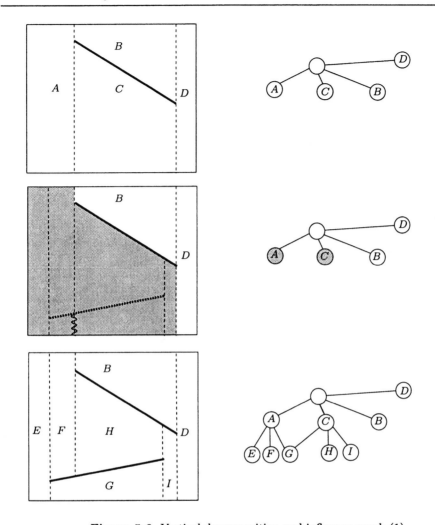

Figure 5.6. Vertical decomposition and influence graph (1).

Each internal node of the influence graph has at most four children, and the running time needed to carry out all the operations described in the previous paragraph is clearly proportional to the number of trapezoids in $\mathcal{L}(S)$. The update condition is therefore satisfied.

From lemma 5.2.4, we know that the expected number of trapezoids in the vertical decomposition of a random r-sample is $O(n + ar^2/n^2)$, if n is the number of segments in S and a is the number of intersecting pairs of segments. Theorem 5.3.4 therefore shows that the on-line algorithm just described has an expected complexity of $O(n \log n + a)$ and uses expected storage $O(n + a)$. The average complexity of the n-th insertion is $O(\log n + a/n)$.

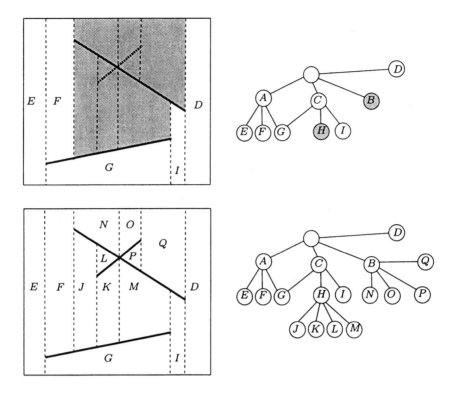

Figure 5.7. Vertical decomposition and influence graph (2).

5.4   Accelerated incremental algorithms

One of the problems encountered in solving a problem using the incremental method, that is, in identifying the set $\mathcal{F}_0(\mathcal{S})$ of those regions defined and without conflict over a given set $\mathcal{S}$ of objects, is the detection of conflicts between a new object and a region defined and without conflict over the current subset. Algorithms that use a conflict graph are static, as opposed to on-line algorithms which use an influence graph. In this section, which may be skipped in a first reading, we show how to combine both data structures to transform an on-line algorithm into a static one that has a lower asymptotic average complexity.

5.4.1   The general method

Theorem 5.3.4 on the influence graph shows that, if the expected number of regions defined and without conflict over $\mathcal{S}$ is $O(r)$, then the complexity of any algorithm that uses an influence graph is dominated by the cost of the locating phases in the incremental steps. (This cost is $O(n \log n)$, whereas the cost of the updates is only $O(n)$.) The idea is to lower the complexity of the locating phase

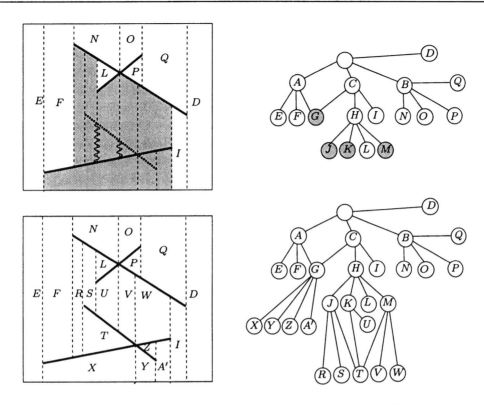

Figure 5.8. Vertical decomposition and influence graph (3).

by using a conflict graph, in addition to the influence graph. We cannot afford to maintain this conflict graph at every step, however, so we update it only at certain steps of the algorithm.

Let $\mathcal{S}_k$ be the current subset immediately after step k. The *conflict graph at step k* is the conflict graph that stores all the conflicts between the regions in $\mathcal{F}_0(\mathcal{S}_k)$ and the objects in $\mathcal{S} \setminus \mathcal{S}_k$. The following theorem shows how to use a knowledge of this conflict graph at step k to speed up the subsequent locating phases.

Theorem 5.4.1 *If an on-line algorithm satisfies the update condition, a knowledge of the conflict graph at step k can be used to perform the locating phase in a subsequent step l, with an average complexity of*

$$O\left(\sum_{r=k+1}^{l-1} \frac{f_0(\lfloor r/2 \rfloor, \mathcal{S})}{r^2} \right).$$

In particular, if $f_0(r, \mathcal{S})$ is $O(r)$, the cost of a locating phase at step l is $O\left(\log(\frac{l}{k})\right)$ on the average.

Proof. The conflict graph at step k can be augmented, for each object O in $S \setminus S_k$, by a list of pointers to the nodes of the influence graph which correspond to a region of $\mathcal{F}_0(S_k)$ that conflicts with O. In order to locate the object O_l at step l, the algorithm may start to traverse the influence graph not from the root, but from the nodes of the influence graph which correspond to a region of $\mathcal{F}_0(S_k)$ that conflicts with O_l. If the update condition is satisfied, the number of children of each node is bounded by a constant, and the number of nodes visited is proportional to the number of regions F created between steps $k+1$ and $l-1$ that conflict with O_l. A region F in $\mathcal{F}_j^i(S)$ is created at step r with probability $\frac{i}{r}p_j^i(r)$. Given this, the conditional probability that F conflicts with a given object O_l is $\frac{j}{n-r}$. The expected number $w(l, S)$ of nodes visited while locating O_l is thus

$$
w(l, S) = O\left(\sum_{r=k+1}^{l-1} \sum_{i=1}^{b} \sum_{j=0}^{n-b} |\mathcal{F}_j^i(S)| \, p_j^i(r) \, \frac{i}{r} \, \frac{j}{n-r} \right).
$$

In this expression we recognize the first order moment (lemma 4.2.5). Using corollary 4.2.7 to the moment theorem, we obtain

$$
w(l, S) = O\left(\sum_{r=k+1}^{l-1} \frac{b}{r(n-r)} \, m_1(\mathcal{R}, S) \right) = O\left(\sum_{r=k+1}^{l-1} \frac{f_0(\lfloor r/2 \rfloor, S)}{r^2} \right). \quad \square
$$

The accelerated algorithm proceeds as follows. At each step, the algorithm updates the influence graph. At certain steps, numbered $n_1, n_2, \ldots, n_k, \ldots$, the algorithm builds the conflict graph between the regions in $\mathcal{F}_0(S_{n_k})$ and the objects in $S \setminus S_{n_k}$. This conflict graph is used to speed up the locating phases in the steps whose numbers range from $(n_k + 1)$ to n_{k+1}.

Of course, this design is useful only if the computation of the conflict graph at steps $n_1, n_2, \ldots, n_k, \ldots$ is not too unwieldy. The following theorem presents a general situation when an on-line algorithm that runs in expected time $O(n \log n)$ can be speeded up into a static algorithm that runs in expected time $O(n \log^* n)$.

Theorem 5.4.2 *Let S be a set of n objects, and $\mathcal{R}$ be a random r-sample of S. Suppose that the expected number $f_0(r, S)$ of regions defined and without conflict over $\mathcal{R}$ is $O(r)$. If the conflict graph at any step k can be built in expected time $O(n)$, then the randomized accelerated algorithm runs in expected time $O(n \log^* n)$.*

Proof. The conflict graph is computed at steps $n_k = \left\lfloor n/\log^{(k)} n \right\rfloor$, for $k = 1, \ldots,$ $\log^* n$. The conflict graph is therefore computed $\log^* n$ times overall, accounting for an expected complexity of $O(n \log^* n)$. The locating phases, between step n_k and step n_{k+1}, have a total average complexity of

$$
\sum_{l=n_k+1}^{n_{k+1}} O\left(\log\left(\frac{l}{n_k}\right)\right) = \sum_{l=n_k+1}^{n_{k+1}} O\left(\log\left(\frac{l}{n}\log^{(k)} n\right)\right)
$$
$$
= (n_{k+1} - n_k)O\left(\log\log^{(k)} n\right)
$$
$$
= O(n).
$$

The total contribution of the locating phases to the running time is therefore, on the average, $O(n \log^* n)$. This fact combined with theorem 5.3.4 proves that the average complexity of the accelerated algorithm is $O(n \log^* n)$. $\qquad\square$

5.4.2 An example: vertical decomposition of a polygon

The vertical decomposition of a polygon $\mathcal{P}$, which we denote by $\mathcal{Dec}(\mathcal{P})$, is the decomposition $\mathcal{Dec}(\mathcal{S_P})$ of the set of line segments $\mathcal{S_P}$ that are the edges on the boundary of the polygon $\mathcal{P}$. The decomposition of a simple polygon is a very interesting structure since we can derive from it a triangulation of the polygon, as we explain in chapter 12. The previous method helps to compute the decomposition of a polygon with n sides in expected time $O(n \log^* n)$, which therefore leads to a randomized algorithm of better average complexity than most of its deterministic counterparts.

The algorithm processes the segments in $\mathcal{S_P}$ in a random order, builds an influence graph as described in subsection 5.3.2, and maintains the simplified decomposition of the current set of edges. In accordance with the preceding idea, it computes a conflict graph at steps $n_k = \left\lfloor n/\log^{(k)} n \right\rfloor$, $k = 1, \ldots, \log^* n$, which is then used to locate the subsequent edges between step n_k and step n_{k+1}.

The segments in $\mathcal{S_P}$ may intersect only at their endpoints. Lemma 5.2.4 shows that the number $f_0(r, \mathcal{S_P})$ of trapezoids in the decomposition of a random r-sample of $\mathcal{S_P}$ is $O(r)$.

The conflict graph at step n_k can be computed using the following method in expected time $O(n)$. Let $\mathcal{Dec}_s(\mathcal{S}(n_k))$ be the current decomposition immediately after step n_k. A simplified decomposition will suffice for our purposes. We first begin by locating a given vertex of the polygon in the decomposition $\mathcal{Dec}_s(\mathcal{S}(n_k))$, using brute force and $O(n)$ operations. We then follow the boundary of the polygon, reporting all the conflicts between the edges and the trapezoids of the decomposition $\mathcal{Dec}_s(\mathcal{S}(n_k))$. Immediately after step n_k, an edge has either been inserted already, or it conflicts with some trapezoids in the decomposition $\mathcal{Dec}_s(\mathcal{S}(n_k))$. In the former case, it has been split into possibly many edges of this decomposition, and the total complexity of following these edges is dominated by the size $O(n_k) = O(n)$ of this decomposition. In the latter case, the cost of following these edges is proportional to the number of conflicts between these edges and the trapezoid of the decompositions $\mathcal{D}ec_s(\mathcal{S}(n_k))$.[2] From theorem 4.2.6 and its corollary 4.2.7, the expected number of conflicts reported at step n_k is exactly the first order moment of the current subset of edges at step n_k. From corollary 4.2.7, this number is $O\left(f_0(\lfloor n_k/2 \rfloor, \mathcal{S})\right)$, which is $O(n)$ for non-intersecting segments, as is the case for the edges of a polygon.

The hypotheses of theorem 5.4.2 are thus satisfied, which yields:

Theorem 5.4.3 *A randomized incremental algorithm can build the vertical decomposition of a simple polygon with n edges in expected time $O(n \log^* n)$.*

Remark. The algorithm relies on two facts: the edges are connected, and do not intersect except possibly at common endpoints. The same algorithm therefore works as well in the more general cases of a polygonal line, or a connected set of segments whose pairwise interiors are disjoint.

5.5   Exercises

Exercise 5.1 (Probabilities) Prove that

$$\sum_{r=i}^{n-j} \frac{i}{r} \frac{\dbinom{n-i-j}{r-i}}{\dbinom{n}{r}} = \frac{i! \, j!}{(i+j)!}.$$

Then show that the probabilities p'^i_j and $p'^i_j(r)$ defined in section 5.2 satisfy the following relation:

$$p'^i_j = \sum_{r=1}^{n} p'^i_j(r).$$

Exercise 5.2 (Non-amortized analysis of the conflict graph) Consider a randomized incremental algorithm that processes a set $\mathcal{S}$ of n objects by using a conflict graph. Show that if the update condition is satisfied, then the complexity of step k is on the average

$$O\left(\frac{f_0(k,\mathcal{S})}{k} + \frac{n}{k^2} f_0\left(\left\lfloor \frac{k}{2} \right\rfloor, \mathcal{S}\right) + \frac{n}{(k-1)^2} f_0\left(\left\lfloor \frac{k-1}{2} \right\rfloor, \mathcal{S}\right) \right).$$

---

[2] We may recall that each trapezoid of the decomposition is adjacent to at most four trapezoids through vertical walls and these adjacencies are encoded through additional pointers in the influence graph. Since an edge may not intersect a floor or ceiling, we can trace its conflicts in the current decomposition using constant time per conflict.

Hint: The expected number of regions killed or created during step k of a randomized incremental algorithm is estimated in the proof of theorem 5.3.4. It remains to estimate the expected number of conflict arcs added or removed during step k.

Exercise 5.3 (Complete description of the vertical decomposition) Let S be a set of line segments in the plane, with a intersecting pairs, and let $\mathcal{R}$ be a random r-sample of S. We consider the complete description of the decomposition $Dec(\mathcal{R})$ of $\mathcal{R}$, and particularly the expected number of edges on the boundaries of the trapezoids of $Dec(\mathcal{R})$ which are cut by a random segment of $S \setminus \mathcal{R}$. Show that this expectation is $O(1 + ar/n^2)$.

From this, show that the randomized incremental algorithm described in subsection 5.2.3 can be slightly modified to compute, within the same complexity bounds, a complete description of the decomposition of the line segments, which in particular includes all the adjacency relationships between the trapezoids.

Hint: We may redefine the notions of regions and conflicts as follows: A region defined on $\mathcal{R}$ is a *paddle* with two components, a trapezoid F in the decomposition $Dec(\mathcal{R})$, and a wall butting on the floor or ceiling of F. A paddle is determined by at most six segments. It conflicts with a segment if the interior of the trapezoid intersects the segment. The problem is now to find an upper bound on the number of paddles defined and without conflict with a segment of $S \setminus \mathcal{R}$.

Exercise 5.4 (Storage) Consider the incremental algorithm that uses a conflict graph as in subsection 5.2.2 in order to compute the decomposition of a set S of n segments. Show that if a is the number of intersecting pairs, the storage needed by the algorithm at step k is, on the average,

$$m_1(k, S) = O\left(n + a\frac{k}{n}\right).$$

Using the result of the previous exercise, show that we may reduce the storage to $O(n)$ by storing only one conflict for each non-inserted segment, say with the trapezoid that contains its left endpoint, without affecting the running time of the algorithm.

Exercise 5.5 (Decomposing a set of curves) Show how to generalize the notion of a decomposition for a set of curves supported by algebraic curves of bounded degree. Two such curves intersect at only a constant number of points, which we assume may be computed in constant time. Show that both algorithms given in subsections 5.2.2 and 5.3.2 may be extended to build the decomposition of a set of such curves.

Hint: Do not forget to trace walls from each point where the curves have a vertical tangent.

Exercise 5.6 (Backward analysis) Backward analysis (see also exercises 4.1 and 4.2) gives an alternative proof of the results of this chapter without using the explicit expressions for $p_j^i(r)$ and $p_j'^i(r)$.

For instance, we show how backward analysis can be used to estimate the number $v(k, \mathcal{S})$ of regions created at step k by an incremental algorithm. Note that if $\mathcal{S}_k$ is the current subset immediately after inserting object O_k at step k, the regions created by O_k during this step are the regions of $\mathcal{F}_0(\mathcal{S}_k)$ determined by a subset of $\mathcal{S}_k$ that contains O_k. Since O_k, which has chronological rank k, may be any of the objects in $\mathcal{S}_k$ with uniform probability $\frac{1}{k}$, a region of $\mathcal{F}_0(\mathcal{S}_k)$ is created at step k with probability at most b/k. Therefore, $v(k, \mathcal{S})$ is at most the expectation of $\frac{b}{k}|\mathcal{F}_0(\mathcal{S}_k)|$ over all possible $\mathcal{S}_k$.

Similarly, a region that is killed during step k is a region of $\mathcal{F}_1(\mathcal{S}_k)$ that conflicts with O_k. Any region of $\mathcal{F}_1(\mathcal{S}_k)$ conflicts with O_k with probability $1/k$. The expected number $v'(k, \mathcal{S})$ of regions killed during step k is therefore at most the expectation of $\frac{1}{k}|\mathcal{F}_1(\mathcal{S}_k)|$ over all possible $\mathcal{S}_k$.

It is possible to compute in this fashion the expected numbers $v(k, \mathcal{S})$ and $v'(k, \mathcal{S})$ of regions created or killed during step k. Show how to use backward analysis to prove the other results in this chapter, for instance, to bound the number of conflict arcs that are added to or removed from the conflict graph, or to bound the number of conflicts detected during a locating phase by an algorithm that uses an influence graph.

Exercise 5.7 (Biregions) The notion of *biregion* introduced in this exercise can be used to analyze the average complexity of some algorithms that use an influence graph, but do not satisfy the update condition 5.3.3. A *biregion* is pair of regions which can have a parent–child relationship in the influence graph for at least one permutation of the data. A biregion is determined by a set of at most $2b$ objects, those that determine the parent region and those that determine the child region. Exactly one of the objects that determine the child region conflicts with the parent region. We can extend the notion of conflict to biregions in the following way: an object conflicts with a biregion if it conflicts with at least one of its two regions and does not belong to the set of objects that determine the biregion. A biregion can then be considered as a region in the framework described in chapter 4.

1. Let $\mathcal{S}$ be a set of n objects. Show that a biregion, determined by i objects of $\mathcal{S}$ and in conflict with j objects of $\mathcal{S}$, is defined and with k conflicts over a random r-sample of $\mathcal{S}$ with the probability $p_{j,k}^i(r)$ given by lemma 4.2.1.

2. From this, extend both the sampling theorem and the moment theorem to the case of biregions.

3. In essence, the difference between biregions and regions resides in the following fact. Let FF be a biregion determined by i objects and conflicting with j objects of $\mathcal{S}$. For FF to correspond to an arc in the influence graph built for $\mathcal{S}$, it is not enough that the i objects that determine FF be inserted before any of the j objects that conflict with FF; it must also be the case that the i objects that determine i be processed in a certain order. This order has to meet several criteria. These criteria depend on the algorithm. At the very least, one of the objects that determine the child region, more precisely the one that conflicts with the parent region, must be inserted after all the objects that determine the parent region.

Show that the probability that FF correspond to an arc of the influence graph is $\alpha p_j'^i$, where $p_j'^i$ is given in lemma 5.2.2, and α is a constant that satisfies $\frac{1}{(2b)!} \leq \alpha \leq \frac{1}{i}$ and that depends only on the particular criteria that the insertion order has to meet. Then show that the probability that the biregion FF correspond to an arc of the influence graph that is created at step r is $\alpha \frac{i}{r} p_j^i(r)$, where $p_j^i(r)$ is defined in subsection 4.2.1.

4. Our goal is now to give a randomized analysis of an on-line algorithm that uses an influence graph in which a node can have arbitrarily many children. We thus forget about the second clause in the update condition 5.3.3, and relax the third one by assuming that the parents of a region created by O are either killed by O, or still have no conflict after the insertion of O. In this way, regions defined and without conflict with the current subset may not be leaves of the influence graph, but could have many children before they are killed. The complexity of the update phase is still assumed to take time proportional to the number of arcs added to or removed from the influence graph. For instance, the algorithm that computes convex hulls described in exercise 8.5 meets these conditions.

Let $ff_0(r, S)$ stand for the expected number of biregions defined and without conflict over a random r-sample of S. Show that the number of arcs in the influence graph built for S is, on the average,

$$\Theta\left(\sum_{r=1}^{n} \frac{ff_0(r, S)}{r}\right).$$

Show that the cost of the locating phases is

$$O\left(n \sum_{r=1}^{n} \frac{ff_0(r, S)}{r^2}\right).$$

5. Assume now that the influence graph built for a random r-sample of S has an expected number of arcs at most $g(r, S)$, where g is a known function. For instance, when each node of the influence graph has at most a bounded number of children, we may choose

$$g(r, S) = O\left(\sum_{j=1}^{n} \frac{f_0(j, S)}{j}\right).$$

Show that the n-th incremental step of the on-line algorithm has an average complexity of

$$O\left(\frac{g(n, S)}{n} + \sum_{r=1}^{n} \frac{g(r, S)}{r^2}\right).$$

Exercise 5.8 (Decomposing a polygon) This exercise presents another randomized algorithm that builds the vertical decomposition of a simple polygon with n edges in expected time $O(n \log^* n)$.

The algorithm is incremental but inserts a number of edges of the polygon at a time. Let $\mathcal{P}$ be a polygon, and S the set of its n edges. Assume that the segments in S are ordered in a random order, and let S_i be the subset containing the first r_i segments of S, with $r_i = \left\lfloor n / \left\lceil \log^{(i)} n \right\rceil \right\rfloor$. The subset S_i is thus a random r_i-sample of S, and

$$S_1 \subset S_2 \subset \ldots \subset S_{\log^* n} = S.$$

The algorithm computes a simplified description of the decomposition $\mathcal{Dec}_s(\mathcal{P})$, using $\log^* n$ steps. Step i computes the decomposition $\mathcal{Dec}_s(S_i)$ from $\mathcal{Dec}_s(S_{i-1})$.

In the initial step, we build $Dec_s(S_1)$ using the plane sweep algorithm of subsection 3.2.2, in time $O(r_1 \log r_1)$. (Any algorithm that runs in time $O(r_1 \log r_1)$ would do.)

In a subsequent step i, $i > 1$:

1. We locate the segments of S in $Dec_s(S_{i-1})$. In other words, for each region F in $Dec_s(S_{i-1})$, we compute the set $S(F)$ of segments in S which intersect F.

2. For each region F of $Dec_s(S_{i-1})$, we compute the decomposition of $S(F) \cup S_i$, and the portion of it that lies inside F. To do this, we simply compute the total decomposition $Dec_s(S(F) \cup S_i)$, using the plane sweep algorithm of subsection 3.2.2. (Again, any algorithm that runs in time $O(m \log m)$ for m segments would do.)

3. We obtain $Dec_s(S_i)$ by putting together all the portions $Dec_s(S(F) \cup S_i) \cap F$ inside the trapezoids F of $Dec_s(S_{i-1})$, and merging the regions that share a wall of $Dec_s(S_{i-1})$ which disappears in $Dec_s(S_i)$.

Show that all three phases 1, 2, and 3 can be performed using $O(n)$ operations. To analyze phase 2, note that S_{i-1} is a random r_{i-1}-sample of S_i, then use the extension of the moment theorem given in exercise 4.3 for the function $g(x) = x \log x$.

Exercise 5.9 (Querying the influence graph) The influence graph built by an on-line algorithm can be used to answer conflict queries on a set of objects. For instance, the influence graph built for a vertical decomposition can answer location queries for a point inside this decomposition. Show that, if n segments are stored in the influence graph, answering a given location query takes time $O(\log n)$, on the average over all possible insertion orders of the n segments into the influence graph. More generally, show that the same time bound holds for any conflict query which, on any subset $\mathcal{R}$ of objects, answers with a single region of $\mathcal{F}_0(\mathcal{R})$.

5.6   Bibliographical notes

The first non-trivial (that is, sub-quadratic) algorithm that computes all the intersecting pairs in a set of segments in the plane is that of Bentley and Ottmann [23], which uses a plane sweep method. This algorithm, described in chapter 3, computes all a intersecting pairs in $O((n + a) \log n)$, which falls short of being optimal. About ten years later, Chazelle and Edelsbrunner proposed in [48, 49] a deterministic algorithm that runs in optimal time $O(n \log n + a)$ to compute all the a intersections. The description and implementation of their algorithm is rather complicated, however. At about the same time, Clarkson and Shor [71] and independently Mulmuley [171, 173] proposed randomized incremental algorithms for the same problem that have an optimal average complexity.

The algorithm by Clarkson and Shor that uses a conflict graph is described in section 5.2 in this chapter. In the same paper [71], they also set up the formalism of objects, regions, and conflicts, and introduce the conflict graph in these terms; they give other algorithms that use the conflict graph (computing the intersection of n half-spaces, the diameter of a point set in 3 dimensions), and show how to compute a complete description of the decomposition of line segments and how to lower the storage requirements of their algorithm (see exercises 5.3 and 5.4). Mulmuley's algorithm is very similar to that of Clarkson and Shor, yet its analysis is based on probabilistic games and combinatorial series, and is much less immediate.

The influence graph was first introduced in a paper by Boissonnat and Teillaud [31, 32] where it was called the *Delaunay tree*, and was used there to compute on-line the Delaunay triangulation of a set of points. Guibas, Knuth, and Sharir [117] proposed a similar algorithm to solve the same problem. How to use the influence graph in an abstract setting is described by Boissonnat, Devillers, Schott, Teillaud, and Yvinec in [28] and applied to other problems, especially to compute convex hulls or to decompose a set of segments in the plane. The method was later used to solve numerous other problems. The influence graph is sometimes called the *history* of the incremental construction.

The accelerated algorithm that builds the vertical decomposition of a simple polygon is due to Seidel [204]. This method was subsequently extended to solve other problems by Devillers [80], for instance to compute the Voronoi diagram of a polygonal line or of a closed simple polygon (see section 19.2). The algorithm described in exercise 5.8 that computes the decomposition of a polygon in time $O(n \log^* n)$ is due to Clarkson, Cole and Tarjan [69].

The method called *backward analysis* used in exercise 5.6 was first used by Chew in [59] to analyze an algorithm that computes the Voronoi diagram of a convex polygon (see exercise 19.4). It was subsequently used in a systematic fashion by Seidel in [203] and Devillers in [80].

Mehlhorn, Sharir, and Welzl [167, 168] gave a finer analysis of randomized incremental analysis by bounding the probability that the algorithm exceeds its expected performances by more than a constant multiplicative factor.

Randomized incremental algorithms proved very efficient in solving many geometric problems. The basic methods (using the influence or the conflict graphs) or one of their many variants inspired much work by several researchers such as Mulmuley [172, 174], Mehlhorn, Meiser and Ó'Dúnlaing [164], Seidel [205], Clarkson and Shor [71], and Aurenhammer and Schwarzkopf [18].

There is a class of randomized algorithms which work not by the incremental method, but rather by the divide-and-conquer paradigm. The subdividing step is achieved using a sample of the objects to process. Randomization is used for choosing the sample, and the method can be proved efficient using the probabilistic theorems given in exercises 4.5 and 4.6. Randomized divide-and-conquer is mainly used for building hierarchical data structures that support repeated range queries. Typically, these queries can be expressed in terms of locating a point in the arrangement of a collection of hyperplanes, simplices, or other geometric objects. In a dual situation, the data set is a set of points and the queries ask for those points which lie in a given region (half-space, simplex, ...). Haussler and Welzl [123] spurred new interest in the field with their notion of an ϵ-net. Later, Matoušek introduced the related notion of ϵ-approximations [150]. Chazelle and Friedman [53] showed how to compute these objects in a deterministic fashion using the method of conditional probabilities. The resulting deterministic method is called a *derandomization* of the randomized divide-and-conquer method. This method was then widely used, for instance by Matoušek [150, 151, 152, 153, 154, 155], Matoušek and Schwarzkopf [156], or Agarwal and Matoušek [4]. In his thesis [35], Brönnimann studies the derandomization of geometric algorithms and the related concept of the Vapnik–Chervonenkis dimension. Randomized divide-and-conquer is also used by Clarkson, Tarjan, and Van Wyk in [65] to build the vertical decomposition of a simple polygon.

Last but not least, the book by Mulmuley [177] is entirely devoted to randomized geometric algorithms, and serves as a very comprehensive reference on the topic.

Chapter 6

Dynamic randomized algorithms

The geometric problems encountered in this chapter are again stated in the abstract framework of objects, regions, and conflicts introduced in chapter 4. A *dynamic* algorithm maintains the set of regions defined and without conflict over the current set of objects, when the objects can be removed from the current set as well as added. In contrast, on-line algorithms that support insertions but not deletions are sometimes called *semi-dynamic*.

Throughout this chapter, we denote by S the current set of objects and use the notation introduced in the previous two chapters to denote the different subsets of regions defined over S. In particular, $\mathcal{F}_0(S)$ stands for the set of regions defined and without conflict over S. To design a dynamic algorithm that maintains the set $\mathcal{F}_0(S)$ is a much more delicate problem than its static counterpart. In the previous chapter, we have shown how randomized incremental methods provide simple solutions to static problems. In addition, the influence graph techniques naturally lead to the design of semi-dynamic algorithms. In this chapter, we propose to show how the combined use of both conflict and influence graphs can yield fully dynamic algorithms.

The general idea behind our approach is to maintain a data structure that meets the following two requirements:
- It allows conflicts to be detected between any object and the regions defined and without conflict over the current subset.
- After deleting an object, the structure is identical to what it would have been, had the deleted object never been inserted.

Such a structure is called an *augmented influence graph*, and can be implemented using an influence graph together with a conflict graph between the regions stored in the influence graph and the current set of objects. In some cases, we might be able to do without the conflict graph.

In section 6.2, we describe the augmented influence graph and how to perform insertions and deletions. The randomized analysis of these operations is given in section 6.3. This analysis assumes a probabilistic model which is made precise and unambiguous in section 6.1. The general method is used in section 6.4 to design a dynamic algorithm that builds the vertical decomposition of a set of segments in the plane.

This chapter also uses the terminology and notation introduced in the previous two chapters. To ease the reading process, some definitions are recalled in the text or in the footnotes.

6.1 The probabilistic model

The current set of objects, denoted by $\mathcal{S}$, is the result of a sequence of insertions and deletions. Due to the second requirement which we stated earlier, the data structure does not keep track of the deleted objects. Consequently, at any given time, the data structure depends only on $\mathcal{S}$ and on the order in which the objects in $\mathcal{S}$ were introduced. In fact, an object stored in the data structure may have been inserted and removed several times, yet the current state of the data structure only keeps track of the last insertion.

At any given time, the data structure only depends on the *chronological sequence* $\Sigma = \{O_1, O_2, \ldots, O_n\}$ which is the sequence of objects in $\mathcal{S}$ in the order of their last insertion.

The randomized analysis of a dynamic algorithm assumes that:

- If the last operation is an insertion, each object in the current set $\mathcal{S}$ is equally likely to have been inserted in this operation.

- If the last operation is a deletion, each object present prior to the deletion is equally likely to be deleted in this operation.

It follows from these two assumptions that the chronological sequence Σ is random, and that every permutation of the objects in $\mathcal{S}$ is equally likely to occur in Σ. Let the current set of object $\mathcal{S}$ be of size n, and let i be an integer in $\{1, \ldots, n\}$. Each object in $\mathcal{S}$ is the object O_i of rank i in Σ with uniform probability $1/n$. Moreover, in a deletion, the object O_i of rank i in Σ is deleted with uniform probability $1/n$.

We let $\mathcal{S}_i$ be the subset $\{O_1, O_2, \ldots, O_i\}$ of the first i objects in the chronological sequence. The probabilistic model implies that for i, $1 \leq i \leq n$, $\mathcal{S}_i$ is a random i-sample[1] of $\mathcal{S}$ and, for each pair (i, j) such that $1 \leq i \leq j \leq n$, $\mathcal{S}_i$ is a

---

[1]We may recall that a random i-sample is a random subset of size i of $\mathcal{S}$. Its elements are chosen in a way that makes all the $\binom{n}{i}$ possible subsets of size i of S equally likely.

random i-sample of $\mathcal{S}_j$.

6.2 The augmented influence graph

The augmented influence graph obtained after a sequence of insertions and deletions that results in a set $\mathcal{S}$, is determined only by the chronological sequence Σ of the objects in $\mathcal{S}$ and is denoted $\mathcal{I}a(\Sigma)$. The augmented influence graph $\mathcal{I}a(\Sigma)$ is connected, directed, and acyclic. It has the same nodes and arcs as the graph built by an on-line algorithm which inserts the objects of the sequence Σ in the order of Σ. There is a node in the graph for each region that belongs to $\bigcup_{i=1}^{n} \mathcal{F}_0(\mathcal{S}_i)$, where $\mathcal{F}_0(\mathcal{S}_i)$ denotes the set of regions defined and without conflict over $\mathcal{S}_i$. Let us recall that a region is characterized by two subsets of objects: the subset of objects with which it conflicts, called the influence domain of the region, and the subset of objects with bounded size that determine the region. In the following, we call each object that belongs to this second subset a *determinant* of the region. We call *creator* of a region the object of highest chronological rank among the determinants of the region. As in the preceding chapter, we use the terminology of trees to describe the structure of the augmented influence graph, and often identify a node with the region that is stored therein. This lets us speak for instance of the parent or children of a region, or of the influence domain of a node. The arcs of the influence graph maintain the *inclusion property* that the influence domain of a node is a subset of the union of the influence domains of its parents. For dynamic algorithms, we demand that arcs of the augmented influence graph also ensure a *second inclusion property* stating that, apart from its creator, the set of determinants of a region is contained in the set of determinants of its parent regions.

In addition to the usual information stored in the influence graph, the augmented influence graph stores a conflict graph between the objects in the current set $\mathcal{S}$ and the regions stored in the nodes of the influence graph. This conflict graph is represented as in the preceding chapter by a system of interconnected lists: To each region F stored at a node of the influence graph corresponds a list $L'(F)$ of objects of $\mathcal{S}$ with which it conflicts. To each object O in the current set $\mathcal{S}$ corresponds a list $L(O)$ of regions stored in the entire influence graph that conflict with O. There is a bidirectional pointer between the entry corresponding to a region F in the list $L(O)$ of an object O and the entry corresponding to O in the list $L'(F)$.

Inserting an object

Inserting an object O_n into a structure built for a set $\mathcal{S}_{n-1}$ is very similar to the operation of inserting an object in an on-line algorithm that uses an influence graph. The only difference is that, in addition to the insertion into the influence graph, we must also take care of updating the conflict lists. This can be done in two phases: a locating phase, and an updating phase.

Locating. The algorithm searches for all the nodes in the influence graph of $\mathcal{I}a(\Sigma)$ that conflict with O_n. Each time a conflict is detected, we add a conflict arc to the conflict graph, add O_n to the conflict list of the region that conflicts with it, and add this region to the list $L(O_n)$.

Updating. A node of the influence graph is created for each region in $\mathcal{F}_0(\mathcal{S}_n)$ determined by a set of objects that contains O_n. This node is also linked to parent nodes so that the two inclusion properties hold.

We may recall that a region in $\mathcal{F}_0(\mathcal{S}_n)$ is said to be created by O_n if it is determined by a set of objects that contains O_n. Similarly, a region of $\mathcal{F}_0(\mathcal{S}_{n-1})$ is said to be killed by O_n if it conflicts with O_n. More generally, a region stored in a node of the influence graph $\mathcal{I}a(\Sigma)$ has a *creator* in Σ, and a *killer* if it is not a leaf. The *creator* of F is, among all the objects that determine F, the one that has the highest rank in Σ. The *killer* of F is, among all the objects in Σ that conflict with F, the one with the lowest chronological rank.

For the rest of this chapter, we assume that the augmented influence graph satisfies the update condition 5.3.3. In particular, a node of the graph that stores a region created by O_n is linked only to nodes storing regions killed by O_n.

Deleting an object

To simplify the discussion, assume that the current set $\mathcal{S}$ has n objects, and that the current data structure is the augmented influence graph $\mathcal{I}a(\Sigma)$ corresponding to the chronological sequence $\Sigma = \{O_1, \ldots, O_n\}$. The object to be deleted is O_k, the object that has chronological rank k. The algorithm must modify the augmented influence graph to look as if O_k had never been inserted into Σ. The augmented graph must therefore correspond to the chronological sequence $\Sigma' = \{O_1, \ldots, O_{k-1}, O_{k+1}, \ldots O_n\}$.

For any integer l, $k \leq l \leq n$, let us denote by $\mathcal{S}'_l$ the subset $\mathcal{S}_l \setminus \{O_k\}$ of $\mathcal{S}$. In particular, observe that $\mathcal{S}'_k = \mathcal{S}_{k-1}$.

In what follows, an object is called a *determinant* of a region if it belongs to the set of objects that determine that region. The symmetric difference between the nodes of $\mathcal{I}a(\Sigma)$ and those of $\mathcal{I}a(\Sigma')$ can be described as follows.

1. The nodes of $\mathcal{I}a(\Sigma)$ that do not belong to $\mathcal{I}a(\Sigma')$ are determined by a set of objects that contain O_k. Therefore O_k is a determinant of those regions, and we say that such nodes (and the corresponding regions) are *destroyed* when O_k is deleted.

2. The influence graph $\mathcal{I}a(\Sigma')$ has a node that does not belong to $\mathcal{I}a(\Sigma)$ for

each region in $\bigcup_{l=k+1,\ldots,n} \mathcal{F}_0(\mathcal{S}'_l)$ that conflicts with O_k. Let us say that such a node is *new* when O_k is deleted, and so is its corresponding region. A new region has a creator and, occasionally, a killer in the sequence Σ'. If the region belongs to $\mathcal{F}_0(\mathcal{S}'_l)$, conflicts with O_k, and is determined by a set of objects that contain O_l, then it is a new region after O_k is deleted, and its creator is O_l.

Nodes that play a particular role when O_k is deleted include of course the new nodes as well as the destroyed ones, but the nodes killed by O_k also have a special part to play. The nodes killed by O_k should not be mistaken for the nodes destroyed when O_k is deleted. Nodes killed by O_k correspond to regions of $\mathcal{F}_0(\mathcal{S}_{k-1})$ that conflict with O_k, whereas nodes destroyed when O_k is deleted correspond to regions that admit O_k as a determinant. The latter nodes disappear from the whole data structure when O_k is deleted. The former nodes are killed when O_k is inserted but remain in the data structure (occasionally becoming internal nodes), and they still remain after O_k is deleted.

Upon a deletion, the arcs in the influence graph $\mathcal{I}a(\Sigma)$ that are incident to the nodes destroyed by O_k disappear and the graph $\mathcal{I}a(\Sigma')$ has arcs incident to the new nodes. In particular, new nodes must be linked to some parents (which are not necessarily new nodes). Moreover, a few nodes of $\mathcal{I}a(\Sigma)$ that are not destroyed witness the destruction of some of their parents. Let us call these nodes *unhooked*. They must be rehooked to other parents.

Again, deletions can be carried out in two phases: a *locating phase*, and a *rebuilding phase*.

Locating. The algorithm must identify which nodes of the influence graph $\mathcal{I}a(\Sigma)$ are in conflict with O_k, which nodes have to be destroyed, and which are unhooked. Owing to both inclusion properties, this can be done by a traversal of the influence graph. This time, however, we not only visit the nodes that conflict with O_k, but also those which admit O_k as a determinant. The destroyed or unhooked nodes are inserted into a dictionary which will be looked up during the rebuilding phase.

Rebuilding. The first thing to do is to effectively remove all the destroyed nodes. Those nodes can be retrieved from the dictionary, and all the incident arcs in the graph are also removed from the graph. The conflict lists of the nodes which conflict with O_k are also updated accordingly. We shall not detail these low-level operations any further, as they should not raise any problems. Next, we must create the new nodes, as well as their conflict lists; we must also hook these new nodes and rehook the nodes that were previously unhooked. The detail of these operations depends on the nature of the specific problem in hand. The general design is always the same, however: the algorithm *reinserts* one by one, and in chronological order, all the objects O_l whose rank l is higher than k and that are creators of at least one new or unhooked region. To *reinsert* an object involves creating a node for each new region created by O_l, hooking this node into the influence graph, setting up its conflict list, and finally rehooking all the unhooked nodes created by O_l.

To characterize the objects O_l that must be reinserted during the deletion of O_k, we must explain what *critical regions* and the *critical zone* are. For each $l \geq k$, we call *critical* those regions in $\mathcal{F}_0(\mathcal{S}'_{l-1})$ that conflict with O_k. We call *critical zone*, and denote by $\mathcal{Z}_{l-1}$, the set of those regions.

Lemma 6.2.1 *Any object O_l of chronological rank $l > k$ that is the creator of a new or unhooked node when O_k is deleted conflicts with at least one critical region in $\mathcal{Z}_{l-1}$.*

Proof. If O_l is the creator of a new node, then there is a region F in $\mathcal{F}_0(\mathcal{S}'_l)$ that is determined by O_l and conflicts with O_k. In the influence graph $\mathcal{I}a(\Sigma')$, this region is linked to parents which, according to condition 5.3.3, are associated with regions in $\mathcal{F}_0(\mathcal{S}'_{l-1})$ which conflict with O_l. Still according to this condition, at least one of these nodes conflicts with O_k, which proves the existence of a region G in $\mathcal{F}_0(\mathcal{S}'_{l-1})$ that conflicts with both O_l and O_k.

If O_l is the creator of a unhooked node, then there is a region F in $\mathcal{F}_0(\mathcal{S}'_l) \cap \mathcal{F}_0(\mathcal{S}_l)$ whose determinants include O_l. The region F is linked in the influence graph $\mathcal{I}a(\Sigma')$ to parents, at least one of which is either new or killed by O_k (otherwise the region does not need to be rehooked). Update condition 5.3.3 assures us that this parent conflicts with O_l, which proves that there is a region G in $\mathcal{F}_0(\mathcal{S}'_{l-1})$ that conflicts both with O_l and O_k. $\square$

For each $l > k$, we must thus determine whether there is a critical region in $\mathcal{Z}_{l-1}$ that conflicts with O_l. If so, then O_l is reinserted, and we must find all the critical regions that conflict with O_l. Dynamic algorithms are efficient mostly when reinserting O_l involves traversing only a *local* portion of the influence graph that contains all the critical regions which conflict with O_l.

Before starting the rebuilding phase, the critical zone is initialized with those regions killed by O_k. At each subsequent reinsertion, the critical zone changes. To determine the next object that has to be reinserted and the critical regions that conflict with this object, we maintain in a priority queue $\mathcal{Q}$ the set of killers, according to the sequence Σ', of current critical regions. Killers are ordered within $\mathcal{Q}$ by their chronological rank, and each one points to a list of the critical regions that it kills.

The priority queue $\mathcal{Q}$ is first initialized with those regions killed by O_k. For each critical region F in $\mathcal{Z}_{k-1}$, we identify its killer in Σ' as the object, other than O_k, with the lowest rank in the list $L'(F)$.

At each step of the rebuilding process, the object with the smallest chronological rank O_l is extracted from $\mathcal{Q}$, and we also get all the critical regions that conflict with O_l. The object O_l is then reinserted, and the details of this operation depend of course on the problem in hand. The main obstacle is that we might have to change more than the critical zone of the influence graph. Indeed, the new regions created by O_l always have some critical parents, even though they may also have non-critical parents. Moreover, parents of an unhooked region are new, but the unhooked region itself is not. To correctly set up the arcs in the influence graph that are incident to new nodes, the algorithm must find in $\mathcal{I}a(\Sigma)$ the unhooked nodes and the non-critical parents of the new nodes. At this phase, the dictionary set up in the locating phase is used. After reinserting O_l, the priority queue $\mathcal{Q}$ is updated as follows: the regions in $\mathcal{Z}_{l-1}$ that conflict with O_l are not critical any more; however, any new region created by O_l belongs to $\mathcal{Z}_l$. Then for each of these regions F, the killer of F in Σ' is identified as the object in $L'(F)$ with the smallest chronological rank. This object is then searched for in $\mathcal{Q}$ and inserted there if it is not found. Then F is added to the list of regions killed by O_l.

6.3   Randomized analysis of dynamic algorithms

The randomized analysis of the augmented influence graph and the insertion and deletion operations are based on the probabilistic model described in section 6.1. The first three lemmas in this paragraph analyze the expected number of elementary changes to be performed upon a deletion.

Lemma 6.3.1 *Upon deleting an object, the number of nodes that are destroyed, new, or unhooked is, on the average,*

$$O\left(\frac{1}{n}\sum_{l=1}^{n}\frac{f_0(l,\mathcal{S})}{l}\right),$$

where, as usual, $f_0(l,\mathcal{S})$ stands for the number of regions defined and without conflict over a random sample of size l from $\mathcal{S}$.

Proof. We bound the number of destroyed, new, and unhooked nodes separately.
1. The number of destroyed nodes. A node in $\mathcal{I}a(\Sigma)$ corresponding to a region F in $\mathcal{F}_j^i(\mathcal{S})$ is destroyed during a deletion if the object deleted is one of the i objects that determine the region F. Let F be a region in $\mathcal{F}_j^i(\mathcal{S})$. Given that F corresponds to a node in the influence graph built for $\mathcal{S}$, this node is destroyed during a deletion with a conditional probability $i/n \leq b/n$. From theorem 5.3.2, we know that the expected number of nodes in the influence graph is

$$O\left(\sum_{l=1}^{n}\frac{f_0(l,\mathcal{S})}{l}\right),$$

so the number of nodes destroyed when deleting an object is, on the average,

$$O\left(\frac{1}{n}\sum_{l=1}^{n}\frac{f_0(l,\mathcal{S})}{l}\right).$$

2. The number of new nodes. The regions that correspond to the new nodes in the influence graph when O_k is deleted are exactly the regions created by O_l, for some l such that $k < l \le n$, that belong to $\mathcal{F}_0(\mathcal{S}'_l)$ and conflict with O_k. Let F be a region of $\mathcal{F}^i_j(\mathcal{S})$. This region F belongs to $\mathcal{F}_0(\mathcal{S}'_l)$ with the probability $p^i_j(l-1)$ that was given in subsection 5.2.2. Assuming this, F is created by O_l with conditional probability $i/(l-1)$, and F conflicts with O_k with conditional probability $j/(n-l+1)$. Therefore, for a given k, the number of new nodes in the influence graph upon the deletion O_k is, on the average (using corollary 4.2.7 to the moment theorem),

$$\sum_{i=1}^{b}\sum_{j=1}^{n}\sum_{l=k+1}^{n}|\mathcal{F}^i_j(\mathcal{S})|\ p^i_j(l-1)\ \frac{i}{l-1}\ \frac{j}{n-l+1}\ =\ O\left(\sum_{l'=k}^{n-1}\frac{m_1(l',\mathcal{S})}{l'(n-l')}\right)$$

$$=\ O\left(\sum_{l'=k}^{n-1}\frac{f_0(\lfloor l'/2\rfloor,\mathcal{S})}{l'^2}\right).$$

Averaging over all ranks k, the number of new nodes in the influence graph after a deletion is

$$O\left(\frac{1}{n}\sum_{k=1}^{n}\sum_{l'=k}^{n-1}\frac{f_0(\lfloor l'/2\rfloor,\mathcal{S})}{l'^2}\right) = O\left(\frac{1}{n}\sum_{l=1}^{n-1}\frac{f_0(l,\mathcal{S})}{l}\right).$$

3. The number of unhooked nodes. Unhooked nodes are the non-destroyed children of destroyed nodes. If condition 5.3.3 is satisfied, the number of children of each node in the augmented influence graph is bounded by a constant. It follows that the number of unhooked nodes is at most proportional to the number of destroyed nodes.   □

The update condition 5.3.3 assumes that the number of children of a node is bounded by a constant. However, the number of parents of a node is not necessarily bounded by a constant and the following lemma is useful to bound the number of arcs in the influence graph that are removed or added during a deletion.

Lemma 6.3.2 *The number of arcs in the influence graph that are removed or added during a deletion is, on the average,*

$$O\left(\frac{1}{n}\sum_{l=1}^{n-1}\frac{f_0(l,\mathcal{S})}{l}\right).$$

Proof. The simplest proof of this lemma involves the notion of biregion encountered in exercise 5.7. A *biregion* defined over a set of objects S is a pair of regions defined over S which can possibly be related as parent and child in the influence graph, for an appropriate permutation of S. A *biregion* is determined by at most $2b$ objects, and the notion of conflict between objects and regions can be extended to biregions: an object conflicts with a biregion if it is not a determinant of any of the two regions but conflicts with at least one of the two regions. Biregions obey statistical laws similar to those obeyed by regions. In particular, a biregion determined by i objects of S which conflicts with j objects of S is a biregion defined and without conflict over a random l-sample of S, with the probability $p_j^i(l)$ given by lemma 4.2.1. A biregion defined and without conflict over a subset S_l of S corresponds to an arc in the influence graph whenever the objects that determine the parent region are inserted before those that determine the child region and at the same time conflict with the parent region. This only happens with a probability $\alpha \in [0,1]$ (which depends on the number of objects determining the parent and the child, and the number of objects that at the same time determine the child and conflict with the parent).

A biregion determined by i objects in S and conflicting with j objects in S corresponds to an arc in the influence graph $Ia(\Sigma)$ that was created by O_l, with a probability smaller than $\frac{i}{l} p_j^i(l)$ (see also exercise 5.7); this arc, created by O_l, conflicts with O_k with a probability smaller than

$$\frac{1}{l} \frac{j}{n-l} p_j^i(l).$$

A computation similar to that in the proof of lemma 6.3.1 shows that the expected number of arcs in the influence graph that are created or removed during a deletion (which are those adjacent in the influence graph to new nodes or to destroyed nodes) is

$$O\left(\frac{1}{n} \sum_{l=1}^{n-1} \frac{f\!f_0(l,S)}{l} \right),$$

where $f\!f_0(l,S)$ is the expected number of biregions defined and without conflict over a random l-sample of S. It remains to show that $f\!f_0(l,S)$ is proportional to $f_0(l,S)$. Let S_l be a subset of size l of S. The parent region in a biregion that is defined and without conflict over S_l is a region defined over S_l that conflicts with exactly one object in S_l, and is therefore a region in $\mathcal{F}_1(S_l)$. Conversely, if the update condition 5.3.3 is true, every region in $\mathcal{F}_1(S_l)$ is the parent in a bounded number of biregions defined and without conflict over S_l. It follows that $f\!f_0(l,S)$ is within a constant factor of the expectation $f_1(l,S)$ of the number of regions defined and conflicting with one element over a random l-sample. From corollary 4.2.4 to the sampling theorem, this expected number is $O(f_0(l,S))$. $\square$

Lemma 6.3.3 *The total size of all the conflict lists attached to the nodes that are new or destroyed when an object is deleted is, on the average,*

$$O\left(\sum_{l=1}^{n} \frac{f_0(l, \mathcal{S})}{l^2}\right).$$

Proof.

1. Conflict lists of destroyed nodes. A region F of $\mathcal{F}_j^i(\mathcal{S})$ corresponds to a node of the influence graph $\mathcal{I}a(\Sigma)$ with probability

$$\sum_{l=1}^{n} p_j^i(l) \frac{i}{l}$$

as implied by lemma 5.2.2. The conflict list attached to this node has length j and this node is destroyed during the deletion of an object with probability i/n. The total size of the conflict lists attached to destroyed nodes is thus

$$\sum_{l=1}^{n}\sum_{i=1}^{b}\sum_{j=1}^{n} |\mathcal{F}_j^i(\mathcal{S})| \; p_j^i(l) \frac{i}{l}\frac{i}{n} j \;\; = \;\; O\left(\frac{1}{n}\sum_{l=1}^{n} \frac{m_1(l, \mathcal{S})}{l}\right)$$

$$= \; O\left(\frac{1}{n}\sum_{l=1}^{n}(n-l)\,\frac{f_0(\lfloor\frac{l}{2}\rfloor, \mathcal{S})}{l^2}\right)$$

$$= \; O\left(\sum_{l=1}^{n} \frac{f_0(l, \mathcal{S})}{l^2}\right),$$

as follows from corollary 4.2.7 to the moment theorem.

2. Conflict lists attached to new nodes. A region F of $\mathcal{F}_j^i(\mathcal{S})$ is a new region created by O_l when O_k is deleted, if it is a region of $\mathcal{F}_0(\mathcal{S}_l')$ determined by O_l that conflicts with O_k. The conflict list attached to the new node corresponding to F has $j-1$ elements. The total size of the conflict lists attached to new nodes when deleting O_k is thus, on the average,

$$\sum_{i=1}^{b}\sum_{j=1}^{n}\sum_{l=k+1}^{n} |\mathcal{F}_j^i(\mathcal{S})| \; p_j^i(l-1) \frac{i}{(l-1)}\frac{j}{(n-l+1)} (j-1).$$

Applying corollary 4.2.7, this size is

$$O\left(\sum_{l=k}^{n-1} \frac{m_2(l, \mathcal{S})}{l(n-l)}\right) = O\left(\sum_{l=k}^{n-1}(n-l)\,\frac{f_0(\lfloor\frac{l}{2}\rfloor, \mathcal{S})}{l^3}\right).$$

Averaging over all ranks of k, the above sum becomes

$$O\left(\frac{1}{n}\sum_{k=1}^{n}\sum_{l=k}^{n}(n-l)\,\frac{f_0(\lfloor\frac{l}{2}\rfloor, \mathcal{S})}{l^3}\right) = O\left(\sum_{l=1}^{n} \frac{f_0(l, \mathcal{S})}{l^2}\right). \qquad \square$$

Lastly, setting up the priority queue Q of killers of critical regions involves the regions of the influence graph $\mathcal{I}a(\Sigma)$ that are killed by O_k. The conflict lists of these regions are traversed in order to set up the conflict lists of the new children of these nodes. The following lemma is therefore needed in order to fully analyze dynamic algorithms.

Lemma 6.3.4 *The number of nodes in the influence graph $\mathcal{I}a(\Sigma)$ that are killed by a random object in S is, on the average,*

$$O\left(\frac{1}{n}\sum_{l=1}^{n}\frac{f_0(l,S)}{l}\right).$$

The total size of the conflict lists attached to the nodes killed by a random object is, on the average,

$$O\left(\sum_{l=1}^{n}\frac{f_0(l,S)}{l^2}\right).$$

Proof. A node in $\mathcal{I}a(\Sigma)$ that is killed by an object O_k of rank k corresponds to a region in $\mathcal{F}_0(S_{k-1})$ that conflicts with O_k. A region F in $\mathcal{F}_j^i(S)$ is a region in $\mathcal{F}_0(S_{k-1})$ that conflicts with O_k with probability

$$p_j^i(k-1)\frac{j}{n-k+1}.$$

Hence, the average number of nodes in $\mathcal{I}a(\Sigma)$ killed by a random object of S is

$$\frac{1}{n}\sum_{k=2}^{n}\sum_{i=1}^{b}\sum_{j=1}^{n}|\mathcal{F}_j^i(S)|\ p_j^i(k-1)\ \frac{j}{n-k+1} \quad = \quad O\left(\frac{1}{n}\sum_{k=2}^{n}\frac{m_1(k-1,S)}{n-k+1}\right)$$

$$= \quad O\left(\frac{1}{n}\sum_{k=1}^{n}\frac{f_0(k,S)}{k}\right),$$

as can be deduced from corollary 4.2.7 to the moment theorem.

The total size of the conflict lists attached to nodes killed by a random object is, on the average,

$$\frac{1}{n}\sum_{k=2}^{n}\sum_{i=1}^{b}\sum_{j=1}^{n}|\mathcal{F}_j^i(S)|\ p_j^i(k-1)\ \frac{j}{n-k+1}\ (j-1) \quad = \quad O\left(\frac{1}{n}\sum_{k=2}^{n}\frac{m_2(k-1,S)}{n-k+1}\right),$$

$$= \quad O\left(\sum_{k=1}^{n}\frac{f_0(\lfloor k/2\rfloor,S)}{k^2}\right)$$

$$= \quad O\left(\sum_{l=1}^{n}\frac{f_0(l,S)}{l^2}\right). \qquad \square$$

The detailed operations required to insert or delete an object in an augmented influence graph depend upon the problem under consideration. In particular, deletions demand a number of operations (insertions, deletions, or queries) on a dictionary of nodes, or even several dictionaries. To be able to present a general analysis, we introduce here an update condition for dynamic algorithms that use an augmented influence graph. This condition is similar to those introduced in the previous chapter to analyze incremental algorithms, namely condition 5.2.1 for algorithms using a conflict graph and condition 5.3.3 for algorithms using an influence graph. The condition we introduce here is a reasonable one, which will be fulfilled by all the dynamic algorithms described in this book.

Update condition 6.3.5 (for augmented influence graphs) *A dynamic algorithm that uses an augmented influence graph satisfies the* update condition *when:*

1. *The augmented influence graph satisfies the update condition 5.2.1 for algorithms using an influence graph.*

2. *During a deletion:*

    a. *The number of operations on a dictionary of nodes (insertions, deletions, or queries) is at most proportional to the total number of nodes killed (by the deleted object), destroyed, new, or unhooked.*

    b. *The conflict lists of the new nodes are initialized using a time proportional to the total size of the conflict lists of the nodes killed (by the deleted object), destroyed, and new.*

    c. *All the operations performed to update the augmented influence graph that do not pertain to dictionaries, conflict lists, or the priority queue Q, are elementary and their number is proportional to the total number of destroyed or new nodes, and of arcs incident to these nodes.*

The complexity of a deletion depends partly on the data structures used to implement the dictionaries and the priority queue Q of the killers of critical regions. These data structures store a set of elements that belong to a finite, totally ordered universe, whose cardinality is bounded by a polynomial in the number n of current objects. Therefore, we can use the data structures described in section 2.3, or more simply we may use a standard balanced binary tree. In order to take all these cases into account, the analysis given below introduces two parameters. The first parameter, t, is the complexity of a query, insertion, or deletion performed on a dictionary of size $O(n^c)$, where c is some constant. The second parameter, t', is the complexity of a query, insertion, or deletion performed on a priority queue of size $O(n)$. Parameter t is $O(\log n)$ if a balanced tree is used, and $O(1)$ if the perfect dynamic hashing method of section 2.3 is used.

Parameter t' is $O(\log n)$ if we use a balanced binary tree, but it is $O(\log \log n)$ if we use a stratified tree along with perfect dynamic hashing (see section 2.3). Moreover, as we will see further on, if $f_0(l, S)$ grows at least quadratically, then implementing Q with a simple array of size n will suffice, and t' can be ignored in the analysis.

Theorem 6.3.6 *This theorem details the performances of a dynamic algorithm that uses an augmented influence graph and satisfies the update condition 6.3.5. Let S be the current set of objects, and n the size of S.*

1. *The data structure requires an average storage of*

$$O\left(n \sum_{l=1}^{n} \frac{f_0(l, S)}{l^2}\right).$$

2. *Adding an object can be performed in expected time*

$$O\left(\sum_{l=1}^{n} \frac{f_0(l, S)}{l^2}\right).$$

3. *Deleting an object can be performed in expected time*

$$O\left(\min\left(n, \frac{t'}{n} \sum_{l=1}^{n} \frac{f_0(l, S)}{l}\right) + \frac{t}{n} \sum_{l=1}^{n} \frac{f_0(l, S)}{l} + \sum_{l=1}^{n} \frac{f_0(l, S)}{l^2}\right).$$

As always, $f_0(l, S)$ is the number of regions defined and without conflict over a random l-sample of S, t is the complexity of any operation on a dictionary, and t' is the complexity of an operation on the priority queue Q.

Proof.

1. The storage needed by the augmented influence graph $\mathcal{I}a(\Sigma)$ is proportional to the total size of the conflict lists attached to the nodes of $\mathcal{I}a(\Sigma)$. Each element in one of these conflict lists corresponds to a conflict detected by an on-line algorithm processing the objects in S in the chronological order of the sequence Σ. The expected number of conflicts, for a random permutation of Σ, is thus given by theorem 5.2.3 which analyzes the complexity of incremental algorithms that use a conflict graph.

2. The randomized analysis of an insertion into the augmented conflict graph is identical to that of the incremental step in an on-line algorithm that uses an influence graph. Indeed, the two algorithms only differ in that one updates conflict lists. Each conflict between the inserted object and a node in the current graph is detected by both algorithms. In the dynamic algorithm, detecting such a conflict implies adding the inserted object into the conflict list of the conflicting node, which can be carried out in constant time. The expected complexity of an insertion is thus given by theorem 5.3.4.

3. Let us now analyze the average complexity of deleting an object, say O_k. When locating O_k in the augmented influence graph, the nodes that are visited are exactly the destroyed nodes, and their children, and the nodes that conflict with O_k. Since every node has a bounded number of children, the cost of the traversal is proportional to the number of nodes destroyed or conflicting with O_k. The number of nodes in the influence graph that conflict with O_k is, on the average over all possible sequences Σ,

$$O\left(\sum_{l=1}^{k-1} \frac{f_0(\lfloor \frac{l}{2} \rfloor, \mathcal{S})}{l^2}\right),$$

which we know from the proof of theorem 5.3.4. Averaging over the rank of the deleted object, we get

$$O\left(\sum_{l=1}^{n} \frac{f_0(l, \mathcal{S})}{l^2}\right).$$

From lemma 6.3.1, the latter expression is also a bound on the expected number of nodes destroyed and thus on the global cost of traversing the influence graph.

If the update condition 6.3.5 is realized, lemmas 6.3.3 and 6.3.4 show that the conflict lists of the new regions can be set up in time

$$O\left(\sum_{l=1}^{n} \frac{f_0(l, \mathcal{S})}{l^2}\right).$$

Lemma 6.3.1 and the update condition 6.3.5 (2a) assert that the term

$$O\left(\frac{t}{n} \sum_{l=1}^{n} \frac{f_0(l, \mathcal{S})}{l}\right)$$

accounts for the average complexity of all the operations performed on the dictionaries of nodes.

Since t is necessarily $\Omega(1)$, lemmas 6.3.1 and 6.3.2, together with condition 6.3.5 (2c), assert that the former term also accounts for all the operations that update the augmented influence graph, not counting those on the conflict lists or the priority queue.

It remains to analyze the management of the priority queue $\mathcal{Q}$ of critical region killers. The number of insertions and queries in the priority queue is proportional to the total number of critical regions encountered during the rebuilding phase.

These regions are either killed by the deleted object, or they are new regions. Their average number is thus

$$O\left(\frac{1}{n}\sum_{l=1}^{n}\frac{f_0(l,\mathcal{S})}{l}\right),$$

as asserted by lemmas 6.3.1 and 6.3.4. The average number of minimum queries to be performed on the queue $\mathcal{Q}$ is

$$O\left(\min\left(n,\frac{1}{n}\sum_{l=1}^{n}\frac{f_0(l,\mathcal{S})}{l}\right)\right),$$

since the number of objects to be reinserted is bounded from above by n on the one hand, and by the number of unhooked or new nodes (estimated by theorem 6.3.1) on the other hand. $\square$

Consequently,

- If $f_0(l,\mathcal{S})$ grows slower than quadratically (with respect to l), we use a hierarchical structure for the priority queue, characterized by a parameter t' which bounds the complexity of any operation on this structure (insertion, membership or minimum query). Managing the queue has therefore the associated expected cost

$$O\left(\frac{t'}{n}\sum_{l=1}^{n}\frac{f_0(l,\mathcal{S})}{l}\right).$$

- If on the contrary, $f_0(l,\mathcal{S})$ grows at least quadratically, we use for $\mathcal{Q}$ a simple array of size n. This allows insertions and deletions to be performed into the queue in constant time. The cost of finding the minima during the whole rebuilding phase is then $O(n)$, and managing the queue has in this case the associated expected cost

$$O\left(n+\frac{1}{n}\sum_{l=1}^{n}\frac{f_0(l,\mathcal{S})}{l}\right).$$

- When $f_0(l,\mathcal{S})$ is $O(l)$, the expected number of destroyed or new nodes visited during a deletion is $O(1)$. Updating the conflict lists costs $O(\log n)$ anyway. Both the priority queue and the dictionaries can be implemented simply by balanced binary trees ($t = t' = \log n$) to yield a randomized complexity of $O(\log n)$ for a deletion.

6.4   Dynamic decomposition of a set of line segments

The vertical decomposition of a set of line segments in the plane is a structure defined in section 3.3. It can be built using a conflict graph by a static incremental algorithm, as explained in subsection 5.2.2, or by a semi-dynamic incremental algorithm using an influence graph, as detailed in subsection 5.3.2. By combining both structures using the general method explained in section 6.2, we can dynamically maintain the vertical decomposition of the segments under insertion or deletion of a segment. The algorithm described here is a generalization of the decomposition algorithms given in subsections 5.2.2 and 5.3.2. It is advisable to thoroughly understand both these algorithms before reading further.

Let us first recall that, for this problem, the objects are segments and the regions defined over a set $\mathcal{S}$ of segments are the trapezoids appearing in decompositions of subsets of $\mathcal{S}$. A trapezoid is determined by at most four segments. A segment conflicts with a trapezoid if it intersects the interior of the trapezoid.

Let $\mathcal{S}$ be the current set of segments and let $\Sigma = \{O_1, O_2, \ldots, O_n\}$ be the chronological sequence of segments in $\mathcal{S}$. We may also denote by $\mathcal{S}_l$ the subset of $\mathcal{S}$ consisting of the first l segments in Σ. The dynamic algorithm maintains an augmented influence graph $\mathcal{I}a(\Sigma)$, whose nodes correspond to the trapezoids that are defined and without conflict over the current subsets $\mathcal{S}_l$, $l = 1, \ldots, n$. The nodes and arcs in this graph are identical to those built in the influence graph by the on-line algorithm described in subsection 5.3.2. In addition, the augmented influence graph includes a conflict graph between the trapezoids corresponding to the nodes of the influence graph, and the segments in $\mathcal{S}$. The conflict graph is implemented using the interconnected list system, as explained in section 6.2. The structure does not encode all the adjacency relationships between the trapezoids but only those between the trapezoids in the current decomposition (corresponding to leaves of the influence graph).

Adding the n-th segment does not create problems and can be carried out exactly as explained in subsection 5.3.2. The only difference is that the conflict lists are updated when the conflicts are detected during the locating phase when inserting O_n.

Let us now explain how to delete segment O_k, of rank k in the chronological sequence Σ. As before, we denote by $\mathcal{S}'_l$ the subset $\mathcal{S}_l \setminus \{O_k\}$ of $\mathcal{S}$ and by Σ' the chronological sequence $\{O_1, \ldots, O_{k-1}, O_{k+1}, \ldots O_n\}$.

The algorithm proceeds along the usual lines and performs the two phases: locating and rebuilding. The locating phase detects the nodes of $\mathcal{I}a(\Sigma)$ that conflict with O_k, and the destroyed and unhooked nodes. In the rebuilding phase, the algorithm processes the segments of rank $l > k$ that are the creators of new or unhooked nodes. For this, the algorithm manages a priority queue which contains, at each step of the rebuilding process, the killers in Σ' of the critical regions. For each such object O_l, a killer of a critical region, the algorithm builds the new nodes created by O_l and rehooks the unhooked nodes created by O_l. Figure 6.1 shows how the influence graph built for the four segments $\{O_1, O_2, O_3, O_4\}$ is modified when deleting O_3. The reader may observe again how the graph was created incrementally, in figures 5.6, 5.7 and 5.8. In this example, nodes B and H are killed by O_3, nodes J,K,L,M,N,O,P,Q,S,U,V are destroyed, nodes R,T,W are unhooked (they are created by O_4), and B' is a new node (its creator is O_4).

The subsequent paragraphs describe in great detail the specific operations needed.

Locating. This phase is trivial: all the nodes that conflict with the object O_k to be deleted, or that are determined by a subset containing O_k, are visited together with their children. The algorithm builds a dictionary $\mathcal{D}$ of unhooked or destroyed nodes, which will be used during the rebuilding phase.

Rebuilding. The priority queue $\mathcal{Q}$, which contains the killers of critical regions, is initialized with the nodes in $\mathcal{I}a(\Sigma)$ that are killed by O_k.

At each step in the rebuilding process, the algorithm extracts from the priority queue $\mathcal{Q}$ the object O_l of smallest chronological rank. It also retrieves the list of the critical regions that conflict with O_l.

Each of these regions is split into at most four *subregions* by O_l, and the walls stemming from its endpoints and its intersection points. These subregions are not necessarily trapezoids in the decomposition $\mathcal{D}ec(\mathcal{S}'_l)$. Indeed, the walls cut by O_l have to be shortened, keeping only the part that is still connected to the endpoint or intersection point from which it stems. The other part of the wall must be removed and the adjacent subregions separated by this part must be joined. The join can be one of two kinds: *internal* when the portion of wall to be removed separates two critical regions, and *external* when it separates a critical region from a non-critical region (see figure 6.2).

To detect which regions to join,[2] the algorithm visits all the critical regions that conflict with O_l, and stores in a secondary dictionary $\mathcal{D}'_l$ the walls incident to these regions that are intersected by O_l. Any wall in this dictionary that separates two critical regions gives rise to an internal join, and any wall incident to only one critical region gives rise to an external join.

In a first phase, the algorithm creates a temporary node for each subregion resulting from the splitting of a critical region by O_l or the walls stemming from O_l. The node that corresponds to a subregion F_i of the region F is hooked in the graph as a child of F. Its conflict list is obtained by selecting, from the conflict

---

[2]The algorithm cannot traverse the sequence, ordered by O_l, of critical regions for two reasons: (1) it does not maintain the vertical adjacencies between the internal nodes of the influence graph, and the adjacencies between either the trapezoids of the decomposition $\mathcal{D}ec(\mathcal{S}'_{l-1})$ or the critical regions of $\mathcal{Z}_{l-1}$ are not available, and (2) the intersection of O_l with the union of the regions in $\mathcal{Z}_{l-1}$ may not be connected (see for instance figure 6.4).

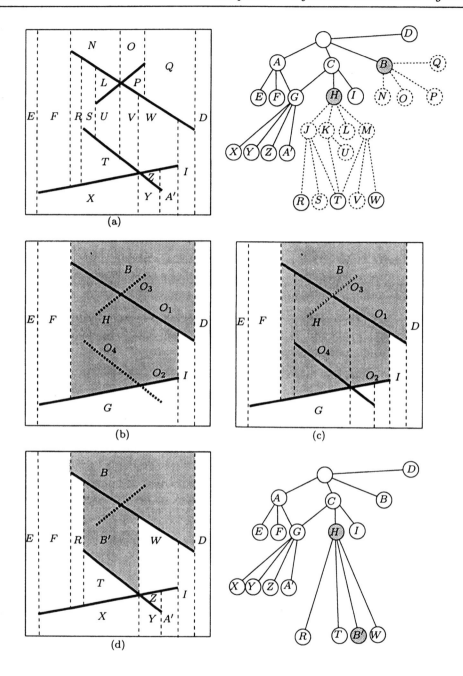

Figure 6.1. Deleting a segment.

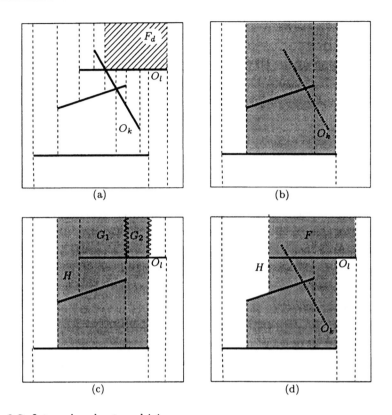

Figure 6.2. Internal and external joins:
(a) The decomposition $\mathcal{D}ec(\mathcal{S}_l)$ before deleting O_k.
(b) The decomposition $\mathcal{D}ec(\mathcal{S}'_{l-1})$, with the critical zone $\mathcal{Z}_{l-1}$ shaded.
(c) Reinserting O_l. Splitting the critical regions and joining: internal join $G = G_1 \cup G_2$, external join $F = G \cup F_d$. Region H is unhooked.
(d) The decomposition $\mathcal{D}ec(\mathcal{S}'_l)$ and the critical zone $\mathcal{Z}_l$.

list of F, the segments intersecting F_i. Then the algorithm processes the internal and the external joins, as explained below.

1. Internal joins. Every maximal set $\{G_1, \ldots, G_h\}$ of subregions, pairwise adjacent and separated by walls to be removed, must be joined together into a single region G. The algorithm creates a temporary node for G. The nodes corresponding to $G_1, G_2, \ldots, G_h$ are removed from the graph and the node corresponding to G inherits all the parents of these nodes. The conflict list of G is obtained by merging the conflict lists of $G_1, G_2, \ldots, G_h$, removing redundancies. For this, we use a procedure similar to that of subsection 5.2.2, but which need not know the order along O_l of the subregions to be joined. By scanning the conflict lists of these subregions successively, the algorithm can build for each segment O in $\mathcal{S}$ a list $L_G(O)$ of the subregions that conflict with O. A bidirectional pointer interconnects the entry in the list $L'(G_i)$ that corresponds to an object O with the entry in $L_G(O)$ corresponding to the subregion G_i. The conflict list of G can be retrieved by scanning again all the conflict lists $L'(G_i)$ of the subregions $G_1, \ldots,$ G_h. This time, each segment O encountered in one of these lists is added to the conflict list of G and removed from the other conflict lists, using the information stored in $L_G(O)$.

Let us call *auxiliary regions* the regions obtained after all the internal joins. These regions are either subregions that needed no internal join, or regions obtained from an internal join of the subregions. An auxiliary region that does not need to undergo any external join is a region of the decomposition $\mathcal{Dec}(\mathcal{S}'_l)$. Let H be such a region. This region is new if it conflicts with O_k, unhooked otherwise. In the former case, the temporary node for H becomes permanent and the killer of H is inserted into the priority queue $\mathcal{Q}$. In the latter case, a node for H already exists in the influence graph $\mathcal{I}a(\Sigma)$. A simple query in the dictionary of unhooked nodes retrieves this node, which can then be rehooked to the parents of the auxiliary node created for H.

2. External joins. In a second phase, the algorithm performs the external joins. An auxiliary region undergoes a left join if its left wall must be removed, and a right join if its right wall must be removed, and a double left–right join if both its vertical walls must be removed. Let G be an auxiliary region undergoing a right join. For instance, this is the case for region $G = G_1 \cup G_2$ in figure 6.2. The right wall of G is on the boundary of the critical zone, since this is an external join. This wall is therefore not cut by the deleted segment O_k. When the decomposition of S is built incrementally according to the order in the sequence Σ, this wall appears at a certain step and is removed when O_l is inserted. Thus, among all the regions in $\mathcal{I}a(\Sigma)$, there is one region F_d created by O_l that contains the right wall of G.[3] The region F_d is necessarily destroyed or unhooked: indeed, F_d is a trapezoid in the decomposition $\mathcal{Dec}(\mathcal{S}_l)$, and has a non-empty intersection with one or more critical regions in $\mathcal{Z}_{l-1}$. As every critical region in $\mathcal{Z}_{l-1}$ is contained in the union of the trapezoids of $\mathcal{Dec}(\mathcal{S}_{l-1})$ of which O_k is a determinant, the region F_d must intersect those trapezoids. Thus at least one of the parents of F_d in the graph $\mathcal{I}a(\Sigma)$ is a destroyed node. Similarly, if the left wall of G must be removed, there is in $\mathcal{I}a(\Sigma)$ one destroyed or unhooked region F_g created by O_l that contains the left wall of G. If the join is double left–right, F_d and F_g may be distinct or identical (see figure 6.3).

Several auxiliary regions may be joined into the same permanent region (see figure 6.4). Let $\{G_1, G_2, \ldots, G_j\}$ be the sequence ordered along O_l of the auxiliary

---

[3]It would have been more desirable to subscript F by l and r for *left* and *right*, but this would have conflicted with the index l for O_l and created confusion. We have kept a French touch with the indices g and r for the French *gauche* and *droit*, meaning respectively left and right. (Translator's note)

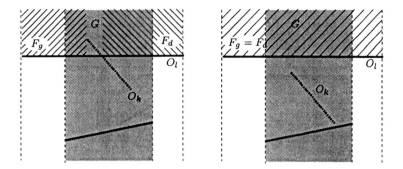

Figure 6.3. External joins: double left–right joins.

regions[4] whose left wall is contained in the same region F_g of $\mathcal{I}a(\Sigma)$ created by O_l. If $j > 1$, then the right walls of these auxiliary regions $\{G_1, G_2, \ldots, G_{j-1}\}$ are also contained in F_g and must be removed as well. If the right wall of G_j is a permanent wall (that does not have to be removed), the join results in a single trapezoid of the decomposition $\mathcal{D}ec(\mathcal{S}'_l)$ that is the same as $F_g \cup G_1 \cup \ldots \cup G_j = F_g \cup G_j$ (see figure 6.4). If the right wall of G_j also has to be removed, then we introduce the ordered sequence of auxiliary regions $\{G_j, G_{j+1}, \ldots, G_h\}$: this sequence consists of regions whose right wall must be removed and which lie in the same region F_d of $\mathcal{I}a(\Sigma)$ created by O_l. The left walls of the regions in $\{G_j, G_{j+1}, \ldots, G_h\}$ then also belong to F_d and have to be removed. The join operates on the auxiliary regions $\{G_1, \ldots, G_j, \ldots, G_h\}$ and results in a unique trapezoid in $\mathcal{D}ec(\mathcal{S}'_l)$ that is the same as $F_g \cup G_1 \cup \ldots \cup G_h \cup F_d = F_g \cup G_j \cup F_d$.

We present below the operations to be performed in the latter case of a double left–right join. The former cases can be handled in a similar manner. Suppose for now that the auxiliary regions $\{G_1, \ldots, G_j, \ldots, G_h\}$ as well as the regions F_g and F_d in the decomposition $\mathcal{I}a(\Sigma)$ that participate in the join are known to the algorithm.

If the trapezoid resulting from the join $F = F_g \cup G_j \cup F_d$ does not conflict with O_k (see figure 6.3, right), it is a trapezoid in the decomposition $\mathcal{D}ec(\mathcal{S}_l)$. Necessarily, the regions F_g, F_d, and F are the same, and the corresponding node in $\mathcal{I}a(\Sigma)$ is unhooked. It then suffices to search for this node in the dictionary of unhooked nodes, to remove the auxiliary nodes created for $G_1, G_2, \ldots, G_h$, and to rehook the node corresponding to F, with the critical nodes in the parents of $G_1, G_2, \ldots, G_h$ as the parents of F.

If the resulting trapezoid $F = F_g \cup G_j \cup F_d$ conflicts with O_k (see figure 6.3, left), then it is a new region of $\mathcal{I}a(\Sigma')$, and the regions F_g and F_d in $\mathcal{I}a(\Sigma)$

---

[4]We must emphasize that even though the given description of the region resulting from an external join refers to the order of the joined auxiliary regions along O_l, the algorithm does not know this order, nor does it need it.

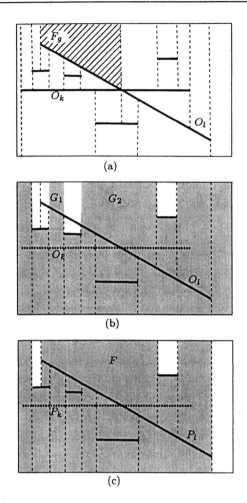

Figure 6.4. External joins.
     (a) The decomposition before deleting O_k.
     (b) Reinserting O_l. The auxiliary regions.
     (c) Joining auxiliary regions G_1 and G_2 into $F = F_g \cup G_1 \cup G_2$.

are destroyed. The auxiliary nodes created for $G_1, G_2, \ldots, G_h$ are removed, and replaced by a single node corresponding to F. This node is then rehooked to the parents of F_g and F_d that are not destroyed, and to all the critical parents of G_1, G_2, $\ldots$, G_h. The conflict list of F is derived from those of $F_g, G_1, G_2, \ldots, G_h$ and F_d, as is the case for internal joins. Lastly, the killer of F is inserted into the priority queue $\mathcal{Q}$.

We now have to explain how to retrieve the unhooked or destroyed nodes corresponding to the regions F_g and F_d involved in the join. Let G be an auxiliary region whose left wall must be removed. The corresponding region F_g is either destroyed or unhooked, created by O_l, and the segments that support its floor and ceiling[5] respectively support the floor and ceiling of G. Any region in the decomposition of a given set of segments is identified uniquely by its floor, its ceiling, and one of its walls. Below, we show that either we can find one of the walls of F_g, or we can identify a destroyed region F_g' which is the unique sibling of F_g in $\mathcal{I}a(\Sigma)$.

- If G conflicts with O_k (as in figure 6.5a), the right wall of F_g is determined by O_k, and can be computed (by looking only at G and O_k).

- If G does not conflict with O_k, but its right wall is permanent (see figure 6.5b), then this right wall is also that of F_g.

- Lastly, if G does not conflict with O_k, and if both its walls must be removed (see figure 6.5c), then segment O_l intersects both walls of a critical region that was subsequently split into G and G'. The other subregion G' also conflicts with O_k but does not undergo any join. In $\mathcal{I}a(\Sigma)$, exactly one node F_g' has O_l for creator, is destroyed, and shares the same floor, same ceiling, and same left wall as G'. This node F_g' has only one parent, and this parent has two children, one of which is F_g' and the other the node F_g that we are looking for: indeed, the parent of F_g' corresponds to a trapezoid in the decomposition $\mathcal{D}ec(\mathcal{S}_{l-1})$ whose two walls are intersected by O_l.

In either case, the region F_g, or its sibling F_g', is known through its creator, its floor, its ceiling, and one of its left or right walls. This information is enough to characterize it. Naturally, the same observation goes for F_d or its sibling F_d'. We can then use the dictionary $\mathcal{D}$ storing all the destroyed or unhooked nodes. This dictionary comes in two parts, $\mathcal{D}_g$ and $\mathcal{D}_d$. In the dictionary $\mathcal{D}_g$, the nodes are labeled with:

- the creator segment,

- the segment supporting the floor of the trapezoid,

- the segment supporting the ceiling of the trapezoid,

- the pair of segments whose intersection determines the right wall of the trapezoid, or the same segment repeated twice if the wall stems from the segment's endpoint.

Similarly, in its counterpart $\mathcal{D}_d$, nodes are labeled the same way, except that in the last component the right wall is replaced by the left wall. Any destroyed or unhooked node is inserted into both dictionaries $\mathcal{D}_g$ and $\mathcal{D}_d$.

---

[5]We recall that the *floor* and *ceiling* of a trapezoid are its two non-vertical sides.

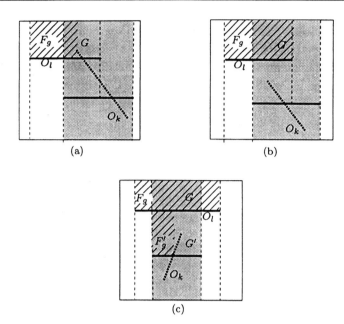

Figure 6.5. External joins:
    (a) G conflicts with O_k.
    (b) the right wall of G is permanent.
    (c) Double left–right join.

Analysis of the algorithm

To analyze this algorithm, we first check that it does satisfy the update conditions 6.3.5. The first condition is satisfied, since the augmented influence graph has the same nodes and arcs as the influence graph built by the on-line algorithm of subsection 5.3.2, which itself satisfies the update condition 5.3.3. Therefore, we need only look at deletions.

1. Number of operations on the dictionaries. Each deletion involves a two-sided dictionary $\mathcal{D}$ of destroyed or unhooked nodes, as well as a dictionary $\mathcal{D}'_l$, for each reinserted segment O_l, of walls in the critical zone intersected by O_l. A destroyed or unhooked node is inserted and queried at most once in $\mathcal{D}$. A critical region in $\mathcal{Z}_{l-1}$ has at most two walls which must be inserted into $\mathcal{D}'_l$, and this region will not be a critical region any more after the reinsertion of O_l. The number of operations on all dictionaries $\mathcal{D}'_l$ is thus at most proportional to the total number of critical regions encountered in the rebuilding phase. Any critical region is either killed or new. The total number of operations is thus at most proportional to the number of nodes that are killed, destroyed, unhooked, or new.

2. Conflict lists of new nodes. The conflict list of a new node is obtained by scanning the conflict lists of the auxiliary or destroyed regions of which it is the union. Similarly, the conflict list of an auxiliary region is obtained by traversing the conflict lists of the subregions of which it is the union, and the conflict lists of those subregions are themselves obtained by consulting the conflict lists of the critical regions cut by the reinserted object. During the rebuilding process, each killed or new region appears at most once as a critical region which conflicts with the reinserted object. Moreover, each subregion is involved in at most one internal join, and each auxiliary or destroyed region in at most one external join. From this, we derive that the conflict lists of the new nodes can be computed in time proportional to the total size of the conflict lists of the nodes that are killed, destroyed, or new.

3. Other operations. Apart from managing the priority queue $\mathcal{Q}$, querying the dictionaries, and setting up the conflict lists of the new nodes, all the remaining operations are elementary. Their number is at most proportional to the number of new or destroyed nodes, and to the number of incident arcs in the augmented influence graph.

If S is a set of n segments, with a intersecting pairs, the mathematical expectation $f_0(l, S)$ of the number of regions defined and without conflict over a random l-sample of S is $O(l + a\frac{l^2}{n^2})$, as given by lemma 5.2.4. Thus,

$$O\left(\frac{1}{n}\sum_{l=1}^{n}\frac{f_0(l, S)}{l}\right) = O\left(1 + \frac{a}{n}\right),$$

$$O\left(\sum_{l=1}^{n}\frac{f_0(l, S)}{l^2}\right) = O\left(\log n + \frac{a}{n}\right).$$

We can now use theorem 6.3.6 to state the following theorem, which summarizes the results so far:

Theorem 6.4.1 *Under the assumptions of dynamic randomized analyses, an augmented influence graph can be used to maintain the vertical decomposition of a set of segments with the following performances. Let S be the current set of segments, n be the size of S, and a be the number of intersecting pairs in S.*

- *The expected storage required by the algorithm is*

$$O(n \log n + a).$$

- *Inserting the n-th segment takes an average time*

$$O(\log n + \frac{a}{n}).$$

- *Deleting a segment takes an average time*

$$O\left(\log n + (1 + \frac{a}{n})(t + t')\right),$$

  where the parameters t and t' stand respectively for the complexities of the operations on dictionaries and priority queues.

Therefore, if we use perfect dynamic hashing together with stratified tree, the expected cost of a deletion is

$$O\left(\log n + \left(1 + \frac{a}{n}\right)\log \log n\right).$$

If we use balanced binary trees, it remains

$$O\left(\log n + \left(1 + \frac{a}{n}\right)\log n\right).$$

For the preceding algorithm, we have merely applied the general principles of the augmented influence graph to the case of computing the vertical decomposition of a set of segments. In fact, in this specific case, we may derive a simpler algorithm, yet one that uses less storage. This algorithm does not need to keep the conflict lists and maintains a non-augmented influence graph. It is outlined in exercises 6.1, 6.2 and 6.3, and its performances are summarized in the following theorem:

Theorem 6.4.2 *Under the assumptions of dynamic randomized analyses, an influence graph can be used to maintain the vertical decomposition of a set of segments with the following performances. Let S be the current set of segments, n be the size of S, and a be the number of intersecting pairs in S.*

- *The expected storage required by the algorithm is*

$$O(n + a).$$

- *Inserting a segment takes an average time of*

$$O(\log n + \frac{a}{n}).$$

- *Deleting a segment takes an average time of*

$$O\left((1 + \frac{a}{n})(t + t')\right),$$

  where the parameters t and t' stand respectively for the complexity of the operations on dictionaries and priority queues.

Therefore, the expected cost of a deletion is $O\left((1 + \frac{a}{n})\log \log n\right)$ if we use perfect dynamic hashing coupled with stratified trees. It remains $O\left((1 + \frac{a}{n})\log n\right)$ if we use balanced binary trees.

6.5   Exercises

Exercise 6.1 (Dynamic decomposition) Let us maintain dynamically the decomposition of a set of segments using an influence graph. Show that the creator of a new trapezoid, or a trapezoid unhooked during a deletion, is also the creator of at least one destroyed trapezoid.

Hint: The proof of this fact relies on the two additional properties possessed by the influence graph of a decomposition:

1. The influence domain of an internal node is contained in the union of the influence domains of its children.

2. If an object O_k is the determinant of an internal node, it necessarily is a determinant of at least one child of this node.

Let O_l be a segment creating a new trapezoid, or a trapezoid unhooked during the deletion of O_k. As in the entire chapter, the segments are indexed by their chronological rank and S_i stands for the set of the first i segments (in chronological order). To prove our assertion, we investigate the addition of O_k and successively O_l to the decomposition $\mathcal{D}ec(S'_{l-1})$. The decompositions obtained are then successively $\mathcal{D}ec(S_{l-1})$ and $\mathcal{D}ec(S_l)$. It can be shown that there is a region H in $\mathcal{D}ec(S_l)$ determined by a set that contains both O_k and O_l.

Exercise 6.2 (Dynamic decomposition) Let us assume that we use an influence graph to dynamically maintain the decomposition of a set S of segments. Here, we consider the deletion of a segment O_k. We still use the notation of section 6.4. In particular, the segments are indexed by their chronological rank. Let O_l be the segment to be reinserted during the deletion of O_k. Show that for any critical region F in the critical zone $\mathcal{Z}_{l-1}$ that conflicts with O_l, there is at least one destroyed region H, created by O_l, that intersects F, and satisfies at least one of the following conditions:

1. F contains a wall of H that stems from an endpoint of O_l, or an intersection point on O_l, and that butts against O_k (see figure 6.6a),

2. F contains a wall of H that stems from an endpoint of O_k, or an intersection point on O_k, and that butts against O_l (see figure 6.6b),

3. F contains a wall of H that stems from the intersection $O_l \cap O_k$ (see figure 6.6c),

4. F is bounded by two walls, one stemming from a point on O_k, the other stemming from a point on O_l, and both walls are contained within F (see figure 6.6d),

5. O_l and O_k support the floor and ceiling of H, both of which intersect F (see figure 6.6e).

Exercise 6.3 (Dynamic decomposition) The aim of this exercise is to show how we may dynamically maintain the decomposition of a set S of segments using a simple influence graph, without the conflict lists.

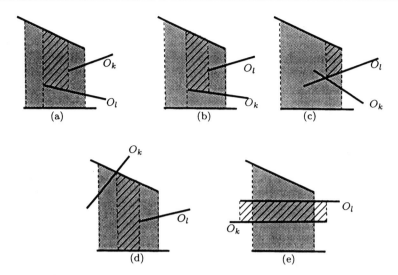

Figure 6.6. Detecting conflicts in critical zone. Region F is shaded, and region H within
is emphasized.

The segments that must be reinserted during a deletion are the creators of destroyed
regions (see exercise 6.1) and can be detected during the locating phase.

Let O_l be one of the segments to be reinserted during the deletion of O_k. To retrieve
all the critical regions that conflict with O_l, the algorithm considers in turn the destroyed
regions H with creator O_l, and selects the critical regions F related to H by one of the
five cases described in exercise 6.2.

For this, the deletion algorithm maintains an augmented dictionary $\mathcal{A}$, storing the
sequence ordered along O_k of critical regions intersected by O_k. Let H be one of the
destroyed regions, created by O_l. If H has a wall that stems from a point on O_l and
butts against O_k, or a wall stemming from $O_l \cap O_k$, this wall is located in the structure
$\mathcal{A}$, and the critical region containing this wall is retrieved. If H has two walls stemming
from a point on O_k and from a point on O_l, the region containing the wall stemming
from the point on O_k is searched for in $\mathcal{A}$, and it is selected if it also contains the wall
of H stemming from the point on O_l. Lastly, if O_l and O_k support the floor and ceiling
of H, the right wall of H is searched for in $\mathcal{A}$, and any critical region that intersects the
floor and the ceiling of H is selected.

1. The selected region obviously conflicts with O_l. As shown in exercise 6.2, any
critical region that conflicts with H is selected. Show that such a region can be selected
at most 16 times.

To speed up the locating phase, the algorithm maintains the lists of nodes killed by
each object stored in the structure. To perform the deletion, the algorithm proceeds
along the following lines.

Locating. The algorithm traverses the influence graph starting on the nodes killed
by O_k, and visits the destroyed or unhooked nodes. During this traversal, the algorithm
sets up a dictionary $\mathcal{D}$ that stores the destroyed and unhooked nodes, and a list $\mathcal{L}$ of the
creators of the destroyed nodes.

Rebuilding. The list $\mathcal{L}$ is sorted by chronological order, for instance by storing the elements in a priority queue, and extracting them in order. The redundant elements are extracted only once. The dictionary $\mathcal{A}$ initially stores the regions killed by O_k.

The objects of $\mathcal{L}$ are processed in chronological order. For each object O_l, the critical regions that conflict with O_l are selected as explained above. The remaining operations are identical to those in the algorithm of section 6.4. The conflict lists of the new regions do not have to be computed. On the other hand, the dictionary $\mathcal{A}$ must be updated.

2. Show that the performances of this algorithm are those given by theorem 6.4.2.

Exercise 6.4 (Lazy dynamic algorithms) In this exercise, we propose a *lazy* method to dynamically maintain the decomposition of a set of segments. For simplicity, let us assume that the segments do not intersect. The algorithm maintains an influence graph in the following lazy fashion:

1. The graph is a mere influence graph, no conflict lists are needed.

2. During an insertion, the nodes corresponding to the new trapezoids are hooked to the nodes corresponding to the killed trapezoids as in the algorithms described in subsection 5.3.2 and section 6.4.

3. During a deletion, the nodes corresponding to the new trapezoids are hooked to leaves of the graph that correspond to destroyed trapezoids. More precisely, a node corresponding to a new trapezoid is hooked to leaves of the graph that correspond to the destroyed nodes that have a non-empty intersection with the new trapezoid. No node is removed from the graph.

4. The algorithm keeps the age of the current graph in a counter, meaning the total number of operations (insertions and deletions) performed on this graph. Each time the number of segments effectively present falls below half the number stored in this counter, the algorithm builds the influence graph anew by inserting the segments effectively present into a brand new structure.

1. Show that when $O(n)$ segments are stored in the structure, the expected cost of an insertion or a location query is still $O(\log n)$.

2. The cost of the periodic recasting of the graph is shared among all the deletions. Show that the amortized complexity of a deletion is still $O(\log n)$ on the average. (Recall that the segments do not intersect, by assumption.)

Hint: It will be noted that the number of children of a node in the influence graph is not bounded any more. The analysis must then have recourse to *biregions* (see exercise 5.7) to estimate the expected complexity of the locating phases.

6.6 Bibliographical notes

The approach discussed in this chapter consists in forgetting deleted objects altogether, and restoring the structure to the exact state which it would have been in, had this object never been inserted. The first algorithm following this approach is that of Devillers, Meiser, and Teillaud [81] which maintains the Delaunay triangulation of a set of points in the plane. The algorithm by Clarkson, Mehlhorn, and Seidel [70] uses the same approach to maintain the convex hull of a set of points in any dimension. The method was then abstracted by Dobrindt and Yvinec [86]. A similar approach is also discussed by Mulmuley [176], whose book is the most comprehensive reference on this topic.

There is another way to dynamize randomized incremental algorithms. This approach, developed by Schwarzkopf [198, 199], can be labeled as *lazy*. As outlined in exercise 6.4, it consists in not removing from the structure the elements that should disappear upon deletions. These elements are marked as destroyed, but remain physically present, and still serve for all subsequent locating phases. Naturally, the structure may only grow. When deletions outnumber insertions, the number of objects still present in the structure is less than half the number of objects still stored, and the algorithm completely rebuilds the structure from scratch, by inserting one by one the objects that were not previously removed.

Finally, we shall only touch the topic of randomized or derandomized dynamic structures which efficiently handle repetitive queries on a given set of objects, while allowing objects to be inserted into or deleted from this set. These structures embody the dynamic version of randomized divide-and-conquer structures, discussed in the notes of the previous chapter. These dynamic versions can be found in the works by Mulmuley [175], Mulmuley and Sen [178], Matoušek and Schwarzkopf [153, 156], Agarwal, Eppstein, and Matoušek [3] and Agarwal and Matoušek [4].

Part II

Convex hulls

Convexity is one of the oldest concepts in mathematics. It already appears in the works of Archimedes, around three centuries B.C. It was not until the 1950s, however, that this theme developed widely in the works of modern mathematicians. Convexity is a fundamental notion for computational geometry, at the core of many computer engineering applications, for instance in robotics, computer graphics, or optimization.

A convex set has the basic property that it contains the segment joining any two of its points. This property guarantees that a convex object has no hole or bump, is not hollow, and always contains its center of gravity. Convexity is a purely affine notion: no norm or distance is needed to express the property of being convex. Any convex set can be expressed as the convex hull of a certain point set, that is, the smallest convex set that contains those points. It can also be expressed as the intersection of a set of half-spaces. In the following chapters, we will be interested in *linear* convex sets. These can be defined as convex hulls of a finite number of points, or intersections of a finite number of half-spaces. Traditionally, a bounded linear convex set is called a *polytope*. We follow the tradition here, but we understand the word polytope as a shorthand for *bounded polytope*. This lets us speak of an *unbounded polytope* for the non-bounded intersection of a finite set of half-spaces.

In chapter 7, we recall the definitions relevant to polytopes, their facial structure, and their combinatorial properties. We introduce the notion of polarity as a dual transform on polytopes, and the notions of projective spaces and oriented projective spaces to extend the above definitions and results to unbounded polytopes. In chapter 8, we present solutions to one of the most fundamental problems of computational geometry, namely that of computing the convex hull of a finite number of points. Chapter 9 contains algorithms which work only in dimension 2 or 3. Lastly, chapter 10 tackles the related linear programming problem, where polytopes are given as intersections of a finite number of half-spaces.

Chapter 7

Polytopes

A polytope is defined as the convex hull of a finite number of points, or also as the bounded intersection of a finite number of half-spaces. Section 7.1 recalls the equivalence of these definitions, and gives the definition of the faces of a polytope. Polarity is also introduced in this section. The polarity centered at O is a dual transform between points and hyperplanes in Euclidean spaces which induces a duality on the set of polytopes containing the center O. Simple and simplicial polytopes are also defined in this section. Section 7.2 takes a close interest in the combinatorics of polytopes. It contains a proof of Euler's relation and the Dehn–Sommerville relations. Euler's relation is the only linear relation between the numbers of faces of each dimension of any polytope, and the Dehn–Sommerville relations are linear relations satisfied by simple polytopes. These relations can be used to show the celebrated upper bound theorem which bounds the number of faces of all dimensions of a d-dimensional polytope as a function of the number of its vertices, or facets. Considering cyclic polytopes shows that the upper bound theorem yields the best possible asymptotic bound. Linear unbounded convex sets enjoy similar properties and are frequently encountered in the rest of this book. Section 7.3 extends these definitions and properties to unbounded polytopes. A simple method to enforce coherence in these definitions is to consider the Euclidean space as embedded in the oriented projective space, an oriented version of the classical projective space.

7.1 Definitions

7.1.1 Convex hulls, polytopes

Let $\mathcal{A}$ be a set of points in $\mathbb{E}^d$. A *linear combination* of points in $\mathcal{A}$ is a sum of the kind $\sum_{i=1}^{k} \lambda_i A_i$, where k is an integer, and for all $i = 1, \ldots, k$, λ_i is a real and A_i a point in $\mathcal{A}$. A linear combination $\sum_{i=1}^{k} \lambda_i A_i$ is *affine* if

$$\sum_{i=1}^{k} \lambda_i = 1.$$

The set of affine linear combinations of points in $\mathcal{A}$ generates an affine subspace of $\mathbb{E}^d$ called the *affine hull* of $\mathcal{A}$. For instance, the affine hull of two distinct points is the line going through these two points. More generally, $k + 1$ points are said to be *affinely independent* if they generate an affine space of dimension k.

A linear combination $\sum_{i=1}^{k} \lambda_i A_i$ is *convex* if

$$\sum_{i=1}^{k} \lambda_i = 1 \quad \text{and} \quad \forall i \in \{1, \ldots, k\}, \ \lambda_i \geq 0.$$

A set $\mathcal{A}$ of points is *convex* if it is stable under convex combinations. Since the set of all convex combinations of two points P and Q is the segment PQ, the convexity of $\mathcal{A}$ is equivalent to the following geometric condition: for any two points P and Q in $\mathcal{A}$, the segment PQ is entirely contained in $\mathcal{A}$. The intersection of two convex sets is also convex.

The dimension of a convex set $\mathcal{A}$ is defined as the dimension of its affine hull $aff(\mathcal{A})$. If $\mathcal{A}$ is convex, its interior as a subset of the topological subspace $aff(\mathcal{A})$ is not empty. It is called the *relative interior* of $\mathcal{A}$.

Let $\mathcal{A}$ be a set of points in $\mathbb{E}^d$. The *convex hull* of $\mathcal{A}$, denoted by $conv(\mathcal{A})$, is formed by the set of all possible linear convex combinations of points in $\mathcal{A}$. Any convex set containing $\mathcal{A}$ must also contain its convex hull: the convex hull of $\mathcal{A}$ is thus the smallest convex set in $\mathbb{E}^d$ that contains $\mathcal{A}$, or equivalently the intersection of all the convex sets that contain $\mathcal{A}$.

The convex hull of a *finite* set of points in $\mathbb{E}^d$ is called a *polytope*. It is a closed bounded subset of $\mathbb{E}^d$. A polytope of dimension k is also called a k-polytope. The convex hull of $k + 1$ affinely independent points is a particular k-polytope called a *simplex* or also k-*simplex*. The convex hull of two affinely independent points A and B is the segment AB; the convex hull of three affinely independent points A, B, and C is the triangle ABC; finally, the convex hull of four affinely independent points A, B, C, and D is the tetrahedron $ABCD$.

7.1.2   Faces of a polytope

Any hyperplane H divides the space $\mathbb{E}^d$ into two half-spaces situated on either side of H. We write H^+ and H^- for these two open half-spaces, and $\overline{H^+}$ and $\overline{H^-}$ for their topological closure. Hence,

$$\mathbb{E}^d = H^+ \cup H \cup H^-$$

$$\overline{H^+} = H^+ \cup H, \ \overline{H^-} = H^- \cup H.$$

Consider a d-polytope $\mathcal{P}$. A hyperplane H *supports* $\mathcal{P}$, and is called a *supporting hyperplane* of P, if $H \cap \mathcal{P}$ is not empty and $\mathcal{P}$ is entirely contained in one of the closed half-spaces $\overline{H^+}$ or $\overline{H^-}$. The intersection $H \cap \mathcal{P}$ of the polytope $\mathcal{P}$ with a supporting hyperplane H is called a *face* of the polytope $\mathcal{P}$. Faces are convex subsets of $\mathbb{E}^d$, with a dimension ranging from 0 to $d-1$. To these faces, called the *proper* faces of $\mathcal{P}$, we add two faces called *improper*: the empty face whose dimension is set to -1 by convention, and the polytope $\mathcal{P}$ itself, of dimension d. A face of dimension j is also called a j-face. A 0-face is called a *vertex*, a 1-face is called an *edge*, and a $(d-1)$-face is called a *facet* of the polytope.

If F is a face of $\mathcal{P}$ and H a supporting hyperplane of $\mathcal{P}$ such that $F = H \cap \mathcal{P}$, H is said to *support* $\mathcal{P}$ *along* F.

Theorem 7.1.1 *The boundary of a polytope is the union of its proper faces.*

Proof. Consider a polytope $\mathcal{P}$. It is easy to show that the union of the faces of $\mathcal{P}$ is included in the boundary of $\mathcal{P}$. Indeed, let F be a face of $\mathcal{P}$, H a hyperplane supporting $\mathcal{P}$ along F, and $\overline{H^+}$ the half-space bounded by H that contains $\mathcal{P}$. Any point X in F belongs to $\mathcal{P}$ and to H, and a neighborhood of this point contains points that do not belong to $\mathcal{P}$. The converse inclusion (of the boundary in the union of the proper faces) results from a general theorem on bounded closed convex sets of $\mathbb{E}^d$, stated in exercise 7.5. It is a consequence of this theorem that there is a supporting hyperplane passing through any point of the boundary of a polytope $\mathcal{P}$; thus every point of the boundary belongs to a supporting hyperplane and hence to a proper face of $\mathcal{P}$. $\square$

Theorem 7.1.2 *A polytope has a finite number of faces. Faces of a polytope are also polytopes.*

Proof. Consider a polytope $\mathcal{P}$, the convex hull $conv(\mathcal{X})$ of a finite set of points $\mathcal{X}$. The theorem can be proved by showing that every proper face of $\mathcal{P}$ is the convex hull of a subset of $\mathcal{X}$. Indeed, let H be a supporting hyperplane of $\mathcal{P}$ and let $\mathcal{X}'$ be the subset of the points of $\mathcal{X}$ that belong to H. We first show that $H \cap \mathcal{P} = conv(\mathcal{X}')$. That $conv(\mathcal{X}') \subset H \cap \mathcal{P}$ is immediate. To prove the converse, we show that any point of $\mathcal{P}$ that does not belong to $conv(\mathcal{X}')$ does not belong to H. Let $H(Y) = 0$ be an equation of H and assume that $\mathcal{P}$ is contained in the half-space $\overline{H^+} = \{Y \in \mathbb{E}^d \ : \ H(Y) \geq 0\}$. For any point X' in $\mathcal{X}'$ or in $conv(\mathcal{X}')$, we have $H(X') = 0$, and for any point X in $\mathcal{X} \setminus \mathcal{X}'$ or in $conv(\mathcal{X} \setminus \mathcal{X}')$, we have $H(X) > 0$. Any point Y in $\mathcal{P}$ is a linear convex combination of points in $\mathcal{X}$. If Y does not belong to $conv(\mathcal{X}')$, at least one of the coefficients of the points in $\mathcal{X} \setminus \mathcal{X}'$ in this combination is strictly positive, and thus $H(Y) > 0$. $\square$

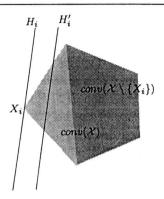

Figure 7.1. For the proof of theorem 7.1.3.

Theorem 7.1.3 *A polytope is the convex hull of its vertices.*

Proof. Let $\mathcal{P}$ be a polytope defined as the convex hull of a finite point set $\mathcal{X}$. By successively removing from $\mathcal{X}$ any point X_i that can be expressed as a linear convex combination of the remaining points, we are left with a minimal subset $\mathcal{X}'$ of $\mathcal{X}$ such that $\mathcal{P} = conv(\mathcal{X}')$. Let us now prove that any point of $\mathcal{X}'$ is a vertex of $\mathcal{P}$. Let X_i be a point of $\mathcal{X}'$. Since $\mathcal{X}'$ is minimal, X_i does not belong to the convex hull $conv(\mathcal{X}' \setminus \{X_i\})$ of the other points, and the theorem stated in exercise 7.4 shows that there is a hyperplane H_i' that separates X_i from $conv(\mathcal{X}' \setminus \{X_i\})$ (see figure 7.1). The hyperplane H_i parallel to H_i' passing through X_i supports $\mathcal{P}$ and contains only X_i among all points of $\mathcal{X}'$. Now theorem 7.1.2 above shows that

$$H_i \cap \mathcal{P} = conv(\{X_i\}) = \{X_i\}.$$

$\square$

The following two theorems are of central importance. They show that a polytope might equivalently be defined as the bounded intersection of a set of closed half-spaces.

Theorem 7.1.4 *Any polytope is the intersection of a finite set of closed half-spaces. More precisely, let $\mathcal{P}$ be a polytope, and $\{F_i : 1 \leq i \leq m\}$ be the set of its $(d-1)$-faces, H_i the hyperplane that supports $\mathcal{P}$ along F_i, and $\overline{H_i^+}$ the closed half-space bounded by H_i that contains $\mathcal{P}$. Then:*

$$\mathcal{P} = \bigcap_{i=1}^{m} \overline{H_i^+}.$$

Proof. The inclusion $\mathcal{P} \subset \bigcap_{i=1}^{m} \overline{H_i^+}$ is trivial, since $\mathcal{P}$ is contained in all the half-spaces $\overline{H_i^+}$. To prove the converse, we show that any point which does not

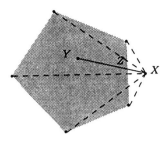

Figure 7.2. For the proof of theorem 7.1.4.

belong to $\mathcal{P}$ does not belong to the intersection $\bigcap_{i=1}^{m} \overline{H_i^+}$. Let X be a point not in $\mathcal{P}$, and Y a point in the interior of $\mathcal{P}$ but not in the hyperplane passing through X and some $d-1$ vertices of $\mathcal{P}$. Such a point exists, since the interior of $\mathcal{P}$ is of dimension d and cannot be contained in the union of a finite number of hyperplanes of dimension $d-1$. Segment XY intersects the boundary of $\mathcal{P}$ in a point Z (see figure 7.2). This point necessarily belongs to a proper face of $\mathcal{P}$ and, from the choice of Y, cannot belong to a face of $\mathcal{P}$ of dimension $j < d-1$. Thus Z belongs to one of the facets F_i of $\mathcal{P}$. Then Z belongs to the hyperplane H_i, Y to the half-space H_i^+ and X to the opposite half-space H_i^-. □

The following theorem is the converse of the previous one.

Theorem 7.1.5 *The intersection of a finite number of closed half-spaces, if it is bounded, is a polytope.*

Proof. The proof goes by induction on the dimension d of the space. In dimension 1, the theorem is trivial. Let

$$Q = \bigcap_{i=1}^{m} \overline{H_i^+} \tag{7.1}$$

be the bounded intersection of a finite number of half-spaces in $\mathbb{E}^d$.

For any j such that $1 \leq j \leq m$, let $F_j = H_j \cap Q$. F_j is thus a bounded intersection of half-spaces in the hyperplane H_j identified with an affine space of dimension $d-1$. By the inductive hypothesis, F_j is a polytope in H_j and thus a polytope of $\mathbb{E}^d$ as well. Let $\mathcal{V}_j$ be the set of vertices of F_j and $\mathcal{V}$ be the union $\bigcup_{j=1}^{m} \mathcal{V}_j$. We can now show that Q is the convex hull of $\mathcal{V}$. Indeed,

- any point X on the boundary of Q belongs to one of the polytopes F_j hence to *conv*$(\mathcal{V})$;

- any point X that belongs to the interior of Q belongs to a segment X_0X_1 which is the intersection of some line passing through X with Q. Since both X_0 and X_1 are on the boundary of Q, they belong to $conv(\mathcal{V})$ and so does X.

From this we may conclude that $Q \subset conv(\mathcal{V})$. And since Q is convex and contains $\mathcal{V}$, the opposite inclusion is trivial. $\square$

Remark. If the intersection Q is of dimension d and if its expression 7.1 is minimal, that is, for any $j = 1, \ldots, m$,

$$Q \neq \bigcap_{i \neq j} \overline{H_i^+},$$

then F_j is a $d-1$ face of the d-polytope Q. To prove this, it suffices to prove that F_j is not empty and that its dimension is $d-1$. The relative interior of F_j inside H_j can be expressed as

$$H_j \cap \left(\bigcap_{i \neq j} H_i^+ \right),$$

and is therefore not empty. Indeed, the intersection $\bigcap_{i \neq j} H_i^+$ is neither empty, nor entirely contained in H_j^-, because Q is not empty. But this intersection is not contained in H_j^+ either, otherwise $\overline{H_j^+}$ could be removed from expression 7.1 without changing the intersection Q.

Theorem 7.1.1 shows that the boundary of a polytope is the union of its proper faces, and the preceding remark shows that the union of the $(d-1)$-faces gives the boundary of a polytope. The following theorem shows more precisely that any proper face of a polytope is entirely contained within a facet of the polytope.

Theorem 7.1.6 *A proper face of a polytope $\mathcal{P}$ is a face of a $(d-1)$-face of $\mathcal{P}$. Conversely, any face of a face of $\mathcal{P}$ is also a face of $\mathcal{P}$.*

Proof. 1. Let F be a proper face of $\mathcal{P}$, H a hyperplane supporting $\mathcal{P}$ along F, and X a point in the relative interior of F. Point X belongs to the boundary of $\mathcal{P}$ and is therefore in a $(d-1)$-face of $\mathcal{P}$, say F_1. Let H_1 be the hyperplane supporting $\mathcal{P}$ along F_1. Point X belongs to the relative interior of face F and to hyperplane H_1 which supports $\mathcal{P}$ along F_1, therefore the whole face F is included in H_1, and thus in the facet F_1.

Moreover, if $\overline{H^+}$ is the half-space bounded by H that contains $\mathcal{P}$, then F_1 is entirely contained in the half-space $H_1 \cap \overline{H^+}$ of H_1. And since F is contained in F_1,

$$F = H \cap \mathcal{P} = H \cap \mathcal{P} \cap H_1 = (H \cap H_1) \cap (\mathcal{P} \cap H_1) = (H \cap H_1) \cap F_1,$$

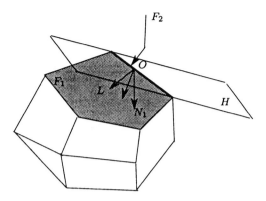

Figure 7.3. For the proof of theorem 7.1.6.

which shows that $H_1 \cap H$ is a hyperplane of H_1 that supports F_1 along F.

2. Let $\mathcal{P}$ be a d-polytope, F_1 be a proper face of $\mathcal{P}$, and F_2 be a face of F_1. Let H_1 be a hyperplane of $\mathbb{E}^d$ supporting $\mathcal{P}$ along F_1 and K a hyperplane of H_1 supporting F_1 along F_2. To show that F_2 is a face of $\mathcal{P}$, we rotate the hyperplane H_1 around K to obtain a hyperplane H of $\mathbb{E}^d$ that supports $\mathcal{P}$ along F_2 (see figure 7.3).

More precisely, suppose that the origin O of $\mathbb{E}^d$ belongs to F_2, and hence to F_1. There is a vector N_1 in $\mathbb{E}^d$ such that

$$H_1 = \{X \in \mathbb{E}^d \ : \ X \cdot N_1 = 0\},$$

$$\overline{H_1^+} = \{X \in \mathbb{E}^d \ : \ X \cdot N_1 \geq 0\},$$

where H_1^+ is the half-space containing $\mathcal{P}$. There is a vector L in H_1 such that

$$K = \{X \in H_1 \ : \ X \cdot L = 0\},$$

$$\overline{K^+} = \{X \in H_1 \ : \ X \cdot L \geq 0\},$$

where K^+ is the half-space containing F_1. Let $\mathcal{V}(\mathcal{P})$ denote the set of vertices of $\mathcal{P}$, $\mathcal{V}(F_1)$ the set of vertices of F_1, and $\mathcal{V}(F_2)$ the set of vertices of F_2. Let

$$\eta_0 = \min\{-\frac{V \cdot L}{V \cdot N_1} \ : \ V \in \mathcal{V}(\mathcal{P}) \setminus \mathcal{V}(F_1)\}$$

and let $\eta < \eta_0$. The hyperplane H of $\mathbb{E}^d$ defined by

$$H = \{X \in \mathbb{E}^d \ : \ X \cdot (\eta N_1 + L) = 0\}$$

supports $\mathcal{P}$ along F_2. Indeed, the hyperplane H contains all the vertices of $\mathcal{V}(F_2)$, and the half-space

$$H^+ = \{X \in \mathbb{E}^d \ : \ X \cdot (\eta N_1 + L) > 0\}$$

contains all the vertices in $\mathcal{V}(\mathcal{P}) \setminus \mathcal{V}(F_1)$ and all the vertices in $\mathcal{V}(F_1) \setminus \mathcal{V}(F_2)$. $\square$

Theorem 7.1.7 *Let $\mathcal{P}$ be a d-polytope.*

1. *The intersection of a family of faces of $\mathcal{P}$ is a face of $\mathcal{P}$.*

2. *Any $(d-2)$-face of $\mathcal{P}$ is the intersection of two $(d-1)$-faces of $\mathcal{P}$.*

3. *For any pair (j,k) of integers such that $0 \leq j \leq k < d$, a j-face of $\mathcal{P}$ is the intersection of all the k-faces of $\mathcal{P}$ that contain it.*

Proof. 1. Let $\{F_1, F_2, \ldots, F_r\}$ be a family of faces of the polytope $\mathcal{P}$. Let F be the intersection $\bigcap_{i=1}^r F_i$. If F is empty, F is trivially a face of $\mathcal{P}$. Otherwise we choose for the origin of $\mathbb{E}^d$ a point O in F. For $i = 1, \ldots, r$, we let H_i be a hyperplane that supports $\mathcal{P}$ along F_i, and N_i be the vector of $\mathbb{E}^d$ such that

$$H_i = \{X \in \mathbb{E}^d \ : \ X \cdot N_i = 0\},$$

and

$$\mathcal{P} \subset \overline{H_i^+} = \{X \in \mathbb{E}^d \ : \ X \cdot N_i \geq 0\}.$$

If $N = \sum_{i=1}^r N_i$, the hyperplane H defined by

$$H = \{X \in \mathbb{E}^d \ : \ X \cdot N = 0\}$$

supports $\mathcal{P}$ along F.

2. Let $\{F_1, F_2, \ldots, F_m\}$ be the facets of polytope $\mathcal{P}$. Let $\{H_1, H_2, \ldots, H_m\}$ be the hyperplanes that support $\mathcal{P}$ along these facets. Let F be a $(d-2)$-face of $\mathcal{P}$. From theorem 7.1.6, F is a $(d-2)$-face of a facet F_j of $\mathcal{P}$. From theorem 7.1.5, facet F_j can be expressed as

$$F_j = H_j \cap \mathcal{P} = H_j \cap \left(\bigcap_{k \neq j} \overline{H_k^+} \right) = \bigcap_{k \neq j} \left(H_j \cap \overline{H_k^+} \right),$$

and any $(d-2)$-face F of F_j can be expressed as

$$F = (H_j \cap H_i) \cap \left(\bigcap_{k \neq \{i,j\}} (H_j \cap \overline{H_k^+}) \right) = H_i \cap H_j \cap \left(\bigcap_{k \neq \{i,j\}} \overline{H_k^+} \right),$$

or equivalently

$$F = H_i \cap H_j \cap \mathcal{P} = F_i \cap F_j.$$

3. Using theorem 7.1.6, a straightforward induction on k (from $k = d-1$ down to $k = j$) shows that a j-face of $\mathcal{P}$ is a face of a k-face of $\mathcal{P}$ for any $j \leq k \leq d-1$.

Using the second assertion in theorem 7.1.7, it is also easy to prove by induction on j (from $j = d - 1$ down to $j = 0$) that any j-face ($0 \leq j \leq d-1$) of a polytope $\mathcal{P}$ is the intersection of all the $(d - 1)$-faces of $\mathcal{P}$ that contain it.

Let then j and k satisfy $0 \leq j \leq k \leq d - 2$. Consider a j-face F of polytope $\mathcal{P}$. The intersection of all the k-faces of $\mathcal{P}$ that contain F is also a face of $\mathcal{P}$ that contains F. To show that this face is precisely F, it suffices to show that F is the intersection of some k-faces of $\mathcal{P}$. From what was said above, F is a face of a $(k + 1)$-face G of $\mathcal{P}$, and thus F is the intersection of all the k-faces of G that contain it. But k-faces of G are also k-faces of $\mathcal{P}$ and therefore F is indeed the intersection of some k-faces of $\mathcal{P}$. $\qquad\square$

Incidences and adjacencies

Two faces F and G of a polytope $\mathcal{P}$ are called *incident* if one is included in the other, and if their respective dimensions differ by one. Two vertices of a polytope are said to be *adjacent* if they are incident to some common edge. Two facets of a polytope are said to be *adjacent* if they are incident to some common $(d-2)$-face.

7.1.3 Polarity, dual of a polytope

The polarity of center O is a geometric one-to-one transformation between points in $\mathbb{E}^d$ (except for the origin) and hyperplanes in $\mathbb{E}^d$ that do not pass through O. Let A be a point in $\mathbb{E}^d$ distinct from the origin O. The polar hyperplane of A, denoted by A^*, is defined by

$$A^* = \{X \in \mathbb{E}^d \ : \ A \cdot X = 1\}.$$

Let H be a hyperplane that does not contain the origin O. The pole of H is the point H^* that satisfies

$$H^* \cdot X = 1, \quad \forall X \in H.$$

This double transformation point-to-hyperplane and hyperplane-to-point is an involution. Indeed, it is a simple task to check that if A differs from O, then A^* does not contain the origin. Similarly, if H does not contain the origin, then H^* differs from O. It is then easy to show that

$$A^{**} = A \quad \text{and} \quad H^{**} = H.$$

Lemma 7.1.8 *The polarity of center O reverses the inclusion relationships between points and hyperplanes: a point A belongs to a hyperplane H if and only if the pole H^* of H belongs to the polar hyperplane A^* of A.*

Proof.

$$A \in H \Longleftrightarrow A \cdot H^* = 1 \Longleftrightarrow H^* \in A^*$$

$\square$

In this subsection, for any hyperplane not containing O, we denote by H^+ the half-space bounded by H that contains O, and by H^- the other half-space bounded by H:

$$H^+ = \{X \in \mathbb{E}^d \ : \ X \cdot H^* < 1\}$$
$$H^- = \{X \in \mathbb{E}^d \ : \ X \cdot H^* > 1\}$$

Lemma 7.1.9 *The polarity centered at O reverses the relative positions of a point and a hyperplane with respect to the center O: a point A belongs to the half-space H^+ if and only if the pole H^* of H belongs to the half-space $(A^*)^+$, and A belongs to H^- if and only if H belongs to $(A^*)^-$.*

Proof.

$$A \in H^+ \quad \Longleftrightarrow \quad A \cdot H^* < 1 \Longleftrightarrow H^* \in A^{*+}$$
$$A \in H^- \quad \Longleftrightarrow \quad A \cdot H^* > 1 \Longleftrightarrow H^* \in A^{*-}$$

$\square$

Generally speaking, we call a *duality* any bijection that reverses inclusion relationships. The preceding relation shows that polarity centered at O is a duality, and the polar hyperplane A^* is often called the dual of A. Similarly, the pole H^* is often called the dual of hyperplane H.

The notion of duality extends naturally to polytopes: a polytope $\mathcal{Q}$ is dual to a polytope $\mathcal{P}$ if there is a bijection between the faces of $\mathcal{P}$ and the faces of $\mathcal{Q}$ that reverses inclusion relationships.

The following theorems show that it is possible to define a *polar image* $\mathcal{P}^{\#}$ for any polytope $\mathcal{P}$ whose interior contains the origin O.

The *polar transformation* centered at O is closely linked to the polarity defined above, but it associates points with half-spaces and not with hyperplanes. Let A be a point of $\mathbb{E}^d$. The polar image $A^{\#}$ of A is the half-space $\overline{A^{*+}}$ bounded by A^* that contains the origin:

$$A^{\#} = \overline{A^{*+}} = \{Y \in \mathbb{E}^d \ : \ Y \cdot A \leq 1\}.$$

For any set $\mathcal{A}$ of points in $\mathbb{E}^d$, we define the *polar image* $\mathcal{A}^{\#}$ of $\mathcal{A}$ as the intersection of the polar images of its points:

$$\mathcal{A}^{\#} = \{Y \in \mathbb{E}^d \ : \ Y \cdot X \leq 1, \ \forall X \in \mathcal{A}\}.$$

Note that this formula allows the definition to be extended to the case where $\mathcal{A}$ contains the origin O.

The two following facts are immediate consequences of the above definition:

1. The polar image $\mathcal{A}^{\#}$ of a set $\mathcal{A}$ is convex.

2. If $\mathcal{A}$ and $\mathcal{B}$ are two sets such that $\mathcal{A} \subset \mathcal{B}$, then $\mathcal{B}^{\#} \subset \mathcal{A}^{\#}$.

In the rest of this subsection, $\mathcal{P}$ stands for a polytope of $\mathbb{E}^d$ whose interior contains the origin O, and $\mathcal{P}^{\#}$ denotes the polar image of $\mathcal{P}$.

Theorem 7.1.10 *The polar polytope $\mathcal{P}^{\#}$ of a polytope $\mathcal{P}$ whose interior contains O is a polytope whose interior contains O.*

Proof. Point O is inside the polar polytope $\mathcal{P}^{\#}$ because $\mathcal{P}$ is bounded. Indeed, if $\mathcal{P}$ is contained within a ball $B(O, \rho)$ centered at O with radius ρ, its polar image $\mathcal{P}^{\#}$ contains the ball $B(O, 1/\rho)$ centered at O with radius $1/\rho$.

A dual argument shows that the image $\mathcal{P}^{\#}$ of $\mathcal{P}$ is bounded because the interior of $\mathcal{P}$ contains O. In fact, $\mathcal{P}$ contains an entire ball $B(O, \epsilon)$ centered at O with radius ϵ. Thus $\mathcal{P}^{\#}$ is entirely contained in the ball $B(O, 1/\epsilon)$ centered at O with radius $1/\epsilon$.

If the polytope $\mathcal{P}$ is the convex hull of n points,

$$\mathcal{P} = conv(\{X_1, \dots, X_n\}),$$

then $\mathcal{P}^{\#}$ is the intersection of the n half-spaces $\overline{X_i^{*+}}$ bounded by the polar hyperplanes X_i^* of the point X_i,

$$\mathcal{P}^{\#} = \bigcap_{i=1}^{n} \overline{X_i^{*+}}.$$

Indeed, the inclusion $\mathcal{P}^{\#} \subset \bigcap_{i=1}^{n} \overline{X_i^{*+}}$ is trivial. To prove the converse, we show that every point that does not belong to $\mathcal{P}^{\#}$ does not belong to $\bigcap_{i=1}^{n} \overline{X_i^{*+}}$. Let Y be a point that does not belong to $\mathcal{P}^{\#}$. There is a point X that belongs to $\mathcal{P}$ such that $Y \cdot X > 1$. Since X is a linear convex combination of $\{X_1, \dots, X_r\}$, its existence implies that $Y \cdot X_i > 1$ for at least one of the points X_i, and thus Y does not belong to $\overline{X_i^{*+}}$.

The polar set $\mathcal{P}^{\#}$ of polytope $\mathcal{P}$ is the bounded intersection of a finite number of half-spaces. It is thus a polytope, by theorem 7.1.4. $\square$

Theorem 7.1.11 *The polar transformation is an involution on the set of polytopes whose interiors contain O, that is for any such polytope $\mathcal{P}$,*

$$\mathcal{P}^{\#\#} = \mathcal{P}.$$

Proof. $\mathcal{P}$ is included in $\mathcal{P}^{\#\#}$, by definition. To prove the converse inclusion, we show that any point Z that does not belong to $\mathcal{P}$ does not belong to $\mathcal{P}^{\#\#}$. Let Z be a point that does not belong to $\mathcal{P}$. Since $\mathcal{P}$ is the intersection $\bigcap_{j=1}^{m} \overline{H_j^+}$ of the half-spaces $\overline{H_j^+}$ bounded by the hyperplanes supporting the facets of $\mathcal{P}$, one of these hyperplanes must separate Z from $\mathcal{P}$. Let us call this hyperplane H_k. The polytope $\mathcal{P}$ is contained in the closed half-space $\overline{H_k^+}$, and Z lies within H_k^-. The point H_k^*, pole of H_k, satisfies

$$\forall X \in \mathcal{P}, \ \ X \cdot H_k^* \leq 1 \ \text{ and } \ Z \cdot H_k^* > 1.$$

Therefore H_k^* belongs to $\mathcal{P}^{\#\#}$, and Z does not belong to $\mathcal{P}^{\#\#}$. $\square$

Lemma 7.1.12 *If A is a point of $\mathbb{E}^d$ on the boundary of a polytope $\mathcal{P}$, the polar hyperplane A^* of A is a hyperplane that supports the polar polytope $\mathcal{P}^{\#}$.*

Proof. For any point A of $\mathcal{P}$, the closed half-space $\overline{A^{*+}}$ contains the polytope $\mathcal{P}^{\#}$. Moreover, if A belongs to the boundary of $\mathcal{P}$, it belongs to one proper face F of $\mathcal{P}$ and there is a supporting hyperplane H of $\mathcal{P}$ that passes through A. The pole H^* of H is a point that belongs to both A^* and $\mathcal{P}^{\#}$. Thus $A^* \cap \mathcal{P}^{\#}$ is not empty and A^* is indeed a supporting hyperplane of $\mathcal{P}^{\#}$. $\square$

Theorem 7.1.13 *There exists a bijection between the faces of $\mathcal{P}$ and those of $\mathcal{P}^{\#}$ which reverses inclusion relationships. This bijection associates the k-faces of $\mathcal{P}$ with the $(d-1-k)$-faces of $\mathcal{P}^{\#}$, for all $k = 0, \ldots, d-1$.*

Proof. With each face F of $\mathcal{P}$, we associate the set

$$F^* = \{Y \in \mathcal{P}^{\#} \ : \ Y \cdot X = 1, \ \forall X \in F\}.$$

The set F^* is a face of $\mathcal{P}^{\#}$. Indeed, F^* can be expressed as

$$F^* = \bigcap_{X \in F} (\mathcal{P}^{\#} \cap X^*),$$

where X^* is the polar hyperplane of X. From lemma 7.1.12, if X is a point of F, then the hyperplane X^* is a supporting hyperplane of $\mathcal{P}^{\#}$, and $X^* \cap \mathcal{P}^{\#}$ is thus a face of $\mathcal{P}^{\#}$. Therefore, F^* is the intersection of a family of faces of $\mathcal{P}^{\#}$, and theorem 7.1.7 proves that it is a face of $\mathcal{P}^{\#}$ as well.

By the definition of the polar image F^* of a face F, if F_1 and F_2 are two faces of $\mathcal{P}$ such that $F_1 \subset F_2$, then $F_2^* \subset F_1^*$.

To prove that the map from F onto F^* that maps a face of $\mathcal{P}$ to a face of $\mathcal{P}^{\#}$ is bijective, we show that it is in fact an involution, that is

$$F^{**} = F.$$

This property is proved below for the proper faces of $\mathcal{P}$ and $\mathcal{P}^{\#}$. In order to extend it to improper faces, we note that the images of $\mathcal{P}$ and $\mathcal{P}^{\#}$ are empty sets because both $\mathcal{P}$ and $\mathcal{P}^{\#}$ have non-empty interiors. Therefore we can make the convention that the d-dimensional face of $\mathcal{P}$ (resp. $\mathcal{P}^{\#}$) corresponds to the empty face of $\mathcal{P}^{\#}$ (resp. $\mathcal{P}$).

Let now F be a proper face of $\mathcal{P}$. Then

$$F^{**} = \{X \in \mathcal{P}^{\#\#} \; : \; X \cdot Y = 1, \; \forall Y \in F^*\}.$$

Since $\mathcal{P}^{\#\#}$ is simply $\mathcal{P}$, the inclusion $F \subset F^{**}$ is immediate. The converse inclusion can be shown by arguing that any point X of $\mathcal{P}$ not in F does not belong to F^{**}. Let X be such a point of $\mathcal{P}$ not in F, and H any hyperplane supporting $\mathcal{P}$ along F. Then X belongs to H^+ and $X \cdot H^* < 1$. Nevertheless, the pole H^* of H lies within the face F^*, which proves that X does not belong to F^{**}.

Finally, let us prove the assertion about the dimensions. Let F be a k-face of $\mathcal{P}$. If F is not a proper face, then $k = -1$ or $k = d$ and the assertion is true. Let F be a k-face of $\mathcal{P}$, with $0 \leq k \leq d-1$. Then F contains $k+1$ affinely independent points, and F^* is contained within the intersection of $k+1$ hyperplanes whose equations are linearly independent. The dimension of F^* is thus at most $d-1-k$. But since $F^{**} = F$, this dimension must equal $d-1-k$ exactly. $\qquad \square$

We have seen that the vertices of $\mathcal{P}$ correspond to facets of $\mathcal{P}^{\#}$, and the converse is true:

- If $\{P_1, P_2, \ldots, P_n\}$ are the vertices of $\mathcal{P}$, then $\{P_1^*, P_2^*, \ldots, P_n^*\}$ are the hyperplanes supporting $\mathcal{P}^{\#}$ along its $(d-1)$-faces.

$$\mathcal{P} = conv(\{P_1, P_2, \ldots, P_n\}) \Longleftrightarrow \mathcal{P}^{\#} = \bigcap_{i=1}^{n} \overline{P_i^{*+}}.$$

- If $\{H_1, H_2, \ldots, H_m\}$ are the hyperplanes supporting $\mathcal{P}$ along its $(d-1)$-faces, then $\{H_1^*, H_2^*, \ldots, H_m^*\}$ are the vertices of $\mathcal{P}^{\#}$.

$$\mathcal{P} = \bigcap_{j=1}^{m} \overline{H_j^+} \Longleftrightarrow \mathcal{P}^{\#} = conv(\{H_1^*, H_2^*, \ldots, H_m^*\}).$$

Finally, the following properties can be easily proved from the preceding ones.

1. If point A belongs to polytope $\mathcal{P}$, the polar hyperplane A^* avoids the polar polytope $\mathcal{P}^{\#}$.

2. If A is a point lying outside $\mathcal{P}$, the polar hyperplane A^* intersects the interior of $\mathcal{P}^{\#}$.

3. When A takes all positions in the relative interior of a face F of $\mathcal{P}$, the polar hyperplane A^* describes the set of hyperplanes supporting $\mathcal{P}^\#$ along F^*.

4. If the origin O lies in the interior of two polytopes $\mathcal{P}_1$ and $\mathcal{P}_2$, then

$$(conv(\mathcal{P}_1 \cup \mathcal{P}_2))^\# = \mathcal{P}_1^\# \cap \mathcal{P}_2^\#.$$

7.1.4   Simple and simplicial polytopes

We recall that a set of $k+1$ points in $\mathbb{E}^d$ is *affinely independent* if its affine linear combinations generate an affine subspace of $\mathbb{E}^d$ of dimension k. A finite set $\mathcal{A}$ of points in $\mathbb{E}^d$ is in *general position* if any subset of $\mathcal{A}$ with at most $d+1$ points is affinely independent.

If $\mathcal{A}$ is a set of points in general position, no affine subspace of dimension j may contain more than $j+1$ points of $\mathcal{A}$.

A set $\mathcal{H}$ of hyperplanes is in *general position* if, for any $j \leq d$, the intersection of any j hyperplanes in $\mathcal{H}$ is of dimension $d-j$, and the intersection of any $d+1$ hyperplanes in $\mathcal{H}$ is empty.

Let $\mathcal{H}$ be a finite set of hyperplanes, O a point not in any of the hyperplanes in $\mathcal{H}$, and $\mathcal{H}^*$ the set of poles of the hyperplanes of $\mathcal{H}$ for the polarity centered at O. The set $\mathcal{H}$ of hyperplanes is in general position if and only if the set $\mathcal{H}^*$ of points is also in general position.

A k-simplex is the convex hull of $k+1$ affinely independent points. Let $\mathcal{S}$ be a k-simplex, the convex hull of a set $\mathcal{A} = \{A_0, \ldots, A_k\}$ of $k+1$ affinely independent points. Any subset $\mathcal{A}'$ of $\mathcal{A}$ of cardinality $k'+1$ $(0 \leq k' \leq k)$ defines a k'-simplex $conv(\mathcal{A}')$ which is a face of $\mathcal{S}$. Therefore, a k-simplex $\mathcal{A}$ has exactly $\binom{k+1}{j+1}$ faces of dimension j $(0 \leq j \leq k)$. A j-face of a k-simplex is incident to $j+1$ $(j-1)$-faces for any $0 \leq j \leq k$, and to $k-j$ $(j+1)$-faces for any $-1 \leq j \leq k-1$.

A polytope is called *simplicial* if all its proper faces are simplices. By virtue of theorem 7.1.6, it is enough to require that its facets have exactly d vertices, and thus are $(d-1)$-simplices. The convex hull of a set of points in general position is a simplicial polytope. Note that this is a sufficient but not necessary condition: the vertices of a simplicial polytope need not be in general position. Indeed, d vertices may lie in the same hyperplane, provided that this hyperplane does not support the polytope along a facet.

A polytope is *simple* if it is dual to a simplicial polytope. Therefore, a polytope $\mathcal{P}$ is simple if any of its vertices belongs to exactly d facets.

Simplices are both simple and simplicial polytopes: they are the only polytopes to possess this property (see exercise 7.6).

By using the bijection between the faces of polytope $\mathcal{P}$ and its dual, we can easily prove the following lemma which will be useful in establishing the Dehn–Sommerville relations (theorem 7.2.2) satisfied by any simple polytope:

Lemma 7.1.14 *For any $0 \leq j \leq k \leq d-1$, any j-face of a simple polytope $\mathcal{P}$ is a face of $\dbinom{d-j}{d-k}$ k-faces of $\mathcal{P}$.*

7.2 The combinatorics of polytopes

Except in the case of simplices, or polytopes in dimensions no greater than 2, the number of faces of a d-polytope is not entirely determined by the number of its vertices, nor by that of its facets. The upper bound theorem, stated and proved in subsection 7.2.3, gives an asymptotic upper bound of $O(n^{\lfloor d/2 \rfloor})$ for the total number of faces of a d-polytope that has n vertices, or respectively n facets. The study of cyclic polytopes (in subsection 7.2.4) shows that this bound is optimal.

If $\mathcal{P}$ is a polytope, we denote by $n_k(\mathcal{P})$ (or simply n_k when $\mathcal{P}$ is unambiguously understood) the number of k-faces of $\mathcal{P}$, for $-1 \leq k \leq d$. In particular, $n_{-1} = n_d = 1$. The proof of the upper bound theorem relies on a set of linear relations satisfied by the numbers $n_k(\mathcal{P})$ of faces of a polytope. One of them is Euler's relation (subsection 7.2.1), and is the only linear relation binding the numbers n_k and satisfied for any polytope (simple or not). The other relations are known as the Dehn–Sommerville relations and are satisfied by all simple polytopes (subsection 7.2.2).

7.2.1 Euler's relation

Theorem 7.2.1 (Euler's relation) *The numbers $n_k(\mathcal{P})$ ($0 \leq k \leq d-1$) of k-faces of a d-polytope $\mathcal{P}$ are bound by the relation*

$$\sum_{k=0}^{d-1}(-1)^k n_k(\mathcal{P}) = 1 - (-1)^d,$$

or, if we also sum over the improper faces of $\mathcal{P}$,

$$\sum_{k=-1}^{d}(-1)^k n_k(\mathcal{P}) = 0.$$

Proof. The proof we present here goes by induction on the dimension d of the polytope. The base case is proved easily since, in one dimension, a polytope has only two proper faces, namely its vertices, and thus satisfies Euler's relation.

Let $\mathcal{P}$ be a polytope of dimension d and let $n = n_0(\mathcal{P})$ be its number of vertices. We may always assume that the x_d-coordinates of any two vertices are distinct, by choosing the coordinate system appropriately. Therefore, a horizontal hyperplane (that is, perpendicular to the x_d-axis) contains at most one vertex of $\mathcal{P}$.

Let $\{P_1, P_2, \ldots, P_n\}$ be the set of vertices of $\mathcal{P}$ sorted by increasing x_d-coordinates. Consider a family of $2n - 1$ horizontal hyperplanes $\{H_1, H_2, \ldots, H_{2n-1}\}$ such that:

- the hyperplane H_{2i-1} $(i = 1, \ldots, n)$ goes through the vertex P_i of $\mathcal{P}$,

- the hyperplane H_{2i} $(i = 1, \ldots, n-1)$ passes between H_{2i-1} and H_{2i+1}.

For each face F of $\mathcal{P}$ and each hyperplane H_j, in this family, we define a signature: $\chi_j(F) = 1$ if H_j intersects the relative interior of F, and $\chi_j(F) = 0$ otherwise.

Consider a face F, and call P_l (resp. P_m) its vertex with minimal (resp. maximal) x_d-coordinate. The horizontal hyperplanes intersecting the relative interior of F lie strictly between horizontal hyperplanes H_{2l-1} and H_{2m-1} that pass through P_l and P_m respectively. If face F is of dimension $k \geq 1$, then l and m are distinct integers, and the number of hyperplanes with even indices that intersect the relative interior of F is one more than the number of hyperplanes with odd indices that intersect the relative interior of F, whence

$$1 = \sum_{j=2}^{2n-2} (-1)^j \chi_j(F).$$

Summing this relation over the set $\mathcal{F}_k(\mathcal{P})$ of k-faces of $\mathcal{P}$, we get

$$n_k(\mathcal{P}) = \sum_{F \in \mathcal{F}_k(\mathcal{P})} \sum_{j=2}^{2n-2} (-1)^j \chi_j(F),$$

and summing over all k,

$$\sum_{k=1}^{d-1} (-1)^k n_k(\mathcal{P}) = \sum_{j=2}^{2n-2} (-1)^j \sum_{k=1}^{d-1} (-1)^k \sum_{F \in \mathcal{F}_k(\mathcal{P})} \chi_j(F). \qquad (7.2)$$

Each hyperplane H_j for $j = 2, \ldots, 2n-2$ intersects polytope $\mathcal{P}$ along a polytope $\mathcal{P}_j = \mathcal{P} \cap H_j$ of dimension $d - 1$, to which we can apply the inductive hypothesis:

$$\sum_{k=0}^{d-2} (-1)^k n_k(\mathcal{P}_j) = \sum_{k=1}^{d-1} (-1)^{k-1} n_{k-1}(\mathcal{P}_j) = 1 - (-1)^{d-1}. \qquad (7.3)$$

When j is even, any $(k-1)$-face of $\mathcal{P}_j$ is the intersection of a k-face of $\mathcal{P}$ with hyperplane H_j, and

$$n_{k-1}(\mathcal{P}_j) = \sum_{F \in \mathcal{F}_k(\mathcal{P})} \chi_j(F), \quad k = 1, \dots, d-1. \tag{7.4}$$

When j is odd, any $(k-1)$-face of $\mathcal{P}_j$ is the intersection of a k-face of $\mathcal{P}$ with hyperplane H_j, except for the vertex P_j that belongs to H_j. Therefore,

$$n_0(\mathcal{P}_j) = 1 + \sum_{F \in \mathcal{F}_1(\mathcal{P})} \chi_j(F), \tag{7.5}$$

$$n_{k-1}(\mathcal{P}_j) = \sum_{F \in \mathcal{F}_k(\mathcal{P})} \chi_j(F), \quad k = 2, \dots, d-1. \tag{7.6}$$

When j is even, we can use equations 7.3 and 7.4 to get

$$\sum_{k=1}^{d-1} (-1)^k \sum_{F \in \mathcal{F}_k(\mathcal{P})} \chi_j(F) = \sum_{k=1}^{d-1} (-1)^k n_{k-1}(\mathcal{P}_j) = -1 + (-1)^{d-1}. \tag{7.7}$$

When j is odd, using equations 7.3, 7.5 and 7.6, we obtain

$$\sum_{k=1}^{d-1} (-1)^k \sum_{F \in \mathcal{F}_k(\mathcal{P})} \chi_j(F) = \sum_{k=1}^{d-1} (-1)^k n_{k-1}(\mathcal{P}_j) + 1 = (-1)^{d-1}. \tag{7.8}$$

It now suffices to multiply relations 7.7 and 7.8 by $(-1)^j$ and to sum over $j = 1, \dots, 2n-1$ to obtain by use of equation 7.2, noticing that there are $n-1$ even and $n-2$ odd relations:

$$\sum_{k=1}^{d-1} (-1)^k n_k(\mathcal{P}) = (-1)^{d-1} - (n-1).$$

Recall now that n is the number $n_0(\mathcal{P})$ of vertices of $\mathcal{P}$. In the last equation, we may now recognize Euler's relation for polytope $\mathcal{P}$. $\square$

In the case of a 2-polytope, Euler's relation, written as

$$n_0(\mathcal{P}) - n_1(\mathcal{P}) = 0,$$

expresses the fact that a polygon has as many vertices as edges. In the case of a 3-polytope, the relation is a bit more interesting and can be written as

$$n_0(\mathcal{P}) - n_1(\mathcal{P}) + n_2(\mathcal{P}) = 2.$$

7.2.2   The Dehn–Sommerville relations

Theorem 7.2.2 (Dehn–Sommerville relations) *The numbers $n_j(\mathcal{P})$ of j-faces of a simple d-polytope satisfy the $d+1$ relations*

$$\sum_{j=0}^{k}(-1)^j\binom{d-j}{d-k}n_j(\mathcal{P}) = n_k(\mathcal{P}),\quad k=0,\dots,d.$$

Proof. Let $\mathcal{P}$ be a simple d-polytope. The Dehn–Sommerville relation for $k=d$ is none other than Euler's relation for $\mathcal{P}$. Suppose now that $k \le d-1$. Any k-face F of $\mathcal{P}$ is a k-polytope, and thus satisfies Euler's relation

$$\sum_{j=-1}^{k}(-1)^j n_j(F) = 0.$$

Summing over the set $\mathcal{F}_k(\mathcal{P})$ of k-faces of $\mathcal{P}$, we get

$$\sum_{F\in\mathcal{F}_k(\mathcal{P})}\sum_{j=-1}^{k}(-1)^j n_j(F) = \sum_{j=-1}^{k}(-1)^j\sum_{F\in\mathcal{F}_k(\mathcal{P})}n_j(F) = 0.$$

The sum

$$\sum_{F\in\mathcal{F}_k(\mathcal{P})}n_j(F)$$

is exactly the number of pairs (F,G) of faces of $\mathcal{P}$, where F is a k-face and G is a j-face entirely contained in F. Since $\mathcal{P}$ is a simple polytope, for each $0 \le j \le k \le d-1$, any j-face of $\mathcal{P}$ is contained in exactly $\binom{d-j}{d-k}$ k-faces of $\mathcal{P}$ (lemma 7.1.14). Therefore,

$$\sum_{F\in\mathcal{F}_k(\mathcal{P})}n_j(F) = \binom{d-j}{d-k}n_j(\mathcal{P}),\quad (0 \le j \le k \le d-1).$$

Finally, we get

$$-n_k(\mathcal{P}) + \sum_{j=0}^{k}(-1)^j\binom{d-j}{d-k}n_j(\mathcal{P}) = 0,$$

which is exactly the $(k+1)$-st Dehn–Sommerville relation for polytope $\mathcal{P}$. $\qquad\square$

The Dehn–Sommerville relations for a simple 3-polytope are

$$
\begin{aligned}
n_0 &= n_0\\
3n_0 - n_1 &= n_1\\
3n_0 - 2n_1 + n_2 &= n_2\\
n_0 - n_1 + n_2 - n_3 &= n_3.
\end{aligned}
$$

The first relation is trivial, the following two are equivalent, and the last is precisely Euler's relation. They can be compacted into two linearly independent equations binding the numbers n_0, n_1, and n_2 of proper faces of a simple 3-polytope. Fixing the number $n = n_2$ of facets, these relations may be expressed as

$$n_0 = 2n - 4, \quad n_1 = 3n - 6.$$

This proves the following theorem:

Theorem 7.2.3 *A simple 3-polytope with n facets has exactly $2n - 4$ vertices and $3n - 6$ edges,*

and its dual counterpart:

Theorem 7.2.4 *A simplicial 3-polytope with n vertices has exactly $2n - 4$ facets and $3n - 6$ edges.*

The following subsection shows that the Dehn–Sommerville relations alone can be used to derive an upper bound on the number of faces of any polytope as a function of its number of vertices or facets.

7.2.3 The upper bound theorem

Theorem 7.2.5 (Upper bound theorem) *Any d-polytope with n facets (or n vertices) has at most $O(n^{\lfloor d/2 \rfloor})$ faces of all dimensions and $O(n^{\lfloor d/2 \rfloor})$ pairs of incident faces of all dimensions.*

Proof. We are interested first in simple polytopes. If $\mathcal{P}$ is a simple polytope of dimension d, the Dehn–Sommerville relations,

$$\sum_{j=0}^{k} (-1)^j \binom{d-j}{d-k} n_j(\mathcal{P}) = n_k(\mathcal{P}), \quad (k = 0, \ldots, d),$$

yield $d+1$ linear relations between the d numbers $n_0, n_1, \ldots, n_{d-1}$ of proper faces of polytope $\mathcal{P}$. The first relation (obtained for $k = 0$) is trivial, and the others are not all linearly independent. But one may prove easily that the odd relations (those that correspond to odd values of k) are linearly independent. Indeed, the coefficients of n_{2p+1} in the equations obtained for $k = 2q+1$, with p and q ranging from 0 to $\lfloor \frac{d-1}{2} \rfloor$, form a triangular matrix. Thus the Dehn–Sommerville relations form a system of rank at least

$$r = \left\lfloor \frac{d+1}{2} \right\rfloor.$$

In fact, it can be shown that there are exactly r linearly independent relations among the Dehn–Sommerville relations (see exercise 7.7). Moreover, it can be shown that the Dehn-Somerville system can be solved for the variables n_j, $j = 0, \ldots, r - 1$, yielding an expression for these variables as a linear combination of the n_j's, $j = r, \ldots, d$ (see exercise 7.8). If the simple polytope has n facets, there is a trivial bound, for $j \geq r$,

$$\binom{n}{d - j} = O(n^{d-j})$$

on its number of j-faces. Indeed, lemma 7.1.14 shows that a j-face of a simple d-polytope is the intersection of $d - j$ facets. We conclude that a simple d-polytope with n facets has $O(n^{\lfloor d/2 \rfloor})$ faces. In a simple polytope, k-faces ($k \leq d$) are incident to $d - k$ ($k + 1$)-faces; thus the number of pairs of incident faces is also $O(n^{\lfloor d/2 \rfloor})$. We therefore have proved the theorem for a simple polytope with n facets.

A dual statement of the theorem also shows that the theorem is true for simplicial d-polytopes with n vertices. To extend this result to arbitrary polytopes, it suffices to show that simple and simplicial polytopes maximize the number of faces and incidences between faces. The following perturbation argument shows that the numbers of faces and incidences of faces of a non-simplicial d-polytope are less than those of some simplicial d-polytope obtained by slightly perturbing the vertices. Let $\mathcal{P}$ be a non-simplicial d-polytope, and n be the number of its vertices. Each face of $\mathcal{P}$ is the convex hull of its vertices and may therefore be triangulated, or in other words decomposed into a union of simplices whose vertices are the vertices of that face. In a triangulation,[1] each face F of $\mathcal{P}$ is expressed as the union of simplices whose relative interiors form a partition of F. A simple scheme to triangulate a d-polytope and its faces is to proceed recursively, or equivalently in a bottom-up fashion, as follows. Let F be a $(k + 1)$-face of $\mathcal{P}$. To triangulate F, we choose a vertex A of F, and consider the $(k + 1)$-simplices $conv(A, T)$, where T ranges over the k-simplices in the recursively obtained triangulation of the k-faces of F which do not contain A. The number of faces of the triangulation is at least the number of faces of $\mathcal{P}$. Slightly perturbing the vertices of P while keeping the union of the simplices in the triangulation convex (and this can always be done, see exercise 7.10) yields a simplicial polytope $\mathcal{P}'$ whose faces are in one-to-one correspondence with the simplices in the triangulation of $\mathcal{P}$. The numbers of faces and incidences of $\mathcal{P}'$ are thus strictly greater than their counterparts for $\mathcal{P}$.   □

---

[1]Triangulations are studied at length in chapters 11, 12, and 13.

7.2.4 Cyclic polytopes

In this subsection, we prove that the bound given in the upper bound theorem (theorem 7.2.5) is optimal. For this, we introduce a particular class of polytopes, and show that their numbers of faces and incidences achieve the bound given in the upper bound theorem.

The *moment curve* is the curve $\mathcal{M}_d$ in $\mathbb{E}^d$ followed by a point M_τ parameterized by a real number τ:

$$\mathcal{M}_d = \{M(\tau) = (\tau^1, \tau^2, \ldots, \tau^d), \tau \in \mathbb{R}\}.$$

Lemma 7.2.6 *Any subset of $k \leq (d+1)$ points on the moment curve is linearly independent.*

Proof. Consider $d+1$ points $\{M_0, M_1, \ldots, M_d\}$ on the moment curve, for the values $\{\tau_0, \tau_1, \ldots, \tau_d\}$ of the parameter. The determinant formed by the coordinates of these points is

$$\begin{vmatrix} 1 & \tau_0 & \tau_0^2 & \cdots & \tau_0^d \\ 1 & \tau_1 & \tau_1^2 & \cdots & \tau_1^d \\ \vdots & \vdots & \vdots & & \vdots \\ 1 & \tau_d & \tau_d^2 & \cdots & \tau_d^d \end{vmatrix} = \prod_{0 \leq i < j \leq d} (\tau_j - \tau_i),$$

the so-called Van Der Monde determinant, and does not vanish when the τ_i's are pairwise distinct. $\square$

A consequence of this lemma is that any hyperplane in $\mathbb{E}^d$ intersects the moment curve in at most d points.

A *cyclic polytope* in $\mathbb{E}^d$ is the convex hull of $n \geq d + 1$ points on the moment curve. By the above lemma, a cyclic polytope is simplicial.

Let now $\mathcal{P}$ be a cyclic polytope in $\mathbb{E}^d$, the convex hull of n points $\{M_1, M_2, \ldots, M_n\}$ of $\mathcal{M}_d$ with respective parameters $\{\tau_1, \tau_2, \ldots, \tau_n\}$. Let $\mathcal{I}$ a subset of the set of indices $\{1, 2, \ldots, n\}$, of cardinality $k \leq d/2$, and consider the polynomial

$$\pi_\mathcal{I}(\tau) = \prod_{i \in \mathcal{I}} (\tau - \tau_i)^2.$$

This polynomial has degree $2k \leq d$, and there is a point H^* in $\mathbb{E}^d$ and a real h_0 which may be used to interpret $\pi_\mathcal{I}(\tau)$ in the form

$$\pi_\mathcal{I}(\tau) = H^* \cdot M(\tau) - h_0, \quad M(\tau) \in \mathcal{M}_d.$$

The polynomial $\pi_\mathcal{I}(\tau)$ is positive and vanishes exactly at $\tau = \tau_i$, $i \in \mathcal{I}$. The hyperplane H defined by the equation

$$H^* \cdot X = h_0$$

is thus a hyperplane that supports $\mathcal{P}$ along the $(k-1)$-face $conv(\{M_i, i \in \mathcal{I}\})$. We conclude that the convex hull of any subset of $k \le d/2$ vertices of $\mathcal{P}$ is a face of $\mathcal{P}$, which proves the following theorem:

Theorem 7.2.7 (Cyclic polytope) *A cyclic polytope with n vertices has* $\begin{pmatrix} n \\ k \end{pmatrix}$ *$(k-1)$-faces, for all $0 \le k \le d/2$.*

Considering the dual of a cyclic polytope, we can also prove:

Theorem 7.2.8 *For any integer $n \ge d+1$, there is a polytope in $\mathbb{E}^d$ that has n facets and exactly* $\begin{pmatrix} n \\ k \end{pmatrix}$ *$(d-k)$-faces, for all $0 \le k \le d/2$.*

7.3 Projective polytopes, unbounded polytopes

The study of polytopes shows the combinatorial equivalence that exists between the convex hull of a finite set of points and the intersection of a finite set of half-spaces, when it is bounded. In the following chapters (and especially in chapter 14 on hyperplane arrangements and in chapter 17 on Voronoi diagrams), convex sets defined as the intersection of a finite set of half-spaces play a special role. These intersections are not necessarily bounded, however. The *projective space* extends the Euclidean space $\mathbb{E}^d$ by adding points at infinity. Using projective geometry yields a unified treatment of bounded and unbounded subsets. One drawback of projective spaces, however, is their lack of orientation: for instance, we cannot define the notion of segment joining two points or the position of a point with respect to a hyperplane without ambiguity. As a consequence of this lack of orientation, the notions of convexity and of half-spaces have no meaning in projective geometry. *Oriented projective geometry* allows us to salvage these notions and at the same time keep a complete duality between convex hulls of a finite set of points and intersections of a finite set of hyperplanes. A presentation that uses oriented projective space makes for an easy extension of the combinatorial results on polytopes to unbounded polytopes of $\mathbb{E}^d$.

7.3.1 Projective spaces

We define a *vector line*, or *line* for short, as a one-dimensional subspace in a vector space. Formally, a projective space of dimension d is the space of all vector lines in a vector space $\mathbb{V}^{d+1}$ of dimension $d+1$. The projective subspaces are formed by the subsets of those vector lines that belong to vector subspaces in $\mathbb{V}^{d+1}$. The subset of those vector lines that belong to a vector subspace of dimension $k+1$ in $\mathbb{V}^{d+1}$ is a *projective subspace* of *dimension k*, also called a *projective k-subspace*.

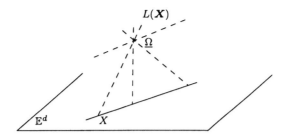

Figure 7.4. Projective subspaces.

More concretely, we can embed the d-dimensional space $\mathbb{E}^d$ in a projective space $\mathbb{P}^d$ by the following construction. Consider, in the $(d+1)$-dimensional space $\mathbb{E}^{d+1}$, the embedding of $\mathbb{E}^d$ as a hyperplane in $\mathbb{E}^{d+1}$. In $\mathbb{E}^{d+1}$, we set the origin at a point $\underline{\Omega}$ that does not belong to the hyperplane $\mathbb{E}^d$. The projective space $\mathbb{P}^d$ of dimension d is defined as the set of all lines in $\mathbb{E}^{d+1}$ that pass through $\underline{\Omega}$. Alternatively, it can be defined as the quotient of $\mathbb{E}^{d+1} \setminus \{\underline{\Omega}\}$ by the equivalence relation $\mathcal{R}$ such that $\underline{X} \,\mathcal{R}\, \underline{X'}$ if there exists a real $\lambda \neq 0$ such that $\underline{X} = \lambda \underline{X'}$. Any affine subspace of dimension $k + 1$ in $\mathbb{E}^{d+1}$ that contains $\underline{\Omega}$ corresponds to a *projective subspace* of dimension k in $\mathbb{P}^d$, consisting of all the lines in this subspace that pass through $\underline{\Omega}$.

To avoid confusion, elements of $\mathbb{E}^{d+1}$ are systematically denoted by an underlined symbol, and elements of $\mathbb{P}^d$ by a boldface symbol. Points in the projective space $\mathbb{P}^d$ are lines in $\mathbb{E}^{d+1}$ that pass through $\underline{\Omega}$. The space $\mathbb{E}^d$ itself is embedded into $\mathbb{E}^{d+1}$, so any point X in $\mathbb{E}^d$ corresponds to a point $\underline{X}$ in $\mathbb{E}^{d+1}$, and the line passing through $\underline{\Omega}$ and $\underline{X}$ is therefore a point $\boldsymbol{X}$ in $\mathbb{P}^d$. Reciprocally, any point $\boldsymbol{X}$ in $\mathbb{P}^d$ is a line $L(\boldsymbol{X})$ in $\mathbb{E}^{d+1}$ that passes through $\underline{\Omega}$. If this line $L(\boldsymbol{X})$ is not parallel to the hyperplane $\mathbb{E}^d$, $\boldsymbol{X}$ corresponds to the point X in $\mathbb{E}^d$ at the intersection of $\mathbb{E}^d$ and $L(\boldsymbol{X})$ (see figure 7.4). The points in $\mathbb{P}^d$ which correspond to lines in $\mathbb{E}^{d+1}$ parallel to $\mathbb{E}^d$ do not correspond to points in $\mathbb{E}^d$; they are called the *points at infinity*. The points at infinity form a subspace of dimension $d - 1$ in $\mathbb{P}^d$, that is a hyperplane in $\mathbb{P}^d$ called the *hyperplane at infinity* and denoted by $\boldsymbol{H}_\infty$. If X is a point in $\mathbb{E}^d$, we systematically denote by $\underline{X}$ the point in $\mathbb{E}^{d+1}$ that corresponds to X and by $\boldsymbol{X}$ the corresponding projective point. For each point $\boldsymbol{X}$ in $\mathbb{P}^d$, we denote by $L(\boldsymbol{X})$ the corresponding line in $\mathbb{E}^{d+1}$.

Any affine subspace F of $\mathbb{E}^d$ can be embedded as a projective subspace in $\mathbb{P}^d$, which is the family of lines passing through $\underline{\Omega}$ in the affine subspace $\mathit{aff}(\{F, \underline{\Omega}\})$ of $\mathbb{E}^{d+1}$. Any line of $\mathbb{E}^d$ can therefore be extended to a projective line by adding the point at infinity, and similarly a subspace of dimension k of $\mathbb{E}^d$ can be extended to a projective subspace of dimension k of $\mathbb{P}^d$ by adding a $(k - 1)$-subspace at infinity.

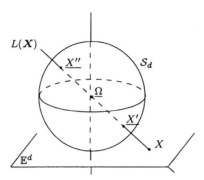

Figure 7.5. Central projection.

Sometimes it can help to represent the projective space $\mathbb{P}^d$ as the set of antipodal (*i.e.* diametrically opposite) pairs of points on a sphere $\mathcal{S}_d$ in $\mathbb{E}^{d+1}$ centered at the origin $\underline{\Omega}$. The point $\boldsymbol{X}$ of $\mathbb{P}^d$ corresponding to the affine line $L(\boldsymbol{X})$ in $\mathbb{E}^{d+1}$ can be represented as the pair of points at the intersection of $L(\boldsymbol{X})$ and $\mathcal{S}_d$. In this representation, k-subspaces of $\mathbb{P}^d$ are represented by great k-spheres of $\mathcal{S}_d$, which are intersections of $\mathcal{S}_d$ with affine $(k+1)$-subspaces of $\mathbb{E}^{d+1}$ that contain $\underline{\Omega}$. The hyperplane at infinity $\boldsymbol{H}_\infty$ corresponds to the great $(d-1)$-sphere of $\mathcal{S}_d$ in a hyperplane parallel to $\mathbb{E}^d$. The function induced by this representation maps a point $\boldsymbol{X}$ in $\mathbb{P}^d$ not in $\boldsymbol{H}_\infty$ to the point $X = L(\boldsymbol{X}) \cap \mathbb{E}^d$ of $\mathbb{E}^d$, and is commonly referred to as the *central projection* (see figure 7.5).

Homogeneous coordinates

We must note at this point that the equivalence relation $\mathcal{R}$ used in the definition of $\mathbb{P}^d$ is compatible neither with the affine structure nor with the vector-space structure of $\mathbb{E}^{d+1}$. Indeed, if $\underline{X}_1, \underline{X}_2, \underline{Y}_1, \underline{Y}_2$ are points in $\mathbb{E}^{d+1}$, it may happen that $\underline{X}_1 \mathcal{R} \underline{X}_2$ and $\underline{Y}_1 \mathcal{R} \underline{Y}_2$, yet $(\underline{X}_1 + \underline{Y}_1) \; \mathcal{R} \; (\underline{X}_2 + \underline{Y}_2)$ does not hold. As a consequence, the projective space $\mathbb{P}^d$ is neither an affine space, nor a vector space.

Nevertheless, any basis of $\mathbb{E}^{d+1}$ can be used as a coordinate system for $\mathbb{P}^d$: we represent point $\boldsymbol{X}$ as a $(d+1)$-tuple $(x_1, \ldots, x_{d+1})$ of reals, the coordinates of some point in $\mathbb{E}^{d+1}$ on the line $L(\boldsymbol{X})$. This $(d+1)$-tuple $(x_1, \ldots, x_{d+1})$ is not uniquely defined, yet it is unique up to a non-null multiplicative factor, and constitutes the *homogeneous coordinates* of $\boldsymbol{X}$. Any projective hyperplane $\boldsymbol{H}$ can be described as the set of projective points whose homogeneous coordinates satisfy a linear equation

$$\sum_{i=1}^{d+1} h_i x_i = 0$$

whose coefficients are unique up to a multiplicative factor.

Let O be a point of $\mathbb{E}^d$ chosen as the origin. If a basis of $\mathbb{E}^{d+1}$ is formed by adding to a basis of $\mathbb{E}^d$ the vector $\underline{\Omega O}$, the point X in $\mathbb{E}^d$ with coordinates $(x_1, \ldots, x_d)$ is mapped to the projective point with homogeneous coordinates $(x_1, \ldots, x_d, 1)$. The hyperplane at infinity has equation $x_{d+1} = 0$.

Below, we use the same notation X for a point of $\mathbb{E}^d$ and its coordinate vector $(x_1, \ldots, x_d)$. Likewise, the notation $\boldsymbol{X}$ denotes either a projective point or (any of) its homogeneous coordinate vectors $(x_1, \ldots, x_{d+1})$.

Let $k \leq d+1$, and $\{\boldsymbol{A_0}, \boldsymbol{A_1}, \ldots, \boldsymbol{A_k}\}$ be a set of $k+1$ points in $\mathbb{P}^d$. These points are said to be *independent* if the smallest projective subspace that contains them has dimension k. Points $\{\boldsymbol{A_0}, \boldsymbol{A_1}, \ldots, \boldsymbol{A_k}\}$ are independent if their coordinate matrix in some basis (which has dimension $(k + 1) \times (d + 1)$) has rank $k + 1$.

Projective mappings

In a projective space, the hyperplane at infinity is like any other hyperplane and plays no particular role. In general, the properties of a projective space $\mathbb{P}^d$ are invariant under any linear map $\boldsymbol{X} \longrightarrow \boldsymbol{XT}$ whose matrix $\boldsymbol{T}$ is non-singular. Such a mapping is called a projective mapping. It transforms a k-dimensional projective subspace into another projective subspace of the same dimension. The hyperplane at infinity may be mapped onto any hyperplane of $\mathbb{P}^d$ by a suitable projective mapping.

Polarity, duality

Any hyperplane $\boldsymbol{H}$ in a projective space has a homogeneous equation of the kind

$$\boldsymbol{H} = \{\boldsymbol{X} \; : \; \sum_{i=1}^{d+1} h_i x_i = 0\}.$$

Let $\boldsymbol{S}$ be the $(d + 1) \times (d + 1)$ matrix

$$\boldsymbol{S} = \begin{pmatrix} \mathbb{I}^d & 0 \\ 0 & -1 \end{pmatrix},$$

where $\mathbb{I}^d$ stands for the $d \times d$ identity matrix. Let $\boldsymbol{H^*}$ be the projective point $(h_1, \ldots, h_d, -h_{d+1})$. The homogeneous equation of H can be rewritten in matrix form

$$\boldsymbol{H} = \{\boldsymbol{X} \; : \; \boldsymbol{H^*SX^t} = 0\}.$$

Point $\boldsymbol{H^*}$ is the pole of hyperplane $\boldsymbol{H}$. Conversely, to any projective point $\boldsymbol{P}$ with homogeneous coordinates $(p_1, \ldots, p_{d+1})$ there corresponds a polar hyperplane $\boldsymbol{P^*}$ with homogeneous equation

$$\boldsymbol{P^*} = \{\boldsymbol{X} \; : \; \boldsymbol{PSX^t} = 0\}.$$

This double correspondence between points and hyperplanes is called the *polarity centered at O*. It is exactly the extension to projective spaces of the polarity centered at O described for Euclidean spaces in subsection 7.1.3. In a Euclidean space, the polarity centered at O maps points other than O to hyperplanes that do not pass through O. In a projective space, the polarity centered at O maps points to hyperplanes, in a one-to-one fashion without restrictions: the projective point $\boldsymbol{O}$ (corresponding to the center O) is mapped to the polar hyperplane at infinity $\boldsymbol{H}_\infty$, and the pole of a hyperplane $\boldsymbol{H}$ that passes through O is the point at infinity in the direction normal to the hyperplane $\boldsymbol{H}$. In the projective space, as in its Euclidean counterpart, the polarity centered at O is an involution, that is,

$$\boldsymbol{P}^{**} = \boldsymbol{P} \quad \text{and} \quad \boldsymbol{H}^{**} = \boldsymbol{H},$$

and reverses inclusion relationships, that is,

$$\boldsymbol{P} \in \boldsymbol{H} \Longleftrightarrow \boldsymbol{H}^* \in \boldsymbol{P}^*.$$

Polarity is therefore a duality.

More generally, for any symmetric non-singular $(d+1) \times (d+1)$ matrix $\Delta_{\mathcal{B}}$, we consider the mapping that maps a point $\boldsymbol{P}$ to the hyperplane $\boldsymbol{P}^*$ satisfying $\boldsymbol{P}^* = \{\boldsymbol{X} \ : \ \boldsymbol{P}\Delta_{\mathcal{B}}\boldsymbol{X}^t = 0,\}$ and a hyperplane $\boldsymbol{H}$ to the point $\boldsymbol{H}^*$ satisfying $\boldsymbol{H} = \{\boldsymbol{X} \ : \ \boldsymbol{H}^*\Delta_{\mathcal{B}}\boldsymbol{X}^t = 0\}$. This mapping is an involution between points and hyperplanes, therefore is one-to-one, and reverses inclusion relationships. The set $\mathcal{B}$ of those projective points $\boldsymbol{X}$ that satisfy

$$\boldsymbol{X}\Delta_{\mathcal{B}}\boldsymbol{X}^t = 0$$

corresponds to a quadric $\mathcal{B}$ in $\mathbb{E}^d$, and the duality just defined is called the *polarity with respect to $\mathcal{B}$*. Using this terminology, the polarity centered at O is the polarity with respect to the unit sphere $\mathcal{S}_{d-1}$ centered at O. The *signature* of the quadric $\mathcal{B}$ is the set of signs of its eigenvalues. In fact, it can be shown that in a projective space, two quadrics with the same signature or with opposite signatures can be derived from one another by a projective mapping. The corresponding polarities are called *equivalent*.

Besides the polarity centered at O, one of the polarities most widely used in computational geometry is that with respect to the unit paraboloid, $\mathcal{P}_{d-1}$, with Cartesian equation in $\mathbb{E}^d$

$$x_d = \sum_{i=1}^{d-1} x_i^2$$

and homogeneous equation in $\mathbb{P}^d$

$$\boldsymbol{X}\Delta_{\mathcal{P}}\boldsymbol{X}^t = 0 \ \text{ with } \ \Delta_{\mathcal{P}} = \begin{pmatrix} \mathbb{I}_{d-1} & 0 & 0 \\ 0 & 0 & -1/2 \\ 0 & -1/2 & 0 \end{pmatrix}.$$

The paraboloid $\mathcal{P}_{d-1}$ can therefore be derived from the unit sphere $\mathcal{S}_{d-1}$ by a projective mapping sending the center O of $\mathcal{S}_{d-1}$ to infinity along the x_d-axis. The polarity with respect to $\mathcal{P}_{d-1}$ is therefore projectively equivalent to the polarity centered at O. For more details on this polarity, see exercises 7.13 and 7.14.

7.3.2 Oriented projective spaces

Motivation

Projective geometry is a powerful and attractive framework for geometry and algorithms. For instance, it allows us to ignore many particular cases arising from the presence of parallel subspaces. Projective geometry also helps in giving a unified presentation of conics and quadrics, and gives the concept of a projective mapping which generalizes that of an affine transformation, while also adding *perspective mappings* which swap points at infinity and points at a finite distance.

Such a bonus does not come without drawbacks, however. The most serious, from the viewpoint of computational geometry, concern convexity and half-spaces, which do not exist in projective geometry. Let us now expand a little on these two points.

About convexity. In a projective space $\mathbb{P}^d$, there is no way to unambiguously define the segment joining two points. Indeed, a projective space is neither a vector space nor an affine space, and even the notion of linear combination of projective points has no meaning. If $\boldsymbol{P}$ and $\boldsymbol{Q}$ are two projective points, we may still let $\lambda \boldsymbol{P} + \mu \boldsymbol{Q}$ be the quotient of the set of points

$$\{\lambda \underline{P} + \mu \underline{Q} \ : \ \underline{P} \in L(\boldsymbol{P}), \ \underline{Q} \in L(\boldsymbol{Q})\}$$

by the relation $\mathcal{R}$ introduced in subsection 7.3.1. This set, however, is not an equivalence class of this relation, but a union of such classes. If λ and μ are fixed non-zero real numbers, $\lambda \boldsymbol{P} + \mu \boldsymbol{Q}$ is the 2-subspace of $\mathbb{E}^{d+1}$ which is the affine hull of the lines $L(\boldsymbol{P})$ and $L(\boldsymbol{Q})$. The same set is obtained if we let λ and μ vary. This remains true even if we require that λ and μ satisfy the convexity condition: $\lambda \geq 0$, $\mu \geq 0$ and $\lambda + \mu = 1$. This condition has no effect on the set generated by $\lambda \boldsymbol{P} + \mu \boldsymbol{Q}$.

The lines in $aff(L(\boldsymbol{P}), L(\boldsymbol{Q}))$ passing through $\underline{\Omega}$ do indeed form a projective line, which is the smallest projective subspace generated by $\boldsymbol{P}$ and $\boldsymbol{Q}$. But since the convexity condition has no effect, there is no means to distinguish a subset of this line which might be the segment $\boldsymbol{P}\boldsymbol{Q}$.

In the spherical model, the points $\boldsymbol{P}$ and $\boldsymbol{Q}$ are represented by two pairs of antipodal points $(\underline{P}, \neg \underline{P})$ and $(\underline{Q}, \neg \underline{Q})$. The line joining them is a great circle of $\mathcal{S}_d$ passing through $\underline{P}, \neg \underline{P}, \underline{Q}, \neg \underline{Q}$. The points $\underline{P}, \neg \underline{P}, \underline{Q}, \neg \underline{Q}$ determine on this circle four arcs pairwise diametrically opposite, or equivalently two projective arcs. There is no way to identify one of these two projective arcs as being the segment that joins P and Q.

Without segments, we certainly cannot define what it means for a set to be convex, nor what the convex hull of a set of points is.

About half-spaces.   Let us consider a hyperplane H in the projective space $\mathbb{P}^d$, with homogeneous equation $H^*SX^t = 0$. If X does not belong to H, the sign of the bilinear homogeneous form H^*SX^t is arbitrary and without significance since the homogeneous coordinates are defined up to a multiplicative factor (of either sign). It is therefore impossible to locate the point P on either side of H. In fact, a projective hyperplane does not separate the space into two disconnected half-spaces. In the spherical model, a hyperplane is represented by a great $(d-1)$-sphere of $\mathcal{S}_d$. Each projective point P is represented as a pair $(\underline{P}, \neg\underline{P})$ of two antipodal points on $\mathcal{S}_d$, and each of these points belongs to a different hemisphere determined by H.

Oriented projective geometry remedies this situation while keeping the advantages of projective geometry.

Definition

For each vector V of a vector space, the set $\{\lambda V \ : \ \lambda \in \mathbb{R}, \lambda \geq 0\}$ is an *oriented vector line*. An *oriented projective space* of dimension d consists of oriented lines of a vector space $\mathbb{V}^{d+1}$ of dimension $d+1$. A subspace of this space consists of the oriented lines lying in a subspace of $\mathbb{V}^{d+1}$.

More concretely, the oriented projective space $\mathbb{P}_o^d$ that extends the affine space $\mathbb{E}^d$ can be described in terms of the embedding of $\mathbb{E}^d$ in the space $\mathbb{E}^{d+1}$. As before, we let the origin $\underline{\Omega}$ be a point of $\mathbb{E}^{d+1}$ not in the hyperplane that we consider as $\mathbb{E}^d$. The oriented projective space $\mathbb{P}_o^d$ is the set of all rays cast from $\underline{\Omega}$ in $\mathbb{E}^{d+1}$, or equivalently the set of equivalence classes of the points in $\mathbb{E}^{d+1} \setminus \{\underline{\Omega}\}$ for the relation $\mathcal{R}_o$ defined by: $\underline{X} \, \mathcal{R}_o \underline{X}'$ if there exists $\lambda > 0$ such that $\underline{X} = \lambda \underline{X}'$. Thus, a point in the projective space corresponds to two points in the oriented projective space, which are then called *opposite points*. In the spherical representation, the oriented projective space amounts to distinguishing the two points in an antipodal pair of $\mathcal{S}_d$.

When a basis of $\mathbb{E}^{d+1}$ is understood, a point in the oriented projective space has a vector of homogeneous coordinates which is defined up to a positive multiplicative factor. In the rest of this chapter, we denote by P either a point in the oriented projective space or its vector $(p_1, p_2, \ldots, p_{d+1})$ of homogeneous coordinates, and by $\neg P$ its opposite point. As in a projective space, $k+1$ points $\{A_0, A_1, \ldots, A_k\}$ in $\mathbb{P}_o^d$ are independent if their coordinate vectors are independent. In particular, points P and $\neg P$ are not independent.

The subspaces of dimension k of $\mathbb{P}_o^d$ are the subsets of points in $\mathbb{P}_o^d$ that correspond to the rays cast from $\underline{\Omega}$ inside a subspace of $\mathbb{E}^{d+1}$ of dimension $k+1$ that contains $\underline{\Omega}$. In this manner, a subspace that contains a point $\boldsymbol{P}$ also contains its opposite $\neg\boldsymbol{P}$, and the subspaces of $\mathbb{P}_o^d$ coincide with those of $\mathbb{P}^d$. One of the main advantages of working in oriented projective geometry is the possibility of orienting the subspaces. Let $\boldsymbol{F}$ be a k-subspace of $\mathbb{P}_o^d$, corresponding to the $k+1$-subspace $\underline{F}$ of $\mathbb{E}^{d+1}$. Any set of vectors in $\mathbb{E}^{d+1}$ that forms a basis for $\underline{F}$ gives a coordinate system in $\boldsymbol{F}$.[2] In such a coordinate system, a point of $\boldsymbol{F}$ is represented by a $(k+1)$-vector of homogeneous coordinates defined up to a positive multiplicative factor. For two $(k+1)$-tuple of independent points in $\boldsymbol{F}$, $\{\boldsymbol{A_0}, \boldsymbol{A_1}, \ldots, \boldsymbol{A_k}\}$ and $\{\boldsymbol{B_0}, \boldsymbol{B_1}, \ldots, \boldsymbol{B_k}\}$, we consider a $(k+1) \times (k+1)$ matrix U that transforms the homogeneous coordinates of $\{\boldsymbol{A_0}, \boldsymbol{A_1}, \ldots, \boldsymbol{A_k}\}$ to those of $\{\boldsymbol{B_0}, \boldsymbol{B_1}, \ldots, \boldsymbol{B_k}\}$ in the same coordinate system. Two $(k+1)$-tuples $\{\boldsymbol{A_0}, \boldsymbol{A_1}, \ldots, \boldsymbol{A_k}\}$ and $\{\boldsymbol{B_0}, \boldsymbol{B_1}, \ldots, \boldsymbol{B_k}\}$ are called *equivalent* if the determinant of U is positive. The sign of the determinant of U does not depend on the choice of the coordinate system or on the choice of the vectors of homogeneous coordinates used to represent each $\boldsymbol{A_i}$ or $\boldsymbol{B_j}$. The $(k+1)$-tuples of points of $\boldsymbol{F}$ fall into two equivalence classes. To give an orientation to $\boldsymbol{F}$ is to choose one of these two classes as the positive orientation. There are thus two possible orientations for an oriented projective space. An oriented projective k-subspace is determined by a $(k+1)$-tuple which at once determines the subspace and its positive orientation. From now on, projective subspaces are supposed to be oriented, and we denote by $\boldsymbol{F}$ and $\neg\boldsymbol{F}$ an oriented subspace and the same subspace with the opposite orientation.

In particular, a projective oriented hyperplane $\boldsymbol{H}$ can be defined by a d-tuple $\{\boldsymbol{A_0}, \boldsymbol{A_1}, \ldots, \boldsymbol{A_{d-1}}\}$ of affinely independent points, by the homogeneous equation

$$\boldsymbol{H} = \{\boldsymbol{X} \in \mathbb{P}_o^d \; : \; [\boldsymbol{A_0}, \boldsymbol{A_1}, \ldots, \boldsymbol{A_{d-1}}, \boldsymbol{X}] = 0\},$$

where $[\boldsymbol{A_0}, \boldsymbol{A_1}, \ldots, \boldsymbol{A_{d-1}}, \boldsymbol{X}]$ is the determinant of the matrix whose coefficients are the homogeneous coordinates of $\{\boldsymbol{A_0}, \boldsymbol{A_1}, \ldots, \boldsymbol{A_{d-1}}, \boldsymbol{X}\}$. The coefficients $(h_1, \ldots, h_{d+1})$ in the homogeneous equation $\sum_{i=1}^{d+1} h_i x_i = 0$ defining $\boldsymbol{H}$ are thus defined up to a positive multiplicative factor. It is possible to determine two classes between the points in $\mathbb{P}_o^d \setminus \boldsymbol{H}$, and an oriented projective hyperplane separates the space into two half-spaces

$$\boldsymbol{H}^+ = \{\boldsymbol{X} \in \mathbb{P}_o^d \; : \; [\boldsymbol{A_0}, \boldsymbol{A_1}, \ldots, \boldsymbol{A_{d-1}}, \boldsymbol{X}] < 0\},$$

$$\boldsymbol{H}^- = \{\boldsymbol{X} \in \mathbb{P}_o^d \; : \; [\boldsymbol{A_0}, \boldsymbol{A_1}, \ldots, \boldsymbol{A_{d-1}}, \boldsymbol{X}] > 0\}.$$

---

[2]For instance, for any $(k+1)$-tuple $\{\boldsymbol{A_0}, \boldsymbol{A_1}, \ldots, \boldsymbol{A_k}\}$ of independent points in $\boldsymbol{F}$, the vectors generating the oriented vector lines of $\mathbb{E}^{d+1}$ corresponding to $\{\boldsymbol{A_0}, \boldsymbol{A_1}, \ldots, \boldsymbol{A_k}\}$ form a coordinate system for $\boldsymbol{F}$.

Duality

The notion of duality can be extended without problems to the oriented projective space. The oriented projective point H^* defined by

$$\forall X \in \mathbb{P}_o^d, \ \ H^*SX^t = [A_0, A_1, \ldots, A_{d-1}, X],$$

is the pole of the oriented projective hyperplane H for the polarity centered at O. Likewise, H is the polar hyperplane of H^*. The pole of the hyperplane $\neg H$ with the opposite orientation is the point opposite to the pole of H. Polarity reverses the inclusion relationships between points and hyperplanes and, moreover, reverses the relative positions of a point and a hyperplane, that is

$$\begin{array}{ccccccc}
P \in H & \Longleftrightarrow & H^*SP^t = 0 & \Longleftrightarrow & PSH^{*t} = 0 & \Longleftrightarrow & H^* \in P^* \\
P \in H^+ & \Longleftrightarrow & H^*SP^t < 0 & \Longleftrightarrow & PSH^{*t} < 0 & \Longleftrightarrow & H^* \in P^{*+} \\
P \in H^- & \Longleftrightarrow & H^*SP^t > 0 & \Longleftrightarrow & PSH^{*t} > 0 & \Longleftrightarrow & H^* \in P^{*-}.
\end{array}$$

In the spherical model, a point in the oriented projective space $\mathbb{P}_o^d$ is represented by a single point of $\mathcal{S}_d$, and an oriented projective hyperplane is represented by an oriented great $(d-1)$-sphere. The two half-spaces determined by an oriented projective hyperplane H correspond to the two hemispheres bounded on $\mathcal{S}_d$ by this great sphere.

Remark. The points at a finite distance of a hyperplane H in $\mathbb{P}_o^d$, which are the points in $H \setminus (H \cap H_\infty)$, project onto a hyperplane H in $\mathbb{E}^d$. The points in the half-spaces H^+ and H^-, however, do not project onto a half-space of $\mathbb{E}^d$. In fact, the projective half-spaces determined by H each project onto the whole Euclidean space $\mathbb{E}^d$. Let us denote by H_∞^+ and H_∞^- the two half-spaces in $\mathbb{P}_o^d$ determined by the hyperplane at infinity:

$$H_\infty^+ = \{X \in \mathbb{P}_o^d \ : \ x_{d+1} > 0\}$$

$$H_\infty^- = \{X \in \mathbb{P}_o^d \ : \ x_{d+1} < 0\}.$$

Now let H be a hyperplane of $\mathbb{P}_o^d$ which projects onto a hyperplane H in $\mathbb{E}^d$. Each of the subsets $H^+ \cap H_\infty^+$ and $H^- \cap H_\infty^-$ projects onto one of the half-spaces of $\mathbb{E}^d$ determined by H, say H^+, and each of the subsets $H^+ \cap H_\infty^-$ and $H^- \cap H_\infty^+$ projects onto the other half-space bounded by H, namely H^- (see figure 7.6).

7.3.3   Projective polytopes, unbounded polytopes

Projective simplices, convexity

We may also define the notion of simplex in an oriented projective space. Let $\{P_0, P_1, \ldots, P_k\}$ be a set of $k+1$ independent points in $\mathbb{P}_o^d$. This set of points

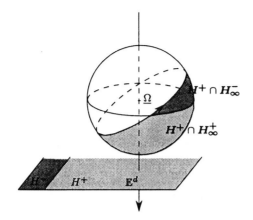

Figure 7.6. Projection onto $\mathbb{E}^d$ of an oriented projective half-space.

determines a k-subspace $\boldsymbol{F}$, and we denote by $[\boldsymbol{P_0}, \boldsymbol{P_1}, \ldots, \boldsymbol{P_k}]$ the determinant of the $(k+1) \times (k+1)$ matrix whose coefficients are the homogeneous coordinates of $\{\boldsymbol{P_0}, \boldsymbol{P_1}, \ldots, \boldsymbol{P_k}\}$ in some coordinate system of $\boldsymbol{F}$. Without loss of generality, we may assume that this determinant is positive.

The interior of the simplex $\boldsymbol{P_0}\boldsymbol{P_1} \ldots \boldsymbol{P_k}$ can now be defined as the set of those points $\boldsymbol{X}$ of $\boldsymbol{F}$ for which the determinants $[\boldsymbol{P_0}, \ldots, \boldsymbol{P_{i-1}}, \boldsymbol{X}, \boldsymbol{P_{i+1}}, \ldots, \boldsymbol{P_k}]$, obtained by substituting $\boldsymbol{X}$ for $\boldsymbol{P_i}$ for every i, $0 \le i \le k$, are all positive. In this manner, the points in the simplex $\boldsymbol{P_0}\boldsymbol{P_1} \ldots \boldsymbol{P_k}$ are those whose homogeneous coordinates can be derived from those of $\{\boldsymbol{P_0}, \boldsymbol{P_1}, \ldots, \boldsymbol{P_k}\}$ by a linear combination with non-negative coefficients.

In particular, the notion of a segment joining two points can be defined for any pair of oriented projective points which are not opposite. In the spherical model, the segment $\boldsymbol{PQ}$ is represented by the shortest arc joining $\boldsymbol{P}$ and $\boldsymbol{Q}$ on the great circle of S_d passing through $\boldsymbol{P}$ and $\boldsymbol{Q}$.

Theorem 7.3.1 *The points on a projective segment can be mapped in a one-to-one fashion to the points of a segment in $\mathbb{E}^d$.*

Proof. Let $\boldsymbol{P}$ and $\boldsymbol{Q}$ be two points in $\mathbb{P}_o^d$, not opposite to one another. If $\boldsymbol{P}$ and $\boldsymbol{Q}$ both belong to one of the half-spaces $\boldsymbol{H_\infty^+}$ or $\boldsymbol{H_\infty^-}$, then segment $\boldsymbol{PQ}$ of $\mathbb{P}_o^d$ projects onto segment PQ in $\mathbb{E}^d$. Otherwise, since $\boldsymbol{P}$ is not opposite to $\boldsymbol{Q}$, there is a hyperplane $\boldsymbol{H}$ such that $\boldsymbol{P}$ and $\boldsymbol{Q}$ lie on the same side of $\boldsymbol{H}$. The projective mapping sending $\boldsymbol{H}$ to $\boldsymbol{H_\infty}$ transforms segment $\boldsymbol{PQ}$ into a segment $\boldsymbol{P'Q'}$ which projects onto segment $P'Q'$ in $\mathbb{E}^d$. $\square$

We may now redefine the notion of convexity in an oriented projective space:

- a subset $\mathcal{X}$ of $\mathbb{P}_o^d$ is called *quasi-convex* if for each pair (P, Q) of non-opposite points in $\mathcal{X}$, the segment joining P and Q is entirely contained in $\mathcal{X}$;

- a subset $\mathcal{X}$ of $\mathbb{P}_o^d$ is called *convex* if it is quasi-convex, and does not contain a pair of opposite points.

Defined this way, convex sets include segments, simplices, and open half-spaces; antipodal pairs of points, closed half-spaces, projective subspaces, and the entire oriented projective space are quasi-convex.

The notions of quasi-convexity and convexity are invariant under projective mapping. The intersection of quasi-convex sets is quasi-convex, and the intersection of convex sets is convex.

Projective polytopes

The quasi-convex hull of a set of points in $\mathbb{P}_o^d$ is the smallest quasi-convex set that contains that set. The quasi-convex hull is not always convex, but it is convex when the set of points is contained within an open half-space of $\mathbb{P}_o^d$. In this case, we may speak of the convex hull of the set of points. Quasi-convex and convex hulls consist of all linear combinations of the points with non-negative coefficients. A *projective polytope* is the convex hull of a finite set of points which is entirely contained in an open half-space.

Theorem 7.3.2 *The points of a projective polytope can be put in one-to-one correspondence with the points of a polytope in $\mathbb{E}^d$.*

Proof. Let $\mathcal{P} = conv(\{P_0, P_1, \ldots, P_n\})$ be a projective polytope, and H a hyperplane of $\mathbb{P}_o^d$ which bounds a half-space containing $\{P_0, P_1, \ldots, P_n\}$. Any projective mapping which transforms H into H_∞ transforms $\mathcal{P}$ into a projective polytope $\mathcal{P}'$ which lies entirely on one side of H_∞, either H_∞^+ or H_∞^-. This polytope projects onto a polytope $\mathcal{P}'$ of $\mathbb{E}^d$. If P_i' is the image of P_i under the projective mapping, then

$$\mathcal{P}' = conv(\{P_0', P_1', \ldots, P_n'\}), \qquad \mathcal{P}' = conv(\{P_0', P_1', \ldots, P_n'\}).$$

$\square$

The notions of supporting hyperplanes and faces can be carried over to the projective setting without problems. The above correspondence therefore establishes a one-to-one correspondence between the faces of the polytopes $\mathcal{P}$ and $\mathcal{P}'$, which also allows us to transfer to projective polytopes the combinatorial properties of Euclidean polytopes. All the theorems in sections 7.1 and 7.2 can therefore be stated for projective polytopes. Here, we only give the projective statements of theorems 7.1.4 and 7.1.5, which concern the polar transformations.

A set $\mathcal{H}$ of projective hyperplanes in $\mathbb{P}_o^d$ is in *general position* if, for any $j \leq d$, the intersection of any j of them is a projective subspace of dimension $d - j$, and if moreover the intersection of any $d + 1$ of them is empty. The intersection of m closed projective half-spaces $\bigcap_{j=1}^m \overline{H_j^+}$ is contained in an open projective half-space if and only if there is a subset of $d + 1$ hyperplanes in general position among the hyperplanes H_j bounding all the half-spaces: such an intersection is called *non-trivial*. Theorems 7.1.4 and 7.1.5 can now be restated in a projective setting:

Theorem 7.3.3 *A projective polytope is the non-trivial intersection of a finite number of closed projective half-spaces. Any non-trivial intersection of a finite number of closed projective half-spaces is a projective polytope.*

In the oriented projective space, the polarity centered at O (or any other polarity, for that matter) can be used to define an involutive one-to-one mapping on the set of all projective polytopes, without <u>any</u> restrictions. The polar image $A^\#$ of a polar point A is the closed half-space $\overline{A^{*+}}$ defined by

$$A^\# = \{Y \in \mathbb{P}_o^d \ : \ ASY^t \leq 0\} = \overline{A^{*+}},$$

and the polar image $\mathcal{A}^\#$ of an oriented projective subset $\mathcal{A}$ is defined as the intersection of all the images of points in $\mathcal{A}$:

$$\mathcal{A}^\# = \{Y \in \mathbb{P}_o^d \ : \ ASY^t \leq 0, \ \forall A \in \mathcal{A}\}.$$

Let now

$$\mathcal{P} = conv(P_1, \ldots, P_n)$$

be a projective polytope. For each $i = 1, \ldots, n$, we denote by P_i^* the polar hyperplane of P_i and by $\overline{P_i^{*+}} = P_i^\#$ the polar half-space of P_i^*. The polar image of the polytope $\mathcal{P}$ is the intersection

$$\mathcal{P}^\# = \bigcap_{i=1}^n \overline{P_i^{*+}}.$$

If $\mathcal{P}$ is a projective polytope of dimension d, the set $\{P_1, \ldots, P_n\}$ contains at least $d + 1$ independent points and the intersection $\mathcal{P}^\#$ is non-trivial. Thus $\mathcal{P}^\#$ is a projective polytope. The proofs of theorem 7.1.11, lemma 7.1.12, and theorem 7.1.13 can now be stated almost *verbatim* for projective spaces and lead to the following theorem:

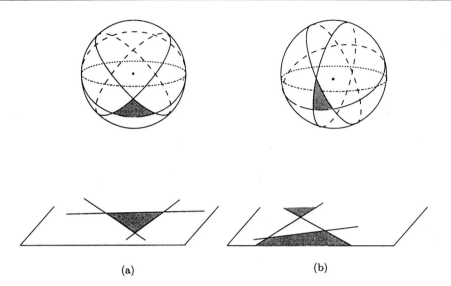

(a)   (b)

Figure 7.7. Projection onto $\mathbb{E}^d$ of a projective polytope.

Theorem 7.3.4 *The polar transformation induces an involutive one-to-one mapping defined on the set of all projective polytopes of dimension d in the oriented projective space $\mathbb{P}_o^d$. There exists a bijection between the k-faces of polytope $\mathcal{P}$ and the $(d-k-1)$-faces of polytope $\mathcal{P}^{\#}$, which reverses inclusion relationships. Moreover, a projective point A is respectively inside, outside, or on the boundary of a polytope $\mathcal{P}$ if its polar hyperplane A^{*} respectively avoids, intersects, or supports the polar polytope $\mathcal{P}^{\#}$.*

In this manner, in oriented projective geometry, the polarity centered at O or, more generally, any polarity, unambiguously defines a perfect duality between convex hulls of finite sets of points and finite intersections of half-spaces.

It should not be forgotten, however, that this duality between convex hulls and intersections of half-spaces is a duality between *projective polytopes*. The corresponding faces $\mathcal{P}$ of $\mathcal{P}^{\#}$ are also projective polytopes, but they do not always project onto polytopes in $\mathbb{E}^d$. A projective polytope projects onto $\mathbb{E}^d$ as a polytope only if it lies entirely within H_∞^- or H_∞^+, or equivalently if it does not intersect the hyperplane at infinity H_∞ (see figure 7.7).

All the other theorems in sections 7.1 and 7.2 can be restated *verbatim* for projective polytopes. In particular, a projective polytope satisfies Euler's relation, and the Dehn–Sommerville relations if it is simple, hence also the upper bound theorem.

Unbounded polytopes

Let us comment again on the projection onto $\mathbb{E}^d$ of a projective polytope. Let $\mathcal{P}$ be a polytope in $\mathbb{P}_o^d$, given as the non-trivial intersection

$$\mathcal{P} = \bigcap_{j=1}^{m} \overline{\boldsymbol{H}_j^+}.$$

The projective hyperplane $\boldsymbol{H}_j$ is the projective extension of the hyperplane H_j in $\mathbb{E}^d$. The intersection $\boldsymbol{H}_j^+ \cap \boldsymbol{H}_\infty^+$ therefore projects as a half-space H_j^+ in $\mathbb{E}^d$ (see figure 7.6). As a consequence, if polytope $\mathcal{P}$ is contained in $\boldsymbol{H}_\infty^+$, then it projects onto the polytope

$$P = \bigcap_{j=1}^{m} \overline{H_j^+}$$

in $\mathbb{E}^d$. On the other hand, if $\mathcal{P}$ intersects $\boldsymbol{H}_\infty$, it projects onto the union of two linear convex unbounded subsets

$$\left(\bigcap_{j=1}^{m} \overline{H_j^+} \right) \cup \left(\bigcap_{j=1}^{m} \overline{H_j^-} \right).$$

In this case, the projective polytope

$$\mathcal{Q} = \left(\bigcap_{j=1}^{m} \overline{\boldsymbol{H}_j^+} \right) \cap \overline{\boldsymbol{H}_\infty^+}$$

projects onto the unbounded intersection of m half-spaces in $\mathbb{E}^d$

$$Q = \left(\bigcap_{j=1}^{m} \overline{H_j^+} \right).$$

Conversely, let

$$Q = \bigcap_{j=1}^{m} \overline{H_j^+}$$

be an unbounded intersection of m half-spaces in $\mathbb{E}^d$. Such an intersection is called *non-trivial* if there is, among the hyperplanes bounding the half-spaces, a subset of d hyperplanes in general position.

Theorem 7.3.5 *Any unbounded and non-trivial intersection Q of closed half-spaces in $\mathbb{E}^d$ is the projection onto $\mathbb{E}^d$ of a projective polytope.*

Proof. Let O be a point lying inside Q, and $\boldsymbol{O}$ the projective point in $\boldsymbol{H}_{\infty}^{+}$ whose projection onto $\mathbb{E}^d$ is O. For every $j = 1, \dots, m$, let $\boldsymbol{H}_j$ be the projective extension of H_j and $\overline{\boldsymbol{H}_j^{+}}$ the closed half-space of $\mathbb{P}_o^d$ bounded by $\boldsymbol{H}_j$ which contains $\boldsymbol{O}$. Then $\boldsymbol{Q} = \left(\bigcap_{j=1}^{m} \overline{\boldsymbol{H}_j^{+}} \right) \cap \overline{\boldsymbol{H}_{\infty}^{+}}$ is a non-trivial intersection of closed projective half-spaces, whose projection onto $\mathbb{E}^d$ is Q. $\square$

This allows us to call any non-trivial unbounded intersection of a finite subset of closed half-spaces, an *unbounded polytope*.

The notions of a supporting hyperplane and faces can therefore be extended to unbounded polytopes. The facial structure and combinatorial properties of unbounded polytopes can be derived from those of projective polytopes by central projection.

7.4   Exercises

Exercise 7.1 (Radon's theorem) Show that any set $\mathcal{X}$ of at least $d + 2$ points of $\mathbb{E}^d$ can be split into two subsets $\mathcal{X}_1$ and $\mathcal{X}_2$, such that $conv(\mathcal{X}_1) \cap conv(\mathcal{X}_2) \neq \emptyset$.

Hint: There exists a non-trivial linear relationship between the points in $\mathcal{X}$

$$\sum_{i=1}^{r} \lambda_i X_i = 0.$$

For $\mathcal{X}_1$ (resp. $\mathcal{X}_2$), choose the points X_i whose coefficients λ_i in the above relation are positive (resp. negative or zero).

Exercise 7.2 (Helly's theorem) Let $\{\mathcal{K}_1, \dots, \mathcal{K}_r\}$ be a family of r convex sets of $\mathbb{E}^d$. Show that if any $d + 1$ convex sets have a non-empty intersection, then so does the whole family.

Hint: One possible proof goes by induction on r. By induction, we know that there is a point X_i in the intersection $\bigcap_{j \neq i} \mathcal{K}_i$, for all $i = 1, \dots, r$. Then use Radon's theorem on the set $\{X_i \, : \, i = 1, \dots, r\}$ to construct a point X that belongs to all the sets $\mathcal{K}_i$.

Exercise 7.3 (Carathéodory's theorem) Show that the convex hull of a subset $\mathcal{X}$ of $\mathbb{E}^d$ can be described as the set of all possible convex linear combinations of $d + 1$ points of $\mathcal{X}$. Use this to show that every polytope is a finite union of simplices.

Hint: Let $conv(\mathcal{X})$ be the convex hull of a subset $\mathcal{X}$ of $\mathbb{E}^d$ and X a point in $conv(\mathcal{X})$ given by a minimal convex linear combination $X = \sum_{i=1}^{r} \lambda_i X_i$. If $r > d + 1$, the points X_i are not independent. Use this to show that we may remove one of the points from the combination, and that it is therefore not minimal.

Exercise 7.4 (Supporting hyperplanes) Let $\mathcal{K}$ be a closed bounded convex set in $\mathbb{E}^d$. A hyperplane H supports $\mathcal{K}$ if and only if $H \cap \mathcal{K}$ is not empty but $\mathcal{K}$ is entirely contained in one of the two half-spaces bounded by H.

1. Show that for any point X not in $\mathcal{K}$, there is a supporting hyperplane of $\mathcal{K}$ which separates $\mathcal{K}$ and X, that is such that $X \in H^-$ and $\mathcal{K} \subset \overline{H^+}$.

2. Show that any closed bounded convex subset of $\mathbb{E}^d$ is the intersection of all the half-spaces that contain it and that are bounded by its supporting hyperplanes.

Hint: First show that for any point $X \notin \mathcal{K}$, there is a unique point $\Phi(X)$ of $\mathcal{K}$ such that

$$d(X, \Phi(X)) = \min\{d(X, Y) \ : \ Y \in \mathcal{K}\};$$

here, $d(X, Y)$ denotes the Euclidean distance between X and Y.

For each $X \notin \mathcal{K}$, let $D^+(\Phi(X), X)$ be the infinite ray originating at $\Phi(X)$ towards X. The hyperplane H passing through $\Phi(X)$ and normal to $D^+(\Phi(X), X)$ is a supporting hyperplane of $\mathcal{K}$ and separates X from $\mathcal{K}$.

Exercise 7.5 (Supporting hyperplanes) Let $\mathcal{K}$ be a closed bounded convex subset of $\mathbb{E}^d$. Show that every point on the boundary of $\mathcal{K}$ belongs to a supporting hyperplane of $\mathcal{K}$.

Hint: Consider the mapping Φ from $\mathbb{E}^d$ to $\mathcal{K}$ defined by

$$\forall X \in \mathcal{K}, \ \ \Phi(X) = X,$$

$$\forall X \notin \mathcal{K}, \ \ \Phi(X) = X',$$

where X' is the unique point of $\mathcal{K}$ such that

$$d(X, X') = \min\{d(X, Y) \ : \ Y \in \mathcal{K}\}.$$

1. Show that for every X and Y in $\mathbb{E}^d \setminus \mathcal{K}$,

$$d(\Phi(X), \Phi(Y)) \le d(X, Y).$$

From this, deduce that Φ is continuous.

2. Let $\mathcal{S}_{d-1}$ be a $(d-1)$-sphere of $\mathbb{E}^d$, the boundary of a ball in $\mathbb{E}^d$ that contains $\mathcal{K}$. Show that the image under Φ of $\mathcal{S}_{d-1}$ is the whole boundary of $\mathcal{K}$.

3. Show that every point on the boundary of $\mathcal{K}$ is the image under Φ of at least one point of $\mathcal{S}_{d-1}$. Use the previous exercise to show that there is a supporting hyperplane of $\mathcal{K}$ through this point.

Exercise 7.6 (Simplices) Show that simplices are the only polytopes which are both simple and simplicial.

Exercise 7.7 (The Dehn–Sommerville relations) Prove the Dehn–Sommerville relations for a d-polytope form a system of linear relations whose rank is exactly $\lfloor \frac{d+1}{2} \rfloor$.

Hint: Consider a d-polytope $\mathcal{P}$. The *face-vector* of $\mathcal{P}$ is the d-dimensional vector whose components are $\{n_0(\mathcal{P}), \ldots, n_{d-1}(\mathcal{P})\}$. Show that the face-vectors of cyclic d-polytopes with $d+1, d+2, \ldots, d+\lfloor \frac{d+1}{2} \rfloor$ vertices are independent.

Exercise 7.8 (The upper bound theorem) Let $\mathcal{P}$ be a simplicial d-polytope and let $n_j = n_j(\mathcal{P})$ denote the number of j-faces of $\mathcal{P}$. Show that the Dehn–Sommerville relations on the numbers n_k can be solved for the numbers n_j, $j = 0, \ldots, \lfloor \frac{d-1}{2} \rfloor$, yielding those numbers as linear combinations of the numbers n_j with $j = \lceil \frac{d}{2} \rceil, \ldots, d$.

Hint: Given integers $r \geq 1$, $d \geq 2r - 2$, let $D(r,d)$ be the $r \times r$ determinant

$$
\begin{vmatrix}
\binom{d}{r-1} & \binom{d-1}{r-1} & \cdots & \binom{d-r+1}{r-1} \\
\binom{d}{r-2} & \binom{d-1}{r-2} & \cdots & \binom{d-r+1}{r-2} \\
\vdots & \vdots & & \vdots \\
\binom{d}{0} & \binom{d-1}{0} & \cdots & \binom{d-r+1}{0}
\end{vmatrix}
$$

Show by induction on r that $D(r,d) = 1$. This is trivial if $r = 1$. If $r > 1$, one may subtract column $i+1$ from column i for $i = 1, \ldots, r-1$ in that order. This yields

$$
D(r,d) = \begin{vmatrix} D(r-1,d-1) & C(r-1,d-1) \\ 0 & 1 \end{vmatrix} = D(r-1,d-1)
$$

for some $(r-1)$-vector $C(r-1,d-1)$. This proves the inductive step.

Now let d be as in the exercise, and put $2r = d$ if d is even, $d = 2r - 1$ if d is odd. For each $k = 1, \ldots, r$, there is a Dehn–Sommerville relation

$$
\binom{d}{r-k} n_0 - \binom{d-1}{r-k} n_1 + \cdots + (-1)^{r-1} \binom{d-r+1}{r-k} n_{r-1} + L_k = n_{d-r+k},
$$

where L_k is some integral linear combination of $n_r, \ldots, n_{d-r+k}$. Regard these relations as a system of r linear equations for $n_0, \ldots, n_{r-1}$. The matrix of coefficients of this system is integral with determinant $\pm D(r,d) = \pm 1$, and therefore has an integral inverse. Hence each of $n_0, \ldots, n_{r-1}$ can be expressed as an integral linear combination of $n_r, \ldots, n_d$.

Exercise 7.9 (Euler's relation) Show that Euler's relation is the only non-trivial linear relation satisfied by the numbers $n_k(\mathcal{P})$ ($0 \leq k \leq d-1$) of faces of any d-polytope.

Hint: By induction on the dimension d, we show that any linear relation

$$
\sum_{j=0}^{d-1} \lambda_j n_j(\mathcal{P}) = \lambda_d
$$

satisfied by all d-polytopes $\mathcal{P}$ is proportional to Euler's relation. For this, it suffices to consider any $(d-1)$-polytope $\mathcal{Q}$ and, from it, to build two d-polytopes: a pyramid $\mathcal{P} = conv(\mathcal{Q}, P)$ where P is a point of $\mathbb{E}^d$ that does not belong to the affine hull $aff(\mathcal{Q})$ of $\mathcal{Q}$, and a bipyramid $\mathcal{R} = conv(\mathcal{Q}, \{P, P'\})$ where P and P' are two points on opposite sides of $aff(\mathcal{Q})$ such that the segment PP' intersects $\mathcal{Q}$.

Exercise 7.10 (Canonical triangulation) A triangulation of a polytope $\mathcal{P}$ is a set of simplices whose relative interiors partition the faces of $\mathcal{P}$. The canonical triangulation $\mathcal{T}_c(\mathcal{P})$ of a polytope $\mathcal{P}$ is defined by the following bottom-up construction process:

- The 0-simplices of $\mathcal{T}_c(\mathcal{P})$ are the vertices of $\mathcal{P}$.

- For $0 < k \le d$, the k-simplices of $\mathcal{T}_c(\mathcal{P})$ are the simplices $conv(V_F, S)$ where V_F is the lexicographic smallest vertex of a k-face F of $\mathcal{P}$ and S is a $(k-1)$-simplex of $\mathcal{T}_c(\mathcal{P})$ included in the boundary of F and not including V_F.

The aim of this exercise is to show that for $0 \le k < d$ the k-simplices of $\mathcal{T}_c(\mathcal{P})$ are in one-to-one correspondence with the k-faces of a simplicial polytope with the same number of vertices as $\mathcal{P}$. Therefore the number of simplices in the canonical triangulation $\mathcal{T}_c(\mathcal{P})$ of any d-polytope $\mathcal{P}$ with n vertices is bounded above by $O(n^{\lfloor \frac{d}{2} \rfloor})$.

For this we describe a transformation on polytopes called *pulling a vertex*. Let $\mathcal{P}$ be a d-polytope and V a vertex of $\mathcal{P}$. Let V' be a point in $\mathbb{E}^d$ such that (i) V' belongs to the intersection $\bigcap_{V \in F} H_F^-$ involving, for all the facets F of $\mathcal{P}$ that contain V, the half-space H_F^- bounded by the hyperplane supporting $\mathcal{P}$ along F and disjoint from $\mathcal{P}$ and (ii) the segment VV' does not meet any of the hyperplanes supporting P along a facet except in vertex V. The polytope $\mathcal{P}' = conv(V' \cup \mathcal{P})$ is said to be obtained from $\mathcal{P}$ by pulling vertex V to V'.

1. Show that the k-faces of $\mathcal{P}'$ are (i) the k-faces of $\mathcal{P}$ which do not contain V and (ii) faces of the form $conv(V' \cup G)$ where G is a $(k-1)$-subface not containing V of a k-face F containing V.

2. Show that the number of k-faces of $\mathcal{P}$, is less than the number of k-faces of $\mathcal{P}'$.

3. Consider the polytope $\mathcal{P}_c$ obtained from $\mathcal{P}$ by pulling successively, in lexicographic order, the vertices of $\mathcal{P}$. Show that $\mathcal{P}_c$ is a simplicial polytope with the same number of vertices as $\mathcal{P}$ and that, for all $0 \le k \le d$, the number $n_k(\mathcal{P})$ of k-faces of $\mathcal{P}$ is less than the number $n_k(\mathcal{P}_c)$ of k-faces of $\mathcal{P}_c$. Conclude.

Define the operation *pushing a face* of a polytope $\mathcal{P}$ as the dual of the operation *pulling a vertex* and show that repeated applications of this operation allow us to build a simple polygon $\mathcal{P}'_c$ with n facets from any polytope $\mathcal{P}$ with n facets, such that $n_k(\mathcal{P}) \le n_k(\mathcal{P}'_c)$ for each k, $0 \le k \le d$.

Exercise 7.11 (Maximal polytope) Show that there exists a polytope with n vertices on a sphere, or on a paraboloid, with maximal complexity $\Omega(n^{\lfloor d/2 \rfloor})$.

Hint: In the Euclidean space $\mathbb{E}^d$, when d is even ($d = 2p$), we consider the curve $\mathcal{M}'_d$ on the unit sphere, parameterized by

$$\mathcal{M}'_d = \{M(\tau) = \frac{1}{p}\left(\sin(\tau), \cos(\tau), \sin(2\tau), \cos(2\tau), \ldots, \sin(p\tau), \cos(p\tau)\right), \tau \in [0, \pi/2]\}.$$

Using the identity:

$$\begin{vmatrix} 1 & \cos(\tau_0) & \sin(\tau_0) & \ldots & \cos(p\tau_0) & \sin(p\tau_0) \\ 1 & \cos(\tau_1) & \sin(\tau_1) & \ldots & \cos(p\tau_1) & \sin(p\tau_1) \\ \vdots & \vdots & \vdots & & \vdots & \vdots \\ 1 & \cos(\tau_{2p}) & \sin(\tau_{2p}) & \ldots & \cos(p\tau_{2p}) & \sin(p\tau_{2p}) \end{vmatrix} = 4^{n^2} \prod_{0 \le i < j \le d} \sin\left(\frac{1}{2}(\tau_j - \tau_i)\right),$$

show that the convex hull of n points on this curve, parameterized respectively by $\{\tau_1, \tau_2, \ldots, \tau_n\}$, is a polytope whose faces are in one-to-one correspondence with those of a cyclic polytope (introduced in subsection 7.2.4). Conclude that it is possible to build a maximal polytope whose vertices lie on the unit sphere of $\mathbb{E}^d$, with exactly $\binom{n}{k}$ $(k-1)$-faces for any k, $1 \le k \le d/2$.

By considering the projective mapping that sends the unit sphere of $\mathbb{E}^d$ onto the unit paraboloid in $\mathbb{E}^d$, with equation

$$x_d = \sum_{i=1}^{d-1} x_i^2,$$

show that it is possible to build a maximal polytope whose vertices are on the unit paraboloid of $\mathbb{E}^d$.

Exercise 7.12 (Upper bound theorem) This exercise presents a very simple proof of the upper bound theorem. This proof considers a polytope, given as the intersection of n half-spaces in $\mathbb{E}^d$ bounded by hyperplanes in general position, and shows that the number of vertices of this polytope is $O(n^{\lfloor \frac{d}{2} \rfloor})$.

1. Show that any vertex of the polytope is the vertex that has the minimal or maximal x_d-coordinate in a k-face, for some $k \ge \lceil \frac{d}{2} \rceil$. For this, we consider a vertex P of the polytope. This vertex is incident to d edges, at least $\lceil \frac{d}{2} \rceil$ of which are contained in the half-space $x_d > x_d(P)$ or the half-space $x_d < x_d(P)$.

2. Note that a face has a unique vertex with maximal x_d-coordinate, and a unique vertex with minimal x_d-coordinate. Recall the bound of $\binom{n}{d-k}$ on the number of faces of dimension k of a polytope given by the intersection of n half-spaces in $\mathbb{E}^d$ and conclude.

Exercise 7.13 (Polarity with respect to a paraboloid) Consider the polarity with respect to the unit paraboloid $\mathcal{P}$ with homogeneous equation

$$X \Delta_{\mathcal{P}} X^t = 0 \quad \text{with} \quad \Delta_{\mathcal{P}} = \begin{pmatrix} \mathbb{I}^{d-1} & 0 & 0 \\ 0 & 0 & -1/2 \\ 0 & -1/2 & 0 \end{pmatrix},$$

where $\mathbb{I}^{d-1}$ is the $(d-1) \times (d-1)$ identity matrix. Show that the restriction of this polarity to the Euclidean space maps a point P in $\mathbb{E}^d$ with coordinates $(p_1, p_2, \ldots, p_d)$ to the hyperplane P^* in $\mathbb{E}^d$ with equation

$$x_d = 2 \sum_{i=1}^{d-1} p_i x_i - p_d,$$

and a non-vertical hyperplane H with equation $x_d = 2 \sum_{i=1}^{d-1} h_i x_i - h_d$ to the point $H^* = (h_1, h_2, \ldots, h_d)$.

Show that this transformation is a one-to-one mapping between the points of $\mathbb{E}^d$ and the non-vertical hyperplanes in $\mathbb{E}^d$.

Show that this bijection reverses inclusion relationships, that is,

$$P \in H \iff H^* \in P^*.$$

For a non-vertical hyperplane H, with equation $\sum_{i=1}^{d} h_i x_i + h_{d+1} = 0$, we denote by H^+ (resp. H^-) the half-space above (resp. below) H, given by the inequality $x_d > \sum_{i=1}^{d-1} h_i x_i - h_d$ (resp. $x_d < \sum_{i=1}^{d-1} h_i x_i - h_d$). Show that the polarity with respect to paraboloid $\mathcal{P}$ reverses the relative vertical positions of a point and a hyperplane, that is,

$$P \in H^+ \iff H^* \in P^{*+}$$
$$P \in H^- \iff H^* \in P^{*-}.$$

Exercise 7.14 (Lower convex hull) Let $\{P_1, P_2, \ldots, P_n\}$ be a set of points in $\mathbb{E}^d$ and O' be a point on the x_d-axis, with $x_d > 0$ large enough such that the facial structure of $conv(O', P_1, P_2, \ldots, P_n)$ is stable as O' goes to infinity along the x_d-axis. We call *lower convex hull* of $\{P_1, P_2, \ldots, P_n\}$, and we denote by $conv^-(P_1, P_2, \ldots, P_n)$, the set of faces of $conv(O', P_1, P_2, \ldots, P_n)$ which do not contain O'. Using the oriented projective space and the polarity with respect to the unit paraboloid $\mathcal{P}$ studied in exercise 7.13, show that there is a one-to-one correspondence between the faces of $conv^-(P_1, P_2, \ldots, P_n)$ and those of the unbounded intersection $\bigcap_{i=1}^{n} \overline{P_i^{*+}}$, where the half-spaces P_i^{*+} are defined as in exercise 7.13.

Exercise 7.15 (Euler's relation) Show that Euler's relation for an unbounded polytope of $\mathbb{E}^d$ can be expressed as

$$\sum_{k=0}^{d} (-1)^k n_k(\mathcal{P}) = 0,$$

where $n_k(\mathcal{P})$ is the number of k-faces of the unbounded polytope $\mathcal{P}$.

Exercise 7.16 (Half-space intersection) Let $Q = \bigcap_{j=1}^{m} \overline{H_j^+}$ be the intersection of m half-spaces in $\mathbb{E}^d$. Suppose that Q is not empty and that a point O inside Q is known.

1. Show that Q is bounded if and only if O is in the interior of the convex hull $conv(\{O, H_1^*, \ldots, H_m^*\})$.

2. Show that if O is a vertex of $conv(\{O, H_1^*, \ldots, H_m^*\})$, the faces of Q are in one-to-one correspondence with the faces of $conv(\{O, H_1^*, \ldots, H_m^*\})$ that do not contain O.

Exercise 7.17 (Zonotopes) The Minkowski sum $\mathcal{A} \oplus \mathcal{B}$ of two polytopes $\mathcal{A}$ and $\mathcal{B}$ is the polytope defined by

$$\mathcal{A} \oplus \mathcal{B} = \{A + B : A \in \mathcal{A}, B \in \mathcal{B}\}.$$

Clearly, this operator is associative. A *zonotope* is a polytope that can be expressed as the Minkowski sum of a finite set of line segments. Let $\{S_1, \ldots, S_n\}$ be a set of n line segments in $\mathbb{E}^d$, and $\mathcal{Z}$ the zonotope $S_1 \oplus \cdots \oplus S_n$. A translation brings the midpoint of each segment to the origin while simply translating the zonotope. The endpoints of segment S_i are denoted by A_i and $-A_i$.

1. Show that the polar transform $\mathcal{Z}^{\#}$ of $\mathcal{Z}$ is the polytope given by

$$\mathcal{Z}^{\#} = \{X : \sum_{i=1}^{n} |X \cdot A_i| \leq 1\}.$$

2. Show that the faces of $\mathcal{Z}$ are of the form

$$F = S_{i_1} \oplus \cdots \oplus S_{i_r} + \varepsilon_{i_r+1} A_{i_r+1} + \cdots + \varepsilon_{i_n} A_{i_n}, \qquad (7.9)$$

where $\varepsilon_{i_j} = \pm 1$, $j = r + 1, \ldots, n$.

There is a close relationship between zonotopes of $\mathbb{E}^d$ and arrangements of hyperplanes in $\mathbb{E}^{d+1}$ (see also exercise 14.8).

7.5   Bibliographical notes

The reader interested further in polytopes will do well to turn to the books by McMullen and Shephard [159], Grünbaum[114], Brønsted [37], or Berger [24]. Each of these works covers and goes well beyond all the material in sections 7.1 and 7.2. The reader will find in particular solutions to exercises 7.1 to 7.7. The presentation of polytopes given in sections 7.1 and 7.2 follows the same order as that by McMullen and Shephard.

The upper bound theorem 7.2.5 which we give here is an asymptotic version of the true upper bound theorem, which gives an exact and optimal bound on the number of k-faces of a polytope with n facets or vertices. Exercise 7.12 presents another proof of the upper bound theorem that does not use Euler's or the Dehn–Sommerville relations. This proof is due to Seidel. The oriented projective space was introduced by Stolfi in [210, 211].

Chapter 8

Incremental convex hulls

To compute the convex hull of a finite set of points is a classical problem in computational geometry. In two dimensions, there are several algorithms that solve this problem in an optimal way. In three dimensions, the problem is considerably more difficult. As for the general case of any dimension, it was not until 1991 that a deterministic optimal algorithm was designed. In dimensions higher than 3, the method most commonly used is the incremental method. The algorithms described in this chapter are also incremental and work in any dimension. Methods specific to two or three dimensions will be given in the next chapter.

Before presenting the algorithms, section 8.1 details the representation of polytopes as data structures. Section 8.2 shows a lower bound of $\Omega(n \log n + n^{\lfloor d/2 \rfloor})$ for computing the convex hull of n points in d dimensions. The basic operation used by an incremental algorithm is: given a polytope $\mathcal{C}$ and a point P, derive the representation of the polytope $conv(\mathcal{C} \cup \{P\}\}$ assuming the representation of $\mathcal{C}$ has already been computed. Section 8.3 studies the geometric part of this problem. Section 8.4 shows a deterministic algorithm to compute the convex hull of n points in d dimensions. This algorithm requires preliminary knowledge of all the points: it is an *off-line* algorithm. Its complexity is $O(n \log n + n^{\lfloor (d+1)/2 \rfloor})$, which is optimal only in even dimensions. In section 8.5, the influence graph method explained in section 5.3 is used to obtain a semi-dynamic algorithm which allows the points to be inserted on-line. The randomized analysis of this algorithm shows that its average complexity is optimal in all dimensions. Finally, section 8.6 shows how to adapt the augmented influence graph method of chapter 6 to yield a fully dynamic algorithm for the convex hull problem, allowing points to be inserted or deleted on-line. The expected complexity of an insertion or deletion is $O(\log n + n^{\lfloor d/2 \rfloor - 1})$, which is optimal.

Throughout this chapter, we assume that the set of points whose convex hull is to be computed is in *general position*. This means that any subset of $k+1 \leq d+1$ points generates an affine subspace of dimension k. This hypothesis is not crucial

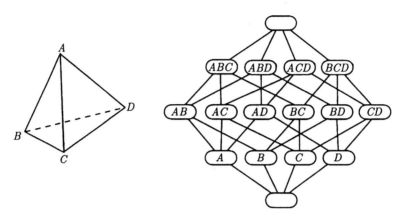

Figure 8.1. A tetrahedron and its incidence graph.

for the deterministic algorithm (see exercise 8.4), but it allows us to simplify the description of the algorithm and to focus on the central ideas. It becomes an essential assumption, however, for the randomized analyses of the on-line and dynamic algorithms.

8.1   Representation of polytopes

To compute the convex hull of a set of points amounts to setting up a data structure that represents the polytope which is the convex hull of the set. A polytope is generally represented by the *incidence graph* of its faces, which stores a node for each face and an arc for each pair of incident faces. Recall that two faces are incident if their dimensions differ by one and if one is contained in the other. Figure 8.1 shows the incidence graph of a tetrahedron.

Using the upper bound theorem 7.2.5, the incidence graph of a d-polytope can be stored using $O(n^{\lfloor d/2 \rfloor})$ space. This graph describes the entire combinatorial structure of the polytope. In order to describe its geometric structure, some additional information has to be stored: for instance, the node storing a vertex contains the coordinates of that vertex, and the node storing a facet contains the coefficients in an equation of the hyperplane that supports the polytope along that facet.

Sometimes, it may be enough to store subgraphs of the incidence graph. The *j-skeleton* of a polytope is the subgraph of the incidence faces of dimension at most j. The 1-skeleton of a polytope is simply made up of the vertices and edges of that polytope.

In a d-polytope, every $(d-2)$-face is incident to exactly two $(d-1)$-faces (theorem 7.1.7); two $(d-1)$-faces of a polytope are said to be *adjacent* if they are incident to a common $(d-2)$-face. Thus, the incidence graph of a polytope also encodes the *adjacency graph*, which has a node for each facet and an arc for each pair of adjacent facets. The arcs of the adjacency graph are in one-to-one correspondence with the $(d-2)$-faces of the polytope. If the polytope is simplicial, the full incidence graph can be retrieved from the adjacency graph in time linear in the number of faces (see exercise 8.2).

8.2   Lower bounds

Theorem 8.2.1 *The complexity of computing the convex hull of n points in d dimensions is $\Omega(n \log n + n^{\lfloor d/2 \rfloor})$.*

Proof. Subsection 7.2.4 shows that the convex hull of n points in the Euclidean space $\mathbb{E}^d$ may have $\Omega(n^{\lfloor d/2 \rfloor})$ faces. In any dimension, $\Omega(n^{\lfloor d/2 \rfloor})$ is thus a trivial lower bound for the complexity of computing convex hulls. In two dimensions, the lower bound $\Omega(n \log n)$ is a consequence of theorem 8.2.2 proved below. Finally, any set of points in $\mathbb{E}^2$ can be embedded into $\mathbb{E}^3$, so the complexity of computing convex hulls in $\mathbb{E}^3$ cannot be smaller than in $\mathbb{E}^2$. □

Theorem 8.2.2 *The problem of sorting n real numbers can be transformed in linear time into the problem of computing the convex hull of n points in $\mathbb{E}^2$.*

Proof. Consider n real numbers $x_1, x_2, \ldots, x_n$, which we want to sort. One way to do this is to map the number x_i to the point A_i with coordinates (x_i, x_i^2) on the parabola with equation $y = x^2$ (see figure 8.2). The convex hull of the set of points $\{A_i : i = 1, \ldots, n\}$ is a cyclic 2-polytope, and the list of its vertices is exactly the list of the vertices $\{A_i : i = 1, \ldots, n\}$ ordered according to their increasing abscissae. □

8.3   Geometric preliminaries

The incremental method for computing the convex hull of a set $\mathcal{A}$ of n points in $\mathbb{E}^d$ consists in maintaining the succession of convex hulls of the consecutive sets obtained by adding the points in $\mathcal{A}$ one by one. Each convex hull is represented by its incidence graph. Let $\mathcal{C}$ be the convex hull of the current subset and P the point to be inserted next into the structure, at a given step in the algorithm. The problem is thus to obtain the incidence graph of $conv(\mathcal{C} \cup \{P\})$, once we know that of $\mathcal{C}$. The following lemmas clarify the relations existing between these two graphs.

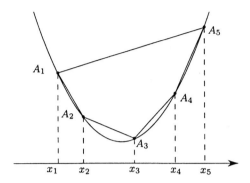

Figure 8.2. Transforming a sorting problem into a convex hull problem in two dimensions.

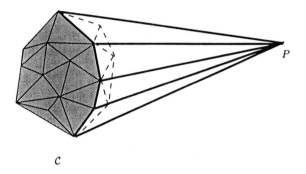

Figure 8.3. The incremental construction of a convex hull.

Suppose that point P and polytope C are in *general position*, meaning that P and the vertices of C form a set of points in general position. The facets of C can then be separated into two classes with respect to P. Let F be a facet of C, H_F the hyperplane that supports C along F, and H_F^+ (resp. H_F^-) the half-space bounded by H_F that contains (resp. does not contain) C. The facet F is *red* with respect to P if it is visible from point P, that is if P belongs to the half-space H_F^-. It is colored *blue* if P belongs to H_F^+. From the general position assumption, it follows that P never belongs to the supporting hyperplane H_F and therefore every facet of C is either red or blue with respect to P.

Using theorem 7.1.7, any face of C is the intersection of the facets of C which contain it. The faces of C of dimension strictly smaller than $d-1$ can be separated into three categories with respect to P: a face of C is *red* if it is the intersection of red facets only, *blue* if it is the intersection of blue facets only, or *purple* if it is the intersection of red and blue facets.

Intuitively, the red faces are those that would be lit if a point source of light was shining from P, the blue faces are those that would remain in the shadow, and the purple faces would be lit by rays tangent to C. In figure 8.3, the blue faces of C are shaded, the red edges are outlined in dashed lines, and the purple edges are shown in bold.

Lemma 8.3.1 *Let C be a polytope and P a point in general position with respect to C. Every face of $conv(C \cup \{P\})$ is either a blue or purple face of C, or the convex hull $conv(G \cup \{P\})$ of P and a purple face G of C.*

Proof. Note that if P belongs to C, all the facets of C are blue with respect to C (theorem 7.1.4) and the content of the lemma is trivial.

In the other case, we first show that a blue face of C is a face of $conv(C \cup \{P\})$. Let F be a facet of C that is blue with respect to P. Since P belongs to the half-space H_F^+, the hyperplane H_F which supports C along F also supports $conv(C \cup \{P\})$ and $conv(C \cup \{P\}) \cap H_F = F$, which proves that F is indeed a facet of $conv(C \cup \{P\})$. Any blue facet of C is thus a facet of $conv(C \cup \{P\})$. Any blue face of C, being the intersection of blue facets of C, is also the intersection of facets of $conv(C \cup \{P\})$: therefore a blue face of C is also a face of $conv(C \cup \{P\})$ (theorem 7.1.7).

Next we show that, for any purple face G of C, G and $conv(G \cup \{P\})$ are faces of $conv(C \cup \{P\})$. If G is a purple face of C, then there is at least one red facet of C, say F_1, and one blue facet of C, say F_2, that both contain G (see figure 8.4). Let H_1 (resp. H_2) be the hyperplane supporting C along F_1 (resp. F_2). Point P belongs to the half-space H_1^+ which contains C, and since $H_1 \cap conv(C \cup \{P\}) = G$ we have shown that G is a face of $conv(C \cup \{P\})$. Point P also belongs to the half-space H_2^- that does not contain C. Imagine a hyperplane that rotates around $H_1 \cap H_2$ while supporting C along G. There is a position H for which this hyperplane passes through point P. Hyperplane H supports $conv(C \cup \{P\})$, and since $conv(C \cup \{P\}) \cap H = conv(G \cup \{P\})$, we have proved that $conv(G \cup \{P\})$ is a face of $conv(C \cup \{P\})$.

Finally, let us show that every face of $conv(C \cup \{P\})$ is either a blue or a purple face of C, or the convex hull $conv(G \cup \{P\})$ of P and of a purple face G of C. Indeed, a hyperplane that supports $conv(C \cup \{P\})$ is also a supporting hyperplane of C, unless it intersects $conv(C \cup \{P\})$ only at point P. As a consequence, any face of $conv(C \cup \{P\})$ that does not contain P is a (blue or purple) face of C, and any face $conv(C \cup \{P\})$ that contains P is of the form $conv(G \cup \{P\})$ where G is a purple face of C. Note that the vertex P of $conv(C \cup \{P\})$ is also a face of the form $conv(G \cup \{P\})$ obtained when G is the empty face of C. Indeed, when

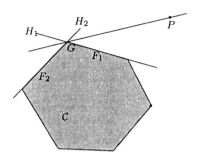

Figure 8.4. Faces of $conv(G \cup \{P\})$.

P does not belong to $\mathcal{C}$, $\mathcal{C}$ necessarily has some facets that are blue and some facets that are red with respect to P. The empty face, being the intersection of all faces of $\mathcal{C}$, is therefore purple. $\Box$

The following lemma, whose proof is straightforward, investigates the incidence relationships between the faces of $\mathcal{C}$ and those of $conv(\mathcal{C} \cup \{P\})$.

Lemma 8.3.2 *Let $\mathcal{C}$ be a polytope and P a point in general position with respect to $\mathcal{C}$.*

- *If F and G are two incident faces of polytope $\mathcal{C}$, either blue or purple with respect to P, then F and G are incident faces of $conv(\mathcal{C} \cup \{P\})$.*

- *If G is a purple face of $\mathcal{C}$, then G and $conv(G \cup \{P\})$ are incident faces of $conv(\mathcal{C} \cup \{P\})$.*

- *Finally, if F and G are incident purple faces of $\mathcal{F}$, then $conv(F \cup \{P\})$ and $conv(G \cup \{P\})$ are incident faces of $conv(\mathcal{C} \cup \{P\})$.*

Recall that two facets of a polytope $\mathcal{C}$ are adjacent if they are incident to the same $(d-2)$-face and that the adjacency graph of a polytope stores a node for each facet and an arc for each pair of adjacent facets.[1] We say that a subset of facets of a polytope $\mathcal{C}$ is *connected* if it induces a connected subgraph of the adjacency graph of $\mathcal{C}$.

Lemma 8.3.3 *Consider a polytope $\mathcal{C}$ and a point P in general position. The set of facets of $\mathcal{C}$ that are red with respect to P is connected, and the set of facets of $\mathcal{C}$ that are blue with respect to P is also connected.*

---

[1]Two facets sharing a common k-face, $k < d - 2$, may be not adjacent, even though they are connected as a topological subset of the boundary of the polytope. Such a situation is only possible in dimension $d \geq 3$.

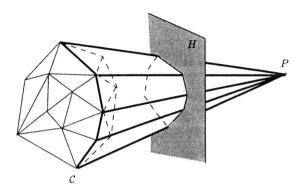

Figure 8.5. Isomorphism between the purple faces and the faces of a $(d-1)$-polytope.

Proof. If P belongs to $\mathcal{C}$, the set of the red facets is empty, any facet is blue, and the lemma is trivial. We will therefore assume that P does not belong to $\mathcal{C}$.

The connectedness of the set of red facets can be proved easily in two dimensions. Indeed, the polytope $conv(\mathcal{C} \cup \{P\})$ has two edges incident to P. By lemma 8.3.1, there are exactly two purple vertices of $\mathcal{C}$ with respect to P. Hence, the adjacency graph of the 2-polytope $\mathcal{C}$ is a cycle that has exactly two arcs connecting a blue and a red facet.

Let us now discuss the case of dimension d, and suppose for a contradiction that the set of facets of $\mathcal{C}$ that are red with respect to P is not connected. Therefore, we may choose two points Q and R on two facets of $\mathcal{C}$ that belong to two distinct connected components of the set of red facets of $\mathcal{C}$. Let H be the affine 2-space passing through points P, Q, and R. This plane intersects polytope $\mathcal{C}$ along a 2-polytope $\mathcal{H} \cap H$. The edges of $\mathcal{C} \cap H$ that are red with respect to P are exactly the intersections of the red facets of $\mathcal{C}$ with H. The points Q and R belong to two separate connected components of the set of red edges of $\mathcal{C} \cap H$. Connectedness of the set of red faces of a 2-polytope would then not hold, a contradiction.

Analogous arguments prove the connectedness of the set of facets of $conv(\mathcal{C} \cup \{P\})$ that are blue with respect to P. $\square$

Finally, the lemma below completely characterizes the subgraph of the incidence graph induced on the faces of $\mathcal{C}$ that are purple with respect to P.

Lemma 8.3.4 *Let $\mathcal{C}$ be a polytope and P a point in general position with respect to $\mathcal{C}$. If $\mathcal{C}$ has n vertices and does not contain P, then the set of the proper faces of $\mathcal{C}$ that are purple with respect to P is isomorphic, for the incidence relationship, to the set of faces of a $(d-1)$-polytope whose number of vertices is at most n.*

Proof. From lemma 8.3.1, we know that the faces of polytope $\mathcal{C}$ that are purple with respect to P are in one-to-one correspondence with the faces of $conv(\mathcal{C} \cup \{P\})$

that do not contain P. Since point P does not belong to C, there must be a hyperplane H which separates P from C (see exercise 7.4). Hyperplane H intersects all the faces of $conv(C \cup \{P\})$ that contain P except for the vertex P, and those faces only. Moreover, the traces in H of the faces of $conv(C \cup \{P\})$ are the proper faces of the $(d-1)$-polytope $conv(C \cup \{P\}) \cap H$, and the traces in H of incident faces of $conv(C \cup \{P\})$ are incident faces of $conv(C \cup \{P\}) \cap H$. Thus, the incidence graph of the $(d-1)$-polytope $conv(C \cup \{P\}) \cap H$ is isomorphic to the subgraph of the incidence graph of $conv(C \cup \{P\})$ induced by the faces that contain vertex P. Lemmas 8.3.1 and 8.3.2 show that this subgraph is isomorphic to the subgraph of the incidence graph of $\mathcal{P}$ induced by the faces of C that are purple with respect to P. Lastly, the vertices of polytope $conv(C \cup \{P\}) \cap H$ are the traces in H of the edges of $conv(C \cup \{P\})$ incident to vertex P, and their number is at most n. $\square$

8.4   A deterministic algorithm

In this section we describe an incremental deterministic algorithm to build the convex hull of a set $\mathcal{A}$ of n points. The points in $\mathcal{A}$ are processed in increasing lexicographic order of their coordinates. To simplify the description of the algorithm, we assume below that the set is in general position. We denote by $\{A_1, A_2, \ldots, A_n\}$ the points of $\mathcal{A}$ indexed by lexicographic order. Let $\mathcal{A}_i$ be the set of the first i points of $\mathcal{A}$.

The general idea of the algorithm is as follows:

1. Sort the points of $\mathcal{A}$ in increasing lexicographic order of their coordinates.

2. Initialize the convex hull to the simplex $conv(\mathcal{A}_{d+1})$, the convex hull of the first $d+1$ points of $\mathcal{A}$.

3. In the incremental step: the convex hull of $conv(\mathcal{A}_i)$ is built knowing the convex hull $conv(\mathcal{A}_{i-1})$ and the point A_i to be inserted.

Details of the incremental step

Because of the lexicographic order on the points of $\mathcal{A}$, point A_i never belongs to the convex hull $conv(\mathcal{A}_{i-1})$, and is therefore a vertex of $conv(\mathcal{A}_i)$. The preceding lemmas show that the subgraph of the incidence graph of $conv(\mathcal{A}_{i-1})$ restricted to the faces that are blue with respect to A_i is also a subgraph of the incidence graph of $conv(\mathcal{A}_i)$. All the efficiency of the incremental algorithm stems from the fact that the incidence graph of the current convex hull can be updated in an incremental step without looking at the blue faces or at their incidences.

To perform this incremental step, we proceed in four phases:

Phase 1. We first identify a facet of $conv(\mathcal{A}_{i-1})$ that is red with respect to A_i.

Phase 2. The red facets and the red or purple $(d-2)$-faces of $conv(\mathcal{A}_{i-1})$ are traversed. A separate list is set up for the red facets, the red $(d-2)$-faces, and the purple $(d-2)$-faces.

Phase 3. Using the information gathered in phase 2, we identify all the other red or purple faces of $conv(\mathcal{A}_{i-1})$. For each dimension k, $d-3 \geq k \geq 0$, a list $\mathcal{R}_k$ of the red k-faces is computed, as well as a list $\mathcal{P}_k$ of the purple k-faces.

Phase 4. The incidence graph is updated.

Before giving all the details for each phase, let us first describe precisely the data structure that stores the incidence graph. For each face F of dimension k $(0 \leq k \leq d-1)$ of the convex hull, this data structure stores:

- the list of the *sub-faces* of F, which are the faces of dimension $k-1$ incident to F,

- the list of the *super-faces* of F, which are the faces of dimension $k+1$ incident to F,

- the color of the face (red, blue, purple) in the current step, and

- a pointer $p(F)$ whose use will very soon be clarified.

If F is a super-face of G, then a bidirectional pointer links the record for F in the list of super-faces of G to the record for G in the list of sub-faces of F.

    Phase 1. To find an initial red facet in $conv(\mathcal{A}_{i-1})$, we take advantage of the lexicographic order on the points in $\mathcal{A}$. Because of this order, A_{i-1} is always a vertex of $conv(\mathcal{A}_{i-1})$ and there is at least one facet of $conv(\mathcal{A}_{i-1})$ containing A_{i-1} which is red with respect to A_i. Indeed, let $\mathcal{F}_{i-1}$ be the set of facets of $conv(\mathcal{A}_{i-1})$ that contain A_{i-1} as a vertex. Let also H be the hyperplane whose equation is $x_1 = x_1(A_{i-1})$, and H^+ the half-space bounded by H that contains $conv(\mathcal{A}_{i-1})$, and H^- the other half-space bounded by H. Since A_{i-1} is a vertex of $conv(\mathcal{A}_{i-1})$ with maximal abscissa, $H^+ \cup A_{i-1}$ contains the intersection of all the half-spaces H_F^+ when $F \in \mathcal{F}_{i-1}$. Point A_i belongs to $\overline{H^-}$, and therefore cannot belong to this intersection of half-spaces (see figure 8.6). Thus, at least one facet F in $\mathcal{F}_{i-1}$ must be red with respect to A_i. All the facets of $\mathcal{F}_{i-1}$ were created at the previous incremental step, so it suffices to store the list of facets created during an incremental step and to traverse this list during the next incremental step in order to find an initial red facet.

    Phase 2. In the second phase, we use the connectedness of the set of red facets (lemma 8.3.3). A depth-first traversal of the subgraph of red facets in

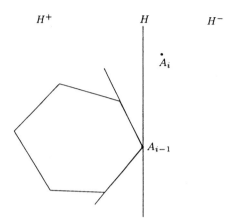

Figure 8.6. One of the facets of $conv(\mathcal{A}_{i-1})$ containing A_{i-1} must be red with respect to A_i.

the adjacency graph[2] of $conv(\mathcal{A}_{i-1})$, starting with the initial red facet that was found in phase 1, visits all the facets visible from A_i, which we color red, and their $(d-2)$-faces, which we color red if they are incident to two red facets, or purple if they are incident to a blue facet. The traversal backtracks whenever the facet encountered was already colored red, or if it is a blue facet.

Phase 3. We now know all the red and purple $(d-2)$-faces, and the red facets. In this phase, all the remaining red and purple faces are colored, and their lists are set up in order of decreasing dimensions. Assume inductively that all the red and purple faces of dimension $k' \geq k+1$ have already been identified and colored, and that the lists $\mathcal{R}_{k'}$ and $\mathcal{P}_{k'}$ have already been set up. We process the k-faces in the following way. Each sub-face of a face of $\mathcal{P}_{k+1}$ that has not yet been colored is colored purple and added to the list $\mathcal{P}_k$. Afterwards, each sub-face of $\mathcal{R}_{k+1}$ that has not yet been colored is added to the list $\mathcal{R}_k$.

Phase 4. To update the incidence graph, we proceed as follows. All the red faces are removed from the incidence graph, and so are all the arcs adjacent to these faces in the graph. The purple faces are processed in order of increasing dimension k. If F is a k-face purple with respect to P, a new node is created for the $(k+1)$-face $conv(F \cup \{A_i\})$ and linked by an arc to the node for F in the incidence graph. Also the pointer $p(F)$ is set to point to the new node created for $conv(F \cup \{A_i\})$. It remains to link this node to all the incident k-faces of the form $conv(G \cup \{A_i\})$, where G is a $(k-1)$-face incident to F. For each sub-face G of F, its pointer $p(G)$ gives a direct access to the node corresponding to $conv(G \cup \{A_i\})$, and the incidence arc can be created.

---

[2]The adjacency graph is already stored in the incidence graph, and need not be stored separately (see subsection 8.1).

Analysis of the algorithm

Phase 1 of each incremental step can be carried out in time proportional to the number of facets created at the previous step. The total cost of phase 1 over all the incremental steps is thus dominated by the total number of facets created.

At step i that sees the insertion of A_i, the cost of phase 2 is proportional to the number of nodes visited during the traversal of the adjacency graph. The nodes visited correspond to red facets of $conv(\mathcal{A}_{i-1})$, and to the blue facets adjacent to these red facets. The total cost of this phase is thus at most proportional to the number of red facets of $conv(\mathcal{A}_{i-1})$ and of their incidences.

The cost of phase 3 is bounded by (a constant factor times) the number of arcs in the incidence graph that are visited, and this number is the same as the number of incidences between red or purple faces of $conv(\mathcal{A}_{i-1})$.

Lastly, the cost of phase 4 is proportional to the total number of red faces and of their incidences, plus the number of purple faces and of their incidences to purple faces.

In short, when incrementally adding a point to the convex hull, the cost of phases 2, 3, and 4 is proportional to the number of red or purple faces, plus the number of faces incident to a red face, plus the number of incident purple faces. Red faces and their incidences correspond to the nodes and arcs of the incidence graph that are removed from the graph. The purple faces and the incidences between two purple faces correspond to nodes and arcs of the incidence graph that are added to the graph. The total cost of phases 2, 3, and 4 is thus proportional to the number of changes undergone by the incidence graph. Since a node or arc that is removed will not be inserted again (red faces will remain inside the convex hull for the rest of the algorithm), this total number of changes is proportional to the number of arcs and nodes of the incidence graph that are created throughout the execution of the algorithm, which also takes care of the cost of phase 1. The following lemma bounds this number.

Lemma 8.4.1 *The number of faces and incidences created during the execution of an incremental algorithm building the convex hull of n points in d dimensions is $O(n^{\lfloor (d+1)/2 \rfloor})$.*

Proof. Lemma 8.3.1 shows that the subgraph of the incidence graph of $conv(\mathcal{A}_i)$ induced by the faces created upon the insertion of A_i is isomorphic to the set of faces of $conv(\mathcal{A}_{i-1})$ that are purple with respect to A_i. The number of incidences between a new face and a purple face of $conv(\mathcal{A}_{i-1})$ is also proportional to the number of purple faces of $conv(\mathcal{A}_{i-1})$. Finally, lemma 8.3.4 shows that the set of purple faces of $conv(\mathcal{A}_{i-1})$ is isomorphic to a $(d-1)$-polytope that has at most $i-1$ vertices. The upper bound theorem 7.2.5 shows that the number of these faces and incidences between these faces, is $O(i^{\lfloor (d-1)/2 \rfloor})$. This is thus a bound on the number of faces and incidences created upon inserting A_i. Summing over all i, $i = 1, \ldots, n$, the total number of facets and incidences created by the algorithm is:

$$\sum_{i=1}^{n} O(i^{\lfloor (d-1)/2 \rfloor}) = O(n^{\lfloor (d+1)/2 \rfloor}).$$

$\square$

The storage needed by this algorithm is proportional to the maximum size of the incidence graph stored at any step, which is $O(n^{\lfloor d/2 \rfloor})$. Taking into account the initial sorting of the vertices, we conclude with the following result:

Theorem 8.4.2 *The incremental algorithm builds the convex hull of n points in d dimensions in time $O(n \log n + n^{\lfloor (d+1)/2 \rfloor})$ and storage $O(n^{\lfloor d/2 \rfloor})$.*

This algorithm is optimal in the worst case when the dimension of the space is even.

8.5   On-line convex hulls

Computing the convex hull of a set of points is one of the geometric problems to which the randomization techniques developed in chapter 5 apply. Randomized algorithms compute the convex hull of n points in optimal expected time, in any dimension: $O(n \log n)$ in dimension 2 or 3, and $O(n^{\lfloor d/2 \rfloor})$ in dimension $d > 3$. Let us once again recall that the average value involved here is over all the possible random choices of the algorithm, not over some spatial distribution of the points. The only assumption we make on the points is that they are in general position.

The algorithm which we present here is an incremental on-line algorithm (or semi-dynamic) that uses the influence graph method described in section 5.3, to which we refer the reader if need be. The term "on-line" means that the algorithm is able to maintain the convex hull of a set of points as the points are added one by one without preliminary knowledge of the whole set. This algorithm is in fact deterministic. Only the analysis is randomized and assumes that the order in which the points are inserted is random.

Convex hulls in terms of objects, regions, and conflicts

This section applies the formalism described in chapter 4. In order to do so, we must first recast the convex hull problem in terms of objects, regions, and conflicts.

The *objects* are naturally the points of $\mathbb{E}^d$. A *region* is defined as the union of two open half-spaces. Such a region is determined by a set of $d + 1$ points in general position. Let $\{P_0, P_1, \ldots, P_{d-1}, P_d\}$ stand for such a $(d + 1)$-tuple. Let H_d be the hyperplane containing $\{P_0, P_1, \ldots, P_{d-1}\}$ and H_d^- be the half-space bounded by H_d that does not contain P_d. Similarly let H_0 be the hyperplane containing $\{P_1, \ldots, P_{d-1}, P_d\}$ and let H_0^- be the half-space bounded by H_0 that does not contain P_d. The region determined by the $(d+1)$-tuple is the union of the two open half-spaces H_d^- and H_0^-. A point conflicts with a region if it belongs to at least one of the two open half-spaces that make up the region. In this case, the *influence domain* of a region is simply the region itself.

With this definition of regions and conflicts, the convex hull of a set $\mathcal{S}$ of n affinely independent points can be described as the set of regions defined and without conflict over $\mathcal{S}$. In fact, the regions defined and without conflict over $\mathcal{S}$ are in bijection with the $(d-2)$-faces of $conv(\mathcal{S})$. Indeed, let a region be determined by the $(d+1)$-tuple $\{P_0, P_1, \ldots, P_{d-1}, P_d\}$ of points in $\mathcal{S}$. Because the points in $\mathcal{S}$ are assumed to be in general position, if this region is without conflict over $\mathcal{S}$, the two $d-1$ simplices $F_d = conv(\{P_0, P_1, \ldots, P_{d-1}\})$ and $F_0 = conv(\{P_1, \ldots, P_{d-1}, P_d\})$ are facets of $conv(\mathcal{S})$, and the $(d-2)$-simplex $G = F_0 \cap F_d = conv(\{P_1, \ldots, P_{d-1}\})$ is the $(d-2)$-face of $conv(\mathcal{S})$ that is incident to both these facets. This region will be denoted below by (F_0, F_d) or sometimes by (F_d, F_0). The set of regions defined and without conflict over a set $\mathcal{S}$ therefore not only gives the facets of $conv(\mathcal{S})$, but also their adjacency graph. Using this information, it is an easy exercise to build the complete incidence graph of $conv(\mathcal{S})$ in time proportional to the number of faces of all dimensions of $conv(\mathcal{S})$ (see exercise 8.2).[3]

The algorithm

The algorithm is incremental, and in fact closely resembles that which is described in section 8.4. The convex hull $conv(\mathcal{S})$ of the current set $\mathcal{S}$ is represented by its incidence graph. At each step, a new point P is inserted. The faces of $conv(\mathcal{S})$ can be sorted into three categories according to their color with respect to P, as explained in section 8.3: red faces, blue faces, and purple faces. The on-line algorithm, like the incremental algorithm, identifies the faces that are red and purple with respect to P, then updates the incidence graph. The main difference resides in the order with which the points are inserted. The on-line algorithm processes the points in the order given by the input, and therefore cannot take advantage of the lexicographic order to detect the red facets. For this reason, the algorithm maintains an *influence graph*. As we may recall, the influence graph

---

[3]It would certainly be more natural to define a region as a open half-space determined by d affinely independent points. In this case the region is one of the half-spaces bounded by the hyperplane generated by these d affinely independent points, and a point conflicts with such a region if it lies in this half-space. With these definitions, the facets of the convex hull $conv(\mathcal{S})$ of a set $\mathcal{S}$ of n points in $\mathbf{E}^d$ are in bijection with the regions defined and without conflict over $\mathcal{S}$.

In fact, such a definition of regions is perfectly acceptable and so is an incremental algorithm based on these definitions (see exercise 8.5). Such an algorithm, however, does not satisfy the update conditions 5.2.1 and 5.3.3, and its analysis calls for the notion of biregion introduced in exercise 5.7.

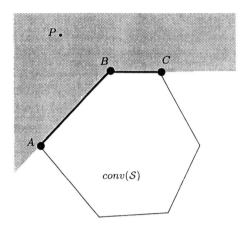

Figure 8.7. On-line convex hull: regions and conflicts.
The influence domain of region (AB, AC) is shaded, and the $(d-2)$-faces
corresponding to regions conflicting with P are represented in bold.

is used mainly to detect the conflicts between the point to be inserted and the regions defined and without conflict over the points inserted so far. The influence graph is an oriented acyclic graph that has a node for each region that, at some previous step in the algorithm, appeared as a region defined and without conflict over the current subset of points. At each step of the algorithm, the regions defined and without conflict over the current subset correspond to the leaves of the influence graph. The arcs in this graph link these nodes such that the following *inclusion* property is always satisfied: the influence domain of a node is always contained in the union of the influence domains of its parents.[4] A depth-first traversal of the influence graph can detect all the conflicts between the new point P and the nodes in the graph. With a knowledge of the conflicts between points P and the regions defined and without conflict over $\mathcal{S}$, it is easy to find the facets of $conv(\mathcal{S})$ that are red with respect to P. Indeed:

- A region defined and without conflict over $\mathcal{S}$ that conflicts with P corresponds to a red or purple $(d-2)$-face of $conv(\mathcal{S})$, since it is incident to two $(d-1)$-faces of $conv(\mathcal{S})$, at least one of which is red (see figure 8.7).

- A region defined and without conflict over $\mathcal{S}$ that does not conflict with P corresponds to a $(d-2)$-face of $conv(\mathcal{S})$ that is blue with respect to P.

In an initial step, the algorithm processes the first $d+1$ points that are inserted into the convex hull. The incidence graph is set to that of the d-simplex formed

---

[4]Recall also that we frequently identify a node in the influence graph with the region that it corresponds to, which for instance lets us speak of conflicts with a node, of the influence domain of a node, or of the children of a region.

by these points, and the influence graph is initialized by creating a node for each of the regions that correspond to the $(d-2)$-faces of this simplex.

To describe the current step, we denote by S the current set of points already inserted, and by P the new point that is being inserted. The current step consists of a location phase and an update phase.

Locating. The location phase aims at detecting the regions *killed* by the new point P. These are the regions defined and without conflict over S that conflict with P. For this, the algorithm recursively visits all the nodes that conflict with P, starting from the root.

Updating. If none of the regions defined and without conflict over S is found to conflict with P, then P must lie inside the convex hull $conv(S)$, and there is nothing to update: the algorithm may proceed to the next insertion. If a region corresponding to a $(d-2)$-face of $conv(S)$ is found to conflict with P, however, then at least one of the two incident $(d-1)$-faces is red with respect to P. Starting from this red face, the incidence graph of $conv(S)$ can be updated into that of $conv(S \cup \{P\})$ by executing phases 2, 3, and 4 of the incremental algorithm described above in section 8.4.

Its remains to show how to update the influence graph. Let us recall that the nodes of the influence graph are in bijection with the $(d-2)$-faces of the successive convex hulls, and that the corresponding regions are determined by a pair of adjacent facets, or also by the $d+1$ vertices that belong to these facets. To update the influence graph, the algorithm considers in turn each of the purple $(d-2)$-faces of $conv(S)$, and each of the $(d-3)$-faces incident to these faces.

1. Consider a $(d-2)$-face G_1 of $conv(S)$ that is purple with respect to P, and let (F_1, F_1') be the corresponding region; F_1 and F_1' are two $(d-1)$-faces of $conv(S)$ that are incident to G_1. We may assume that F_1 is blue with respect to P and F_1' is red (see figure 8.8). The face G_1 is a $(d-2)$-face of $conv(S \cup \{P\})$ that corresponds to the new region (F_1, F_1''), where F_1'' is the convex hull $conv(G_1 \cup \{P\})$. A new node of the influence graph is created for region (F_1, F_1'') and this node is hooked into the influence graph as the child of (F_1, F_1'). In this way, the inclusion property is satisfied. Indeed, let H_1 and H_1' be the hyperplanes supporting $conv(S)$ along F_1 and F_1', respectively. The hyperplane H_1'' supporting $conv(S \cup \{P\})$ along F_1'' is also a hyperplane supporting $conv(S)$ along G_1. As a consequence, the half-space $H_1''^-$ that does not contain $conv(S \cup \{P\})$ is contained in the union of the half-spaces H_1^- and $H_1'^-$, which do not contain $conv(S)$. The influence domain of region (F_1, F_1'') is therefore contained within that of (F_1, F_1').

2. Let K be a $(d-3)$-face of $conv(S)$, purple with respect to P, and let G_1 and G_2 be the purple $(d-2)$-faces of $conv(S)$ that are incident to K.[5] Let (F_1, F_1') and (F_2, F_2') be the two regions corresponding to G_1 and G_2, the faces F_1 and F_2

---

[5] The set of purple faces of $conv(S)$ being isomorphic to a $(d-1)$-polytope (lemma 8.3.4), any purple $(d-3)$-face of $conv(S)$ is incident to exactly two purple $(d-2)$-faces (theorem 7.1.7).

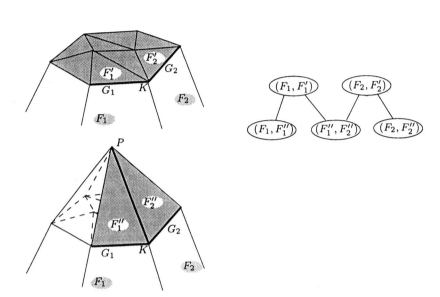

Figure 8.8. On-line convex hull: new regions when inserting a point P.

being blue with respect to P while faces F_1' and F_2' are red (see figure 8.8). The convex hull $conv(K \cup \{P\})$ is a $(d-2)$-face of $conv(\mathcal{S} \cup \{P\})$, and is incident to the $(d-1)$-faces $F_1'' = conv(G_1 \cup \{P\})$ and $F_2'' = conv(G_2 \cup \{P\}))$. In the influence graph, a new node is created for the region (F_1'', F_2''), and hooked into the graph to two parents which are the nodes corresponding to regions (F_1, F_1') and (F_2, F_2'). Let us verify that the inclusion property is satisfied. Indeed, the influence domain of (F_1'', F_2'') is the union $H_1''^- \cup H_2''^-$, where $H_1''^-$ (resp. $H_2''^-$) is the half-space bounded by hyperplane H_1'' (resp. H_2'') that supports $conv(\mathcal{S} \cup \{P\})$ along F_1'' (resp. F_2'') and does not contain $conv(\mathcal{S} \cup \{P\})$. The half-space $H_1''^-$ is contained in the the influence domain of region (F_1, F_1'), and similarly $H_2''^-$ is contained in the influence domain of (F_2, F_2'). Consequently, the influence domain of (F_1'', F_2'') is contained in the union of the influence domains of (F_1, F_1') and (F_2, F_2').

This description can be carried over almost *verbatim* to the case of dimension 2. We need only remember that the polytope $conv(\mathcal{S})$ has an empty face of dimension -1, incident to all of its vertices. If P is not contained within $conv(\mathcal{S})$, the empty face is purple and incident to the two purple vertices of $conv(\mathcal{S})$ (see also figure 8.9).

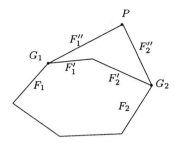

Figure 8.9. On-line convex hull in two dimensions.

Randomized analysis of the algorithm

In this randomized analysis, we assume that the points are inserted in a random order. The performances of the algorithm are then estimated on the average, assuming that all $n!$ permutations are equally likely.

To apply the results in chapter 5, we must verify that the algorithm satisfies the update condition 5.3.3 for algorithms that use an influence graph.

1. Testing conflict between a point and a region boils down to testing whether a point belongs to two half-spaces, and can be performed in constant time.

2. The number of children of each node in the influence graph is bounded. In fact, each node has d children, or none. Indeed, when inserting a point P, the node corresponding to a purple $(d-2)$-face G of $conv(S)$ receives d children: one for the $(d-2)$ face G of $conv(S \cup \{P\})$, and $d-1$ corresponding to $conv(K \cup \{P\})$ for each $(d-3)$-subface K of G. The nodes corresponding to red or blue $(d-2)$-faces of $conv(S)$ do not receive children. The nodes corresponding to the red or purple $(d-2)$-faces of $conv(S)$ are killed by P: they no longer correspond to regions without conflict and will not receive children after the insertion of P.

3. The parents of a region created by a point P are recruited among the regions killed by P. From the analysis of phases 2, 3, and 4 of the incremental step in section 8.4, we can deduce that updating the incidence graph takes time proportional to the total number of red and purple faces of $conv(S)$ and of their incidences. If every $(d-2)$-face of the convex hull is linked by a bidirectional pointer with the corresponding node in the influence graph, it is easy to see that updating the influence graph takes about the same time as updating the incidence graph. The set of points being in general position, the facets of $conv(S)$ are simplices; thus the number of red or purple faces and of their incidences is proportional to the number of red facets of $conv(S)$. Each of these red facets is incident to $d-1$ red or purple

$(d-2)$-faces of $conv(S)$, each of which corresponds to a region that conflicts with P. Each region defined and without conflict over S that conflicts with $\mathcal{P}$ corresponds to a $(d-2)$-face of $conv(S)$ that is incident to one or two red facets. As a result, the number of red facets of $conv(S)$, and therefore the complexity of the update phase, is proportional to the number of regions killed by the new point P.

Since the update conditions are satisfied, the randomized analysis of the on-line convex hull computation can now be established readily by theorem 5.3.4 which analyzes algorithms that use an influence graph. The number of regions without conflict defined over a set S of n points in a d-dimensional space is exactly the number of $(d-2)$-faces of the convex hull $conv(S)$, which is $O(n^{\lfloor d/2 \rfloor})$ according to the upper bound theorem 7.2.5.

Theorem 8.5.1 *An on-line algorithm that uses the influence graph method to build the convex hull of n points in d dimensions requires expected time $O(n \log n + n^{\lfloor d/2 \rfloor})$, and storage $O(n^{\lfloor d/2 \rfloor})$. The expected time required to perform the n-th insertion is $O(\log n + n^{\lfloor d/2 \rfloor - 1})$.*

8.6   Dynamic convex hulls

The previous section shows that it is possible to build on-line the convex hull of a set of points in optimal expected time and storage, using an influence graph. Such an algorithm is called *semi-dynamic*, since it can handle insertions of new points. Fully dynamic algorithms, however, handle not only insertions but also deletions.

The possibility of deleting points makes the task of maintaining the convex hull much more complex. Indeed, during an insertion, the current convex hull and the new point entirely determine the new convex hull. After a deletion, however, points that were hidden inside the convex hull may appear as vertices of the new convex hull. A fully dynamic algorithm must keep, in one way or another, some information for all the points in the current set, be they vertices of the current convex hull or not.

The goal of this section is to show that the augmented influence graph method described in chapter 6 allows the convex hull to be maintained dynamically.

The algorithm which we now present uses again the notions of objects, regions, and conflicts as defined in the preceding section. It conforms to the general scheme of dynamic algorithms described in chapter 6, to which the reader is referred should the need arise. Besides the current convex hull (described by the incidence graph of its faces), the algorithm maintains an augmented influence graph whose nodes correspond to regions defined over the current set. After each deletion, the structure is rebuilt into the exact state it would have been in, had the deleted point never been inserted. Consequently, the augmented influence graph only depends on the sequence $\Sigma = \{P_1, P_2, \ldots, P_n\}$ of points in the current set, sorted by chronological order: P_i occurs before P_j if the last insertion of P_i occurred before the last insertion of P_j.

Let us denote by $\mathcal{I}a(\Sigma)$ the augmented influence graph obtained for the chronological sequence Σ. The nodes and arcs of $\mathcal{I}a(\Sigma)$ are exactly the same as those of the influence graph built by the incremental algorithm of the preceding section, when the objects are inserted in the order given by Σ. We denote by $\mathcal{S}_l$ the subset of $\mathcal{S}$ formed by the first l objects in Σ. The nodes of $\mathcal{I}a(\Sigma)$ correspond to the regions defined and without conflict over the subsets $\mathcal{S}_l$, for $l = 1, \ldots, n$. The arcs of $\mathcal{I}a(\Sigma)$ ensure both *inclusion properties*: that the domain of influence of a node is contained in the union of the domains of influence of its parents, and that a determinant of this node is either the creator of this node or is contained in the union of the sets of determinants of its parents. Moreover, the augmented influence graph contains a conflict graph between the regions that correspond to nodes in the influence graph, and the objects in $\mathcal{S}$. This conflict graph is implemented by a system of interconnected lists such as that described in section 6.2: each node of the conflict graph has a list (sorted in chronological order) of the objects that conflict with the corresponding region; also, for each object we maintain a list of pointers to the nodes in the influence graph that conflict with that object. The record corresponding to an object in the conflict list of a node is interconnected with the record corresponding to that node in the conflict list of the object.

Insertion

Inserting the n-th point into the convex hull is carried out exactly as in the on-line algorithm described in section 8.5, except that while we are locating the object in the influence graph, each detected conflict is added to the interconnected conflict lists.

Deletion

Let us now consider the deletion of point P_k. For $l = k, \ldots, n$, we denote by $\mathcal{S}'_l$ the subset $\mathcal{S}_l \setminus \{P_k\}$ of $\mathcal{S}$, and by Σ' the chronological sequence $\{P_1, \ldots, P_{k-1}, P_{k+1}, \ldots P_n\}$. When deleting P_k, the algorithm rebuilds the augmented influence graph, resulting in $\mathcal{I}a(\Sigma')$. For this, we must:

1. remove from the graph $\mathcal{I}a(\Sigma)$ the *destroyed* nodes, which correspond to regions having P_k as a determinant,[6]

---

[6]Recall that an object is a determinant of a region if it belongs to the set of objects that determine this region.

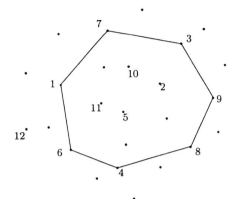

Figure 8.10. Convex hull: creator and killer of a region.
Points are numbered by chronological ranks. Unnumbered points have rank
greater than 12. Region $(6\,1,6\,4)$ has point 6 as its creator and point 12 as its
killer.

2. create a *new* node for each region defined and without conflict over one of
   the subsets $\mathcal{S}'_l$, $l = k+1, \ldots, n$ that conflicts with P_k,

3. set up the new arcs that are incident to the new nodes. The new nodes must
   be hooked to their parents which may or may not be new. The unhooked
   nodes, which are nodes of $\mathcal{I}a(\Sigma)$ that are not destroyed but have destroyed
   parents, must be rehooked.

Before we describe the deletion algorithm, it is useful to recall a few definitions.
A region G of $\mathcal{I}a(\Sigma)$ is *created* by P_l or also P_l is the *creator* of G, if P_l is among
all determinants of G the one with highest chronological rank. A region G of
$\mathcal{I}a(\Sigma)$ is *killed* by P_l or also P_l is the *killer* of G if P_l has the lowest rank among
the points that conflict with G (see figure 8.10).

The deletion algorithm proceeds in two substeps: the location phase and the
rebuilding phase.

Locating. During this phase, the algorithm identifies the nodes in $\mathcal{I}a(\Sigma)$ that
are killed by P_k, and the destroyed and unhooked nodes. For this, the algorithm
recursively visits all the nodes that conflict with P_k or have P_k as a determinant,
starting at the root. During the traversal, the algorithm removes P_k from the
conflict lists, and builds a dictionary of the destroyed or unhooked nodes for use
during the rebuilding phase.

Rebuilding. During this phase, the algorithm creates the new nodes, hooks
them to the graph, builds their conflict lists and rehooks the unhooked nodes.

For this, the algorithm considers in turn all the objects P_l of rank $l > k$ that
are the creators of some new or unhooked node. A point P_l of rank $l > k$ is the

creator of some new or unhooked node if and only if there exists a region defined and without conflict over $\mathcal{S}'_{l-1}$ which conflicts with both P_l and P_k (lemma 6.2.1). When processing P_l, we call a region *critical* if it is defined and without conflict over $\mathcal{S}'_{l-1}$ but conflicts with P_k. The *critical zone* is the set of all critical regions. The critical zone evolves as we consider the objects P_l in turn. At the beginning of the rebuilding phase, the critical regions are the regions of $\mathcal{I}a(\Sigma)$ that are killed by P_k. Subsequently, the critical regions are either regions in $\mathcal{I}a(\Sigma)$ that are killed by P_k, or new regions in $\mathcal{I}a(\Sigma')$. At each substep in the rebuilding phase, the next point to be processed is the point of smallest rank among all the points that conflict with one or more of the currently critical regions. To find this point, the algorithm maintains a priority queue $\mathcal{Q}$ of the points in Σ' that are the killers of critical regions. Each point P_l in $\mathcal{Q}$ also stores the list of the current critical regions that it kills. The priority queue $\mathcal{Q}$ is initialized with the killers in Σ' of the regions in $\mathcal{I}a(\Sigma)$ that were killed by P_k.

At each substep in the rebuilding phase, the algorithm extracts the point P_l of smallest rank in $\mathcal{Q}$, and this point is then *reinserted* into the data structure. To reinsert a point means to create new nodes for the new regions created by P_l, to hook them to the influence graph, and to rehook the unhooked nodes created by P_l. The $(d-2)$-faces of $conv(\mathcal{S}'_{l-1})$ that are red or purple with respect to the point P_k that is removed correspond to critical regions and are, below, called *critical faces*. Unless explicitly stated, the color *blue*, *red*, or *purple*, is now given with respect to the point P_l that is being reinserted. The regions that are unhooked or new and created by P_l can be derived from the critical purple $(d-2)$-faces and their $(d-3)$-subfaces, which will be considered in turn by the algorithm.

1. Processing the critical purple $(d-2)$-faces

Along with point P_l, we know the list of critical regions with which it conflicts. These regions correspond to the critical red or purple $(d-2)$-faces, and a linear traversal of this list allows the sublist of its critical purple $(d-2)$-faces to be extracted.

Let G be a critical purple $(d-2)$-face, and (F, F') be the corresponding region; F and F' are $(d-1)$-faces of $conv(\mathcal{S}'_{l-1})$, both incident to G, and we may assume that F is blue with respect to P_l while F' is red (see figure 8.11 in dimension 3 and figure 8.12 in dimension 2.)

In the convex hull $conv(\mathcal{S}'_l)$, G is a $(d-2)$-face that corresponds to (F, F''), a region defined and without conflict over $\mathcal{S}'_l$, where F'' is the convex hull $conv(G \cup \{P_l\})$ (see figure 8.11 in dimension 3 and figure 8.12 in dimension 2.)

If region (F, F'') conflicts with P_k (see figures 8.11a and 8.12a), then it is a new region created by P_l. In the augmented influence graph, a new node is created for this region, with node (F, F') as parent. The conflict list of (F, F'') can be

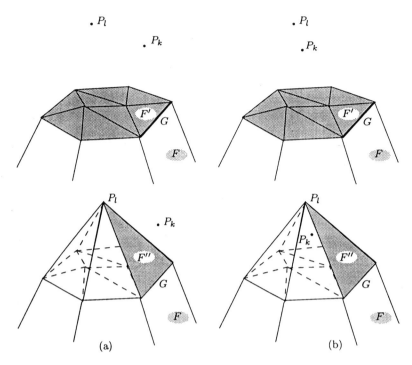

Figure 8.11. Deleting from a 3-dimensional convex hull: handling critical purple $(d-2)$-faces.
(a) (F, F'') is a new region.
(b) (F, F'') is an unhooked region.

set up by selecting the objects in conflict with (F, F'') from the conflict list of (F, F'). The killer of (F, F'') in Σ' is inserted in the priority queue $\mathcal{Q}$ if it was not found there. Finally, region (F, F'') is added to the list of critical regions killed by this point.

If region (F, F'') does not conflict with P_k (see figures 8.11b and 8.12b), then it corresponds to an unhooked node created by P_l. This node is found by using the dictionary $\mathcal{D}$ of destroyed and unhooked nodes, and hooked as a child of (F, F').

2. Handling the critical purple $(d-3)$-faces

Critical purple $(d-3)$-faces are subfaces of critical purple $(d-2)$-faces.[7] For each such $(d-3)$-face, we must know the at most two critical purple $(d-2)$-faces incident to it. To find them, we build an auxiliary dictionary $\mathcal{D}'$ of the $(d-3)$-subfaces of critical purple $(d-2)$-faces. Each entry in the dictionary $\mathcal{D}'$ for a

---

[7]According to lemma 8.3.4, each critical purple $(d-3)$-face is incident to two purple $(d-2)$-faces, at least one of which is critical.

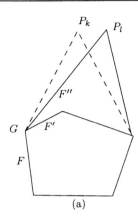 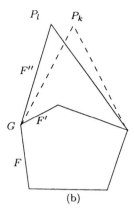

(a) (b)

Figure 8.12. Deleting from a 2-dimensional convex hull: handling critical purple $(d-2)$-faces.
(a) (F, F'') is a new region.
(b) (F, F'') is an unhooked region.

$(d-3)$-face K has two pointers for keeping track of the critical purple $(d-2)$-faces incident to K.

Let K be such a $(d-3)$-face (see figure 8.13 in dimension 3 and figure 8.14 in dimension 2). We denote by G_1 and G_2 the two purple $(d-2)$-faces incident to K. At least one of them is a critical face, but not always both. We denote by (F_1, F_1') and (F_2, F_2') the regions corresponding to faces G_1 and G_2 of the convex hull $conv(\mathcal{S}_{l-1}')$. We may assume that facets F_1 and F_2 are blue, while F_1' and F_2' are red.

The $(d-2)$-face $conv(K \cup \{P_l\})$ of $conv(\mathcal{S}_l')$ corresponds to some region (F_1'', F_2''), where $F_1'' = conv(G_1 \cup \{P_l\})$ and $F_2'' = conv(G_2 \cup \{P_l\})$ (see figure 8.13; see also figure 8.14, in dimension 2, in which K is the empty face of dimension -1, and G_1 and G_2 are the two vertices of $conv(\mathcal{S}_{l-1}')$, both purple with respect to P_l).

2.a If both G_1 and G_2 are critical faces, the corresponding nodes in $\mathcal{I}a(\Sigma')$ may be retrieved through dictionary $\mathcal{D}'$.

2.a.1 If region (F_1'', F_2'') conflicts with P_k (see figure 8.14a), it is a new region created by P_l; a node is created for this region, and inserted into the influence graph with both (F_1, F_1') and (F_2, F_2') as parents. The conflict list of (F_1'', F_2'') may be obtained by merging the conflict lists of (F_1, F_1') and (F_2, F_2'), and then selecting from the resulting list the objects that conflict with (F_1'', F_2''). Merging the conflict lists can be carried out in time proportional to the total length, because these lists are ordered chronologically.[8] The killer of (F_1'', F_2'') in the sequence Σ' is inserted into the priority queue $\mathcal{Q}$ if not found there, and region (F_1'', F_2'') is added to the list of critical regions killed by this point.

---

[8]An alternative to this solution is to forget about ordering the conflict lists and to resort to

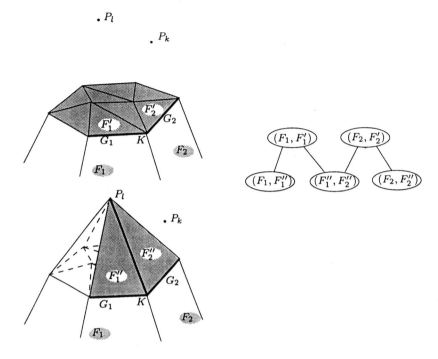

Figure 8.13. Deleting from a 3-dimensional convex hull: handling critical purple $(d-3)$-faces.

2.a.2 If region (F_1'', F_2'') does not conflict with P_k (see figure 8.14b), then this region is an unhooked region created by P_l. It suffices to find the corresponding node using dictionary $\mathcal{D}$ and hook it back to the nodes corresponding to (F_1, F_1') and (F_2, F_2').

2.b When only one of the purple $(d-2)$-faces G_1 and G_2 incident to K is critical, say G_1, the algorithm must find in the influence graph the node corresponding to G_2, the other purple $(d-2)$-face incident to K. Lemma 8.6.1 below proves that, in this case, $conv(K, P_l)$ is a $(d-2)$-face of $conv(\mathcal{S}_l)$ which corresponds to a destroyed or unhooked node of $\mathcal{I}a(\Sigma)$, whose parents include precisely the node corresponding to region (F_2, F_2'). To find (F_2, F_2'), we may therefore search in the dictionary $\mathcal{D}$ of destroyed or unhooked nodes, created by P_l, corresponding to the $(d-2)$-face $conv(K, P_l)$ of $conv(\mathcal{S}_l)$. This node is uniquely known from this criterion, because we know not only the $(d-2)$-face $conv(K, P_l)$ of its corresponding region, but also its creator P_l.

Lemma 8.6.1 *Let K be a $(d-3)$-face of $conv(\mathcal{S}_{l-1}')$ incident to two purple faces G_1 and G_2, only one of which is critical, say G_1. Then $conv(K, P_l)$ is a*

---

the method used in section 6.4 for merging the conflicts lists of trapezoids.

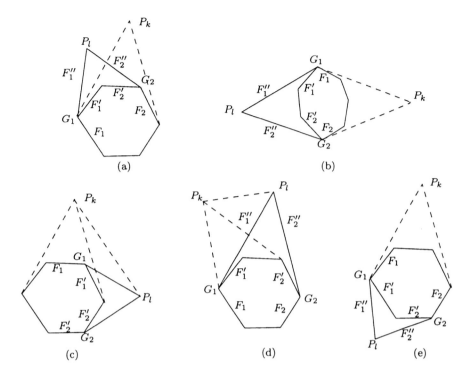

Figure 8.14. Deleting from a 2-dimensional convex hull: handling the critical purple $(d-3)$-faces. Critical purple $(d-3)$-face K here is the empty face of dimension -1. G_1 and G_2 are its two purple vertices.
(a) G_1 and G_2 are critical, (F_1'', F_2'') is new.
(b) G_1 and G_2 are critical, (F_1'', F_2'') is unhooked.
(c) G_1 is critical, G_2 is not, and G_1 is not a face of $conv(\mathcal{S}_{l-1})$.
(d) G_1 is critical, G_2 is not, and G_1 is a face of $conv(\mathcal{S}_{l-1})$, but not purple with respect to P_l.
(e) G_1 is critical, G_2 is not, and G_1 is a face of $conv(\mathcal{S}_{l-1})$, this time purple with respect to P_l.

$(d-2)$-face of $conv(\mathcal{S}_l)$, its corresponding node in $\mathcal{I}a(\Sigma)$ is destroyed or unhooked, and one of its parents is the region (F_2, F_2') that corresponds to the face G_2 of $conv(\mathcal{S}_{l-1}')$.

Proof. For the proof, imagine that P_k then P_l are inserted into $\mathcal{S}_{l-1}'$: then we obtain successively $\mathcal{S}_{l-1}$ and $\mathcal{S}_l$.

The $(d-2)$-face K of $conv(\mathcal{S}_{l-1}')$ is purple with respect to P_k since it belongs to a critical $(d-2)$-face as well as to a non-critical $(d-2)$-face. As a result, both K and $conv(K, P_k)$ are faces of $conv(\mathcal{S}_{l-1})$.

Since it is not critical, the $(d-2)$-face G_2 is also a $(d-2)$-face of $conv(\mathcal{S}_{l-1})$, and its corresponding region is still (F_2, F_2'), hence face G_2 of $conv(\mathcal{S}_{l-1})$ is purple with respect to P_l.

The $(d-3)$-face K of $conv(\mathcal{S}_{l-1})$ is purple with respect to P_l since it is incident to G_2. As a result, $conv(K, P_l)$ is a $(d-2)$-face of $conv(\mathcal{S}_l)$, incident to the $(d-1)$-face $conv(G_2, P_l)$. In the graph $\mathcal{I}a(\Sigma)$, one of the parents of the node corresponding to the $(d-2)$-face $conv(K, P_l)$ of $conv(\mathcal{S}_l)$, is region (F_2, F_2').

We now have to show that the $(d-2)$-face $conv(K, P_l)$ of $conv(\mathcal{S}_l)$ corresponds to a region that is either destroyed or unhooked when P_k is deleted. For this, we consider the face G_1 of $conv(\mathcal{S}'_{l-1})$. The situation is one of three (see figure 8.14c, d, e). If this face is red with respect to P_k (see figure 8.14c), then it is not a face of $conv(\mathcal{S}_{l-1})$ any more. If this face is purple with respect to P_k, it remains a face of $conv(\mathcal{S}_{l-1})$, but it may be blue (see figure 8.14d) or remain purple (see figure 8.14e) with respect to P_l.

In the first two cases, $conv(K, P_k)$ is necessarily a $(d-2)$-face of $conv(\mathcal{S}_{l-1})$, purple with respect to P_l. Indeed, the set of those purple faces is isomorphic to a $(d-1)$-polytope, and since G_1 is not purple with respect to P_l, it must be replaced by another $(d-2)$-face incident to K which can only be $conv(K, P_k)$. Consequently, the region corresponding to the $(d-2)$-face $conv(K, P_l)$ of $conv(\mathcal{S}_l)$ is region $(conv(K, P_l, P_k), F_2'')$ which is destroyed during the deletion of P_k.

In the third case, $F_1'' = conv(G_1, P_l)$ must be a facet of $conv(\mathcal{S}_l)$, and the region that corresponds to the $(d-2)$-face $conv(K, P_l)$ of $conv(\mathcal{S}_l)$ is region (F_1'', F_2''), which is an unhooked region created by P_l.   □

Once the node corresponding to the $(d-2)$-face G_2 of $conv(\mathcal{S}'_l)$ has been found, operations can resume as before, apart from a simple detail. If the region (F_1'', F_2'') that corresponds to the $(d-2)$-face $conv(K, P_l)$ of $conv(\mathcal{S}'_l)$ is new, then its conflict list may be obtained by merging that of the critical region (F_1, F_1') and that of the destroyed region $(conv(K, P_l, P_k), F_2'')$. (We do this in order to avoid traversing the conflict list of region (F_2, F_2') corresponding to face G_2, which is neither new nor destroyed.)

Randomized analysis of the algorithm

The algorithm is deterministic. Yet the analysis given here is randomized and assumes the following probabilistic model:

- the chronological sequence Σ is a random sequence, each of the $n!$ permutations being equally likely;

- each insertion concerns, with equal probability, any of the objects present in the current set immediately after the insertion;

- each deletion concerns, with equal probability, any of the objects present in the current set immediately before the deletion.

Theorem 8.6.2 *Using an augmented influence graph allows the fully dynamic maintenance of the convex hull of points in $\mathbb{E}^d$, under insertion or deletion of points. If the current set has n points:*

- *the structure requires expected storage $O(n \log n + n^{\lfloor d/2 \rfloor})$,*

- *inserting a point takes expected time $O(\log n + n^{\lfloor d/2 \rfloor - 1})$,*

- *deleting a point takes expected time $O(\log n)$ in dimension 2 or 3 and time $O(tn^{\lfloor d/2 \rfloor - 1})$ in dimension $d > 3$. The parameter t represents the complexity of an operation on the dictionaries used by the algorithm ($t = O(\log n)$ if balanced binary trees are used, $t = O(1)$ if perfect dynamic hashing is used.)*

Proof. During the rebuilding phase in a deletion, the number of queries into the dictionary of destroyed or unhooked nodes is at most proportional to the number of destroyed or unhooked nodes. For each point P_l that is reinserted, the number of updates or queries on the dictionary of $(d-3)$-faces incident to critical purple $(d-2)$-faces is proportional to the number of these critical purple $(d-3)$-faces. Thus, the total number of accesses to the dictionaries is proportional to the total number of critical faces encountered that correspond to new or killed nodes. The conflict lists of new nodes can be set up in time at most proportional to the total sizes of the conflict lists of new or killed nodes. All the other operations performed during a deletion, except handling the priority queue, take constant time, and their number is proportional to the number of destroyed, new, or unhooked nodes.

As a result, the algorithm indeed satisfies the update condition 6.3.5 for algorithms that use an augmented conflict graph. Its randomized analysis is therefore the same as in section 6.3, and is given in theorem 6.3.6 in terms of $f_0(l, S)$, the expected number of regions defined and without conflict over a random l-sample of S. For the case of convex hulls, since the number of such regions for any sample is bounded in the worst case by $O(l^{\lfloor d/2 \rfloor})$ (upper bound theorem 7.2.5), so is their expectation $f_0(l, S)$. This results in the performance given in the statement of theorem 6.3.6. In dimension 2 or 3, the number of operations to be performed on the dictionaries and on the priority queue is $O(1)$ whereas handling the conflict lists always takes $O(\log n)$ time. Therefore, it suffices to implement dictionaries and priority queues with balanced binary trees. In dimensions higher than 3, deletions have supra-linear complexity, and the priority queue may be implemented using a simple array. $\square$

8.7 Exercises

Exercise 8.1 (Extreme points) *Extreme points in a set of points are those which are vertices of the convex hull. Show that to determine the extreme points of n points in $\mathbb{E}^2$ is a problem of complexity $\Theta(n \log n)$.*

Hint: You may use the notion of an algebraic decision tree: an algebraic tree of degree a is a decision tree where the test at any node evaluates the sign of some algebraic function of degree a for the inputs. Loosely stated, a result by Ben-Or (see also subsection 1.2.2) says that any algebraic decision tree that decides whether a point in $\mathbb{E}^k$ belongs to some connected component W of $\mathbb{E}^k$ must have a height $h = \Omega(\log c(W) - k)$, where $c(W)$ is the number of connected components of W.

Exercise 8.2 (Adjacency graph) Let a simplicial d-polytope be defined as the convex hull of n points. Show that knowledge of the facets of the graph (given by their vertices), along with their adjacencies, suffices to reconstruct the whole incidence graph of the polytope in time linear in the size of the adjacency graph, which is $O(n^{\lfloor d/2 \rfloor})$.

Exercise 8.3 (1-skeleton) This problem is the dual version of its predecessor. Let a simple d-polytope be defined as the intersection of n half-spaces. Suppose that the 1-skeleton is known, that is the set of its vertices and the arcs joining them. Each vertex is given as the intersection of d bounding hyperplanes. Show that the whole incidence graph of the polytope may be reconstructed in time $O(n^{\lfloor d/2 \rfloor})$.

Exercise 8.4 (Degenerate cases) Generalize the incremental algorithm described in section 8.4 to build the convex hull of a set of points which is not assumed to be in general position.

Exercise 8.5 (On-line convex hulls) Give an algorithm to compute on-line the convex hull of a set of points in $\mathbb{E}^d$, by using an influence graph whose nodes correspond to regions which are half-spaces. Give the randomized analysis of this algorithm.

Hint: Each region, or half-space, is now determined by a subset of d affinely independent points that generates its bounding hyperplane. A point conflicts with a half-space if it lies inside. The regions defined and without conflict over a set S are in bijection with the facets, or $(d-1)$-faces, of the convex hull $conv(S)$ of S.

Upon inserting a point P into S, the regions killed by P correspond to the facets of $conv(S)$ that are red with respect to P, and the regions created by P correspond to the facets $conv(G \cup P)$ of $conv(S \cup P)$ where G is any $(d-2)$-face of $conv(S)$ that is purple with respect to P.

1. Let F_1 and F_2 be the facets of $conv(S)$ incident to a $(d-2)$-face G, which is purple with respect to P. Show that the node of the influence graph that corresponds to the facet $conv(G \cup P)$ must have both nodes corresponding to F_1 and F_2 as parents.

2. In this manner, a node in the graph may receive a child without being killed, therefore the number of children of a node is not bounded any more. The maximum number of parents is two, however. For this particular problem, define and use the notion of a biregion that was introduced in exercise 5.7, and show that the expected complexity of the algorithm is $O(n \log n + n^{\lfloor \frac{d}{2} \rfloor})$.

Exercise 8.6 (Intersection of half-spaces) Give a randomized incremental algorithm that uses a conflict graph to build the intersection of n half-spaces in $\mathbb{E}^d$ whose bounding hyperplanes are in general position. Try to achieve an expected running time of

$O(n \log n + n^{\lfloor \frac{d}{2} \rfloor})$. Give an on-line version of the preceding algorithm that uses an influence graph.

Show that in the version of the algorithm that uses a conflict graph, the storage requirements may be lowered if only one conflict is stored for each half-space.

Hint: Objects are half-spaces, regions are segments. A segment is determined by $d + 1$ half-spaces, or rather by the $d + 1$ hyperplanes which bound these half-spaces. The line that supports the segment is the intersection of $d-1$ hyperplanes, and the endpoints of the segment are the intersections of this line with the two remaining hyperplanes. A segment conflicts with a half-space if it has an intersection with the (open) complementary half-space. The segments defined and without conflict over these half-spaces are precisely the edges of the polytope, obtained as the intersection of the half-spaces.

When a new half-space $\overline{H^+}$ bounded by a hyperplane H is inserted, the conflict graph identifies all the edges that lie in H^-, which disappear, and those that intersect H. An edge E that intersects H gives a shorter edge $E' \subset E$, and the conflict list of E' is set up by traversing that of E. To obtain the new edges that lie in H, it suffices to follow, for each 2-face F incident to each edge E that intersects H, the edges of F that conflict with $\overline{H^+}$ until the second edge E' of F that intersects H is found. The new edge $F \cap H$ has vertices $E \cap H$ and $E' \cap H$. Its conflict list can be obtained by traversing the conflict lists of the edges of F killed by $\overline{H^+}$. Knowing the 1-skeleton, the whole incidence graph of the intersection may be updated.

8.8 Bibliographical notes

The incremental deterministic convex hull algorithm described in section 8.4 is due to Seidel [201]. This algorithm is also described in detail in Edelsbrunner's book [89] where degenerate cases are also handled (see exercise 8.4).

The first randomized algorithm to build convex hulls was due to Clarkson and Shor [71]. This algorithm uses a conflict graph and in fact solves the dual problem of computing the intersection of half-spaces (see exercise 8.6). The on-line algorithm that uses an influence graph is due to Boissonnat, Devillers, Teillaud, Schott and Yvinec [28]. The dynamic algorithm presented in section 8.6 is due to Dobrindt and Yvinec [86]. Clarkson, Mehlhorn, and Seidel [70] and independently Mulmuley [176, 177] proposed similar solutions for dynamically maintaining convex hulls.

Chazelle [46] proposed a deterministic algorithm that is optimal in any dimension greater than 3. This algorithm is a *derandomized* incremental algorithm, and uses the method of conditional probabilities to determine which point must be inserted next. Brönnimann, Chazelle, and Matoušek [36] and Brönnimann [35] give a simpler version which works in any dimension.

The lower bound proposed in exercise 8.1 to identify the extreme points in a set of points in the plane is due to Yao [220]. The solution to exercise 8.1 can be found in the book by Preparata and Shamos [192].

Chapter 9

Convex hulls
in two and three dimensions

There are many algorithms that compute the convex hull of a set of points in two and three dimensions, and the present chapter does not claim to give a comprehensive survey. In fact, our goal is mainly to explore the possibilities offered by the divide-and-conquer method in two and three dimensions, and to expand on the incremental method in the case of a planar polygonal line.

In dimension 2, the divide-and-conquer method leads, like many other methods, to a convex hull algorithm that is optimal in the worst case. The main advantage of this method is that it also generalizes to three dimensions while still leading to an algorithm that is optimal in the worst case, which is not the case for the incremental method described in chapter 8. The performances of this divide-and-conquer algorithm rely on the existence of a circular order on the edges incident to a given vertex. In dimensions higher than three, such an order does not exist, and the divide-and-conquer method is no longer efficient for computing convex hulls. The 2-dimensional divide-and-conquer algorithm is described in section 9.2, and generalized to dimension 3 in section 9.3. But before these descriptions, we must comment on the representation of polytopes in dimensions 2 and 3, and describe a data structure that explicitly provides the circular order of the edges or facets around a vertex of a 3-dimensional polytope.

The problem of computing the convex hull of a polygonal line is interesting from the point of view of its complexity. Indeed, the lower bound of $\Omega(n \log n)$ on the complexity of computing the convex hull of n points does not hold if the points are assumed to be the vertices of a simple polygonal line. In fact, any simple polygonal line that links the points in a given set determines an order on those points which is not completely unrelated to the order of the vertices on the boundary of the convex hull. In section 9.4, we show how it is possible to compute in time $O(n)$ the convex hull of a set of n points given as the vertices

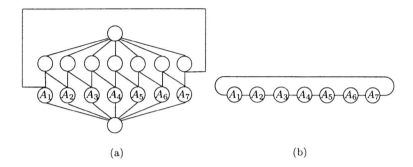

Figure 9.1. Representation of a 2-polytope: (a) the incidence graph, (b) the circular list of its vertices.

of a simple polygonal line, using an incremental algorithm that takes advantage of the order of the points along the polygonal line.

As in chapter 8, and to simplify the presentation of the algorithms, we assume that the sets of points to be processed are in general position. We leave it to the reader to work out how to modify the algorithms so that they can handle any sets of points without increasing the complexity.

9.1   Representation of 2- and 3-polytopes

Representation of 2-polytopes

The proper faces of a 2-polytope consist of its vertices and edges. Each edge is incident to two vertices and each vertex to two edges. In fact, the incidence graph of a 2-polytope is a cyclic graph that alternates vertices and edges (see figure 9.1a). Without losing information, a 2-polytope may be represented by the doubly-linked circular list of its vertices. Either direction in this list corresponds to an order on the boundary of the polytope. If the plane that contains the 2-polytope has an orientation, it induces an order on this boundary that is called the *direct* (or *counter-clockwise*) order of the vertices, and the reverse order is called the *indirect* (or *clockwise*) order of the vertices.

Representation of 3-polytopes

Generally speaking, any 2-face of a d-polytope is a 2-polytope and there is a circular ordering on the set of edges and vertices contained in a given 2-face. By duality, there is also a circular ordering on the set of $(d-1)$- and $(d-2)$-faces of a d-polytope containing a given $(d-3)$-face.

In particular, for a 3-polytope, there is circular order on the set of edges and

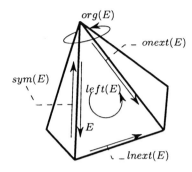

Figure 9.2. Representation of a 3-polytope.

vertices contained in any given facet, and also on the set of edges and facets containing any given vertex. Let us agree that supporting hyperplanes are oriented by the outward normal, pointing into the half-space that does not contain the polytope. This orientation induces a circular order on the edges and vertices contained in a facet, which we again call the *direct* (or *counter-clockwise*) order; the other orientation induces the *indirect* (or *clockwise*) order.

Cycles of edges of a 3-polytope, around a vertex or a facet, are not stored in the incidence graph of the polytope. These cycles are commonly used by algorithms that deal with 3-polytopes, however, and for this reason an alternative data structure is often preferred: the *edge-list* representation stores the order of the edges incident to a given vertex or to a given facet of the 3-polytope.

In this structure, vertices and facets are represented by a single node, whereas an edge is stored in a double node, one for each possible orientation of the edge. To orient an edge is to choose an order on its two vertices: the *origin* is the first vertex of the edge while the *end* is the last one. We can now make a distinction between the two facets incident to an oriented edge: the facet incident on the left, or *left incident facet*, is the one whose direct orientation traverses the edge from origin to end, and the *right incident facet* is the one whose indirect orientation traverses the edge from origin to end. In this data structure, each edge node stores five pointers displayed in figure 9.2:

$org(E)$ points towards the origin vertex of the oriented edge E,

$left(E)$ points towards the facet incident to E on the left,

$sym(E)$ points towards the node for the reverse edge. In this way, $org(sym(E))$
    points towards the end of E and $left(sym(E))$ towards the facet incident
    to E on the right,

$onext(E)$ points towards the edge E' that shares the same origin as E, and whose facet incident on the right is the same as the facet incident to E on the left:

$$org(E') \;\;=\;\; org(E)$$
$$left(sym(E')) \;\;=\;\; left(E),$$

$lnext(E)$ points towards the edge E'' that follows E in the circular order of edges on the boundary of $left(E)$:

$$left(E'') \;\;=\;\; left(E)$$
$$org(E'') \;\;=\;\; org(sym(E)).$$

Conversely, each facet keeps a pointer to one oriented edge that has the facet as its left incident facet. The entire edge cycle on the boundary of a facet may be obtained in direct (resp. indirect) order by repeated applications of the functor $lnext()$ (resp. $sym(onext())$). The time taken for this operation is constant per edge on the boundary.

Each vertex node also keeps a pointer to one of the edges originating at that vertex. We define the order of edges around a vertex as follows: All the edges originating at that vertex may be obtained in direct (resp. indirect) order by repeated applications of the functor $onext()$ (resp. $lnext(sym())$). Again, the time needed to enumerate these edges is constant per each edge.

9.2 Divide-and-conquer convex hulls in dimension 2

Building the convex hull of a set $\mathcal{A}$ of points using the divide-and-conquer method consists of dividing the set $\mathcal{A}$ into two subsets $\mathcal{A}_1$ and $\mathcal{A}_2$ of equal size, recursively computing the convex hull $conv(\mathcal{A}_1)$ and $conv(\mathcal{A}_2)$ of each set, and merging them into a single convex hull $conv(\mathcal{A}) = conv(conv(\mathcal{A}_1) \cup conv(\mathcal{A}_2))$.

The divide-and-conquer algorithm which we present here divides the set $\mathcal{A}$ into two sets $\mathcal{A}_1$ and $\mathcal{A}_2$ separated by a vertical line. To efficiently split the subsets in the recursive steps, the algorithm begins by sorting once and for all the set $\mathcal{A}$ in order of increasing abscissae. Again, we may assume that $\mathcal{A}$ is in general position.

The only subtlety of the algorithm lies in the method used to merge the convex hulls $conv(\mathcal{A}_1)$ and $conv(\mathcal{A}_2)$ into the convex hull $conv(\mathcal{A})$. We assume that both $conv(\mathcal{A}_1)$ and $conv(\mathcal{A}_2)$ are represented by the doubly linked circular lists of their vertices.

The edges of $conv(\mathcal{A}_1)$ and $conv(\mathcal{A}_2)$ can be split into two categories, which it is helpful to color again *red* or *blue* as follows:

- an edge of $conv(\mathcal{A}_1)$ is *red* with respect to $conv(\mathcal{A}_2)$ if it is not an edge of $conv(\mathcal{A})$,

- an edge of $conv(\mathcal{A}_1)$ is *blue* with respect to $conv(\mathcal{A}_2)$ if it is an edge of $conv(\mathcal{A})$.

The color of an edge of $conv(\mathcal{A}_2)$ with respect to $conv(\mathcal{A}_1)$ is defined symmetrically. Intuitively, the red edges of $conv(\mathcal{A}_1)$ are those that would be lit if $conv(\mathcal{A}_2)$ was an extended source of light. Blue edges would remain in the shadow of $conv(\mathcal{A}_1)$. Using the terminology of chapter 8, an edge E of $\mathcal{P}$ is red with respect to a point A if it is visible from A, meaning that A belongs to the half-plane H_E^- that is bounded by the line supporting $\mathcal{P}$ along E and that does not contain $\mathcal{P}$. Thus, an edge of $conv(\mathcal{A}_1)$ is blue with respect to $conv(\mathcal{A}_2)$ if it is blue with respect to all the vertices of $conv(\mathcal{A}_2)$, and red with respect to $conv(\mathcal{A}_2)$ if it is red with respect to at least one vertex of $conv(\mathcal{A}_2)$. The edges of $conv(\mathcal{A}_2)$ that are red or blue with respect to $conv(\mathcal{A}_1)$ are defined symmetrically. In the rest of this section, provided there is no ambiguity, we say *red* or *blue* for short instead of red or blue with respect to $conv(\mathcal{A}_1)$ or to $conv(\mathcal{A}_2)$.

The vertices of $conv(\mathcal{A}_1)$ can fall into one of the following three categories with respect to $conv(\mathcal{A}_2)$:

- a vertex of $conv(\mathcal{A}_1)$ is *red* with respect to $conv(\mathcal{A}_2)$ if it is not a vertex of $conv(\mathcal{A})$,

- a vertex of $conv(\mathcal{A}_1)$ is *blue* with respect to $conv(\mathcal{A}_2)$ if it is a vertex of $conv(\mathcal{A})$ that is not incident to a red edge,

- a vertex of $conv(\mathcal{A}_1)$ is *purple* with respect to $conv(\mathcal{A}_2)$ if it is a vertex of $conv(\mathcal{A})$ that is incident to at least one red edge.

The color of a vertex of $conv(\mathcal{A}_2)$ with respect to $conv(\mathcal{A}_1)$ is defined symmetrically. For convenience, we will say *red, blue,* or *purple* instead of red, blue, or purple with respect to $conv(\mathcal{A}_1)$ or to $conv(\mathcal{A}_2)$.

It follows from lemma 8.3.3 and section 8.4 that the set of edges of $conv(\mathcal{A}_1)$ that are red with respect to some vertex A of $conv(\mathcal{A}_2)$ is connected among the set of edges of $conv(\mathcal{A}_1)$ and contains at least one edge incident to the vertex of $conv(\mathcal{A}_1)$ that has maximal abscissa. As a consequence, the set of edges of $conv(\mathcal{A}_1)$ that are red with respect to $conv(\mathcal{A}_2)$ is connected and not empty, and the set of edges of $conv(\mathcal{A}_1)$ that are blue with respect to $conv(\mathcal{A}_2)$ is also connected. If there is such a blue edge, then $conv(\mathcal{A}_1)$ has exactly two purple vertices, each adjacent to a blue edge and to a red edge. If there is no blue edge on the boundary of $conv(\mathcal{A}_1)$, then $conv(\mathcal{A}_1)$ has only one purple vertex (see figure 9.3). Symmetrically, $conv(\mathcal{A}_2)$ also has one or two purple vertices.

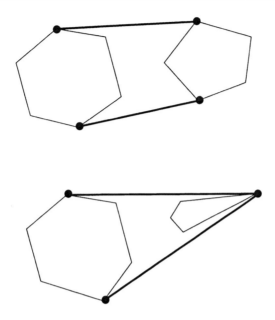

Figure 9.3. Blue, red, and purple faces.
Purple vertices are shown in bold.

Moreover, the total number of purple vertices on both convex hulls must be at least three.

The edges of $conv(\mathcal{A})$ that are neither edges of $conv(\mathcal{A}_1)$ nor of $conv(\mathcal{A}_2)$ must intersect the separating vertical line H_0, and there are exactly two such edges. They must connect one purple vertex of $conv(\mathcal{A}_1)$ to a purple vertex of $conv(\mathcal{A}_2)$: they are the exterior bitangents to polytopes $conv(\mathcal{A}_1)$ and $conv(\mathcal{A}_2)$. We call the *upper bitangent* the one that intersects the separating line H_0 above the other, which is called the *lower bitangent*. Much of the work in the merging process is to identify these two bitangents.

Let A_k be the vertex of $conv(\mathcal{A}_1)$ with the greatest abscissa, and A_{k+1} be the vertex of $conv(\mathcal{A}_2)$ with the smallest abscissa. The segment $A_k A_{k+1}$ lies outside both $conv(\mathcal{A}_1)$ and $conv(\mathcal{A}_2)$. Both vertices A_k and A_{k+1} are incident to a red edge. To find the upper bitangent to $conv(\mathcal{A}_1)$ and $conv(\mathcal{A}_2)$, the merging step moves a segment $U_1 U_2$ upwards from position $A_k A_{k+1}$, while staying outside $conv(\mathcal{A}_1)$ and $conv(\mathcal{A}_2)$. The left endpoint U_1 moves counter-clockwise on the boundary of $conv(\mathcal{A}_1)$, taking position at vertices of $conv(\mathcal{A}_1)$ that are incident to a red edge. Likewise, the right endpoint U_2 moves clockwise on the boundary of $conv(\mathcal{A}_2)$, taking position at vertices of $conv(\mathcal{A}_2)$ that are incident to a red edge. More precisely, let $succ(U_1)$ denote the successor of U_1 along the oriented boundary of $conv(\mathcal{A}_1)$, and $pred(U_2)$ the predecessor of U_2 along the

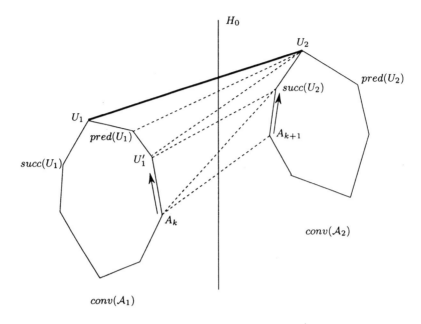

Figure 9.4. Divide-and-conquer convex hull in dimension 2.

oriented boundary of $conv(\mathcal{A}_2)$. A move of U_1 or U_2 is computed by the following procedure:

$$\textbf{while}\quad U_1 \in H^-_{pred(U_2)U_2} \textbf{ or } U_2 \in H^-_{U_1 succ(U_1)}$$
$$\textbf{if}\quad U_1 \in H^-_{pred(U_2)U_2} \textbf{ then } U_2 = pred(U_2)$$
$$\textbf{else } U_1 = succ(U_1)$$

In this manner, the endpoints U_1 and U_2 of segment U_1U_2 only traverse red edges of $conv(\mathcal{A}_1)$ and $conv(\mathcal{A}_2)$, and both U_1 and U_2 keep in contact with a red edge. It remains to show that, when the loop is exited, the line joining U_1 and U_2 is a line supporting both $conv(\mathcal{A}_1)$ and $conv(\mathcal{A}_2)$, and therefore is the desired upper bitangent. When U_1U_2 has reached its final position, the edge $U_2 pred(U_2)$ of $conv(\mathcal{A}_2)$ is blue with respect to U_1 and the edge $U_1 succ(U_1)$ of $conv(\mathcal{A}_1)$ is blue with respect to U_2. Without loss of generality, we may assume that the last move was that of U_1, on the boundary of $conv(\mathcal{A}_1)$ (the proof is entirely symmetrical in the converse situation). Then the edge $pred(U_1)U_1$ of $conv(\mathcal{A}_1)$ is red with respect to U_2, and as a result vertex U_1 is purple with respect to U_2, so that line U_1U_2 supports $conv(\mathcal{A}_1)$ (lemma 8.3.1). It remains to show that U_1U_2 also supports $conv(\mathcal{A}_2)$, or in other words that vertex U_2 of $conv(\mathcal{A}_2)$ is purple with respect to U_1. The edge $U_2 pred(U_2)$ of $conv(\mathcal{A}_2)$ is blue with respect to U_1, however, so we only have to show that edge $succ(U_2)U_2$ is red with respect to U_1. Let U_1' be the position of U_1 on the boundary of $conv(\mathcal{A}_1)$ during the last move of the other endpoint on the boundary of $conv(\mathcal{A}_2)$. The edge $succ(U_2)U_2$ of $conv(\mathcal{A}_2)$ is red with respect to U_1'. All the vertices between U_1' and U_1 on $conv(\mathcal{A}_1)$ lie on the same side of $succ(U_2)U_2$. As a result, the edge $succ(U_2)U_2$ is also red with respect to U_1.

We obtain the lower bitangent in the same fashion, only U_1 moves on the boundary of $conv(\mathcal{A}_1)$, starting at position A_k and passing clockwise over the vertices of $conv(\mathcal{A}_1)$. Likewise, U_2 moves on the boundary of $conv(\mathcal{A}_2)$, starting at position A_{k+1} and passing counter-clockwise over the vertices of $conv(\mathcal{A}_2)$.

To analyze this algorithm, it suffices to notice that each test between a vertex and an edge simply consists in evaluating the sign of a 3×3 determinant, and can therefore be performed in constant time. Moreover, only two tests are performed at each step, to follow a red edge of either $conv(\mathcal{A}_1)$ or $conv(\mathcal{A}_2)$, or to discover that a bitangent has been found. A red edge of $conv(\mathcal{A}_1)$ or $conv(\mathcal{A}_2)$ is not part of the convex hull $conv(\mathcal{A})$, and therefore will never be tested again in the entire algorithm. The total time needed in the recursion for these operations is therefore at most proportional to the number of edges of convex hulls created by the algorithm. At each merging step, two new edges are created, and the total number of these steps is $O(n)$ if the size of the original set $\mathcal{A}$ is n. The total number of edges created is thus linear, and the complexity of the operations in the divide-and-conquer recursive calls is $O(n)$. In two dimensions, the total complexity of the algorithm is therefore dominated by the cost of the initial sorting. Notice that if the points are sorted along one axis, the algorithm runs in time $O(n)$.

Theorem 9.2.1 *A divide-and-conquer algorithm computes the convex hull of a set of n points in $\mathbb{E}^2$ in optimal time $O(n \log n)$.*

9.3 Divide-and-conquer convex hulls in dimension 3

The divide-and-conquer algorithm described in section 9.2 can be made to work in dimension 3. Below, we describe this algorithm, which is deterministic and optimal.

The algorithm in three dimensions follows the same paradigm as its counterpart in two dimensions. The points are first sorted in order of increasing abscissae, then the recursive procedure is performed.

Dividing. The set $\mathcal{A}$ is split into two almost equal-sized subsets $\mathcal{A}_1$ and $\mathcal{A}_2$, separated by a vertical line.

Solving. Compute the convex hulls $conv(\mathcal{A}_1)$ and $conv(\mathcal{A}_2)$ of each subset separately, using recursive calls to the procedure.

Merging. The convex hull $conv(\mathcal{A})$ of $\mathcal{A}$ can be computed by forming the convex hull of the two polytopes $conv(\mathcal{A}_1)$ and $conv(\mathcal{A}_2)$.

The only delicate step is the merging step, which amounts to computing the convex hull of two polytopes separated by a plane. The discussion below shows that this computation can be carried out in time $O(n)$ if n is the total number of vertices of both polytopes. The analysis of the algorithm is then immediate: the initial sorting takes time $O(n \log n)$ and the complexity $t(n)$ of the recursive calls to the procedure satisfies

$$t(n) = 2t(n/2) + O(n),$$

which solves to $t(n) = O(n \log n)$. This proves the theorem below. Notice that in the three-dimensional case, the divide-and-conquer algorithm has complexity $\Theta(n \log n)$ even if the points are sorted along one axis.

Theorem 9.3.1 *In three dimensions, a divide-and-conquer algorithm computes the convex hull of a set of n points in optimal time $O(n \log n)$.*

Convex hull of two polytopes separated by a plane

Let $\mathcal{C}_1$ and $\mathcal{C}_2$ be two polytopes separated by a plane. We want to compute their convex hull $\mathcal{C} = conv(\mathcal{C}_1 \cup \mathcal{C}_2)$. To keep things simple, we assume that the interiors of $\mathcal{C}_1$ and $\mathcal{C}_2$ are both non-empty, and that their vertices are in general position. Both assumptions are satisfied by the convex hulls $conv(\mathcal{A}_1)$ and $conv(\mathcal{A}_2)$ computed by the recursive procedure described above if we assume, as usual, that the set $\mathcal{A}$ of points is in general position. We may also assume that the two polytopes $\mathcal{C}_1$ and $\mathcal{C}_2$ are separated by the plane H_0 with equation $x = 0$, and that the abscissae of the vertices of $\mathcal{C}_1$ are negative, while those of $\mathcal{C}_2$ are positive.

Overview of the algorithm

As in the incremental algorithm of section 8.3, it helps to color the facets of polytopes $\mathcal{C}_1$ and $\mathcal{C}_2$ *red* or *blue*, and the faces of dimension $d - 2$ or less either *red*, *blue*, or *purple*. Intuitively, the faces of polytope $\mathcal{C}_1$ that are red with respect to $\mathcal{C}_2$ are those that would be lit if $\mathcal{C}_2$ was an extended light source. The blue faces of $\mathcal{C}_1$ are the ones in the shadow, and the purple faces are lit by tangent beams. The colors of faces of $\mathcal{C}_2$ can similarly be explained by imagining that $\mathcal{C}_1$ is an extended light source. Throughout this section, we call a face of $\mathcal{C}_1$ (resp. $\mathcal{C}_2$) *red*, *blue*, or *purple*, if it is *red*, *blue*, or *purple* with respect to the opposite polytope $\mathcal{C}_2$ (resp. $\mathcal{C}_1$). Since the light sources $\mathcal{C}_1$ or $\mathcal{C}_2$ are polytopes and not points any longer, the distinction between blue and purple faces is considerably

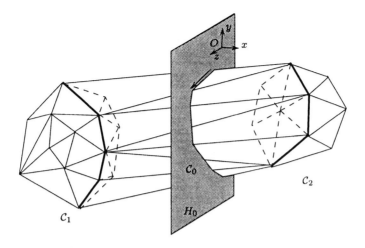

Figure 9.5. Convex hull of two polytopes separated by a plane.

more subtle than in the incremental algorithm. The colors can be attributed as follows:

- A face (facet, edge, or vertex) of C_1 or C_2 that is not a face of C is *red*.

- A facet of C_1 or C_2 that is a facet of C is *blue*.

- An edge of C_1 or C_2 that is an edge of C is *purple* if it is incident to at least one red facet, and *blue* otherwise.

- A vertex of C_1 or C_2 that is a vertex of C is *purple* if it is incident to at least one red or purple edge, and *blue* otherwise.

It is very tempting to believe that the purple vertices and edges of C_1 (resp. C_2) form a cycle in the incidence graph of C_1 (resp. C_2) as is the case for the set of edges and vertices that are purple with respect to a point. This is not true, however. Indeed, a purple vertex can be incident to an arbitrarily high number of purple edges (for instance, vertex A in figure 9.6a is incident to three purple edges). This number may even be zero when the purple vertex is the only non-red face of polytope C_1 (consider for instance vertex A in figure 9.6b). A purple edge may be incident to two red facets: this happens for instance to an edge of C_2 whose affine hull is a line that does not intersect C_1, but whose incident facets have their supporting planes intersecting C_1 (for instance, edge AB in figure 9.6a).

The faces of C that are neither faces of C_1 nor of C_2 are the *new* faces, and they necessarily intersect the separating plane H_0. A new edge is the convex hull of a purple vertex of C_1 and a purple vertex of C_2. A new facet is a triangle, the convex hull of a purple edge of C_1 and of a purple vertex of C_2, or conversely the

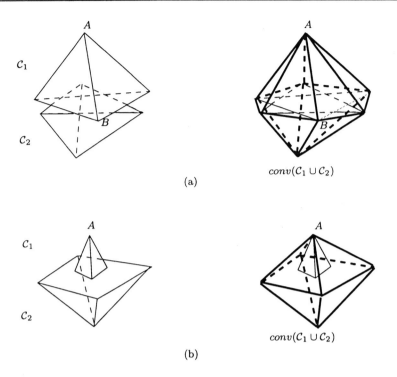

Figure 9.6. Convex hull of two 3-polytopes. Peculiar cases.

convex hull of a purple edge of C_2 and of a purple vertex of C_1. Let C_0 be the 2-polytope formed by the intersection of C with the plane H_0 (see figure 9.5). The new edges of C intersect H_0 at the vertices of C_0, and the circular order on the faces of the 2-polytope C_0 induces a circular order on the new faces (edges and facets) of C.

The main idea is then to build the new facets of C in turn, in the order given by the edges of C_0. For instance, the plane H_0 may be oriented by the x-axis, and the boundary of the 2-polytope C_0 will be followed in counter-clockwise order. As we will show below, the algorithm takes advantage on the order of edges of a 3-polytope incident to a given vertex. So we choose to represent polytopes by the edge-list structure, which explicitly encodes this order (see section 9.1).

The overview of the merging algorithm is then:

1. We first find a new edge of C.

2. The algorithm then discovers the other new faces (facets and edges) of C in the order induced by C_0. At the same time, the purple faces (edges and vertices) of C_1 and C_2 are found.

3. In a third stage, all the red faces (facets, edges, and vertices) of C_1 and C_2 are found and the edge-list representation of C is built from those of C_1 and of C_2.

Finding the first new edge

A first new edge of C can be found by applying the two-dimensional bitangent-finding algorithm to the projections C_1' and C_2' of polytopes C_1 and C_2 on the plane $z = 0$. Vertex A_1 of greatest abscissa of C_1 projects onto vertex A_1' of greatest abscissa of C_1'; starting from this vertex, its successor in the counter-clockwise order on the boundary of C_1' may be found by looking at all the vertices of C_1 incident to A_1. From one vertex to the next, we can build the vertices of C_1'. The projection C_2' of C_2 can be obtained in a similar way, starting at A_2', the projection of the vertex A_2 of C_1 that has the least abscissa. Polytopes C_1' and C_2' are disjoint and separated by the line $x = 0$ in the plane $z = 0$. The two-dimensional merging algorithm, starting with segment $A_1'A_2'$ which lies outside C_1' and C_2', yields a bitangent $U_1'U_2'$ to both polytopes (U_1' belonging to C_1' and U_2' to C_2'). The vertices U_1 and U_2 of C_1 and C_2 that project onto U_1' and U_2' yield a new edge U_1U_2 of C. Moreover, we know that the vertical plane that contains U_1U_2 is a supporting plane of both C_1 and C_2.

Finding the other new faces

To find the other new faces of C, the algorithm uses the *gift-wrapping* method which consists in pivoting a plane around the current new edge, so that it supports both C_1 and C_2, as if we were trying to wrap both C_1 and C_2 with a single sheet of paper. More precisely, let A_1A_2, $A_1 \in C_1$, $A_2 \in C_2$, be the new edge that was most recently discovered. The algorithm knows a plane H_{12} that supports C along its edge A_1A_2: take for H_{12} the vertical plane passing through A_1A_2 if this edge is the first edge found in the previous stage, or otherwise take the affine hull of the most recently discovered new facet AA_1A_2, which is incident to the oriented edge A_1A_2 on its left. At this point, the algorithm must discover the new facet of C that is right incident to the oriented edge A_1A_2. This facet is a triangle A_1A_2A' where A' is either a vertex of C_1 or a vertex of C_2. Consider a plane H that pivots around edge A_1A_2, starting at position H_{12} and moving counter-clockwise, meaning that its trace $H_{12} \cap H_0$ in the separating plane H_0 pivots counter-clockwise around the vertex $A_1A_2 \cap H_0$ of C_0. Vertex A' is the first vertex of either C_1 or C_2 that is touched by H. We say that the *winner* of C_1 for the pivot A_1A_2 is the first vertex A_1' of C_1 that is touched by H. Similarly, the *winner* of C_2 for the pivot A_1A_2 is the first vertex A_2' of C_2 that is touched by H. Necessarily, A is one of A_1' or A_2', and which one can be decided in the following

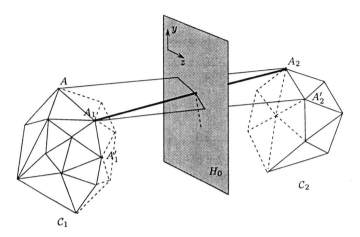

Figure 9.7. How to find the winner for the pivot $A_1 A_2$.

way. Let N be the unit vector normal to plane H_{12}, and directed outside $\mathcal{C}$.[1]
Likewise, for each $i = 1, 2$, let H_i' be the affine plane of the triangle $A_1 A_2 A_i'$, and
N_i' its unit normal vector directed outside $\mathcal{C}_i$.[2] Vertex A' is that A_i' $(i = 1, 2)$ for
which the dihedral angle between H_{12} and H_i' is minimal, or equivalently that for
which the dot product $N \cdot N_i$ is minimal.

We must now explain how to find the winners A_1' and A_2'. These problems
being exactly symmetrical, we will restrict our attention the problem of finding
the winner of $\mathcal{C}_1$.

If A_1' is a vertex of $\mathcal{C}_1$ that is adjacent to A_1, then we denote by $pred(A_1')$
and $succ(A_1')$ the vertices of $\mathcal{C}_1$ adjacent to A_1' that respectively precede and
follow A_1' in the counter-clockwise order around A_1. Denote by H_1' the planar
affine hull of the triangle $A_1 A_1' A_2$ and $H_1'^+$ the half-space bounded by this plane
that is opposite to the wedge product $-A_1 A_2 \wedge A_1 A_1'$, which induces a clockwise
orientation of the triangle $A_1 A_2 A_1'$.

Lemma 9.3.2 *The winner of $\mathcal{C}_1$ for the pivot $A_1 A_2$ is the unique vertex A_1' of
$\mathcal{C}_1$ adjacent to A_1 such that $pred(A_1')$ and $succ(A_1')$ both belong to the half-space
$H_1'^+$.*

Proof. A_1' is the winner of pivot $A_1 A_2$ if and only if triangle $A_1' A_1 A_2$ is the face
of polytope $conv(\mathcal{C}_1 \cup \{A_2\})$ incident to the oriented edge $A_1 A_2$ on its right. Then
H_1' is a supporting plane of $conv(\mathcal{C}_1 \cup \{A_2\})$ and therefore of $\mathcal{C}_1$, so that $A_1 A_1'$ is
an edge of $\mathcal{C}_1$, and both $pred(A_1')$ and $succ(A_1')$ belong to half-space $H_1'^+$.

---

[1]The direction of vector N is the one that orients the triangle (A_1, A_2, A) counter-clockwise.
It is the direction of the wedge product $A_1 A_2 \wedge A_1 A$.

[2]On the other hand, the direction of vector N_i is the one that orients the triangle (A_1, A_2, A_i')
clockwise. It is the direction of the wedge product $-A_1 A_2 \wedge A_1 A_i'$.

Reciprocally, let A_1' be a vertex of $\mathcal{C}_1$ adjacent to A_1 such that both $pred(A_1')$ and $succ(A_1')$ belong to half-space $H_1'^+$. Then polytope $\mathcal{C}_1$, which is contained in the intersection of two half-spaces bounded by the planes supporting $\mathcal{C}_1$ along its facets $A_1 A_1' succ(A_1')$ and $A_1 A_1' pred(A_1')$, is therefore contained in $\overline{H_1'^+}$. As a result, H_1' supports $\mathcal{C}_1$ and $A_1' A_1 A_2$ is the facet of polytope $conv(\mathcal{C}_1 \cup \{A_2\})$ incident to the oriented edge $A_1 A_2$ on its right, which shows that A_1' is the winner of the pivot $A_1 A_2$. $\qquad\square$

During the algorithm, we may occasionally encounter several pivots incident to the same vertex A_1 of $\mathcal{C}_1$. The following lemma shows that it is not necessary to test each of the vertices of $\mathcal{C}_1$ adjacent to A_1, for each of these pivots.

Lemma 9.3.3 *When a vertex A_1 of $\mathcal{C}_1$ is incident to several pivots, the algorithm encounters these pivots in such an order that their winners are ordered clockwise around this vertex A_1 of $\mathcal{C}_1$. Likewise, the pivots incident to a vertex A_2 of $\mathcal{C}_2$ are encountered in an order such that their winners are ordered counter-clockwise around this vertex A_2 of $\mathcal{C}_2$.*

Proof. Here we prove the assertion concerning polytope $\mathcal{C}_1$. A proof for polytope $\mathcal{C}_2$ is entirely symmetrical. Since A_1 is a vertex of the convex hull $\mathcal{C}$, there is a plane H_1 that separates vertex A_1 from all the other vertices of $\mathcal{C}_1$ and $\mathcal{C}_2$ (see exercise 7.4). Such a plane intersects all the edges of $\mathcal{C}$ and $\mathcal{C}_1$ incident to A_1 (see figure 9.8). Plane H_1 intersects polytope $\mathcal{C}$ along the 2-polytope $H_1 \cap \mathcal{C}$ and polytope $\mathcal{C}_1$ along the 2-polytope $H_1 \cap \mathcal{C}_1$ contained inside $H_1 \cap \mathcal{C}$.

Let us orient the plane H_1 by a normal unit vector N_1 directed towards the half-space that does not contain A_1. The counter-clockwise order on the edges of the 2-polytope $H_1 \cap \mathcal{C}_1$ corresponds to the indirect order on the facets of $\mathcal{C}_1$ that contain vertex A_1.

The order in which the new faces of $\mathcal{C}$ that contain A_1 are discovered is consistent with the counter-clockwise order of the faces of the 2-polytope $H_1 \cap \mathcal{C}$. To see this, it suffices to consider a plane H that pivots around $H_0 \cap H_1$ from H_0 towards H_1. As H pivots from H_0 to H_1, the 2-polytope $H \cap \mathcal{C}$ changes. Nevertheless, each new face of $\mathcal{C}$ that contains A_1 always keeps a trace in H corresponding to a face of $H \cap \mathcal{C}$, and all these traces remain in the same order along the boundary of $H \cap \mathcal{C}$.

Any purple edge of $\mathcal{C}_1$ is also an edge of $\mathcal{C}$, and the trace on plane H_1 of a purple edge of $\mathcal{C}$ that is incident to A_1 is a common vertex of both polytopes $H_1 \cap \mathcal{C}$ and $H_1 \cap \mathcal{C}_1$. The trace on H_1 of a pivot $A_1 A_2$, however, is a vertex of $H_1 \cap \mathcal{C}$ but not of $H_1 \cap \mathcal{C}_1$. The trace on H_1 of the plane that pivots around the edge $A_1 A_2$ is a line L that pivots around the point $A_1 A_2 \cap H_1$. The pivoting plane touches $\mathcal{C}_1$ at the winner of A_1' of $\mathcal{C}_1$ for the pivot $A_1 A_2$ whenever L becomes a supporting line of the polytope $H_1 \cap \mathcal{C}_1$. The point along which L supports

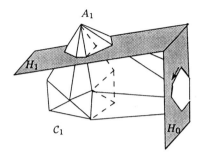

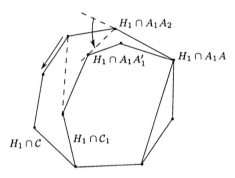

Figure 9.8. Successive pivots incident to a vertex A_1 of C_1.

$H_1 \cap C_1$ is the trace on H_1 of the edge $A_1 A_1'$. During the course of the algorithm, the trace on H_1 of the successive pivots incident to A_1 moves counter-clockwise on the boundary of $H_1 \cap C$; as a result, the point at which the line L touches $H_1 \cap C_1$ moves counter-clockwise on the boundary of $H_1 \cap C_1$ (see figure 9.8). The edge $A_1 A_1'$ therefore traverses the list of edges of C incident to A_1 in indirect (or clockwise) order.   □

In order to find the winner in C_1 for a pivot incident to a vertex A_1, the algorithm need only consider the edges of C_1 incident to vertex A_1 in clockwise order. When the algorithm considers the first pivot incident to A_1, the algorithm starts searching at any edge incident to A_1. If the algorithm has already encountered one or more pivots incident to A_1, however, then it starts the search at the winner of C_1 for the last encountered pivot incident to A_1.

When the algorithm discovers a new face ($A_1 A_2 A_1'$ or $A_1 A_2 A_2'$) while pivoting around edge $A_1 A_2$, it also exhibits a new edge of C ($A_2 A_1'$ or $A_1 A_2'$). The pivoting process may be started again around this new pivot. Moreover, a purple vertex A_1' (or A_2') and a purple edge $A_1 A_1'$ (or $A_2 A_2'$) are also identified. The pivoting process is terminated when the pivot is back at the initial edge $U_1 U_2$. All the new facets have been discovered, and the purple faces of C_1 and C_2 (vertices and edges) have been sorted out. Note that a purple vertex can be enumerated several times in the algorithm, and that some purple edges (those incident to two red facets) may be enumerated twice.

Reconstruction of C

The third stage of the merging process must identify all the red faces of C_1 and C_2, and also build the edge-list representation of the convex hull C of C_1 and C_2. For this, we may begin by traversing the list of purple edges and color the incident facets: a facet of C_i ($i = 1, 2$) incident to a purple oriented edge $A_i A_i'$ is red if there is a new facet of C incident to $A_i A_i'$ on the same side. It is blue otherwise. The red facets are then colored by propagating the red color, as every facet incident to a red facet and that is not colored blue must be a red facet as well. The red edges and vertices are then easily determined: an edge incident to a red facet must be red, unless it was colored purple in the previous stage (finding the purple edges and vertices); likewise, a vertex incident to a red or purple edge is red, unless it was colored purple in the previous stage.

When the new faces of C are discovered and the red and purple faces of polytopes C_1 and C_2 have been determined, it is easy to create the edge-list representation of C from those of C_1 and C_2.

Analysis of the algorithm

During the initial stage, each edge of C_1 or C_2 is examined at most twice, once for each endpoint, in order to build the projections C_1' and C_2'. In dimension 2, the complexity of the algorithm that finds the tangent of two convex polytopes is linear (see section 9.2), and hence the first new edge is found in time $O(n_1 + n_2)$, if n_1 and n_2 stand for the number of vertices of C_1 and C_2 respectively.

The complexity of the second stage is clearly proportional to the number of edges of C_1 and C_2 that are considered when searching the winners of each pivot. From lemma 9.3.3, the list of edges incident to a purple vertex is traversed at most twice during the entire algorithm, once to find the winner of the first pivot incident to this vertex, and another time to find the winners of all the other pivots incident to this vertex. The complexity of this stage is therefore $O(n_1 + n_2)$ as well.

Finally, finding the red faces of C_1 and C_2 in the third stage can be carried out in time proportional to the number of these faces and of new faces of C, which is again $O(n_1 + n_2)$. We have proved:

Theorem 9.3.4 *Let C_1 and C_2 be two polytopes in $\mathbb{E}^3$ separated by a plane, and having respectively n_1 and n_2 vertices. It is possible to compute the convex hull $C = conv(C_1 \cup C_2)$ in time $O(n_1 + n_2)$.*

9.4   Convex hull of a polygonal line

The reader will have noticed that, in dimension 2, the complexity of the incremental algorithm and that of the divide-and-conquer algorithm are both dominated by the cost of the initial sorting. In fact, the costs of both algorithms are only linear if the points are already sorted along some direction. Now a set of points that is sorted along, say, the x-axis, forms a simple polygonal line, meaning that it does not intersect itself. It is therefore tempting to conjecture that the convex hull of *any* simple polygonal line in $\mathbb{E}^2$ can be obtained in linear time. It is the goal of this section to show that this is indeed the case.

Simple polygonal lines and polygons are defined in a general way in chapter 12. For this reason, here we only give a brief summary of definitions and results which we will use in the forthcoming algorithm. A polygonal line is an ordered sequence of points, called the *vertices* of the polygonal line. The segments that join two consecutive vertices are called the *edges* of the polygonal line. The polygonal line is *simple* if:

1. its vertices are all distinct, except perhaps the first and last which may be identical, and

2. two edges do not intersect, except perhaps at a common endpoint.

A polygonal line is *closed* if the first and last vertices are identical. A simple and closed polygonal line is also called a *polygon*. Thus a polygon may be defined entirely by its circular sequence of vertices. A deep theorem of Jordan (a proof of the theorem for polygons is given in exercise 11.1) states that any simple closed curve separates the plane $\mathbb{E}^2$ into two connected components, exactly one of which is bounded. Thus a polygon $\mathcal{P}$ separates the points in $\mathbb{E}^2 \setminus \mathcal{P}$ into two connected regions, only one of which is bounded. This bounded region is called the *interior* of polygon $\mathcal{P}$, and denoted by $int(\mathcal{P})$. The other (unbounded) region is called the *exterior* of the polygon and is denoted by $ext(\mathcal{P})$. Regions $int(\mathcal{P})$ and $ext(\mathcal{P})$ are topological open subsets[3] of $\mathbb{E}^2$, and the topological closure $\overline{int(\mathcal{P})}$ of $int(\mathcal{P})$ is the union $int(\mathcal{P}) \cup \mathcal{P}$. In this section, the Euclidean space $\mathbb{E}^2$ is oriented and a polygon will be described as a circular list of vertices in direct (or counterclockwise) order. This defines an orientation on the edges. By convention, we agree that the direct orientation of a polygon is such that the interior of the polygon is to the left of each oriented edge.

Let $\mathcal{A}$ be a set of n points in $\mathbb{E}^2$. One may wonder why knowing a simple polygonal line $\mathcal{L}(\mathcal{A})$ joining these points would help in computing the convex hull of $conv(\mathcal{A})$. The following theorem shows a deep connection between the

---

[3]For a brief survey of the topological notions of open, closed, and connected subsets, see chapter 11.

order of the vertices along the boundary of the convex hull $conv(\mathcal{A})$ and the order of these points along the polygonal line $\mathcal{L}(\mathcal{A})$.

Theorem 9.4.1 *Consider two polygons $\mathcal{P}$ and $\mathcal{Q}$, such that the interior of $\mathcal{Q}$ is entirely contained inside the interior of $\mathcal{P}$. The common vertices of $\mathcal{P}$ and $\mathcal{Q}$ are encountered in the same order when both polygons are traversed in a counter-clockwise order.*

Let $\Sigma_{\mathcal{P}}$ and $\Sigma_{\mathcal{Q}}$ and be the circular sequences of vertices of $\mathcal{P}$ and $\mathcal{Q}$. Let $\Sigma'_{\mathcal{P}}$ be the subsequence of $\Sigma_{\mathcal{P}}$ that corresponds to vertices common to both $\mathcal{P}$ and $\mathcal{Q}$, and similarly let $\Sigma'_{\mathcal{Q}}$ be the subsequence of $\Sigma_{\mathcal{Q}}$ that corresponds to vertices common to both $\mathcal{P}$ and $\mathcal{Q}$. The theorem states that the two sequences $\Sigma'_{\mathcal{P}}$ and $\Sigma'_{\mathcal{Q}}$ are identical. In particular, if $\mathcal{P}$ is the convex hull of $\mathcal{Q}$, then the theorem states that the sequence of vertices of $conv(\mathcal{Q})$ is a subsequence of the sequence of vertices of $\mathcal{Q}$.

Proof. Because the interior of $\mathcal{Q}$ is entirely contained inside the interior of $\mathcal{P}$, edges of $\mathcal{P}$ and $\mathcal{Q}$ cannot intersect in a point which is not a vertex of $\mathcal{P}$ or $\mathcal{Q}$. In the following, we first assume that the edges of $\mathcal{P}$ and $\mathcal{Q}$ intersect only in points which are vertices of both $\mathcal{P}$ and $\mathcal{Q}$, and then remove this assumption at the end of the proof.

A *chord* of a topological 2-ball is a simple curve contained in the interior of the 2-ball except for its endpoints which lie on the boundary. The proof of the theorem relies on a consequence of Jordan's theorem that states that any closed topological 2-ball is separated by a chord into two connected components. (A proof of this consequence is given in exercise 11.2.) Both subsequences $\Sigma'_{\mathcal{P}}$ and $\Sigma'_{\mathcal{Q}}$ have the same vertices, namely those common to both $\mathcal{P}$ and $\mathcal{Q}$, but not necessarily in the same order. To prove that they are in fact identical, it suffices to show that two consecutive vertices in $\Sigma'_{\mathcal{Q}}$ are also consecutive in $\Sigma'_{\mathcal{P}}$. Let then A_1 and A_2 be two consecutive vertices of $\Sigma'_{\mathcal{Q}}$. Let $\mathcal{Q}_{12}$ be the portion of $\mathcal{Q}$ that joins A_1 to A_2 in counter-clockwise order, and $\mathcal{Q}_{21}$ be the portion that joins A_2 to A_1 in counter-clockwise order. Likewise, let $\mathcal{P}_{12}$ be the portion of $\mathcal{P}$ that joins A_1 to A_2 in counter-clockwise order, and $\mathcal{P}_{21}$ be the portion that joins A_2 to A_1 in counter-clockwise order (see figure 9.9). The vertices of $\mathcal{P}$ are distinct from those of $\mathcal{Q}_{12}$ (except for the endpoints A_1 and A_2) because A_1 and A_2 are consecutive in the subsequence $\Sigma'_{\mathcal{Q}}$. Hence $\mathcal{P}_{12}$ and $\mathcal{Q}_{12}$ cannot share a common vertex except for A_1 and A_2. Furthermore, there are no intersections between edges of $\mathcal{P}$ and $\mathcal{Q}$ except at a common vertex. Hence $\mathcal{Q}_{12}$ is a chord of the topological 2-ball $int(\mathcal{P})$. Moreover, $\mathcal{Q}_{21}$ is contained in $\overline{int(\mathcal{P})}$ and does not intersect the chord $\mathcal{Q}_{12}$ except at its endpoints A_1 and A_2. Hence $\mathcal{Q}_{21}$ is entirely contained in one of the two connected components of $\overline{int(\mathcal{P})} \setminus \mathcal{Q}_{12}$. Clearly, $\mathcal{P}_{12}$ is entirely contained in the other component except for its endpoints A_1 and A_2. Therefore, $\mathcal{P}_{12}$ and $\mathcal{Q}_{21}$ cannot share a common vertex except for A_1 and A_2. This shows that A_1

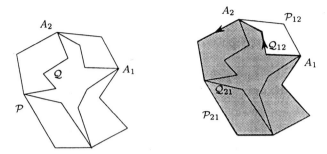

Figure 9.9. For the proof of theorem 9.4.1.

and A_2 are consecutive in the subsequence $\Sigma'_{\mathcal{P}}$, and proves the theorem when $\mathcal{P}$ and $\mathcal{Q}$ intersect only at common vertices.

The case where some vertices of $\mathcal{P}$ can lie on edges of $\mathcal{Q}$ is handled by adding those vertices of $\mathcal{P}$ as vertices of $\mathcal{Q}$, splitting the corresponding edges of $\mathcal{Q}$. Similarly, when some vertices of $\mathcal{Q}$ lie on edges of $\mathcal{P}$, we add those vertices of $\mathcal{Q}$ as vertices of $\mathcal{P}$ and split the corresponding edges of $\mathcal{P}$. The same vertices are added to the subsequences $\Sigma'_{\mathcal{P}}$ and $\Sigma'_{\mathcal{Q}}$, producing two subsequences $\Sigma''_{\mathcal{P}}$ and $\Sigma''_{\mathcal{Q}}$ which are identical by the proof above. Thus $\Sigma'_{\mathcal{P}}$ and $\Sigma'_{\mathcal{Q}}$, obtained by removing the same elements, are also identical. $\square$

The following corollary takes more interest in the convex hull of polygonal lines. Let $\mathcal{A}$ be a set of n points in the plane and $\mathcal{L}(\mathcal{A}) = (A_1, A_2, \ldots, A_n)$ be a simple polygonal line joining the points in $\mathcal{A}$. Consider the ranks in $\mathcal{L}(\mathcal{A})$ of the vertices on the convex hull $conv(\mathcal{A})$, and denote by A_m and A_M the vertices of $conv(\mathcal{A})$ with respectively lowest and highest rank (see figure 9.10).

Corollary 9.4.2 *When the boundary of $conv(\mathcal{A})$ is oriented counter-clockwise, the vertices encountered between A_m and A_M form a subsequence of vertices of $\mathcal{L}(\mathcal{A})$, whereas the vertices encountered between A_M and A_m form a subsequence of the polygonal line that is the reverse of $\mathcal{L}(\mathcal{A})$.*

Proof. We denote by $\mathcal{C}(\mathcal{A})$ the polygon that constitutes the boundary of the convex hull $conv(\mathcal{A})$, by $\mathcal{C}_{mM}$ the portion of $\mathcal{C}(\mathcal{A})$ that joins A_m to A_M in counter-clockwise order, and by $\mathcal{C}_{Mm}$ the portion of $\mathcal{C}(\mathcal{A})$ that joins A_M to A_m in counter-clockwise order. To prove the first part of this corollary, it suffices to apply the previous theorem to the polygons $\mathcal{P}$ and $\mathcal{Q}$ defined as follows. Polygon $\mathcal{P}$ is the concatenation of $\mathcal{C}_{mM}$ and a polygonal line $\mathcal{P}_{Mm}$ that joins A_M to A_m such that $\mathcal{L}(\mathcal{A})$ and $\mathcal{C}(\mathcal{A})$ are both contained within $\overline{int(\mathcal{P})}$ (see figure 9.10). Polygon $\mathcal{Q}$ is the concatenation of $\mathcal{L}_{mM}$ (the portion of $\mathcal{L}(\mathcal{A})$ that joins A_m to A_M) and of $\mathcal{P}_{Mm}$. The second part of the theorem can be proved very similarly. $\square$

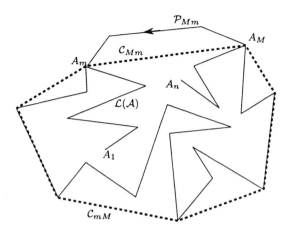

Figure 9.10. Convex hull of a polygonal line.

The remainder of this section presents an algorithm that builds the convex hull of a polygonal line in $\mathbb{E}^2$ in linear time.

Let $\mathcal{A}$ be a set of n points in $\mathbb{E}^2$ and $\mathcal{L}(\mathcal{A})$ a simple polygonal line whose vertices are the points of $\mathcal{A}$. The algorithm we present here is an incremental algorithm that processes the points of $\mathcal{A}$ in the order of $\mathcal{L}(\mathcal{A})$.

Let $(A_1, A_2, \ldots, A_n)$ be the sequence $\mathcal{L}(\mathcal{A})$ and $\mathcal{A}_i = \{A_1, A_2, \ldots, A_i\}$ be the set of the first i points of $\mathcal{L}(\mathcal{A})$. The convex hull $conv(\mathcal{A}_i)$ of $\mathcal{A}_i$ is maintained as a doubly connected circular list of vertices. The algorithm maintains a pointer to the vertex of $conv(\mathcal{A}_i)$ with the highest rank in $\mathcal{L}(\mathcal{A})$.

The initial step builds a circular list for the triangle $A_1 A_2 A_3$ and the pointer points to A_3. The current step inserts point A_i in the structure, and updates the data structure that stores $conv(\mathcal{A}_{i-1})$ so that it represents $conv(\mathcal{A}_i)$. The algorithm works in two phases.

First phase. The algorithm determines whether point A_i belongs to the interior or exterior of $conv(\mathcal{A}_{i-1})$. Lemma 9.4.3 below shows that this reduces to evaluating the signs of the two determinants $[pred(A_M)A_M A_i]$ and $[A_M succ(A_M)A_i]$, where A_M has the highest rank among the vertices of $conv(\mathcal{A}_{i-1})$, and $pred(A_M)$ and $succ(A_M)$ are respectively the predecessor and successor of A_M in a counterclockwise enumeration of the vertices on the boundary of $conv(\mathcal{A}_{i-1})$ (see figure 9.11).

Lemma 9.4.3 *Let H_p^+ (resp. H_s^+) be the half-space bounded by the line supporting $conv(\mathcal{A}_{i-1})$ along the edge $pred(A_M)A_M$ (resp. $A_M succ(A_M)$), and that contains $conv(\mathcal{A}_{i-1})$. Point A_i is interior to polytope $conv(\mathcal{A}_{i-1})$ if and only if A_i belongs to the intersection of half-spaces $H_p^+ \cap H_s^+$.*

Proof. The condition is obviously required. To show that it also suffices, we

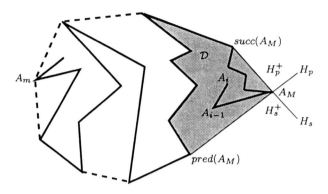

Figure 9.11. For the proof of lemma 9.4.3.

see that the simple polygonal line $\mathcal{L}(\mathcal{A})$ connects the vertices $pred(A_M)$ and $succ(A_M)$ of $conv(\mathcal{A}_{i-1})$, and that the portion of $\mathcal{L}(\mathcal{A})$ that joins $pred(A_M)$ to $succ(A_M)$ together with the edges $pred(A_M)A_M$ and $A_M succ(A_M)$ of $conv(\mathcal{A}_{i-1})$ form a simple closed polygonal line that bounds a region $\mathcal{D}$ of $\mathbb{E}^2$ entirely contained in $conv(\mathcal{A}_{i-1})$. This region is shaded in figure 9.11. The simple polygonal line $\mathcal{L}(\mathcal{A})$ also connects vertex A_M to A_i, and the portion of $\mathcal{L}(\mathcal{A})$ that joins A_M to A_i cannot intersect the portion of $\mathcal{L}(\mathcal{A})$ that joins $pred(A_M)$ to $succ(A_M)$. This guarantees that if point A_i belongs to both half-spaces H_p^+ and H_s^+, then it also belongs to $\mathcal{D}$ and thus to $conv(\mathcal{A}_{i-1})$.   □

Second phase. The algorithm now updates the convex hull if point A_i does not belong to the interior of polytope $conv(\mathcal{A}_{i-1})$. In this case, the previous lemma shows that at least one of the edges $pred(A_M)A_M$ and $A_M succ(A_M)$ is *red* with respect to A_i, if we use the terminology of the incremental algorithm of chapter 8. To update the convex hull, the algorithm need only perform steps 2, 3, and 4 of the incremental algorithm described in section 8.3.

Theorem 9.4.4 *The algorithm described previously builds the convex hull of a simple polygonal line in $\mathbb{E}^2$ in linear time.*

Proof. For each vertex of the polygonal line $\mathcal{L}(\mathcal{A})$, phase 1 requires constant time, since it only involves the computation of the signs of two 2×2 determinants. As to the second phase, if it is performed, its complexity is shown in section 8.3 to require time that is proportional to the number of the edges of polytope $conv(\mathcal{A}_{i-1})$ that are red with respect to A_i. The total contribution of this phase to the complexity of the algorithm is thus proportional to the number of edges created by the algorithm, which is $O(n)$.   □

9.5 Exercises

Exercise 9.1 (Common bitangents) Let C_1 and C_2 be two 2-polytopes, separated by a vertical line Δ. Let $C = conv(C_1 \cup C_2)$. Show that the edges of C intersecting Δ may be found in time $O(\log n)$, where n is the total number of vertices of C_1 and C_2.

Hint: For each polytope C, let us denote by C^+ the upper hull of C, which is the convex polygonal line whose vertices are vertices of C and that joins the vertex of highest abscissa to the vertex of lowest abscissa of C in the counter-clockwise order of the vertices along the boundary of C (see exercise 7.14). Similarly, we define the lower hull C^-. Note that the boundary of C is a concatenation of C^+ and C^- and that any vertical line that intersects C also intersects an edge of C^+ and an edge of C^-. If C is the convex hull $conv(C_1 \cup C_2)$ of two polytopes C_1 and C_2 separated by a vertical line Δ, we call *bridges* the two edges of C intersected by Δ. We separately search for the *upper bridge*, which is the edge of C^+ intersected by Δ, and the *lower bridge*, which is the edge of C^- intersected by Δ. The upper bridge is an edge of C that joins a vertex of C_1^+ to a vertex of C_2^+. It is possible to find it by a binary search on each of C_1^+ and C_2^+. Indeed, consider a vertex U_1 of C_1^+ and a vertex U_2 of C_2^+, and look at the color with respect to U_1 of the edges of C_2^+ incident to U_2, and the color with respect to U_2 of the edges of C_1^+ incident to U_1. There are nine possible cases, and in each case, at least one of the four chains determined by U_1 and U_2 on C_1^+ and C_2^+ can be discarded from further consideration.

Exercise 9.2 (Dynamic convex hulls) We present an algorithm for maintaining the convex hull of a set of points in the plane under insertion and deletion. In fact, the algorithm maintains the upper and lower hulls separately (see exercise 9.1).

The data structure used to represent the upper hull is a balanced binary tree whose leaves correspond to the points in the set ordered by increasing abscissae. At each internal node N in the tree, we store a secondary data structure which allows us to efficiently restore the convex hull of the points stored in the subtree rooted at N. More precisely, this structure is a catenable queue which maintains a list under the following operations: insertion and deletion of list items at either end of the list, splitting the list at a given item, and concatenation of two lists. Each of these operations may be performed in time logarithmic in the size of the lists. Let N be a node of the primary tree. We denote by $conv^+(N)$ the upper hull of the points stored in the subtree rooted at N. The catenable queue stored at N contains the portion of $conv^+(N)$ that is not on the boundary of the convex hull $conv^+(M)$, where M is the parent of N in the tree. Moreover the position of the first vertex of $conv^+(M)$ that does not belong to $conv^+(N)$ is stored at node N in an integer $j(N)$. The catenable queue stored at the root maintains the upper hull of all the points stored in the tree.

1. Show that the data structure requires a storage $O(n)$ if n is the number of points stored in the structure.

2. Show that each insertion or deletion takes time $O(\log^2 n)$ when the structure stores n points.

Hint: Let us consider, for instance, the insertion of a point P into the structure that maintains the upper hull. We follow the path in the tree that leads, from the root, to the leaf that is the closest to P (in the x-order). At each node N on this path, we update the catenable queues of N and of its sibling N' so that they store the upper convex hulls $conv^+(N)$ and $conv^+(N')$, which may be done in time $O(\log n)$. Indeed, it suffices to split the catenable queue of the parent M of N and N' at position $j(N)$, and to concatenate the sublists with those stored at N and N'. A node is created for P. Then, while going the other way on this path, the primary structure is re-balanced and the correct chains can be computed for each node. Let N be a node on the reverse path, and N' be its sibling. We can compute the upper hull $conv^+(M)$ of their parent M because the upper hulls $conv^+(N)$ and $conv^+(N')$ are known, and the bridge joining them can be computed in time $O(\log n)$ (see exercise 9.1). The lists $conv^+(N)$ and $conv^+(N')$ may thus be split at the bridge and we keep only the portion on the right for the left sibling and on the left for the right sibling. All these operations may be carried out in time $O(\log n)$ for each node on the reverse path from the new node P to the root of the primary tree. During a rotation (simple or double) of the tree, only a constant number of nodes switch children and a similar operation restores the correct chains stored in these nodes, in time $O(\log n)$ for each node.

Exercise 9.3 (Onion peeling) Given a set S of n points in the plane, we consider the subsets

$$
\begin{aligned}
S_0 &= S, \\
S_1 &= S_0 \setminus \{\text{set of vertices of } conv(S)\} \\
&\;\;\vdots \\
S_i &= S_{i-1} \setminus \{\text{set of vertices of } conv(S_i)\}
\end{aligned}
$$

until S_k has at most three elements. Give an algorithm that computes the iterated convex hulls $conv(S_0), conv(S_1), \ldots, conv(S_k), \ldots$ in total time $O(n \log n)$.

Exercise 9.4 (Diameter, antipodal pairs) Let $\mathcal{P}$ be a set of points of $\mathbb{E}^d$. The *diameter* of S, denoted by $d(S)$, is the maximal distance between two points of S. A pair (P_i, P_j) of vertices of the convex hull $conv(S)$ is said to be *antipodal* if $conv(S)$ admits two parallel hyperplanes supporting $\mathcal{P}$ along P_i and P_j respectively.

1. Show that if the diameter $d(S)$ occurs for P_i and P_j, that is $d(P_i, P_j) = d(S)$, then P_i and P_j are vertices of the convex hull $conv(S)$, and (P_i, P_j) is an antipodal pair.

2. Derive an algorithm in $\mathbb{E}^2$ that enumerates all the antipodal pairs to find the diameter in time $O(n \log n)$.

Exercise 9.5 (Hierarchical representation of a 3-polytope) Let $\mathcal{P}$ be a 3-polytope with n vertices. The degree of a vertex is the number of edges incident to that vertex. Two vertices of $\mathcal{P}$ are adjacent if they are incident to a common edge.

1. Show that $\mathcal{P}$ has at least $n/3$ vertices whose degree is less than 9. (Use Euler's relation to count the number of adjacencies between edges and vertices of $\mathcal{P}$.)

2. Show that any maximal subset of non-adjacent vertices of $\mathcal{P}$ with degree at most 8 has at least $n/27$ vertices.

Denote by $\mathcal{V}(\mathcal{P})$ the set of vertices of $\mathcal{P}$. A hierarchical representation of $\mathcal{P}$ is a nested sequence of polytopes $(\mathcal{P}_0 \supset \mathcal{P}_1 \supset \ldots \supset \mathcal{P}_k)$ such that

- $\mathcal{P}_0 = \mathcal{P}$,

- $\mathcal{P}_k$ is a simplex,

- for any $0 \le i \le k$, $\mathcal{V}(\mathcal{P}_{i+1}) \subset \mathcal{V}(\mathcal{P}_i)$ and $\mathcal{V}(\mathcal{P}_i) \setminus \mathcal{V}(\mathcal{P}_{i+1})$ is a maximal subset of non-adjacent vertices of $\mathcal{P}_i$ that have degree at most 8 in $\mathcal{P}_i$.

3. Show that $k = O(\log n)$ and that a hierarchical representation of $\mathcal{P}$ may be built in time $O(n)$.

Exercise 9.6 (Querying 3-polytopes) Let $\mathcal{P}$ be a 3-polytope with n vertices. Using the hierarchical representation of $\mathcal{P}$ described in exercise 9.5, show that the following queries can be answered in time $O(\log n)$:

1. Computing the intersection with a line: compute the intersection of $\mathcal{P}$ with a line L, or if this intersection $L \cap \mathcal{P}$ is empty, find the point P of $\mathcal{P}$ that minimizes the distance to L.

2. Detecting the intersection with a plane: if the intersection of $\mathcal{P}$ with a plane H is not empty, then output a point on the boundary of $\mathcal{P} \cap H$, otherwise return a point P of $\mathcal{P}$ that minimizes the distance to H.

3. Finding a plane of support: given a line L, and a plane H containing L that does not intersect $\mathcal{P}$, compute the first vertex of $\mathcal{P}$ encountered as H pivots around L.

4. Ray shooting: given a point O inside the polytope, and a direction U, find the face of $\mathcal{P}$ intersected by the ray originating at O along the direction U.

Hint: In the first three cases, the algorithm can take advantage of the solution for $\mathcal{P}_{i+1}$ to compute the solution for $\mathcal{P}_i$, in constant time. Note that the fourth query type is dual to the third.

Exercise 9.7 (Intersection of two convex polygons) Give an algorithm that computes the intersection of two convex polygons in the Euclidean plane $\mathbb{E}^2$, in linear time $O(n + m)$ if m and n are the respective number of vertices of either polygon.

Exercise 9.8 (Union of tricolored triangles) Consider a set of n points in $\mathbb{E}^2$ and the set of triangles having these points as vertices. Show that if each point is colored with one of k given colors, it is possible to compute the union of all tricolored triangles (no two vertices have the same color) in time $O(n \log n)$.

9.6   Bibliographical notes

There are several algorithms that compute the convex hull of a set of points, especially in dimension 2. It is commonly accepted that the first optimal algorithm in dimension 2 was given by Graham [111] in 1972. This algorithm sorts the points by polar angle around a point inside the convex hull (for instance any convex combination of three original points), then determines the edges of the convex hull by scanning this list. In fact, Toussaint [215] shed new light on the history of the problem by pointing out a paper by

Bass and Schubert [21], published in 1967, that proposed an optimal algorithm for the convex hull in dimension 2, although the paper had a slight error and gave no complexity analysis. Another well known algorithm is Jarvis' "gift-wrapping" algorithm [130] which computes the convex hull by successively finding all supporting hyperplanes. These algorithms and others are detailed in the book by Preparata and Shamos [192], which also gives a solution to exercise 9.4.

If the convex hull of n points in dimension d may have $\Omega(n^{\lfloor d/2 \rfloor})$ faces in the worst case, the number of faces may be much less. It is thus important to have output-sensitive algorithms, meaning that their complexities depend on the size of the output. Jarvis' algorithm is output-sensitive, as it runs in time $O(nh)$ for a set of n points whose convex hull has h vertices. Kirkpatrick and Seidel [137] gave an algorithm in dimension 2 that runs in time $O(n \log h)$. This algorithm uses a curious variant of the divide-and-conquer algorithm, which may appropriately be called *marriage-before-conquest*, since the upper and lower bridges connecting the convex hull of two subsets are computed before the convex hull of either subset is known. Edelsbrunner and Shi [99] generalized the idea to yield a convex hull algorithm in $\mathbb{E}^3$ that runs in $O(n \log^2 h)$.

The on-line algorithm presented in section 9.4 is due to Melkman [169]. The algorithms by Lee [146] and Graham and Yao [112] are both correct and use only one stack. The problem of dynamically maintaining the convex hull of a planar set of points (see exercise 9.2) was solved by Overmars and Van Leeuwen [184] in time $O(\log^2 n)$ for each operation. If only insertions or only deletions are to be performed, time $O(\log n)$ for each operation may be achieved. The case of insertions was studied by Preparata [190] and that of deletions by Chazelle [43] and Hershberger and Suri [125]. Chazelle's algorithm [43] computes the onion peeling described in exercise 9.3. Hershberger and Suri [126] also showed that a sequence of n insertions and deletions can be performed in amortized time $O(\log n)$ for each operation if the sequence of operations is known in advance. Moreover, their data structure can also handle queries (tangent line passing through a given point, intersection with a line, finding a vertex, and more generally any query that can be handled with binary search) in time $O(\log n)$, and this after any number of operations are performed.

The divide-and-conquer algorithm in dimension 3 was first proposed by Preparata and Hong in 1977 [191], but the first entirely correct description of the algorithm was given in 1987 by Edelsbrunner [89].

The hierarchical decomposition of 3-polytopes (see exercise 9.5) was invented by Dobkin and Kirkpatrick [85] who used it to compute the distance between two polyhedra. Edelsbrunner and Maurer [94] also used it to answer different types of queries on 3-polytopes (see exercise 9.6). Chazelle [45] also used the same kind of decomposition to compute the intersection of two 3-polytopes in linear time.

The algorithm in exercise 9.8 that computes the union of tricolored triangles is due to Boissonnat, Devillers, and Preparata [27].

Chapter 10

Linear programming

Several problems, geometric or of other kinds, use the notion of a polytope in d-dimensional space more or less implicitly. The preceding chapters show how to efficiently build the incidence graph which encodes the whole facial structure of a polytope given as the convex hull of a set of points. Using duality, the same algorithms allow one to build the incidence graph of a polytope defined as the intersection of a finite number of half-spaces. It is not always necessary, however, to explicitly enumerate all the faces of the polytope that underlies a problem. This is the case in linear programming problems, which are the topic of this chapter.

Section 10.1 defines what a linear programming problem is, and sets up the terminology commonly used in optimization. Section 10.2 gives a truly simple algorithm that solves this class of problem. Finally, section 10.3 shows how linear programming may be used as an auxiliary for other geometric problems. A linear programming problem may be seen as a shortcut to avoid computing the whole facial structure of some convex hull. Paradoxically, the application we give here is an algorithm that computes the convex hull of n points in dimension d. Besides its simplicity, the interest of the algorithm is mostly that its complexity depends on the output size as well as on the input size. Here, the output size is the number f of faces of all dimensions of the convex hull, and thus ranges widely from $O(1)$ (size of a simplex) to $\Theta(n^{\lfloor d/2 \rfloor})$ (size of a maximal polytope). It is therefore often useful to use an algorithm whose running time depends on the number of faces f effectively computed, rather than an algorithm whose complexity can only be bounded by the worst-case complexity of a polytope. The algorithm presented in section 10.3 runs in time $O(n^2 + f \log n)$ on a set of n points whose convex hull has f faces, which is competitive in dimensions d higher than 5. This algorithm relies on the existence of a shelling order of the polytope, which is defined and proved to exist in section 10.3.

10.1   Definitions

A linear programming problem consists of optimizing a linear function of d variables, where the variables must satisfy a given set of n linear constraints. The linear function to be minimized may be written as a dot product:

$$f(X) = V \cdot X,$$

where V is a given vector of $\mathbb{E}^d$ and X a variable vector in the same space. The linear constraints that X must satisfy may be written as

$$A_i \cdot X \leq a_i, \quad 1 \leq i = 1, \ldots, n,$$

where, for each $i = 1, \ldots, n$, A_i is a vector in $\mathbb{E}^d$ and a_i a real constant. Geometrically, each constraint can be expressed by the fact that the point X lies in a closed half-space. Denote by $\overline{H_i^+}$ the closed half-space that corresponds to the i-th constraint:

$$\overline{H_i^+} = \{X \in \mathbb{E}^d \ : \ A_i \cdot X \leq a_i\}.$$

The intersection $\bigcap_{i=1}^{n} \overline{H_i^+}$ is called the *feasible domain* of the problem. If the feasible domain is bounded and not empty, then it is a polytope $\mathcal{P}$ and the solution to the linear programming problem is a vertex of $\mathcal{P}$, or occasionally the set of points on a higher-dimensional face of $\mathcal{P}$. This face F of $\mathcal{P}$ is characterized as follows: the set of supporting hyperplanes of $\mathcal{P}$ along F includes a hyperplane normal to the direction V, and $\mathcal{P}$ is contained in the half-space bounded by this supporting hyperplane that contains the vector V. If the feasible domain is empty, the linear programming problem is termed *unfeasible*. On the contrary, if the feasible domain is infinite in the direction $-V$, the linear programming problem is termed *unbounded*. To have a bounded problem, one may always restrict the feasible domain to a large box by adding $2d$ constraints. The size of this box can be chosen in such a way that if the problem is bounded, the solution is guaranteed to lie in the box (see exercise 10.1).

For example, the following location problem about convex hulls is a linear programming problem in disguise: given n points $\{P_1, P_2, \ldots, P_n\}$ in $\mathbb{E}^d$, and a query point Q, the problem asks if Q belongs to the convex hull $conv(\{P_1, P_2, \ldots, P_n\})$. If an origin O is chosen inside this convex hull (for instance as the centroid of $d+1$ of the points), the polarity of center O (see section 7.1.3) lets the dual version of the problem be expressed as follows: given n half-spaces $\{\overline{P_i^{*+}} \ : \ i = 1, \ldots, n\}$ whose intersection $\mathcal{P}^\#$ is bounded, and a query hyperplane Q^*, determine whether Q^* intersects the polytope $\mathcal{P}^\#$. The half-space $\overline{P_i^{*+}}$ is the half-space bounded by the hyperplane P_i^* polar to P_i and that contains O, and Q^* is the polar hyperplane of Q. Point Q lies inside $\mathcal{P}$ if and only Q^* avoids the dual polytope $\mathcal{P}^\#$ (see section 7.1.3). The equation of Q^* is

$$Q \cdot X = 1,$$

and to locate point Q inside or outside $\mathcal{P}$ reduces to solving the following two problems:

1. Minimize the function $f_1(X) = Q \cdot X$ with the constraints that $X \in \overline{P_i^{*+}}$ for $i = 1, \ldots, n$.

2. Minimize the function $f_2(X) = -Q \cdot X$ with the same constraints.

Point Q is inside polytope $\mathcal{P}$ if and only if $1 < u_1$ or $1 > -u_2$ where u_1 and u_2 are the respective minima of $f_1(X)$ and $f_2(X)$.

10.2 Randomized linear programming

Several algorithms have been devised to solve linear programming problems. The deterministic ones with best asymptotic complexity perform the minimization in time linear in the number n of constraints, but exponential in the number d of variables. Some randomized algorithms have sub-exponential complexity in n and d. The algorithm described here is a randomized algorithm that runs in expected time $O(d^4 d! n)$. Its main advantage is its simplicity.

The algorithm

This algorithm is an incremental algorithm that adds the constraints one by one while maintaining the solution to the current linear programming problem. For the randomization, we assume that the constraints are inserted in random order.

To make the description simpler, assume for now that all linear programming problems at any incremental step are feasible, bounded, and have a unique solution. These assumptions will be relaxed afterwards.

Here is the algorithm: In an initial step, compute the optimal vertex of the polytope $\mathcal{P}_d$, given as the intersection of the half-spaces that correspond to the first d constraints:

$$\mathcal{P}_d = \bigcap_{j=1}^{d} \overline{H_j^+}.$$

Subsequently, the constraints $\overline{H_i^+}$ are added one by one. The algorithm computes the optimal vertex X_i of the polytope $\mathcal{P}_i = \bigcap_{j=1}^{i} \overline{H_j^+}$ knowing the optimal vertex X_{i-1} of the polytope $\mathcal{P}_{i-1}$. It proceeds as follows. If vertex X_{i-1} belongs to the half-space $\overline{H_i^+}$, then $X_i = X_{i-1}$, and nothing else is done for this step. Otherwise, vertex X_i necessarily belongs to hyperplane H_i, so we know one of the hyperplanes incident to the optimal vertex. To find the other hyperplanes, the algorithm recursively solves a $(d-1)$-dimensional linear programming problem with $d-1$ variables and $i-1$ constraints. The optimizing function for this problem is $f'(X) = V' \cdot X$, where V' is the projection of V onto the hyperplane H_i, and the constraints are the half-spaces $\overline{H'^+_j} = \overline{H^+_j} \cap H_i$ in H_i, for each $j = 1, \ldots, i-1$. The optimal vertex is therefore obtained by recursively solving problems of smaller dimension. The recursion stops when $d = 1$, since the feasible domain is an interval which may be computed easily in time linear in the number of constraints.

Let us now indicate how to relax the assumptions stated at the beginning of our description of the algorithm.

First of all, one of the sub-problems $\mathcal{P}_i$ may be unfeasible. But this will be detected when adding the i-th constraint H_i, since at the bottom of the recursion the feasible interval will be found empty. If so, the entire problem is also unfeasible and we may stop the algorithm without even adding the remaining constraints.

We also assumed that the linear programming problems under consideration are bounded. To guarantee this, we add constraints that bound a hyper-rectangular domain $\mathcal{P}_0$ known to contain the solution if the problem is bounded (see exercise 10.1). This limits the search space to the hyper-rectangle. In the initial step, the algorithm computes the optimal vertex of the hyper-rectangle $\mathcal{P}_0$, and the algorithm continues with the constraints $\overline{H^+_i}$, $i = 1, \ldots, n$. If the optimal vertex is found on the boundary of $\mathcal{P}_0$, then the problem is unbounded.

Finally, the algorithm demands that the solution to each sub-problem $\mathcal{P}_i$ be unique, so that this solution does not depend on the order in which the constraints are inserted. To enforce this, when the optimal solution is a face of $\mathcal{P}_i$ of dimension $k > 0$, we let the algorithm systematically choose the vertex of this face which has the highest lexicographic order: if the x_d-coordinates of several vertices are equal, we compare their x_{d-1}-coordinates, etc., until the optimal vertex is found.

Randomized analysis of the algorithm

The analysis of the algorithm is also particularly simple. The order in which the constraints are added is assumed to be random, so we evaluate the performances of the algorithm on the average over all possible permutations. We prove that the expected running time of the algorithm is $O(d^4 d! n)$ when the linear programming problem has d variables and n constraints. More precisely, let $t(d, i)$ be the expected time to insert the i-th constraint into a linear programming problem with d variables and i constraints. We prove by induction on d and i that $t(d, i)$ is $O(d^4 d!)$. Indeed, if d equals 1 each insertion is processed in constant time. When $d > 1$, inserting the i-th constraint reduces to testing if X_{i-1} belongs to the half-space $\overline{H^+_i}$, except when X_i differs from X_{i-1}, in which case a linear programming problem with $d-1$ variables and $i-1$ constraints has to be solved. The latter case occurs only when the i-th constraint is bounded by one of the d constraint hyperplanes intersecting in X_i. Knowing the first i constraints, this happens with conditional probability d/i. Furthermore, since this conditional probability does not depend upon the subset formed by the first i constraints, d/i is simply the probability that X_i differs from X_{i-1}. Testing whether X_{i-1} belongs to the half-space $\overline{H_i^+}$ entails computing the sign of a $d \times d$ determinant. The cost of this operation, which we denote by $f(d)$, is $O(d^3)$. Thus $t(d, i)$ obeys the recursive equation

$$t(d, i) = f(d) + \frac{d}{i} \left((i - 1)t(d - 1, i - 1) \right),$$

which yields

$$
\begin{aligned}
t(d, i) &\leq f(d) + dt(d - 1, i - 1) \\
&\leq f(d) + df(d - 1) + \cdots + d! f(1) \\
&= O(d^4 d!)
\end{aligned}
$$

Note finally that the hyperplanes being in general position is not a requirement for the algorithm. This assumption is only needed in the analysis of the algorithm. In fact, a perturbation argument clearly shows that when more than d hyperplanes intersect in one point, the probability that the optimal vertex X_i of sub-problem $\mathcal{P}_i$ is no longer the optimal vertex of sub-problem $\mathcal{P}_{i+1}$ is only smaller, so that the average running time only decreases.

The following theorem summarizes the results of this section.

Theorem 10.2.1 *A linear programming problem with d variables and n constraints may be solved by an incremental randomized algorithm in expected time $O(d^4 d! n)$.*

10.3 Convex hulls using a shelling

In dimension d, the size of the convex hull may vary greatly between $(d + 1)!$ (number of faces of a d-simplex) and $\Omega(n^{\lfloor d/2 \rfloor})$ (complexity of a cyclic polytope, see subsection 7.2.7). It is conceivable that there might be a real advantage in using an algorithm whose running time is a function of the size of the effectively computed convex hull rather than the maximal size of this convex hull in the worst case. The algorithm we present here computes the convex hull of a set of n points in a d-dimensional space in time $O(n^2 + f \log n)$ where f is the number of facets of the convex hull. This algorithm is therefore of great interest in dimensions higher than 5. The main notion involved in its design is that of a *shelling* of a polytope by a line. This notion, which is of interest in its own right, is described under the next heading. A shelling of a polytope is merely a linear order on the facets of that polytope and the algorithm we propose here computes the facets of a polytope in such an order. Let us call *extreme points* the vertices of the convex hull. The algorithm needs, in a first phase, to identify the extreme points among all the points in the set, and for each extreme point, to determine the first facet that contains this point in the shelling order. This is where linear programming helps.

Shelling of a polytope

Let $\mathcal{P}$ be a polytope. A *shelling* of $\mathcal{P}$ is a sequence $(F_1, F_2, \ldots, F_m)$ enumerating all the facets of $\mathcal{P}$ such that, for all i, $1 < i < m$, the set $\{F_1, F_2, \ldots, F_i\}$ of the first i facets is a connected set of facets.[1]

A line that intersects the interior of a polytope $\mathcal{P}$ is said to be *acceptable* for $\mathcal{P}$ if it intersects all the hyperplanes supporting $\mathcal{P}$ along a facet at pairwise different points. The following discussion shows that any oriented acceptable line defines a linear order on the facets of $\mathcal{P}$ which is a shelling.

Let then $\mathcal{P}$ be a polytope, L an oriented line, acceptable for $\mathcal{P}$, and O a point on L and in the interior of $\mathcal{P}$. Imagine a point V that moves along L, from O towards infinity in the direction given by the orientation of the line, then reappearing from the opposite infinity and back towards O. The hyperplanes supporting $\mathcal{P}$ along its facets are encountered in a certain order by V as it moves along L, and this order is entirely determined since L is acceptable (see figure 10.1). This order gives the shelling of $\mathcal{P}$ induced by L. Let U be the directing vector of L. Then L may be parameterized by

$$L = \left\{ V(t) = \frac{-U}{t} \; : \; t \in \,]-\infty, +\infty[\,\setminus\, \{0\} \right\}$$

which simulates the motion of V. To each facet F_i of $\mathcal{P}$ there corresponds a parameter t_i such that the hyperplane H_i supporting $\mathcal{P}$ along F_i is intersected by L at $V(t_i)$. The oriented acceptable line orders the facets of $\mathcal{P}$ in increasing order of their parameters t_i.

Lemma 10.3.1 *The sequence $(F_1, F_2, \ldots, F_m)$ of facets of a polytope $\mathcal{P}$ ordered by an acceptable oriented line is a shelling of the polytope $\mathcal{P}$.*

Proof. For each value $t \in \,]t_1, t_m[$ of the parameter, the point $V(t)$ on L is exterior to the polytope $\mathcal{P}$. We may color the facets of $\mathcal{P}$ with respect to $V(t)$ as in section 8.3.[2] If t is negative and belongs to $]t_i, t_{i+1}[$, then the facets $\{F_1, F_2, \ldots, F_i\}$

---

[1] A set of facets of a polytope $\mathcal{P}$ is said to be connected if it determines a connected subgraph of the adjacency graph of $\mathcal{P}$.

[2] It is helpful to recall the terminology of section 8.3 here. Let $\mathcal{P}$ be a polytope, F a facet of this polytope, H the hyperplane that supports $\mathcal{P}$ along F, H^+ the half-space bounded by H that contains the interior of $\mathcal{P}$, and H^- the other half-space bounded by H. Facet F of $\mathcal{P}$ is blue with respect to a point V if V belongs to H^+, and red if V belongs to H^-.

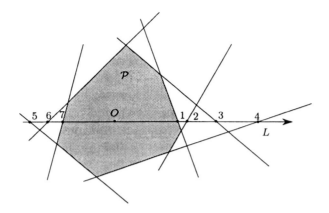

Figure 10.1. The shelling of a polytope $\mathcal{P}$ induced by a line L.

whose corresponding parameters are smaller than t are red with respect to $V(t)$, whereas the facets whose corresponding parameters are greater than t are blue with respect to $V(t)$. If t is positive and belongs to $]t_i, t_{i+1}[$, then the facets $\{F_1, F_2, \ldots, F_i\}$ whose corresponding parameter is smaller than t are blue with respect to $V(t)$, whereas the facets whose corresponding parameter is greater than t are red with respect to $V(t)$. Therefore, the connectedness of the sets of facets $\{F_1, F_2, \ldots, F_i\}$ for every $i = 1, \ldots, m$ is a consequence of lemma 8.3.3. $\square$

Shelling for convex hulls

The algorithm described here builds the convex hull $conv(\mathcal{S})$ of a set $\mathcal{S}$ of n points in the space $\mathbb{E}^d$. The underlying principle of this algorithm is to enumerate the facets of $\mathcal{P}$ in turn, in the order given by the shelling of $conv(\mathcal{S})$ induced by an acceptable oriented line.

The set $\mathcal{S} = \{P_1, P_2, \ldots, P_n\}$ of points is supposed to be in general position. Each facet of $conv(\mathcal{S})$ is thus a $(d-1)$-simplex and is entirely characterized by the set of its d vertices. The algorithm not only builds the facets of $conv(\mathcal{S})$ but also their adjacency graph. From this graph, it is easy to reconstruct the entire incidence graph of the polytope $conv(\mathcal{S})$ (see exercise 8.2).

Before explaining the details of the algorithm, we must clarify how we discover the facets of $conv(\mathcal{S})$ at all, and moreover how this can be done in the order of the shelling induced by a line L. Suppose therefore that there is a line L acceptable for $conv(\mathcal{S})$. As before, we agree that L is given a parametric representation:

$$L = \left\{ V(t) = \frac{-U}{t} \ : \ t \in \ ]-\infty, +\infty[\ \backslash \ \{0\} \ \right\}.$$

Let $(F_1, F_2, \ldots, F_m)$ be the sequence of facets of $conv(\mathcal{S})$ in the shelling of $conv(\mathcal{S})$ induced by L. We also denote by t_i the parameter of the intersection point $H_i \cap L$ of L with the hyperplane H_i that contains facet F_i. At a given stage of the algorithm, the first i facets $(F_1, F_2, \ldots, F_i)$ have been discovered, together with their adjacency relationships. We call *i-horizon* (or horizon for short when i is clearly understood) the boundary of the union $\bigcup_{j=1}^{i} F_j$ of these facets. The horizon is made of the $(d-2)$-faces of $conv(\mathcal{S})$ that are incident to only one of the facets in $(F_1, F_2, \ldots, F_i)$ together with the faces of all dimensions contained in these $(d-2)$-faces.

The algorithm uses the following lemma.

Lemma 10.3.2 *Suppose that the first i facets $(F_1, F_2, \ldots, F_i)$ of the shelling have been discovered, together with their adjacency relationships.*

- *The horizon is made of the faces of $conv(\mathcal{S})$ that are purple[3] with respect to $V(t)$ for any $t \in]t_i, t_{i+1}[$.*

- *The horizon is isomorphic to the set of faces of a $(d-1)$-polytope with respect to the incidence relationships among the faces of $conv(\mathcal{S})$.*

- *For any $(d-2)$-face G of the horizon and any point $V(t)$ on L with parameter $t \in [t_i, t_{i+1}]$, the affine hull of $\{G \cup V(t)\}$ is a hyperplane that supports $conv(\mathcal{S})$ along G.*

Proof. A face G of $conv(\mathcal{S})$ belongs to the horizon if and only if it is contained in at least one facet F_k with parameter $t_k \leq t_i$, and one facet F_l with parameter $t_l \geq t_{i+1}$. It follows from the proof of lemma 10.3.1 that G is purple with respect to any point $V(t)$ on L with parameter $t \in]t_k, t_l[$, and in particular with respect to a point $V(t)$ with parameter $t \in]t_i, t_{i+1}[$. The second statement in the lemma is a direct consequence of lemma 8.3.4. Let us prove the third statement. Let G be a $(d-2)$-face G of the horizon. G is the intersection of two facets F_l and F_k of $conv(\mathcal{S})$ such that $t_k \leq t_i \leq t_{i+1} \leq t_l$ and G is purple with respect to any point $V(t)$ of L with parameter $t \in]t_k, t_l[$. Lemma 8.3.1 thus shows that, for any $t \in]t_k, t_l[$, $conv(\{G \cup V(t)\})$ is a facet of $conv(\mathcal{S} \cup V(t))$, which also proves that the affine hull $aff(\{G \cup V(t)\})$ is a hyperplane that supports $conv(\mathcal{S})$ along G. If $t = t_k$ (resp. $t = t_l$), $conv(\{G \cup V(t)\})$ is exactly F_k (resp. F_l) and the affine hull $aff(\{G \cup V(t)\})$ is a hyperplane that supports $conv(\mathcal{S})$ along F_k (resp. F_l). $\square$

Suppose therefore that the algorithm has already built the portion of the adjacency graph that contains the first i facets of the shelling of $conv(\mathcal{S})$ induced by L. Lemma 10.3.1 implies that the $(i+1)$-st facet F_{i+1} is necessarily incident to one (or several) $(d-2)$-faces of the horizon. Two cases may arise (see figure 10.2).

---

[3]A face of a polytope $\mathcal{P}$ is purple with respect to a point V if it is contained in a blue facet and in a red facet.

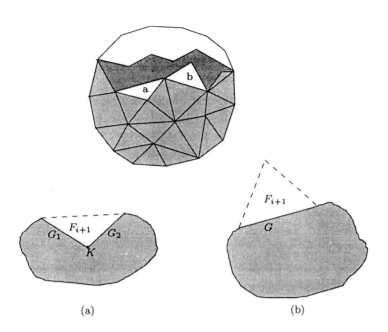

Figure 10.2. Shelling of a polytope :
   (a) Facet F_{i+1} is incident to several $(d-2)$-faces on the horizon.
   (b) Facet F_{i+1} is incident to only a single $(d-2)$-face on the horizon.

First case. Facet F_{i+1} is incident to several $(d-2)$-faces on the horizon. Let G_1 and G_2 be any two $(d-2)$-faces on the horizon that are incident to F_{i+1}. Since F_{i+1} is a $(d-1)$-simplex, G_1 and G_2 must be incident to a common $(d-3)$-face $K = G_1 \cap G_2$. Both faces G_1 and G_2 therefore have $d-1$ vertices, of which $d-2$ belong to K. Together, D_1 and G_2 have $d+1$ distinct vertices. Facet F_{i+1} is the convex hull of $G_1 \cup G_2$.

Second case. The intersection between facet F_{i+1} and the union $\bigcup_{j=1}^{i} F_j$ of the previous facets only yields a single $(d-2)$-face on the horizon. Then facet F_{i+1} is necessarily the convex hull $conv(G \cup \{P\})$ of G and a vertex P of $conv(\mathcal{S})$ that belongs neither to G nor to any of the already discovered facets F_j, $j \leq i$. This implies that F_{i+1} is the first facet in the shelling that has admits P as a vertex.

To summarize, either F_{i+1} is the convex hull of two $(d-2)$-faces on the horizon that share a common $(d-3)$-face, or it is the first facet in the shelling that contains some vertex P of $\mathcal{S}$.

Suppose for now that, in an initial phase, the algorithm has determined for every point P in $\mathcal{S}$ whether it is a vertex of the convex hull $conv(\mathcal{S})$ and, if so, which facet F_P is the first facet in the shelling that contains P. Each vertex of $conv(S)$ has a pointer to a record that contains this $(d-1)$-simplex F_P, the corresponding hyperplane H_P (the affine hull of F_P) and the corresponding value t_P of the parameter of the intersection point $H_P \cap L$. Let $\mathcal{F}'_i$ be the set of simplices F_P whose parameter t_P is greater than t_i.

Since the horizon is isomorphic to a $(d-1)$-polytope (lemma 10.3.2), each $(d-3)$-face on the horizon is incident to two $(d-2)$-faces of the horizon (theorem 7.1.7). For each $(d-3)$-face K on the horizon we store a pointer to a record that contains the simplex F_K which is the convex hull of the $(d-2)$-faces on the horizon that are incident to K, and also store the corresponding hyperplane H_K (the affine hull of F_K) and the corresponding value t_K of the parameter of the intersection point $H_K \cap L$. Let $\mathcal{F}''_i$ be the set of simplices F_K whose parameter t_K is greater than t_i.

Lemma 10.3.3 *The next facet F_{i+1} in the shelling is the simplex F^* of $\mathcal{F}_i =$ $\mathcal{F}'_i \cup \mathcal{F}''_i$ that has the smallest parameter t^*.*

Proof.   The previous comments show that the face F_{i+1} is in $\mathcal{F}_i$. Also, the affine hull H_{i+1} of F_{i+1} is a hyperplane that supports $conv(S)$ and, among all the simplices of $\mathcal{F}_i$ whose affine hull supports $conv(S)$, F_{i+1} is defined as the one with the smallest parameter. Thus it suffices to show that the affine hull H^* of the simplex F^* in $\mathcal{F}_i$ whose parameter is minimal is a hyperplane that supports $conv(S)$. Now if F^* is a simplex of $\mathcal{F}'_i$ stored in a record that is pointed to by a vertex of $conv(S)$, then H^* is indeed a hyperplane that supports $conv(S)$, by its definition. Thus suppose that F^* is a simplex of $\mathcal{F}''_i$ pointed to by some $(d-3)$-faces on the horizon, and let K be such a $(d-3)$-face. Let G_1 and G_2 be the two $(d-2)$-faces of the horizon incident to K. From lemma 10.3.2, the hyperplane $aff(\{G_1 \cup V(t)\})$ supports $conv(S)$, for any $t \in [t_i, t_{i+1}]$. In particular, since $t_i < t^* \leq t_{i+1}$, $H^* = aff(\{G_1 \cup V(t^*)\})$ supports $conv(S)$. $\square$

To build the facets of the polytope $conv(S)$ in the order of its shelling induced by a line L, the preceding lemma suggests that the set $\mathcal{F}_i$ of potential facets be maintained and ordered according to their parameters.

But before this, we must explain how to determine, for each point P in S, whether it is a vertex of the convex hull and, if so, how to find the first facet F_P in the shelling that contains P. This is where linear programming comes in.

Lemma 10.3.4 *To determine whether a point P of S is a vertex of the convex hull $conv(S)$ and, if so, to find the first facet F_P in the shelling that contains P can be cast into a linear programming problem with $n-1$ constraints and $d-1$ variables.*

Proof. Line L is acceptable for $conv(\mathcal{S})$, so there is a point O in L inside $conv(\mathcal{S})$. Let H be a hyperplane in the space $\mathbb{E}^d$ that does not pass through the origin O and H^* be the pole of H in the polarity centered at O (see section 7.1.3):

$$H = \{X \in \mathbb{E}^d \ : \ H^* \cdot X = 1\}.$$

Then H passes through P if and only if H^* satisfies:

$$H^* \cdot P = 1, \tag{10.1}$$

and H^* supports $\mathcal{P} = conv(\mathcal{S})$ if and only if

$$\forall P' \in \mathcal{S}, \ H^* \cdot P' \le 1. \tag{10.2}$$

Also, H intersects line L at a point with parameter $t(H)$ defined by

$$t(H) = -U \cdot H^*.$$

Finding the hyperplane that supports $\mathcal{P}$, contains P, and minimizes the parameter t of its intersection point with L reduces to finding the vector H^* that satisfies equation 10.1 and inequality 10.2, and that also minimizes the linear functional $t(H^*)$. By changing the coordinate system, segment OP can be made to lie on the x_d-axis. Equation 10.1 determines the d-th component of vector H^* and the system given by inequalities 10.2 appears as a set of $n - 1$ linear constraints over the remaining $d - 1$ components of H^*. This system is thus a linear programming problem with $n - 1$ constraints and $d - 1$ variables. If the problem is unfeasible, then there is no supporting hyperplane that passes through P and this means that P lies inside the convex hull $\mathcal{P}$. If not, the general position assumption means that the solution of the linear programming problem is uniquely defined when line L is acceptable. This unique solution is therefore a vertex of the feasibility domain that belongs to $d - 1$ hyperplanes corresponding to $d - 1$ constraints in the system (10.2). Let $\mathcal{S}_P$ be the subset of $d - 1$ points in $\mathcal{S} \setminus \{P\}$ that correspond to these constraints. The first facet F_P in the shelling that contains P is then $F_P = conv(\mathcal{S}_P \cup \{P\})$. □

We can now explain the algorithm in its entirety.

The algorithm

First of all, we must choose an origin O in the interior of $conv(\mathcal{S})$ and an oriented line L that contains O. We assume that L is acceptable for $conv(\mathcal{S})$. If at any stage of the algorithm it appears that L is not acceptable, the vector U directing L can always be perturbed a little to make L acceptable without modifying the result of the previous computations.

The algorithm builds the facets of the convex hull $conv(\mathcal{S})$ in the order of the shelling of the polytope $conv(\mathcal{S})$ induced by L. It also builds the adjacency graph of the facets. Each facet is described by the list of its vertices. In addition to the adjacency graph of the current set of facets, the algorithm maintains the following three data structures.

- The *horizon graph* denoted by $\mathcal{H}$ contains a node for each $(d-2)$-face on the horizon and an edge for each pair of $(d-2)$-faces on the horizon that share a common $(d-3)$-face. Each $(d-2)$-face on the horizon is incident to exactly one facet that is already stored in the adjacency graph, so the corresponding node in the horizon graph contains a pointer to this facet.

- A dictionary $\mathcal{D}$ contains, for each vertex P of $conv(\mathcal{S})$, the $(d-1)$-tuple $\mathcal{S}_P$ of points of $\mathcal{S}$ such that $F_P = conv(\mathcal{S}_P \cup P)$ is the first facet of the shelling that contains P. Each $(d-1)$-tuple $\mathcal{S}_P$ corresponds to a $(d-2)$-face of $conv(\mathcal{S})$ that will be on the horizon at some point during the execution of the algorithm. Each item in the dictionary $\mathcal{D}$ contains a pointer that will, at that time, point to the corresponding node in the horizon graph.

- A priority queue $\mathcal{Q}$ maintains the set $\mathcal{F}_i$ of facets that, when the first i facets $(F_1, \ldots, F_i)$ of the shelling have been computed, are candidates for the next facet F_{i+1}. To each simplex of $\mathcal{F}_i$ corresponds a value of the parameter that is the parameter of the intersection point of the affine hull of this simplex with line L. The simplices in $\mathcal{F}_i$ are ordered in the priority queue by increasing values of their parameters. The priority queue stores the d vertices of each simplex, and a pointer. If the simplex is the first face F_P incident to a vertex P of $conv(\mathcal{S})$, the pointer gives the entry in the dictionary $\mathcal{D}$ corresponding to the $(d-1)$-tuple $\mathcal{S}_P$. If the simplex is associated with a $(d-3)$-face K on the horizon, the pointer gives the edge of the horizon graph that corresponds to K.

Initialization. In a starting phase, the algorithm initializes the structures $\mathcal{Q}$ and $\mathcal{D}$. For this, it must solve the linear programming problem corresponding to each point P. If the problem is unfeasible, then P is not a vertex of the convex hull and can be discarded from further consideration. Otherwise, the solution to the linear programming problem gives a facet F_P which is the first facet incident to P in the shelling of $conv(\mathcal{S})$ by L. This facet is the convex hull $conv(\mathcal{S}_P \cup \{P\})$ of P and a subset $\mathcal{S}_P$ of $\mathcal{S}$ with $d-1$ elements. The priority queue is initialized with the facet F_P for each $P \in \mathcal{S}$ and the subsets $\mathcal{S}_P$ are inserted in the dictionary $\mathcal{D}$.

First phase. Let t_1 be the smallest parameter of the facets present in the priority queue $\mathcal{Q}$ immediately after the initial phase. This value t_1 is achieved by the simplices F_P corresponding to the points P in some subset $\mathcal{S}_1$ of d points in $\mathcal{S}$.

The convex hull $conv(\mathcal{S}_1)$ is the first face F_1 in the shelling. The horizon graph initially contains the complete graph on d nodes corresponding to all the $(d-2)$-faces of the $(d-1)$-simplex F_1. Each subset of $\mathcal{S}_1$ of size $d-1$ is in the dictionary $\mathcal{D}$ and corresponds to a node in the horizon graph. The item in the dictionary $\mathcal{D}$ that corresponds to this subset is located and its corresponding pointer updated. All the simplices of parameter t_1 are retrieved from the priority queue.

Current phase. As long as the priority queue $\mathcal{Q}$ is not empty, the algorithm extracts from $\mathcal{Q}$ the set of candidates with the smallest parameter t^*, uses it to determine the next facet in the shelling, and updates the adjacency graph of the facets and the data structures $\mathcal{H}$, $\mathcal{D}$ and $\mathcal{Q}$.

If t^* is the parameter of a simplex F_P that corresponds to a point P in $\mathcal{S}$, the subset $\mathcal{S}_P$ of vertices of F_P minus P itself is located in the dictionary $\mathcal{D}$. The pointer associated with this item allows the retrieval of the node in the horizon graph that corresponds to $\mathcal{S}_P$. The face F_P may then be added to the adjacency graph. In the horizon graph, the node that corresponds to $\mathcal{S}_P$ is replaced by a complete graph with $d-1$ nodes, each of which corresponds to a subset of $\mathcal{S}_P \cup \{P\}$ with $d-1$ points other than $\mathcal{S}_P$ itself.

In the opposite case, t^* is the parameter of one or more $(d-3)$-faces on the horizon. Let $\mathcal{K}$ be the set of $(d-3)$-faces K for which $t_K = t^*$, and let $\mathcal{G}$ be the set of $(d-2)$-faces on the horizon that are incident to the faces of $\mathcal{K}$. The facet that must be added to the shelling is the $(d-1)$-simplex F_G, the convex hull of the vertices of any two faces in $\mathcal{G}$. This facet F_G is adjacent to all the facets that have already been built and that are incident to the $(d-2)$-facets of $\mathcal{G}$. Let g be the number of faces in $\mathcal{G}$. The pair $(\mathcal{G}, \mathcal{K})$ is a complete subgraph with g nodes in the horizon graph. This subgraph is replaced by the complete subgraph with $d-g$ nodes that correspond to the $(d-2)$-faces of F_G that do not belong to $\mathcal{G}$.

In either case, all the hyperplanes of parameter t^* are extracted from the priority queue. Each new edge in the horizon graph (resp. each edge that was removed from the horizon graph) corresponds to a $(d-3)$-face K' and the convex hull $F_{K'}$ of the two $(d-2)$-faces on the horizon that are incident to K' are inserted into $\mathcal{Q}$ (resp. removed from $\mathcal{Q}$) if its parameter t' is greater than t^*. Also, for each new node in the horizon graph, the algorithm checks whether its set of vertices is stored in the dictionary $\mathcal{D}$ and, if so, updates the corresponding pointer.

When the priority queue $\mathcal{Q}$ is empty, the algorithm has discovered all the $(d-1)$-facets of the convex hull and their adjacency graph. Using this, it may build the entire incidence graph of the polytope $conv(\mathcal{S})$ (see exercise 8.2).

Analysis of the algorithm

Theorem 10.3.5 *Let $\mathcal{S}$ be a set of n points in general position in $\mathbb{E}^d$. The algorithm above computes the convex hull $conv(\mathcal{S})$ in expected time $O(n^2 + f \log n)$, where f stands for the number of facets of $conv(\mathcal{S})$.*

Proof. If the dimension of the space is considered as a constant, the n linear programming problems in the initial phase may be solved in time $O(n^2)$. Let us count the number of operations performed for each facet discovered by the algorithm. The number of nodes added to or removed from the horizon graph is d, and the number of queries and updates into the dictionary $\mathcal{D}$ is $O(d)$. The number of edges created or removed in the horizon graph is $d(d-1)$ and the number of queries and updates into the priority queue $\mathcal{Q}$ is $d(d-1)$ as well. The size of the dictionary $\mathcal{D}$ is at most $O(n^{\lfloor d/2 \rfloor})$ (by the upper bound theorem 7.2.5) and the same bound is valid for the size of $\mathcal{Q}$. Each operation on the data structures can therefore be carried out in time $O(\log n)$. For each facet discovered by the algorithm, the running time is therefore $O(\log n)$. Finally, the computation of the incidence graph from the adjacency graph can be performed in time $O(f)$ (see exercise 8.2).   □

10.4   Exercises

Exercise 10.1 (Unbounded linear programming problems) Consider a (possibly unbounded) linear programming problem, with the constraints expressed as

$$X \cdot A_i \leq a_i, \quad 1 \leq i = 1, \dots, n,$$

where, for each $i = 1, \dots, n$, A_i is a vector in $\mathbb{E}^d$ and a_i a constant. By scaling if necessary, one may assume that all the components of A_i, $i = 1, \dots, n$, are integers. Let $a = \max_{1 \leq i \leq n}(a_i)$, and

$$A = max_{\substack{1 \leq i \leq n \\ 1 \leq j \leq d}} (A_{i,j})$$

if $A_{i,j}$ denotes the j-th component of the vector A_i. Prove that, if a solution X to the linear programming problem exists, any of its components X_j, $j = 1, \dots, d$, satisfies

$$X_j \leq d^{d/2} A^{d-1} a$$

Hint: The point X is the solution of a $d \times d$ system of linear equations whose coefficients are bounded by A on one side and by a for the constant side. The determinant D of this system is a non-zero integer, and thus $|D|$ is at least 1. Cramer's rules imply that X_j is the quotient by D of a $d \times d$ determinant that has the coefficients of $d-1$ columns bounded by A and the coefficients of one column bounded by a.

Exercise 10.2 (Separability) Let $\mathcal{S}_1$ and $\mathcal{S}_2$ be two sets of points in $\mathbb{E}^d$. Show that there is an algorithm that decides in linear time whether the sets can be separated by a hyperplane, that is whether there exists H such that $\mathcal{S}_1$ is contained in one of the hyperplanes bounded by H and $\mathcal{S}_2$ in the other.

Exercise 10.3 (Ray shooting) Let $\mathcal{S}$ be a set of points in general position in $\mathbb{E}^d$. Choose an origin O inside the convex hull $conv(\mathcal{S})$, and a vector V in $\mathbb{E}^d$. Show how to compute in linear time the facet of $conv(\mathcal{P})$ that is intersected by the ray originating at O in the direction V.

Exercise 10.4 (Intersection of half-spaces) Let $Q = \bigcap_{j=1}^{m} H_j^+$ be an intersection of m half-spaces in $\mathbb{E}^d$. Determine whether Q is empty and, if not, find a point O inside the intersection Q in linear time.

Exercise 10.5 (Minimum area annulus) An *annulus* is the portion of the plane contained between two concentric circles. Let n points in the plane be given. Find the annulus of minimal area that contains all these points.

Hint: This can be shown to be a linear programming problem if we use the space of spheres that is introduced in chapter 17. More directly, let $\{P_i(u_i, v_i) : 1 \leq i \leq n\}$ be the set of n points in the plane. The problem can be expressed as deciding whether there is a center (x, y) and two radii r_1 and r_2 such that $r_2^2 - r_1^2$ is minimal subject to the $2n$ constraints

$$r_1^2 \leq (x - u_i)^2 + (y - u_i)^2 \leq r_2^2.$$

This optimization problem can be cast into a linear programming problem if instead of the variables x, y, r_1, and r_2, we express the constraints in terms of the variables x, y, $x^2 + y^2 - r_1^2$ and $x^2 + y^2 - r_2^2$.

Exercise 10.6 (Convex hull in time $O(n \log h)$) Let S be a set of n points in $\mathbb{E}^2$. Show that if S has only h extreme points, the following algorithm computes the lower hull $conv^-(S)$ of S in time $O(n \log h)$. (The lower hull is defined in exercise 7.14).

If $n \leq 2$, then return the segment $conv(S)$.
Otherwise

1. Split S into two balanced subsets by a vertical line Δ.

2. Find the edge E of $conv^-(P)$ that intersects Δ (see exercise 10.3).

3. Recursively compute the lower hulls of each subset $P \cap \overline{\Delta_l^-}$ and $P \cap \overline{\Delta_r^+}$, where $\overline{\Delta_l^-}$ (resp. $\overline{\Delta_r^+}$) is the closed half-space on the left (resp. on the right) of the vertical line Δ_l (resp. Δ_r) that passes through the left (resp. right) endpoint of E.

Exercise 10.7 (Bisection) We denote by $\text{LP}(\mathcal{H}, V)$ the linear programming problem with d variables, where $\mathcal{H}$ is the set of constraints and V is the vector of $\mathbb{E}^d$ defining the minimization function. Assume that the set of constraints $\mathcal{H}$ has n elements. Show that it is possible to know on which side of a hyperplane H parallel to V the solution to $\text{LP}(\mathcal{H}, V)$ lies (assuming that there is one), by solving at most five linear programming problems with $d - 1$ variables and n constraints.

Hint: The first problem to be solved is the restriction $\text{LP}(H \cap \mathcal{H}, V)$ to H of $\text{LP}(\mathcal{H}, V)$. If this problem is unbounded, then $\text{LP}(\mathcal{H}, V)$ itself is either unbounded or unfeasible (the latter happens when some constraint in $\mathcal{H}$ is parallel to $\mathcal{H}$). If there is an optimal solution to this linear programming problem, however, consider two hyperplanes H' and H'' parallel to H on either side of H. Solve the two lower-dimensional linear programming problems $\text{LP}(H' \cap \mathcal{H}_{opt}, V)$ in H' and $\text{LP}(H'' \cap \mathcal{H}_{opt}, V)$ in H'', where $\mathcal{H}_{opt}$ is the minimal set of constraints that defines the solution to $\text{LP}(H \cap \mathcal{H}, V)$, and compare the optimal values of these sub-problems. Finally, if $\text{LP}(H \cap \mathcal{H}, V)$ is unfeasible, then consider the three following subsets of $\mathcal{H}$ in turn: the subset $\mathcal{H}^0$ of constraints whose hyperplanes are parallel to V, the subset $\mathcal{H}^+$ of constraints whose half-spaces are unbounded in the direction of $-V$, and the subset $\mathcal{H}^-$ of constraints whose half-spaces are unbounded in the direction of V. The amount of unfeasibility of $\mathrm{LP}(H \cap \mathcal{H}, V)$ can be defined as the difference between the optimal value of $\mathrm{LP}(H \cap \mathcal{H}^+, V)$ and that of $\mathrm{LP}(H \cap \mathcal{H}^-, -V)$. It suffices to compare the amount of unfeasibility of $\mathrm{LP}(H' \cap \mathcal{H}, V)$ with that of $\mathrm{LP}(H'' \cap \mathcal{H}, V)$.

Exercise 10.8 (Prune-and-search) Let LP be a linear programming problem with d variables. Let V be the vector that defines the optimizing function and H_0 a hyperplane perpendicular to V. The constraints of LP are put into three categories: $\mathcal{H}^0$, $\mathcal{H}^+$ and $\mathcal{H}^-$ as in the previous exercise.

1. Let H_1 and H_2 be two hyperplanes corresponding to constraints of LP that are in the same category. Let H_{12} be the hyperplane parallel to V that contains $H_1 \cap H_2$. Show that if the solution of LP is known to lie on either side of H_{12}, then at least one of the constraints may be pruned away.

2. In dimension 2, show that it is possible to prune half of the constraints by locating the solution of LP with respect to a line H parallel to V, which is explained in the previous exercise.

3. Generalization to dimension $d > 2$: show that a fixed fraction αn of the constraints may be pruned away (with $\alpha = 2^{1-2^d}$) by locating the solution of LP with respect to 2^{d-1} hyperplanes parallel to V.

4. Using this, devise an algorithm that solves a problem LP with d variables and n constraints in time $c(d)n$ where $c(d) = O(2^{O(2^d)})$.

Hint: (For the second question.) Pair off the lines in $\mathcal{H}^+$, and do the same to the lines in $\mathcal{H}^-$. Then project onto H_0 the intersection of the two lines in each pair, as well as the hyperplanes in $\mathcal{H}^0$, and choose for H the line parallel to V that passes through the median of the projections.

Exercise 10.9 (Minimum enclosing circle) Given n points in $\mathbb{E}^2$, find the circle with the smallest radius that contains all these points.

Hint: In a first step, show that the restricted problem where the center of the circle lies on a given line can be solved in time $O(n)$. Show also that it can be decided in time $O(n)$ on which side of this line the center of the (unrestricted) minimum enclosing circle lies. Then apply to this problem the prune-and-search method described in the previous exercise.

Exercise 10.10 (Maximum inscribed sphere) Let $\mathcal{Q}$ be a polytope given as the intersection of m half-spaces in $\mathbb{E}^d$. Determine the sphere inscribed in $\mathcal{Q}$ that has the greatest possible radius.

Exercise 10.11 (LP-type problems) Let $\mathcal{H}$ be a finite set (whose elements are called the *constraints*) and f a function on $2^{\mathcal{H}}$ that takes its values in a totally ordered set. An optimization problem is to find the minimal subset $\mathcal{B}$ of $\mathcal{H}$ such that $f(\mathcal{B}) = f(\mathcal{H})$. Such a problem is an *LP-type problem* if the following two conditions are true:

Monotonicity: For any two subsets $\mathcal{F}$ and $\mathcal{G}$ of $\mathcal{H}$ such that $\mathcal{F} \subset \mathcal{G} \subset \mathcal{H}$, we have $f(\mathcal{F}) \le f(\mathcal{G})$.

Locality: For any two subsets $\mathcal{F}$ and $\mathcal{G}$ of $\mathcal{H}$ such that $\mathcal{F} \subset \mathcal{G} \subset \mathcal{H}$ and $f(\mathcal{F}) = f(\mathcal{G})$, and for any $H \in \mathcal{H}$, we have $f(\mathcal{G}) < f(\mathcal{G} \cup \{H\})$ if and only if $f(\mathcal{F}) < f(\mathcal{F} \cup \{H\})$.

Show that any linear programming problem is an LP-type problem. Show that the minimum enclosing circle problem (see exercise 10.9) is also an LP-type problem, and so is its generalization to minimum enclosing spheres in any dimension.

10.5 Bibliographical notes

The concept of a linear programming problem was created around 1947 when Dantzig invented the *simplex algorithm* to solve scheduling problems for the U.S. Air Force. This method is still widely used, and performs well in practice. Nevertheless, it has been proved to require in the worst case a time that is exponential in the number of the variables and constraints [138]. Linear programming algorithms devised more recently by Khachiyan [135] and Karmarkar [134] run in time polynomial in the number d of variables, the number n of constraints, and the total number b of bits needed to express the coefficients in the constraints. Khachiyan's algorithm is often referred to as the *ellipsoid method*, and Karmarkar's is the precursor of the so-called *interior-point methods*. Papadimitriou and Steiglitz [187] give a good introduction to linear programming and combinatorial optimization. They give a solution to exercise 10.1.

In many cases and in particular in geometric problems, the number d of variables is relatively small. The works of Megiddo [160, 161] and Dyer [87, 88] provide many algorithms that take advantage of this: the corresponding complexities are linear in the number n of constraints when the dimension d is considered a constant. The method in Megiddo [160, 161] and Dyer [87, 88] is the so-called *prune-and-search* method that is outlined in exercises 10.7 and 10.8. The method consists in reducing the domain of $\mathbb{E}^d$ in which to search for the solution (*bisection* of the problem), so as to prune away a constant fraction of constraints that become redundant on the smaller domain. The solution to exercises 10.7 and 10.8 is also presented in the book by Edelsbrunner [89]. The prune-and-search method may also be used to solve quadratic or convex optimization problems. In particular, it can be used to solve the minimum enclosing circle problem as indicated by exercise 10.9. The complete solution to this exercise can be found in a paper by Meggido [160] or in another by Dyer [88].

The complexities of Meggido's and Dyer's algorithms depend doubly exponentially on the number d of variables. This dependence was lowered to singly-exponential by Clarkson [66]. Later on, faster and simpler randomized algorithms were devised by Kalai [133], Clarkson [68], Sharir and Welzl [208], and Matoušek, Sharir, and Welzl [157]. The one we present in section 10.2 is due to Seidel [205]. Welzl [218] adapts it to yield a simple randomized algorithm for the minimum enclosing circle.

The existence of a shelling of a polytope was first proved by Bruggesser and Mani [39]. They used a shelling induced by a line as we did in section 10.3. The convex hull algorithm described in section 10.3 is due Seidel [202]. More recently, Matoušek and Schwarzkopf [156, 153] showed how to preprocess the set of constraints of the n linear programming problems solved by the algorithm in order to solve them more efficiently.

Their approach leads to an algorithm whose complexity is $O(n^{2-\frac{2}{1+\lfloor d/2 \rfloor}+\epsilon} + f \log n)$ for n points in dimension d if the convex hull has f facets. The notation ϵ stands for a constant that can be made as small as wanted (albeit at the cost of increasing the constant in the $O()$ notation). In dimension 4 or 5, this complexity is $O(n^{4/3+\epsilon} + f \log n)$.

The convex hull algorithm in dimension 2 (described in exercise 10.6), whose complexity $O(n \log h)$ depends on the number h of vertices of the computed convex hull, is due to Kirkpatrick and Seidel [137]. In [99], Edelsbrunner and Shi generalized this algorithm to 3 dimensions, obtaining an algorithm of complexity $O(n \log^2 h)$.

Lastly, the LP-type problems defined in exercise 10.11 generalize the formulation in terms of linear programming. Many geometric problems can be expressed as LP-type problems, such as computing the smallest enclosing ellipsoid, the largest ellipsoid inscribed in a polytope, the smallest circle that intersects n convex objects, etc. The randomized algorithm of Clarkson [68] or those of Sharir and Welzl [208] and Matoušek, Sharir, and Welzl [157] actually solve LP-type problems. Chazelle and Matoušek [57] even explicitly discussed under which conditions a deterministic algorithm to solve an LP-type problem can be obtained by derandomization.

Part III

Triangulations

To triangulate a region is to describe it as the union of a collection of simplices whose interiors are pairwise disjoint. The region is then decomposed into elementary cells of bounded complexity. The words *to triangulate* and *triangulation* originate from the two-dimensional problem, but are commonly used in a broader context for regions and simplices of any dimension.[1]

Triangulations and related meshes are ubiquitous in domains where the ambient space needs to be discretized, for instance in order to interpolate functions of several variables, or to numerically solve multi-dimensional differential equations using finite-element methods. Triangulations are largely used in the context of robotics to decompose the free configuration space of a robot, in the context of artificial vision to perform three-dimensional reconstructions of objects from their cross-sections, or in computer graphics to solve problems related to windows or to compute illuminations in rendering an image. Finally, in the context of computational geometry, the triangulation of a set of points, a planar map, a polygon, a polyhedron, an arrangement, or of any other spatial structures, is often a prerequisite to running another algorithm on the data. For instance, this is the case for algorithms performing point location in a planar map by using a hierarchy of triangulations, or for the numerous applications of triangulations to shortest paths and visibility problems.

Triangulations form the topic of the next three chapters. Chapter 11 recalls the basic definitions related to triangulations, and studies the combinatorics of triangulations in dimensions 2 and 3. Chapters 12 and 13 are concerned with algorithmic problems on triangulations in dimensions 2 and 3 respectively. Essentially, two types of problem are studied. In the first kind of problem, the vertices of the triangulation are given and we seek to decompose their convex hull into simplices. In the second kind, the triangulation is required to include

---

[1]Although usage has consecrated that terminology, it would be more appropriate to use the words *to simpliciate* and *simpliciation*.

several given simplices of positive dimension, such as those on the boundary of a polyhedral region. For the latter, we speak of a constrained triangulation. Problems of this kind always have solutions in dimension 2 and algorithms that solve them efficiently are given in chapter 12. In dimension 3, however, this may not always be the case and chapter 13 exhibits a few pitfalls and describes ways of avoiding them.

Chapter 11

Complexes and triangulations

The notion of a triangulation occurs naturally if we seek to describe the fundamental objects of linear geometry (polygons, polyhedra, etc.) in terms of *elementary* objects that have a bounded complexity. Indeed, the simplest of all such objects are simplices.

In section 11.1, we recall the definitions of a simplicial complex, of a cell complex, and of a triangulation. In particular, this section emphasizes the relations between these notions and the usual objects in linear geometry (polygons, polyhedra, polygonal and polyhedral regions, etc.). Section 11.2 studies the combinatorics of triangulations in dimensions 2 and 3. Finally, section 11.3 gives a representation for these complexes and establishes a duality between them.

11.1   Definitions

11.1.1   Simplices, complexes

In this section, we work in the d-dimensional space $\mathbb{E}^d$. Recall that a simplex of dimension k for $k \leq d$, also called a k-simplex, is a k-polytope with $k+1$ vertices, or equivalently the convex hull of $k+1$ affinely independent points. Let $\mathcal{A} = \{A_0, \ldots, A_k\}$ be a set of $k+1$ affinely independent points and S be the k-simplex defined by $\mathcal{A}$. Any subset of $l+1 \leq k+1$ points in $\mathcal{A}$ defines an l-simplex which is a *face* of S. Simplices of dimension $0, 1, 2,$ and 3 are respectively called points, segments, triangles, and tetrahedra.

A *complex* $\mathcal{C}$ is a finite set of simplices that satisfy the following two properties:

1. any face of a simplex in $\mathcal{C}$ is also a simplex in $\mathcal{C}$, and

2. two simplices in $\mathcal{C}$ either do not intersect, or their intersection is a simplex of smaller dimension which is their common face of maximal dimension.

For convenience, we consider that there is an empty simplex, whose dimension is -1, which is a face of any simplex and hence of any complex. The complex $\mathcal{C}$ is a k-complex, or a complex of dimension k, if the maximal dimension of the simplices in $\mathcal{C}$ is exactly k.

The simplices that constitute a complex are called the *faces* of the complex. A face of dimension l is called an l-face. The faces of dimension 0 are called the *vertices* and the faces of dimension 1 are called the *edges*. In dimension d, the faces of dimension d and $d-1$ are respectively called the *cells* and the *facets*. Two faces of a complex are *incident* if one is included in the other and their dimensions differ by one. Two vertices of a complex are *adjacent* if they share a common incident edge, and two cells are *adjacent* if they share a common incident facet. The *l-skeleton* of a k-complex $\mathcal{C}$ is the l-complex, subcomplex of $\mathcal{C}$, consisting of the faces of $\mathcal{C}$ of dimension at most l. The 1-skeleton of a complex $\mathcal{C}$ is isomorphic to a graph whose nodes are the vertices of $\mathcal{C}$ and whose arcs are the 1-faces of $\mathcal{C}$.

A k-complex $\mathcal{C}$ is *homogeneous* or *pure* if and only if any face of $\mathcal{C}$ is a face of some k-simplex in $\mathcal{C}$.

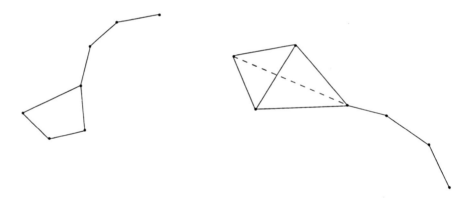

Figure 11.1. A 2-complex and a 3-complex. Neither is homogeneous.

To any complex $\mathcal{C}$ there corresponds a subset of the space $\mathbb{E}^d$ formed by the points in $\mathbb{E}^d$ that belong to the simplices in $\mathcal{C}$. This region is called the *domain* of the complex and is denoted by $dom(\mathcal{C})$. Sometimes, when this does not create ambiguities, we abuse the notation and do not make a distinction between the complex and its corresponding domain.

A k-complex $\mathcal{C}$ is said to be *connected* if its domain is connected. A k-complex $\mathcal{C}$ is connected if and only if its 1-skeleton is connected.

Let $\mathcal{C}$ be a homogeneous k-complex. The *boundary* of $\mathcal{C}$, denoted by $bd(\mathcal{C})$, is the homogeneous $(k-1)$-complex consisting of all the $(k-1)$-faces of $\mathcal{C}$ that belong to only one k-simplex of $\mathcal{C}$, and all the subfaces of these $(k-1)$-faces. A face of $\mathcal{C}$ is said to be *external* if it belongs to the boundary of the complex and *internal* otherwise.

Complexes as they have just been defined are sometimes called *simplicial complexes* since their faces are simplices. For instance, the set of proper faces of a simplicial d-polytope is a canonical example of a simplicial $(d-1)$-complex. Sometimes, our investigations lead us to define a broader kind of complex, called a *cell complex*, whose faces are polytopes or even unbounded polytopes. For instance, it will be very useful to consider the set of faces of a Voronoi diagram (see chapter 17) or of a hyperplane arrangement (see chapter 14) as a cell complex. Formally, a cell complex $\mathcal{C}$ is a set of polytopes, bounded or unbounded, such that

1. any face of a polytope in $\mathcal{C}$ is also a polytope in $\mathcal{C}$, and

2. two polytopes in $\mathcal{C}$ either do not intersect, or their intersection is a polytope of smaller dimension which is their common face of maximal dimension.

For instance, the set of faces of a d-polytope (except for the empty face) is a d-dimensional cell complex, and the set of proper faces of a d-polytope is a $(d-1)$-dimensional cell complex. In the remainder of this chapter, we are mostly concerned with simplicial complexes, which are at the core of the concept of triangulation. Nevertheless, the definitions and properties that are stated for simplicial complexes generalize easily to cell complexes.

11.1.2 Topological balls and spheres, singularities

Triangulations will be defined later on as pure connected complexes without singular faces, and thus we have to give a clear and precise definition of singularities. Examples of singular faces appear in figure 11.2; vertex A of the complex on the left and edge AB of the complex on the right are both singular faces. The notion of singularity is essentially topological and its precise definition calls for that of topological balls and spheres in $\mathbb{E}^d$.

A *topological k-ball* is a set of points in $\mathbb{E}^d$ homeomorphic[1] to the unit ball of $\mathbb{E}^k$ defined as the set of points $X = (x_1, \ldots, x_k)$ such that $\sum_{i=1}^{k} x_i^2 \le 1$. A *topological k-sphere* is a set of points in $\mathbb{E}^{d+1}$ homeomorphic to the unit sphere in $\mathbb{E}^{k+1}$ defined as the set of points $X = (x_1, \ldots, x_{k+1})$ such that $\sum_{i=1}^{k+1} x_i^2 = 1$. For instance, any d-polytope is a topological d-ball and its boundary is a topological $(d-1)$-sphere. With a slight abuse of terminology, we call hereafter a topological ball (resp. topological sphere) a complex $\mathcal{C}$ whose domain $dom(\mathcal{C})$

---

[1]Two subsets of a topological space are homeomorphic if there is a continuous bijection from one to the other whose inverse bijection is also continuous.

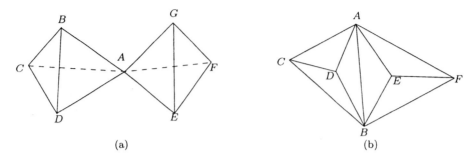

Figure 11.2. Singular faces of a complex:
         (a) A 3-complex showing a singularity at vertex A.
         (b) A 3-complex whose edge AB is singular.

is a topological ball (resp. sphere). Note that the boundary of a topological k-ball is a topological k-sphere, while the boundary of a topological k-sphere is empty.

A 0-complex consisting of two points is a topological 0-sphere, and a 0-complex consisting of a single point is a topological 0-ball. Let $(A_1, A_2, \ldots, A_n)$ be a finite sequence of points in the plane. The set of segments which join two consecutive points in the sequence forms a homogeneous 1-complex if two segments do not intersect except at their common endpoints. Such a 1-complex is called a *polygonal line*. Moreover, if the points are pairwise distinct, then the polygonal line is *simple* and its boundary is a topological 0-sphere formed by its two endpoints $\{A_1, A_n\}$. If the points are pairwise distinct, except for A_1 and A_n which are the same point, the polygonal line is a *polygon* and its boundary is empty. Simple polygonal lines and polygons respectively describe all the topological 1-balls and 1-spheres.

To characterize the singularities of a complex, we introduce the *star* and *shell* operations on a complex. Let $\mathcal{C}$ be a pure connected d-complex and F be a face of $\mathcal{C}$. The *star* of F in $\mathcal{C}$ is the sub-complex of $\mathcal{C}$ consisting of the d-simplices of $\mathcal{C}$ that contain F and all their faces. The *shell* of F in $\mathcal{C}$ is the sub-complex consisting of all the simplices in the star of F in $\mathcal{C}$ that have an empty intersection with F. If $\mathcal{C}$ is a pure d-complex and F is a k-face of $\mathcal{C}$, $k < d$, the shell of F in $\mathcal{C}$ is a pure $(d - k - 1)$-complex. In the complex represented in figure 11.2a, the shell of A is formed by the faces of the triangles BCD and EFG. In the complex of figure 11.2b, the shell of edge AB is formed by the edges CD and EF and all their vertices.

A face F of a complex $\mathcal{C}$ is *singular* if its shell in $\mathcal{C}$ is neither a topological ball nor a topological sphere. For instance, in the 3-complex of figure 11.2a, vertex A is singular because its shell is not connected, nor a topological 0-sphere. For the same reason, edge AB is singular in figure 11.2b.

11.1.3 Triangulations

Here and in the following, we consider d-complexes embedded in a space of dimension d', for some $d' > d$. Hence, d stands for the maximal dimension of the simplices in the complex, not for the dimension of the space.

A *d-triangulation* is defined as a pure connected simplicial d-complex without singular faces. The shell of any k-face of a d-triangulation is thus either a topological $(d - k - 1)$-sphere, or a topological $(d - k - 1)$-ball. More precisely, the following lemma shows that the shell of any internal face is a topological sphere (whose boundary is empty) while the shell of any external face is a topological ball (whose boundary is a sphere).

Lemma 11.1.1 *Let C be a pure simplicial d-complex and F a k-face of C, $k \leq d - 1$. The boundary of the shell of F in C is not empty if and only if F is an external face of C.*

Proof. Let F_1 be a $(d - 1)$-face of C that contains the k-face F, and let F_1' be the $(d - k - 2)$-face of F_1 that is disjoint from F. The face F_1' is a face of the shell of F in C. For each d-simplex $S_1 = conv(F_1 \cup A)$ in C that contains F_1, the simplex $S_1' = conv(F_1' \cup A)$ is a $(d - k - 1)$-simplex in the shell of F that contains F_1'. As a result, F_1' is an external face of the shell of F if and only if F_1 is an external face of C. Note that this proof is perfectly valid for $k = d - 1$, taking $F = F_1$ and F_1'' to be the empty face of dimension -1, which is a face of all the faces of C and hence is in the shell of F in C. $\square$

The notion of a shell also helps to prove the following three lemmas on triangulations.

Lemma 11.1.2 *In a d-triangulation, any $(d - 1)$-face belongs to a most two d-simplices.*

Proof. Let $\mathcal{T}$ be a d-triangulation. If the domain $dom(\mathcal{T})$ of the triangulation is embedded in an affine space of dimension d, then the proposition follows directly from the definition of a complex. Nevertheless, it is conceivable that $\mathcal{T}$ is embedded in a space of higher dimension. In any case, we consider, for any $(d - 1)$-face F of $\mathcal{T}$, the shell of F in $\mathcal{T}$. This shell is a 0-complex formed by one or two vertices of $\mathcal{T}$, so that F belongs to either one or two d-simplices of $\mathcal{T}$. $\square$

Lemma 11.1.3 *For any pair (T, T') of d-simplices of a d-triangulation, there is a sequence $T_1, T_2, \ldots, T_n$ of d-simplices such that $T_1 = T$, $T_n = T'$, and T_i is adjacent to T_{i+1} for all $i = 1, \ldots, n - 1$.*

Proof. The lemma is trivial for 1-triangulations and can be proved by induction on the dimension of the triangulation. Let $A = A_1, A_2, \ldots, A_n = A'$ be a path in the 1-skeleton of T that joins a vertex A of T to a vertex A' of T'. Such a path exists because the complex is connected. We show that for any vertex A_i, $i = 2, \ldots, n-1$ on this path, and any d-simplices T_i and T_{i+1} in T that contain respectively the edges $A_{i-1}A_i$ and A_iA_{i+1}, there exists a sequence of adjacent d-simplices in T that joins T_i to T_{i+1}. Let F_i and F_{i+1} be the $(d-1)$-faces of T_i and T_{i+1} that do not contain A_i. Then F_i and F_{i+1} are $(d-1)$-faces in the shell of A_i in T, and the induction proves that there is a sequence $F_i = G_1, G_2, \ldots, G_m = F_{i+1}$ of adjacent $(d-1)$-simplices in the shell of A_i in T that joins F_i to F_{i+1}. Therefore, the sequence $T_i = conv(G_1 \cup A_i), conv(G_2 \cup A_i), \ldots, conv(G_m \cup A_i) = T_{i+1}$ is a sequence of adjacent d-simplices of T that joins T_i to T_{i+1}.   □

Lemma 11.1.4 *If T is a d-triangulation, the pure $(d-1)$-complex $bd(T)$ that is the boundary of T has itself an empty boundary.*

Proof. We first show that every $(d-2)$-face of $bd(T)$ belongs to two $(d-1)$-simplices of $bd(T)$. Let G be a $(d-2)$-face of $bd(T)$. The shell of G in T is a simple polygonal line whose boundary is formed by two points U and V. Since U belongs to a single edge of the shell of G in T, the $(d-1)$-simplex $conv(G, U)$ belongs to only one d-simplex in T and is thus a $(d-1)$-simplex in $bd(T)$. The same argument applies to the $(d-1)$-simplex $conv(G, V)$.   □

The notion of singularity can be extended to cell complexes. In the rest of this book, unless explicitly mentioned, a complex will denote a pure connected complex without singularities, be it either a simplicial or a cell complex.

11.1.4   Polygons and polyhedra

The 1-triangulations are precisely the simple polygonal lines and the polygons. A triangulation T (of dimension 1 or 2) is said to be planar if its domain $dom(T)$ can be embedded in a space of dimension 2. A planar polygon $\mathcal{P}$ is a closed, simple (that is, not self-intersecting), planar curve, and Jordan's theorem (see exercise 11.1) states that this curves splits $\mathbb{E}^2 \setminus \mathcal{P}$ into two connected regions, exactly one of them being bounded. The *interior* of $\mathcal{P}$ is the bounded region and the *exterior* of $\mathcal{P}$ is the unbounded region.

More generally, we call a *polygonal region* any connected region in the plane whose boundary is one polygon or the union of a finite number of disjoint polygons. Depending on the context, we consider a polygonal region to include its boundary or not. The *edges* and *vertices* of a polygonal region are the edges and vertices of the polygons that bound the region.

A bounded polygonal region whose boundary consists of a single polygon is a topological ball. Such a region is also called *simply connected*. A bounded polygonal region whose boundary consists of $k + 1$ polygons is a polygonal region with k holes. An unbounded polygonal region whose boundary consists of k polygons has k holes.

By lemma 11.1.4, we know that the boundary of a 2-triangulation is a pure 1-complex whose boundary is empty. This complex is not necessarily connected and can be the union of several disjoint polygons. The domain $dom(\mathcal{T})$ of any planar 2-triangulation $\mathcal{T}$ is thus a bounded polygonal region. Reciprocally, we will see in the next chapter that any bounded polygonal region $\mathcal{P}$ is *triangulable*, meaning that the domain it encloses can be decomposed into non-overlapping triangles, or that it can be expressed as the domain of a 2-triangulation $\mathcal{T}$ whose set of vertices is exactly the set of vertices of $\mathcal{P}$.

We will not discuss here the topology of triangulations in dimensions 3 and higher. We need only know that 2-triangulations can be *orientable* or *non-orientable*. To orient a triangle is to choose a circular order on its three vertices. The orientation of a triangle induces an orientation of its three edges. The orientations of two adjacent triangles are *consistent* if they induce an opposite orientation on their common incident edge (see figure 11.3). A 2-triangulation is *orientable* if it is possible to orient each of its triangles in a way such that all adjacent triangles have consistent orientations.

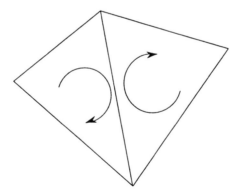

Figure 11.3. Consistent orientations of adjacent triangles.

In the 3-dimensional space $\mathbb{E}^3$, an orientable 2-triangulation whose boundary is empty is a *polyhedron*. A polyhedron may be a topological sphere, but need not be. Generally speaking, it is homeomorphic to a sphere with *handles* (see figure 11.4). The number of handles is called the *genus*. A polyhedron homeomorphic to a topological sphere has *genus* 0. A polyhedron homeomorphic to a torus is a handle and thus has genus 1. More generally, a polyhedron with h

handles is said to have genus h. A polyhedron $\mathcal{P}$ in $\mathbb{E}^3$ separates $\mathbb{E}^3 \setminus \mathcal{P}$ into two connected components, exactly one of which is bounded. We call the bounded component the *interior* of the polyhedron, and the other component the *exterior* of the polyhedron. The polyhedra as they are defined here are often called *simplicial polyhedra* since their faces are simplices. We may extend this definition by defining a *polyhedron* as an oriented cell complex of dimension 2 in $\mathbb{E}^3$ whose boundary is empty.

More generally, we call *polyhedral region* any connected region of $\mathbb{E}^3$ whose boundary is formed by one or several disjoint polyhedra. Depending on the context, a polyhedral region is considered to include its boundary or not. We call *faces* of a polyhedral region the faces (vertices, edges, and facets) of the polyhedra that form the boundary of the region. A bounded polyhedral region whose boundary consists of a single polyhedron of genus 0 is a topological ball. A bounded polyhedral region whose boundary consists of a single polyhedron of genus h is a polyhedral region with h handles. A bounded polyhedral region whose boundary consists of $k + 1$ polyhedra is a polyhedral region with k holes.

The boundary of a 3-triangulation is a pure orientable 2-complex whose boundary is empty. To prove the orientability, it suffices to orient any triangle according to the normal that leaves the 3-simplex of the 3-triangulation that contains this triangle. The boundary of a 3-triangulation is thus formed by one or several polyhedra. Therefore, the domain of a 3-triangulation is a bounded polyhedral region. A polyhedral region $\mathcal{P}$ is *triangulable* if there is a triangulation $\mathcal{T}$ whose domain coincides with $\mathcal{P}$ and whose set of vertices is the same as that of $\mathcal{P}$. We will show in chapter 13 that there are polyhedral regions which are not triangulable.

11.2   Combinatorics of triangulations

The *complexity*, also called the *size*, of a complex is the number of its faces of all dimensions.

11.2.1   Euler's relation for topological balls and spheres

Our study is based on the existence of a linear relation, called Euler's relation, between the faces of different dimensions in a complex. This relation admits an elementary proof in the case of a 2-triangulation or for a 2-complex that can be embedded in a space of dimension 2. Indeed, the 1-skeleton of the complex is then a planar graph (see exercise 11.4). In higher dimensions, this proof does not work. In fact, Euler's relation is one of the most famous results of homology, a theory whose application goes well beyond the scope of this book. We limit ourselves here to proving Euler's relation for topological spheres and balls, basing the proof on a single result of the homology theory. (See the proof of theorem 11.2.1 below for a statement of this result.) In the following subsection we show how to derive from this result all the Euler's relations for 2-complexes, polyhedra, and polyhedral regions in $\mathbb{E}^3$.

Let $\mathcal{C}$ be a d-complex and $n_k(\mathcal{C})$ the number of its k-faces, for $k = 0, \ldots, d$. The *Euler characteristic* $e(\mathcal{C})$ of the d-complex $\mathcal{C}$, is defined as the alternating sum

$$e(\mathcal{C}) = \sum_{k=0}^{d} (-1)^k n_k(\mathcal{C}).$$

Theorem 11.2.1 (Euler's relation) *If the d-complex $\mathcal{C}$ is a topological d-ball, its Euler characteristic is 1:*

$$\sum_{k=0}^{d} (-1)^k n_k(\mathcal{C}) = 1. \tag{11.1}$$

If the d-complex $\mathcal{C}$ is a topological d-sphere, its Euler characteristic is $1 + (-1)^d$:

$$\sum_{k=0}^{d} (-1)^k n_k(\mathcal{C}) = 1 + (-1)^d. \tag{11.2}$$

Proof. The basic result from homology theory mentioned above is that two complexes $\mathcal{C}$ and $\mathcal{C}'$ that have homeomorphic domains also have the same Euler characteristic.

The set of faces of a d-polytope is a topological d-ball. By theorem 7.2.1, its Euler characteristic is 1. By the definition, we know that any topological d-ball is homeomorphic to the domain of a polytope. As a result, the Euler characteristic of any topological d-ball is 1 and it satisfies equation 11.1.

Similarly, the set of proper faces of a d-polytope forms a topological $(d-1)$-sphere. By theorem 7.2.1, its Euler characteristic is $1 + (-1)^{d-1}$. As a result, the Euler characteristic of any topological d-sphere is $1 + (-1)^d$, and it satisfies equation 11.2. □

11.2.2 The complexity of 2-complexes

In this subsection, we study the complexity of 2-triangulations and of cell complexes of dimension 2 that can be embedded into a space of dimension 2, and also of simplicial or cell polyhedra. For complexes that can be embedded into $\mathbb{E}^2$, the number of facets and edges is bounded by the total number of vertices and the number of external vertices of the complex. For polyhedra, the number

of facets and edges can be bounded by the number of vertices and the genus of
the polyhedron.

For any 2-complex $\mathcal{C}$, we denote by $n(\mathcal{C})$ the number of vertices of $\mathcal{C}$, by $m(\mathcal{C})$
the number of its edges, and by $f(\mathcal{C})$ the number of its 2-faces. When the context
is clear, we drop the reference to $\mathcal{C}$ and simply write n, m, and f for $n(\mathcal{C})$, $m(\mathcal{C})$,
and $f(\mathcal{C})$.

Theorem 11.2.2 (Euler's relation in dimension 2)   *Let $\mathcal{C}$ be a 2-complex
whose domain is contained in $\mathbb{E}^2$. If the boundary of $\mathcal{C}$ is the union of $k+1$
polygons (dom($\mathcal{C}$) is thus a polygonal region with k holes), then it satisfies Euler's
relation:*

$$n - m + f = 1 - k. \tag{11.3}$$

Proof. Let us first note that $dom(\mathcal{C})$ is a polygonal region with k holes because
$\mathcal{C}$ is as usual assumed to be pure, connected, and without singularities. We
first prove the theorem when the complex $\mathcal{C}$ is a triangulation $\mathcal{T}$. If $k = 0$,
the triangulation $\mathcal{T}$ is a topological ball and the previous equation is simply
Euler's relation for topological 2-balls that was proved before. For $k \neq 0$, we
invoke a result that we prove independently in the next chapter, showing that
any polygon can be *triangulated*. More precisely, for any polygon $\mathcal{P}$ there exists a
2-triangulation whose vertices are exactly the vertices of $\mathcal{P}$ and whose boundary
is the same as that of $\mathcal{P}$. Let $\mathcal{P}_i$, $i = 1, \ldots, k$, be the polygons forming the
boundaries of the holes of $dom(\mathcal{T})$, and let $\mathcal{T}_i$ be a triangulation of $\mathcal{P}_i$. Each
triangulation $\mathcal{T}_i$ is a topological ball, so it has an Euler characteristic of

$$e(\mathcal{T}_i) = n(\mathcal{T}_i) - m(\mathcal{T}_i) + f(\mathcal{T}_i) = 1. \tag{11.4}$$

The complex $\mathcal{T}' = \mathcal{T} \cup \left(\bigcup_{i=1}^{k} \mathcal{T}_i \right)$ is also a topological ball and satisfies

$$e(\mathcal{T}') = n(\mathcal{T}') - m(\mathcal{T}') + f(\mathcal{T}') = 1. \tag{11.5}$$

The faces common to $\mathcal{T}$ and to $\bigcup_{i=1}^{k} \mathcal{T}_i$ are also faces of $\mathcal{P}_i$, so we can compute
the Euler characteristic of $\mathcal{T}'$ as

$$e(\mathcal{T}') = e(\mathcal{T}) + \sum_{i=1}^{k} e(\mathcal{T}_i) - \sum_{i=1}^{k} e(\mathcal{P}_i). \tag{11.6}$$

Since the Euler characteristic of a polygon is zero, we conclude that

$$e(\mathcal{T}) = 1 - k.$$

Consider now a cell complex $\mathcal{C}$ of dimension 2. The 2-faces of such a complex are
polygonal regions, and can also be triangulated, using the same result as above.

By replacing each non-triangular 2-face of the complex by its triangulation, we obtain a 2-triangulation that also satisfies Euler's relation 11.3. It is then easy to see that this relation is also satisfied by the faces of the cell complex C: each edge added while triangulating a 2-face of C adds an edge but also splits a 2-face into two new 2-faces, so its leaves the linear combination $n - m + f$ unchanged. $\square$

Corollary 11.2.3 *Let C be a 2-complex in $\mathbb{E}^2$, with n vertices, m edges, and f 2-faces and whose domain is a polygonal region with k holes. If C has n_e external vertices (on the boundary of C), then*

$$f \leq 2(n - 1 + k) - n_e,$$
$$m \leq 3(n - 1 + k) - n_e.$$

Equality holds if and only if C is a triangulation.

Proof. Note that n_e is also the number of external edges of C. Each external edge of C is incident to a unique 2-face of C, while each internal edge is shared between exactly two 2-faces. Also, each 2-face of C is incident to at least 3 edges. Counting the number of incidences between an edge and a 2-face, we obtain

$$2m - n_e \geq 3f,$$

and equality holds if and only if C is a triangulation. It now suffices to use Euler's relation 11.3 to prove the corollary. $\square$

Theorem 11.2.4 (Euler's relation for polyhedra) *If C is a polyhedron with n vertices, m edges, f 2-faces, and genus h, then*

$$n - m + f = 2 - 2h. \tag{11.7}$$

Proof. We provide a proof by induction on the genus h of the polyhedron. Let C be a simplicial or cell complex. If the genus h of C is 0, then C is a topological sphere and the relation sought is just Euler's relation for topological 2-spheres, proved in theorem 11.2.1. Now if C is a polyhedron with genus $h > 0$, its Euler characteristic can be derived from that of any polyhedron C_h obtained by adding a handle to a polyhedron C_{h-1} of genus $h - 1$. To add a handle to C_{h-1}, we consider a polyhedron C_0 of genus 0 such that

$$C_{h-1} \cap C_0 = \mathcal{B} \cup \mathcal{B}',$$

where $\mathcal{B}$ and $\mathcal{B}'$ are two disjoint topological 2-balls (which we call *disks*) such that

$$dom(C_{h-1}) \cap dom(C_0) = dom(\mathcal{B}) \cup dom(\mathcal{B}'),$$

(see figure 11.4). The polyhedron C_h can be obtained by removing the internal faces of the two topological disks B and B' from $C_{h-1} \cup C_0$. Let P and P' be the two polygons that form the boundaries of B and B' respectively. The Euler characteristic $e(C_h)$ of C_h is given by

$$e(C_h) = e(C_{h-1}) + e(C_0) - 2(e(B) + e(B')) + e(P) + e(P').$$

The Euler characteristic is 1 for each disk B and B', and 0 for each polygon P and P'. We conclude that

$$e(C_h) = e(C_{h-1}) - 2,$$

and, by induction, that

$$e(C_h) = 2 - 2h.$$

This completes the proof of theorem 11.2.4. $\square$

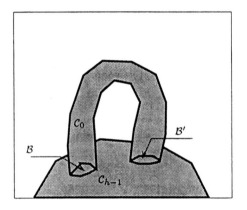

Figure 11.4. A handle.

Corollary 11.2.5 *Let C be a polyhedron of genus h, with n vertices, m edges, and f 2-faces. Then*

$$f \ \leq \ 2n - 4 + 4h \qquad (11.8)$$
$$m \ \leq \ 3n - 6 + 6h. \qquad (11.9)$$

and equality holds if and only if the polyhedron is simplicial.

Proof. By counting the number of incidences between edges and 2-faces of polyhedron C, we obtain the inequality

$$2m \geq 3f$$

which becomes an equality when polyhedron $\mathcal{C}$ is simplicial. Using Euler's relation 11.7 for polyhedra, we derive both relations 11.8 and 11.9.   □

Remark. From corollary 11.2.5, we may also infer that the expected number of edges incident to a random vertex of a polyhedron is at most $6 - \frac{12(1-h)}{n}$. Therefore, any polyhedron of genus h always has a vertex of degree at most 5 when $h = 0$, and of degree at most 6 if $h = 1$.

11.2.3   The complexity of 3-triangulations

Let $\mathcal{T}$ be a 3-triangulation in $\mathbb{E}^3$. We denote by $n(\mathcal{T})$ the number of vertices of $\mathcal{T}$, by $m(\mathcal{T})$ the number of its edges, by $f(\mathcal{T})$ the number of its triangles, and by $t(\mathcal{T})$ the number of its tetrahedra. When there is no ambiguity as to the complex $\mathcal{T}$, we simplify the notation by writing n, m, f and t for $n(\mathcal{T})$, $m(\mathcal{T})$, $f(\mathcal{T})$, and $t(\mathcal{T})$.

Theorem 11.2.6 (Euler's relation in dimension 3) *Consider a 3-triangulation $\mathcal{T}$ whose boundary is a polyhedron of genus h. Euler's relation for $\mathcal{T}$ states that*

$$n - m + f - t = 1 - h. \tag{11.10}$$

Proof. Again, we give a proof by induction on the genus h of the polyhedron that bounds the complex $\mathcal{T}$. If $h = 0$, then $\mathcal{T}$ is a topological 3-ball and the theorem is a consequence of theorem 11.2.1. If $h \neq 0$, then the Euler characteristic of $\mathcal{T}$ is the same as that of any 3-triangulation $\mathcal{T}_h = \mathcal{T}_{h-1} \cup \mathcal{T}_0$, obtained by merging a 3-triangulation $\mathcal{T}_{h-1}$ whose boundary is a polyhedron of genus $h - 1$ and a topological 3-ball $\mathcal{T}_0$ in such a way that

$$bd(\mathcal{T}_{h-1}) \cap bd(\mathcal{T}_0) = \mathcal{B} \cup \mathcal{B}',$$

where $\mathcal{B}$ and $\mathcal{B}'$ are two 2-complexes of disjoint topological balls such that

$$dom(\mathcal{T}_{h-1}) \cap dom(\mathcal{T}_0) = dom(\mathcal{B}) \cup dom(\mathcal{B}').$$

The Euler characteristic $e(\mathcal{T}_h)$ of $\mathcal{T}_h$ is given by

$$e(\mathcal{T}_h) = e(\mathcal{T}_{h-1}) + e(\mathcal{T}_0) - \big(e(\mathcal{B}) + e(\mathcal{B}')\big).$$

From this it follows that

$$e(\mathcal{T}_h) = 1 - h.$$

□

Corollary 11.2.7 *Consider a 3-triangulation T whose boundary is a polyhedron of genus h. The number t of its tetrahedra satisfies*

$$n - 3 + 3h \leq t \leq \frac{n^2}{2} - \frac{3n}{2} - n_e + 3 - 3h.$$

Again, n stands for the number of vertices of T and n_e for the number of its external vertices.

Proof. Denote by n_e, m_e, and f_e the respective numbers of vertices, edges, and triangles on the boundary of T, and by n_i, m_i, and f_i the respective numbers of internal vertices, edges, and triangles of T. Corollary 11.2.5 applied to the boundary of T yields

$$f_e = 2n_e - 4 + 4h \tag{11.11}$$

$$m_e = 3n_e - 6 + 6h. \tag{11.12}$$

Each tetrahedron of T is incident to four triangles. In turn, an internal triangle of T is incident to two tetrahedra, while an external triangle is incident to a unique tetrahedron. Using the same counting argument as in corollary 11.2.3, we obtain

$$4t = 2f_i + f_e. \tag{11.13}$$

Eliminating f_e, f_i and $f = f_e + f_i$ in equations 11.10–11.13, we get

$$t = m - n - n_e + 3 - 3h,$$

which gives the number of tetrahedra in the triangulation as a function of the number of its vertices and edges. The bounds on the number of tetrahedra claimed by the theorem are then an immediate consequence of bounds on the number of edges. On the one hand, the number of edges is trivially bounded above by $n(n-1)/2$. On the other hand, each internal vertex is incident to at least four internal edges, each incident to at most two internal vertices, so the number m_i of internal edges is at least $2n_i$. Thus the total number m of edges must satisfy

$$\frac{n(n-1)}{2} \geq m = m_e + m_i \geq 3n_e - 6 + 6h + 2n_i$$
$$\geq 2n + n_e - 6 + 6h.$$

$\square$

For a 3-triangulation whose boundary is a polyhedron of genus 0, the above bounds can be written as

$$n - 3 \leq t \leq \frac{n^2}{2} - \frac{3n}{2} - n_e + 3.$$

Both upper and lower agree when $n = 4$, and are thus optimal. Below, we show that these bounds may also be matched for any n and at least some values of n_e. For this, we must exhibit a 3-triangulation whose number of tetrahedra is quadratic (resp. linear) in the number of vertices. (See also exercise 11.6).

Lemma 11.2.8 *For any integer n, there is a 3-triangulation with n vertices, all external, and with t tetrahedra, such that*

$$t = \frac{n^2}{2} - \frac{5n}{2} + 3.$$

Proof. Let us choose the vertices of the triangulation on the moment curve. Recall from subsection 7.2.4 that the moment curve is parametrically described as $\{(\tau, \tau^2, \tau^3), -\infty < \tau < +\infty\}$. Lemma 7.2.6 shows that any three points A_1, A_2, A_3 on the moment curve are affinely independent. Moreover, if these points have respective parameters $\tau_1 < \tau_2 < \tau_3$, any point M on the moment curve parameterized by τ lies on a given side of the hyperplane passing through $A_1 A_2 A_3$ when $\tau \in \,]-\infty, \tau_1[\, \cup \,]\tau_2, \tau_3[$, and on the opposite side if $\tau \in \,]\tau_1, \tau_2[\, \cup \,]\tau_3, +\infty[$.

Let $\mathcal{A}_n = \{A_1, \ldots, A_n\}$ be a set of n distinct points on the moment curve Γ, parameterized respectively by $\tau_1 < \cdots < \tau_n$. For each $i = 1, \ldots, n$, we denote by $\mathcal{A}_i$ the set of the first i points in $\mathcal{A}_n$. Consider the convex hull $conv(\mathcal{A}_{n-1})$. Its facets consist of the $2(n-1) - 4$ triangles: $A_1 A_i, A_{i+1}$ for $i = 2, \ldots, n-2$ and $A_i A_{i+1} A_{n-1}$ for $i = 1, \ldots, n-3$. Indeed, the affine hulls of these triangles leave all the points of $\mathcal{A}_{n-1}$ on some given side. For each $i = 1, \ldots, n-3$, the affine hull of the triangle $A_i A_{i+1} A_{n-1}$ separates point A_n from the other points in $\mathcal{A}_{n-1}$ and, according to the terminology set up in chapter 8 (see section 8.3), the triangle $A_i A_{i+1} A_{n-1}$ is a red facet of $conv(\mathcal{A}_{n-1})$ with respect to A_n.

Consider the complex $\mathcal{T}_n$ defined by the following induction:

$$\mathcal{T}_4 = \{A_1 A_2 A_3 A_4\}$$
$$\mathcal{T}_n = \mathcal{T}_{n-1} \cup \{A_i A_{i+1} A_{n-1} A_n : i = 1, \ldots, n-3\},$$

where each complex $\mathcal{T}_i$ is a pure 3-complex described by all the tetrahedra that belong to it (adding all the 0-, 1-, and 2-faces of these tetrahedra). We now show by induction on n that $\mathcal{T}_n$ is a triangulation. Indeed, for $i = 1, \ldots, n-3$, the tetrahedron $A_i A_{i+1} A_{n-1} A_n$ has no points in common with the tetrahedra of $\mathcal{T}_{n-1}$ except for points in the triangles $A_i A_{i+1} A_{n-1}$ on the boundary of $\mathcal{T}_{n-1}$. The complex $\mathcal{T}_n$ is pure, by definition, and connected because its 1-skeleton is a connected graph. Moreover, we can check easily that the shell in $\mathcal{T}_n$ of any vertex A_i is a topological 2-sphere, that for $i = 2, \ldots, n$ the shell of any edge $A_1 A_i$, is a simple polygonal line, that the same holds for edges $A_i A_n$, $i = 1, \ldots, n-1$, and finally that fore any $1 < i < j - 1 < n - 1$ the shell of the edge $A_i A_j$ is a simple polygon. This guarantees that $\mathcal{T}_n$ is indeed a triangulation. The domain of $\mathcal{T}_n$ is the convex hull of $\mathcal{A}$. The number $|\mathcal{T}_n|$ of tetrahedra of $\mathcal{T}_n$ is given by the recurrence

$$|\mathcal{T}_4| = 1 \tag{11.14}$$

$$|\mathcal{T}_n| = |\mathcal{T}_{n-1}| + n - 3. \tag{11.15}$$

Solving this recurrence yields $|\mathcal{T}_n| = n^2/2 - 5n/2 + 3$ which, by corollary 11.2.7, is the maximum number of tetrahedra for a 3-triangulation with n vertices, all external, and whose boundary is a polyhedron of genus 0.   □

The following lemma shows the existence of linear triangulations.

Lemma 11.2.9 *For any pair of integers (n, n_e) such that $4 \le n_e \le n$, there exists a 3-triangulation with n vertices, n_e of which are external, and with t tetrahedra, where*

$$t = n - 3 + 2(n - n_e).$$

Proof. Examining the proof of corollary 11.2.7, we note that the lower bound $n - 3 + 2(n - n_e)$ for the number of tetrahedra can only be achieved when the number of edges itself also achieves its lower bound, $3n_e - 6 + 2(n - n_e)$, and this implies that there can be at most $2(n - n_e)$ internal edges.

We can realize these conditions easily when all the vertices are external, namely when $n = n_e$. Indeed, we may build a triangulation without internal edges incrementally, starting with the tetrahedron defined by the first four vertices. In an incremental step, the next triangulation can be obtained by adding a new tetrahedron adjacent to a single tetrahedron in the previous triangulation through a single facet. For this, we choose the new vertex of the triangulation so that only one facet is red. This implies that all the vertices and edges lie on the convex hull of the set of vertices of the triangulation.

When $n < n_e$, we can build a triangulation with n_e external vertices and edges using the previous construction. We then add $n - n_e$ internal vertices incrementally. In an incremental step, the new vertex A is added inside an existing tetrahedron. Let F_i, $i = 1, \ldots, 4$, be the four facets of this tetrahedron T. In order to make a triangulation of the new set of vertices, we replace T by four new tetrahedra $T_i = conv(F_i, A)$, as shown in figure 11.5. Each new internal vertex therefore adds three tetrahedra to the triangulation, so there are exactly $n_e - 3 + 3(n - n_e) = n - 3 + 2(n - n_e)$ tetrahedra in the resulting triangulation.   □

11.3 Representation of complexes, duality

In the preceding section, we have already emphasized the combinatorial and topological affinities between polytopes and complexes. It is thus natural to use

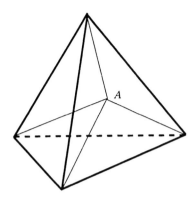

Figure 11.5. Splitting a tetrahedron when adding an internal node.

similar data structures to describe and process these two kinds of objects.

The incidence graph

A d-complex can be encoded by means of its incidence graph, as we did for polytopes. Recall that two faces of the complex are *incident* if one is entirely contained in the other and their dimensions differ by one. The incidence graph of a complex stores a node for each face of the complex and an arc for each pair of incident faces.

The adjacency graph

The adjacency graph of a complex has a node for each cell and an edge for each pair of adjacent cells (meaning that these cells are adjacent to a common facet). Any internal facet is incident to exactly two cells in the complex, so the adjacency graph may be built easily from the incidence graph. This definition is also consistent with the one given for polytopes in section 8.1. Indeed, the adjacency graph of a polytope, as defined in section 8.1, is exactly the adjacency graph of the $(d-1)$-complex formed by the proper faces of this polytope.

The incidence graph of a simplicial complex can be retrieved from its adjacency graph in time linear in the number of faces (see exercise 11.3).

Duality

We may also generalize the concept of a duality from polytopes to complexes (see subsection 7.1.3).

Let $\mathcal{C}$ be a d-complex. A d-complex $\mathcal{C}^*$ is *dual* to $\mathcal{C}$ if there is a bijection between the faces of $\mathcal{C}$ and those of $\mathcal{C}^*$ which reverses inclusion relationships.

Such a bijection associates the k-faces of $\mathcal{C}$ with $(d - k)$-faces of $\mathcal{C}^*$, for any $k = 0, \ldots, d$ (see figure 11.6 for an example).

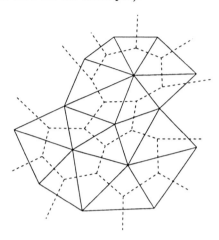

Figure 11.6. A complex (solid edges) and its dual (dashed edges).

Note that in general the dual of a simplicial complex is not a simplicial complex. A complex does not have a unique dual. Nevertheless, all the complexes dual to a given complex $\mathcal{C}$ have isomorphic incidence graphs; we say that they are *combinatorially equivalent*. Moreover, any complex $(\mathcal{C}^*)^*$ dual to the dual of $\mathcal{C}$ is combinatorially equivalent to $\mathcal{C}$ itself.

The adjacency graph of a complex $\mathcal{C}$ is also the 1-skeleton of any dual of $\mathcal{C}$. For this reason, the adjacency graph is also called the *dual graph* of $\mathcal{C}$.

11.4   Exercises

Exercise 11.1 (Jordan's theorem) A *simple curve* in the plane is the image in $\mathbb{E}^2$ of the interval $[0, 1]$ under a continuous bijection f. The endpoints of the curve are $f(0)$ and $f(1)$, and the curve is said to *link* its endpoints. If the mapping f is continuous and bijective over $]0, 1[$, and if $f(0) = f(1)$, then the image $f([0, 1])$ is called a *simple closed curve*. A region $\mathcal{R}$ in the plane is connected if any two of its points can be linked by a simple curve entirely contained within $\mathcal{R}$. Jordan's theorem states that if C is a simple closed curve in $\mathbb{E}^2$, then $\mathbb{E}^2 \setminus C$ has exactly two connected components whose common boundary is C. This exercise presents a simple proof of Jordan's theorem when the simple closed curve C is a polygon.

1. Let C be a polygon. Show that $\mathbb{E}^2 \setminus C$ has at most two connected components. For this, consider a disk D such that $D \cap C$ consists only of a segment. If there are at least three connected components in $\mathbb{E}^2 \setminus C$, then choose three points Q_1, Q_2, Q_3 in distinct components. Show that each of these points can be linked to a point of D by a curve that does not intersect C. Then show that two of the points can be linked by a simple curve entirely contained within $\mathbb{E}^2 \setminus C$.

2. Let Q be a point in $\mathbb{E}^2 \setminus C$ and L any ray extending from Q towards infinity. The intersection $L \cap C$ has connected components which are either points or segments. Each such component S of $L \cap C$ (of zero length or not) is counted twice if C remains on the same side of L just before S and just after S (we say that L *touches* C along S), otherwise it is counted only once (L *goes through* C at S). Show that the parity of this weighted intersection count does not depend on the direction of L, and that it is the same for all points in the same connected component of $\mathbb{E}^2 \setminus C$. By considering a line L that intersects C, show that both parities are possible and thus that $\mathbb{E}^2 \setminus C$ has exactly two connected components, whose common boundary is C.

Exercise 11.2 (Jordan's theorem) The notion of a *simple closed curve* in the plane is defined in the exercise above, which shows that it encloses a region that is a topological 2-ball, called a *disk*, and that the complement of this disk is connected. Prove that this implies that a *chord*, meaning a simple curve that links two points on the boundary of the disk and whose relative interior is entirely contained in the interior of the disk and separates the disk into exactly two distinct connected components.

Hint: Since the complement of the disk is connected, one may join the two endpoints of the chord by a simple curve that lies in the exterior of the disk. The concatenation of this curve and of the chord is a simple closed curve, to which one may again apply Jordan's theorem. One portion of the boundary of the disk lies in the interior, and the other portion in the exterior. Concatenating the chord to these portions yields two simple closed curves, to which we can again apply Jordan's theorem. The bounded regions enclosed by these curves are exactly the two connected components of the disk.

Exercise 11.3 (Incidence graph) Show that the incidence graph of a simplicial d-complex can be retrieved from its adjacency graph in time linear in the number of faces of all dimensions.

Hint: Add all the $(d-1)$-simplices stored in the nodes of the adjacency graph, and for each pair (F, G) of adjacent facets add a $(d-2)$-face incident to F and G. Finally, for $k = d-3, \ldots, 0$, add a node for each k-face of the already constructed $(k+1)$-faces, and merge nodes corresponding to identical k-faces, noticing that such nodes descend from a common $(k+1)$-face.

Exercise 11.4 (Planar maps) A graph G is said to be *planar* if it has a planar embedding: the nodes correspond to points of $\mathbb{E}^2$ and the arcs to simple curves linking two points corresponding to adjacent nodes, such that those curves intersect only at endpoints. The points and simple curves corresponding to the graph for a *planar embedding* of the graph, and the induced subdivision of the plane is commonly called a *planar map* $\mathcal{G}$. The points are called the *vertices* of the map, the curves are the *edges* of the map, and the connected components of $\mathbb{E}^2 \setminus \mathcal{G}$ are the 2-faces (sometimes called regions) of the map.

1. Let n be the number of vertices, m the number of edges, and f the number of 2-faces of a planar map $\mathcal{G}$ and let c be the number of connected components of graph G. Prove Euler's relation:
$$n - m + f = 1 + c.$$

2. Show that if the planar map has f' 2-faces whose boundary consists of only two edges, then the number of edges is bounded by

$$m \leq 3n - 3 - 3c + f'.$$

Hint: Proceed by induction, while analyzing how the sum $n - m + f - c$ varies when a new vertex or a new edge is inserted into the map.

Exercise 11.5 (The Euler characteristic of an unbounded complex) By an unbounded complex we mean a simplicial or cell complex whose cells may be unbounded polytopes (see section 7.3). Show that the Euler characteristic of an unbounded d-complex whose domain is the whole of $\mathbb{E}^d$ is exactly $(-1)^d$.

Exercise 11.6 (Quadratic triangulations) Show that for any pair of integers (n, n_e) such that $4 \leq n_e \leq n$, there exists a triangulation with n vertices, only n_e of which are external, and with t tetrahedra, where

$$t = \frac{n^2}{2} - \frac{3n}{2} - n_e + 3 - 4(n - n_e)(n_e - 4).$$

Hint: For $n = n_e$, see the proof of lemma 11.2.8. For $n_e = 4$, simply choose $n - 2$ points $A_1, \ldots, A_{n-2}$ with respective parameters $\tau_1 < \cdots < \tau_{n-2}$ on the moment curve Γ, and two points B_0 and B_{n-1} such that:

- For $i = 2, \ldots, n - 3$, B_0 belongs to the half-space bounded by the affine hull of triangle $A_1 A_i A_{i+1}$ that does not contain any of the points $A_1, \ldots A_{n-2}$.

- For $i = 1, \ldots, n - 4$, B_{n-1} belongs to the half-space bounded by the affine hull of triangle $A_i A_{i+1} A_{n-2}$ that does not contain any of the points $A_1, \ldots A_{n-2}$.

- The interior of the tetrahedron $B_0 A_1 A_{n-2} B_{n-1}$ contains the points A_i, $i = 2, \ldots, n - 3$.

Build a triangulation of vertices $A_1, \ldots, A_{n-2}$ as in the proof of lemma 11.2.8, then add B_0 and B_{n-1}. For $4 < n_e < n$, choose n_e points on the moment curve, then triangulate them as in the proof of lemma 11.2.8, and finally add the remaining $n - n_e$ points of a smaller scaled moment curve inside one of the previously built tetrahedra.

11.5   Bibliographical notes

The exposition of complexes given in this chapter is voluntarily kept to an elementary level. The reader further interested in homology may find a good introduction in the book by Giblin [110].

Jordan's theorem is a fundamental result; it is intuitively very easy, but its proof is very involved. The simple proof for polygons that is suggested in exercise 11.1 is due to C. Thomassen [214].

The bounds on 3-triangulations were proved by Edelsbrunner, Preparata, and West [95], where the solution of exercise 11.6 can also be found.

Chapter 12

Triangulations in dimension 2

To triangulate a set of points $\mathcal{A}$ is to build a triangulation whose vertices are exactly all the points in $\mathcal{A}$ and whose domain is the convex hull $conv(\mathcal{A})$. In dimension 2, if the number of points in $\mathcal{A}$ is higher than three and no other assumption is made, the problem admits many solutions. It is therefore legitimate to try to obtain the best possible triangulations with respect to some additional criteria. For example, we may want to minimize the total edge-length of the triangulation, or we may also want to avoid long narrow triangles. For the latter criterion, the solution is given by the so-called Delaunay triangulations, which are studied later on (see chapter 17). We may also require that certain given simplices of non-zero dimension be part of the triangulation. We call the corresponding triangulation *constrained*. Obviously, the given simplices must satisfy the intersection property, meaning that they intersect only along their common maximal face, if at all. In dimension 2, constrained triangulations problems always have solutions and there are efficient algorithms to find them. In the next chapter, we will see that here lies a fundamental difference from the case of dimension 3: constrained triangulation problems do not always have a solution in $\mathbb{E}^3$. The most frequently encountered constrained triangulation problem is the *polygon triangulation* problem: the set $\mathcal{A}$ of points is the set of vertices of a polygon $\mathcal{P}$, and the edges of the polygon are constrained to be part of the triangulation. By keeping only the edges and triangles of the resulting constrained triangulation that lie in the polygonal region that is the interior of $\mathcal{P}$, we obtain a *triangulation of the polygon* $\mathcal{P}$, that is, a triangulation whose vertices are exactly those of $\mathcal{P}$ and whose domain is the closed polygonal region interior to $\mathcal{P}$.

Section 12.1 determines the complexity of computing a triangulation for a set of points in $\mathbb{E}^2$, and describes an optimal algorithm. In section 12.2, we prove the existence of solutions for any constrained triangulation problem in $\mathbb{E}^2$. Finally, in section 12.3, we study the more particular problem of computing a triangulation of a given polygon. We clarify the complexity of this problem and describe an algorithm that actually computes a vertical decomposition of the polygon, a decomposition into monotone sub-polygons, then a triangulation. There are many applications of triangulations in dimension 2, some of which can be found in the exercises at the end of this chapter.

12.1   Triangulation of a set of points

12.1.1   The complexity of computing a triangulation

A triangulation $\mathcal{T}$ of a set $\mathcal{A}$ of n points is a triangulation (see previous chapter for the precise definition of a triangulation) such that its vertices are exactly the points in $\mathcal{A}$ and its domain is the convex hull $conv(\mathcal{A})$. The relative interiors of the simplices of this triangulation form a partition of the convex hull $conv(\mathcal{A})$.

Any triangulation $\mathcal{T}(\mathcal{A})$ of $\mathcal{A}$ includes the edges of the convex hull $conv(\mathcal{A})$ as its external edges, and the vertices of $conv(\mathcal{A})$ as its external vertices. Let n_e be the number of extreme points in $\mathcal{A}$ (these are the vertices of the convex hull $conv(\mathcal{A})$). From corollary 11.2.3, we know that any triangulation $\mathcal{T}(\mathcal{A})$ of $\mathcal{A}$ has exactly f triangles and m edges, where

$$\begin{aligned} f &= 2(n-1) - n_e \\ m &= 3(n-1) - n_e. \end{aligned}$$

From a triangulation of $\mathcal{A}$, we easily deduce the external edges, external vertices, and the structure of the boundary, and therefore in $O(n)$ operations we obtain a complete description of the convex hull $conv(\mathcal{A})$. The problem of computing the convex hull of a set $\mathcal{A}$ of n points is thus transformable in time $O(n)$ to the problem of computing a triangulation of the same set $\mathcal{A}$. Since the complexity of computing the convex hull of n points is $\Theta(n \log n)$ (see section 8.2), the complexity of any algorithm that computes a triangulation in dimension 2 must be $\Omega(n \log n)$. Anticipating slightly, we may also mention that there are algorithms that compute a triangulation of a set of n points in time $O(n \log n)$, proving the following theorem:

Theorem 12.1.1 *Computing a triangulation of a set of n points in $\mathbb{E}^2$ is a problem whose complexity is $\Theta(n \log n)$.*

12.1.2   An incremental algorithm

The algorithm we present here uses the incremental method to compute the triangulation of a set of points. This algorithm uses the same scheme as the incremental algorithm that computes the convex hull of a set of points (see section 8.4). The algorithm first sorts the points by increasing lexicographic order on

(x, y), and then maintains a triangulation of the current set obtained by adding the points one by one in that order.

Let $\mathcal{A} = \{A_1, \ldots, A_n\}$ be a set of n points in the plane. To avoid lengthy discussions, we assume as usual that the set of points is in general position. Moreover, we assume that $\mathcal{A}$ has already been sorted by increasing lexicographic order on (x, y), so that $A_1 < A_2 < \cdots < A_n$.

The algorithm not only maintains the triangulation $\mathcal{T}_{i-1}$ built for the subset $\mathcal{A}_{i-1} = \{A_1, \ldots, A_{i-1}\}$ of points already processed, but also the boundary of this triangulation, meaning the boundary of the convex hull $conv(\mathcal{A}_{i-1})$ of the set $\mathcal{A}_{i-1}$. The current triangulation is maintained as a data structure that stores the incidence graph of the triangulation. The convex hull $conv(\mathcal{A}_{i-1})$ is maintained using the doubly linked circular list of its vertices, with a pointer p to the vertex in the list that was last inserted.

In the initial step, we build the triangle formed by the first three points A_1, A_2, A_3 and set the list L to $\{A_1, A_2, A_3\}$, with p pointing to the node that stores A_3.

To describe the current incremental step when A_i is the point to be inserted in the triangulation, we use the same terminology as that of section 8.3 for incremental convex hulls. An edge F of $conv(\mathcal{A}_{i-1})$ is *red* with respect to A_i if the line which is the affine hull of F separates A_i from $conv(\mathcal{A}_{i-1})$, otherwise the edge F is *blue* with respect to A_i. A vertex of $conv(\mathcal{A}_{i-1})$ is *red* with respect to A_i if it is incident to two red edges, *blue* if it is incident to two blue edges, and *purple* it it is incident to both a red and a blue edge. Let us recall that the vertex A_{i-1} is necessarily incident to at least one red edge (see phase 1 of the algorithm described in section 8.3) and that the set of edges on the boundary of $conv(\mathcal{A}_{i-1})$ that are red with respect to A_i is also connected (lemma 8.3.3). Starting at point A_{i-1}, the algorithm traverses the red edges of $conv(\mathcal{A}_{i-1})$, and for each such edge, adds the triangle $conv(F, A_i)$ to the current triangulation. In L, the sub-list of red edges is replaced by the two edges $A_i A_m$ and $A_i A_l$ that connect A_i to the two purple vertices A_m and A_l in $conv(\mathcal{A}_{i-1})$ (see figure 12.1).

The analysis of the algorithm is immediate: Each incremental step of the algorithms takes time proportional to the number of red edges at that step, which is the same as the number of triangles added to the triangulation. A triangle that is added to the triangulation at some step remains a triangle in the subsequent triangulations, and the final triangulation has $O(n)$ triangles. The time needed to perform the $n - 3$ incremental steps is therefore $O(n)$. The initial step can be carried out in constant time. The complexity of the algorithm is thus dominated by the cost of initially sorting the points, which is $O(n \log n)$.

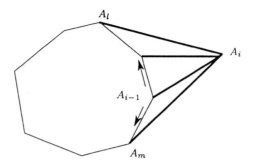

Figure 12.1. Incremental triangulation.

Theorem 12.1.2 *The incremental algorithm described above requires $O(n \log n)$ operations to triangulate a set of n points in the plane. This is optimal in the worst case.*

Notice that the complexity of the incremental algorithm for a set of n points is only $O(n)$ if the points are sorted along some known direction.

12.2   Constrained triangulations

The preceding algorithm outputs a certain triangulation of the set of points. There are many other ways that yield different triangulations, and we may even ask for a triangulation that contains some given edges. The corresponding problem is called a *constrained triangulation* problem.

Let $\mathcal{A}$ be a set of points in the plane. The following theorem shows that any set of edges that do not intersect except at common endpoints and whose endpoints are vertices in $\mathcal{A}$ can be completed into a triangulation of $\mathcal{A}$. In other words, in dimension 2, a constrained triangulation problem always has a solution.

Theorem 12.2.1 *Let $\mathcal{A}$ be a set of points in the plane. Any maximal set of segments that connect the points in $\mathcal{A}$ and have pairwise intersection only at common endpoints is the set of edges of a triangulation of $\mathcal{A}$, and the converse is also true.*

Proof. Let $\mathcal{E}$ be a maximal set of segments that join the points in $\mathcal{A}$ and have pairwise intersection only at common endpoints. The maximality of $\mathcal{E}$ implies that no segment may be added to $\mathcal{E}$ while maintaining this property. The edges in $\mathcal{E}$ must include the edges on the boundary of the convex hull of $\mathcal{A}$, since otherwise adding any of them would contradict the maximality of $\mathcal{E}$. Since the segments in $\mathcal{E}$ are all inside the convex hull of $\mathcal{A}$, they determine a decomposition of this convex hull into polygonal regions. Let us show that any such region must be

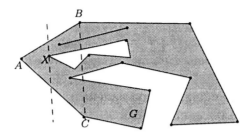

Figure 12.2. For the proof of theorem 12.2.1.

a triangle. Aiming for a contradiction, assume there is a region $\mathcal{G}$ that is not triangular, and let $\mathcal{P}$ denote the polygon that forms the external boundary of $\mathcal{G}$. Consider any vertex A that is convex for $\mathcal{P}$, meaning that the internal angle of the polygon at vertex A is smaller than π (see figure 12.2). Note that any polygon has at least two convex vertices, for instance the vertices of minimal and maximal abscissae. Let AB and AC be the two edges on the boundary of $\mathcal{P}$ that are incident to A. We claim that triangle ABC contains at least some point of $\mathcal{A}$ in its interior. Indeed, since $\mathcal{G}$ is not triangular, either BC is not an edge of $\mathcal{E}$ or the triangle ABC includes the boundary of some hole of $\mathcal{G}$ in its interior. In the former case, BC has to intersect some edge in $\mathcal{E}$, otherwise the maximality of $\mathcal{E}$ is contradicted, and any edge of $\mathcal{E}$ that intersects BC must have an endpoint inside the triangle ABC. Let us choose for X the point closest to A in the direction parallel to BC, such that there is no point of $\mathcal{A}$ in the triangle defined by the intersection of ABC and the half-plane bounded by a line passing through X and parallel to BC which contains A. The edge AX has both endpoints in $\mathcal{A}$ and does not intersect another edge of $\mathcal{E}$ by construction. Again, $\mathcal{E}$ is not maximal, a contradiction.

The converse statement simply states that the set $\mathcal{E}$ of edges in a triangulation is maximal, meaning that any segment having both endpoints in $\mathcal{A}$ and not belonging to $\mathcal{E}$ must intersect the interior of some edge in $\mathcal{E}$. This is trivial, since any segment that does not intersect the segments in $\mathcal{E}$ must lie in the interior of a single triangle, but there cannot be such a segment with endpoints in $\mathcal{A}$ since the interior of any triangle contains no point of $\mathcal{A}$. □

12.3 Vertical decompositions and triangulations of a polygon

12.3.1 Lower bound

Let $\mathcal{P}$ be a polygon in the plane. From the preceding discussion, we know that it is possible to compute a triangulation of the set of vertices of $\mathcal{P}$ constrained to include the edges of the polygon $\mathcal{P}$. Hence the obtained triangulation has, besides the edges and vertices of $\mathcal{P}$, *interior* edges and *interior* triangles that are contained in the interior of $\mathcal{P}$, and occasionally *exterior* edges and *exterior* triangles contained in the exterior of $\mathcal{P}$. The faces of $\mathcal{P}$ and the interior faces of the triangulation form a triangulation whose domain is exactly the interior of $\mathcal{P}$, and whose boundary is the polygon $\mathcal{P}$. Such a triangulation is called a *triangulation of the polygon* $\mathcal{P}$. The following theorem is a straightforward consequence of theorem 12.2.1.

Theorem 12.3.1 *Any polygon $\mathcal{P}$ in the plane can be triangulated, in other words it can be described as the boundary of a 2-triangulation whose vertices are vertices of $\mathcal{P}$.*

If $\mathcal{P}$ is a polygon with n vertices, then any triangulation $\mathcal{T}$ of $\mathcal{P}$ has exactly n vertices which are external. We can use corollary 11.2.3 to show that such a triangulation has exactly $f = n - 2$ triangles, $m = 2n - 3$ edges, and $n - 3$ internal edges. Triangulating a polygon is therefore equivalent to finding the $n - 3$ internal edges that decompose the polygonal region into $n - 2$ triangles.

Incidentally, we note the following property:

Lemma 12.3.2 *The dual graph of a triangulation of a planar polygon is a tree.*

Proof. This graph is obviously connected. Furthermore, it has no cycle. Indeed, the existence of a cycle in the dual graph of a triangulation implies either the existence of a hole in the polygonal region $dom(\mathcal{T})$, or the presence of an internal vertex in the triangulation. $\square$

Knowing a simple polygonal line that joins the points of $\mathcal{A}$ enables us to compute the convex hull $conv(\mathcal{A})$ in linear time (theorem 9.4.4). So the argument that proves a lower bound of $\Omega(n \log n)$ on the complexity of computing a triangulation for a set of n points does not apply to the set of vertices of a polygon. We may legitimately suspect that computing a triangulation of a polygon is a simpler problem than its counterpart for a set of points.

The complexity of computing a triangulation of a simple polygon remained elusive for a long time. Classical algorithms only achieved time $O(n \log n)$ for the general problems, while several algorithms were known to perform in linear time on special kinds of polygons, such as convex, monotone, or star-shaped polygons, or polygons visible from a single segment. In 1986, a deterministic algorithm was proposed whose worst-case complexity is $o(n \log n)$, proving at least that $O(n \log n)$ was not a tight bound. The problem was settled, at least theoretically, in 1990 when a linear-time algorithm that computes the triangulation of any simple polygon in the plane was given. This algorithm is too complex to be presented in this book, or to be of any practical use. Its existence, however, provides a proof of the following theorem:

Theorem 12.3.3 *The complexity of computing a triangulation of a simple polygon with n vertices in the plane is* $\Theta(n)$.

12.3.2 Triangulating monotone polygons

Classical methods used in computational geometry described in chapter 3, the sweep and divide-and-conquer methods, both lead to simple and efficient algorithms that triangulate a simple polygon. Here we choose to develop the method that uses the decomposition of the polygon into trapezoids, which we call the *vertical decomposition* of the polygon. This method leads to some of the currently most efficient deterministic algorithms, and also to the method mentioned above that works in linear time. It is also through vertical decompositions that randomization methods appear in triangulation algorithms.

In principle, this method begins by computing a vertical decomposition of the polygon. From this decomposition, we deduce a decomposition of the interior polygonal region into *monotone* polygonal sub-regions. Finally, the algorithm relies crucially on a method that triangulates a monotone polygon in linear time.

We must begin by defining the monotonicity property of a polygon and how it can be characterized, and describe an algorithm that triangulates a monotone polygon in linear time. In the following discussion, we recall the notion of a vertical decomposition for a polygon, and show how to build a triangulation from the vertical decomposition in linear time, using a decomposition into monotone polygons. Finally, we show how to simply compute the vertical decomposition of a polygon using a sweep algorithm.

Definition and characterization of monotone polygons

A polygonal line $\mathcal{C}$ is *monotone* with respect to a given direction δ if any line perpendicular to the direction δ intersects $\mathcal{C}$ in at most one point. In what follows, the direction δ is supposed to be that of the x-axis. Thus, a polygonal line is monotone if it is the graph of a piecewise linear function $f(x)$, or also if the sequence $P_1, P_2, \ldots P_n$ of its vertices is ordered by increasing abscissae:

$$x(P_1) < x(P_2) < \cdots < x(P_n),$$

or by decreasing abscissae:

$$x(P_1) > x(P_2) > \cdots > x(P_n).$$

A polygon is *monotone* with respect to a given direction if it can be obtained as the concatenation of two polygonal lines which are monotone with respect to that direction.

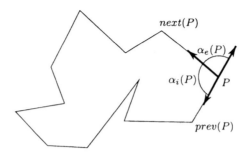

Figure 12.3. Internal and external angles in a polygon.

Before we describe an algorithm that triangulates monotone polygons in linear time, we first give a lemma that allows the monotonicity of a polygon to be easily characterized. A few definitions are needed before we can state the lemma.

Let $\mathcal{P}$ be a polygon, P be a vertex of $\mathcal{P}$, $prev(P)$ and $next(P)$ respectively the predecessor and successor of P in the direct order along the polygon $\mathcal{P}$. Then the vertex P is said to be *convex* if the determinant $[prev(P), P, next(P)]$ is positive, *reflex* if this determinant is negative, and *flat* if the three consecutive vertices are collinear. The notion of a convex vertex can also be defined in terms of *internal* and *external angles* of the polygon. The *internal angle* at a vertex P is the measure $\alpha_i(P)$, in $[0, 2\pi[$, of the angle between the vectors $Pnext(P)$ and $Pprev(P)$. The *external angle* at P is the measure $\alpha_e(P) = \pi - \alpha_i(P)$ of the complementary angle between the vectors $prev(P)P$ and $Pnext(P)$).

A vertex is convex (resp. reflex, flat) if the internal angle of the polygon at that vertex is smaller than (resp. greater than, equal to) π.

A vertex P of the polygon $\mathcal{P}$ that breaks the monotonicity property in the x-direction is necessarily a *start vertex*, that is, a vertex such that

$$x(P) \le x(prev(P)) \quad \text{and} \quad x(P) \le x(next(P)),$$

or an *end vertex*, that is, a vertex such that

$$x(P) \ge x(prev(P)) \quad \text{and} \quad x(P) \ge x(next(P)).$$

A vertex of $\mathcal{P}$ which is neither a start nor end vertex is called a *monotone vertex*. For instance, points P_1, P_{10} and P_{12} on the polygon in figure 12.4 are convex start vertices, while P_4 is a reflex start vertex; similarly, P_3 and P_5 are convex end vertices, while P_{11} and P_{15} are reflex end vertices; all other vertices on this polygon are monotone.

Lemma 12.3.4 *A polygon $\mathcal{P}$ is monotone in the x-direction if and only if it has no start or end reflex vertex.*

Proof. This condition is necessary: indeed, a monotone polygon has only one start vertex, which is the vertex with the minimum abscissa, and only one end vertex, which has maximum abscissa. Both these vertices are convex.

Reciprocally, the following lemma shows that if a polygon has no start or end reflex vertex, then it has only one start vertex and one end vertex, and it is therefore monotone. □

Lemma 12.3.5 *Consider any polygon $\mathcal{P}$. We denote by c_s the number of its convex start vertices, by c_e the number of its convex end vertices, by r_s the number of its reflex start vertices, and by r_e the number of its reflex end vertices. Then*

$$c_s = r_e + 1,$$

$$c_e = r_s + 1.$$

Proof. Let $\mathcal{P}$ be a polygon and $P_1, P_2, \ldots, P_n$ be the circular sequence of its vertices in counter-clockwise order. The sum of all internal and external angles of $\mathcal{P}$ is $n\pi$, since a pair of internal/external angles at any vertex contributes π. The sum of the internal angles is $(n-2)\pi$, because any triangulation of $\mathcal{P}$ has $n-2$ triangles. The sum of the external angles is therefore 2π. Let U_i be the unit vector directed along the vector $prev(P_i)P_i$. Consider a point P that follows the boundary of the polygon in counter-clockwise order and a point U that describes the unit circle centered at the origin. When P is on the edge $prev(P_i)P_i$, U is U_i, and when P is at a vertex P_i, then U describes the circular arc U_iU_{i+1}. Since the sum of the external angles is 2π, when P has described all the boundary, point U has gone through any oriented ray originating at O once more in the counter-clockwise direction than in the clockwise direction. It is easy to see that U crosses the y-axis counter-clockwise each time P is at a convex end vertex, and clockwise each time P is at a reflex start vertex. Similarly, it crosses the opposite $(-y)$-axis counter-clockwise when P is at a convex start vertex, and clockwise when P is at a reflex end vertex (see figure 12.4). □

Incidentally, a polygon is said to be convex if its interior region is convex. Note that a polygon is convex if and only if it has no reflex vertex, and hence if and only if it is monotone for every direction.

Triangulating a monotone polygon

Let $\mathcal{P}$ be a monotone polygon in the direction of the x-axis. Then $\mathcal{P}$ is the concatenation of two monotone polygonal lines that connect the vertices of minimum and maximum abscissae. The algorithm begins by sorting the vertices of $\mathcal{P}$ by increasing abscissae, which can be done by merging the vertices of the upper and

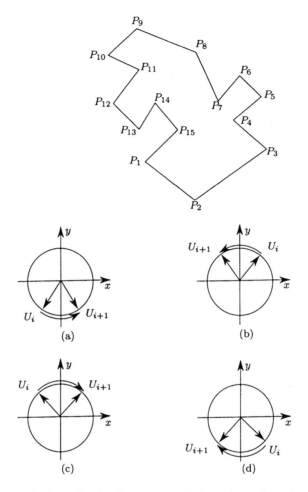

Figure 12.4. Different kinds of motion on the boundary of a polygon:
(a) convex start vertex (for instance P_1)
(b) convex end vertex (for instance P_3)
(c) reflex start vertex (for instance P_4)
(d) reflex end vertex (for instance P_{11}).

lower monotone polygonal lines. Let $Q_0, Q_1, \ldots, Q_{n-1}$ be the resulting ordered sequence of vertices of $\mathcal{P}$. In the course of the algorithm, the vertices of $\mathcal{P}$ are visited in this order, one by one, and the algorithm adds to the edges of $\mathcal{P}$ the internal edges of the triangulation. Each internal edge added by the algorithm separates a triangle in the triangulation from the *remaining* polygon, whose number of vertices decreases by one at each step. Let $Q_0, Q_1, \ldots, Q_{i-1}$ be the vertices already visited by the algorithm before the current step. The algorithm maintains the following invariants:

1. The remaining polygon is monotone. It is the concatenation of the polygonal line formed by the vertices that have not been visited yet, of abscissa greater than $x(Q_i)$, and of a polygonal line (with increasing abscissae) $\{V_0, V_1, \ldots, V_t\}$ whose vertices have already been visited by the algorithm.

2. If $t > 1$, the vertices $\{V_1, \ldots, V_{t-1}\}$ are reflex vertices in the remaining polygon.

The algorithm maintains a stack ordered by increasing abscissae, which contains the vertices of $\{V_0, \ldots, V_t\}$. These vertices are the vertices of the remaining polygon that have already been visited. After the initial sort of the vertices of $\mathcal{P}$, the first vertices Q_0 and Q_1 are put into a stack. The vertex Q_i visited at the current step is a vertex of the remaining polygon which is adjacent either to V_0 or to V_t (see figure 12.5). Only the last vertex Q_{n-1} can be adjacent to both V_0 and V_t. The current step proceeds according to the following scheme.

1. If Q_i is adjacent to V_0 but not to V_t in the remaining polygon (see figure 12.5a), then we add the edges Q_iV_1, Q_iV_2, $\ldots$, Q_iV_t and the triangles $Q_iV_0V_1$, $Q_iV_1V_2$, $\ldots$, $Q_iV_{t-1}V_t$ to the triangulation. The stack is updated so that it contains $\{V_t, Q_i\}$.

2. If Q_i is adjacent to V_t but not to V_0 (see figure 12.5b), as long as the stack contains two vertices and the vertex V_t on top of the stack is a convex vertex in the remaining polygon, the edge Q_iV_{t-1} and the triangle $Q_iV_{t-1}V_t$ are added to the triangulation, and vertex V_t is popped from the stack. Then Q_i is stacked.

3. If the vertex Q_i is the last vertex Q_{n-1} in $\mathcal{P}$, then it is adjacent to both V_0 and V_t (see figure 12.5c), and we add the edges $Q_iV_1, Q_iV_2, \ldots, Q_iV_{t-1}$ and the triangles $Q_iV_0V_1$, $Q_iV_1V_2$, $\ldots$, $Q_iV_{t-1}V_t$ to the triangulation.

A segment or a triangle is said to be *interior* to a polygon if its relative interior is contained within the interior of the polygon. It is easy to prove that each edge and triangle added to the triangulation in the preceding scheme is interior to the remaining polygon. It is also easy to check that the algorithm maintains the invariants at each step. For instance, let us consider the first edge Q_iV_1 and the first triangle $T = Q_iV_0V_1$ added in case 1. The vertices $V_2, \ldots, V_t$ in the stack do not belong to the triangle T because they are on one side of V_0V_1 while Q_i is on the other side. This is guaranteed by the fact that the vertices $\{V_1, \ldots, V_{t-1}\}$ are reflex in the remaining polygon. The vertices in $\mathcal{P}$ that have not yet been visited do not belong to T either because their abscissae are greater than that of Q_i. Thus the triangle T does not contain any vertex of the remaining polygon, except for its three vertices Q_i, V_0 and V_1. The edges Q_iV_0 and V_0V_1 are edges

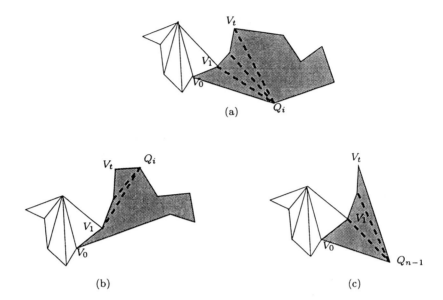

Figure 12.5. Triangulating a monotone polygon:
(a) Q_i is adjacent to V_0 but not to V_t.
(b) Q_i is adjacent to V_t but not to V_0.
(c) Q_i is adjacent to V_0 and to V_t.

of the remaining polygon, so no edge of this polygon can intersect the interior of T, nor the edge Q_iV_1. V_0 is the vertex of minimum abscissa among the vertices of the remaining polygon, it is therefore convex and so T and Q_iV_1 are interior to the remaining polygon.

The analysis of this algorithm is immediate. The initial sort can be performed in linear time since it consists of merging two already sorted lists. Each step in the algorithm can be carried out in time proportional to the number of vertices added to and popped from the stack. Since each vertex is stacked and popped only once, the algorithm has linear complexity, proving that:

Theorem 12.3.6 *It is possible to triangulate a monotone polygon in linear time.*

12.3.3   Vertical decomposition and triangulation of a polygon

From the vertical decomposition to the triangulation

The vertical decomposition $\mathcal{D}ec(\mathcal{P})$ of a polygon $\mathcal{P}$ consists of the vertical decomposition of the set $\mathcal{S}_\mathcal{P}$ of segments formed by the edges of the polygon $\mathcal{P}$. This structure is defined in section 3.3, so we only recall briefly how it is formed. The vertical decomposition (or decomposition for short) depends on a direction in the plane, which we assume is the direction of the y-axis. This direction is called

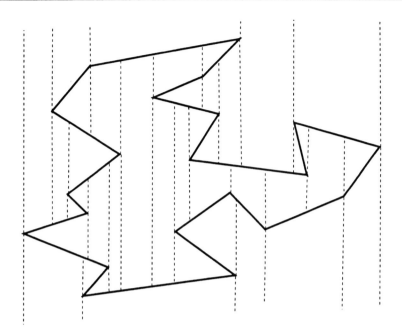

Figure 12.6. The vertical decomposition of a polygon.

vertical and the perpendicular direction is called *horizontal*. To each vertex P of the polygon correspond two *walls*, stemming from P in both directions. These walls are the maximal vertical segments PP_1 and PP_2 such that their relative interiors do not intersect any edge of $\mathcal{P}$. The vertical decomposition $\mathcal{D}ec(\mathcal{P})$ of the polygon is the subdivision of $\mathbb{E}^2$ induced by the edges of the polygon and by the walls stemming from its vertices. The decomposition $\mathcal{D}ec(\mathcal{P})$ can be considered as an (unbounded) cellular complex of size $O(n)$ in the plane. Indeed, the complex has at most $3n$ vertices: the vertices of $\mathcal{P}$ and the $2n$ endpoints of the walls. The cells in the decomposition are trapezoids, with vertical parallel sides, occasionally degenerated into triangles, infinite trapezoids, or half-planes. A wall is said to be *interior* to the polygon if its relative interior is contained in the interior of the polygon, and *exterior* in the opposite case. Similarly, a trapezoid in the decomposition is *interior* to the polygon if its relative interior is contained in the interior of the polygon $\mathcal{P}$, and *exterior* otherwise.

Each trapezoid in the vertical decomposition $\mathcal{D}ec(\mathcal{P})$ has two vertices of $\mathcal{P}$ on its boundary, one on each vertical side. This simple observation allows us to classify the trapezoids into two categories. Let P and Q be the vertices of $\mathcal{P}$ that are on the boundary of a trapezoid (see figure 12.7):

- The trapezoid is said to be *non-diagonal* if P and Q are adjacent vertices on the boundary of the trapezoid.

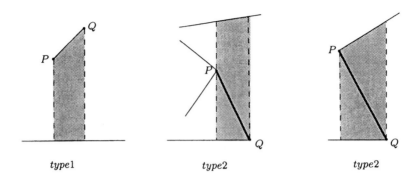

Figure 12.7. The two types of trapezoid in the vertical decomposition of a polygon.

- The trapezoid is said to be *diagonal* if P and Q are not adjacent vertices on the boundary of the trapezoid. This happens when one of P or Q is not a vertex of the trapezoid, or else if P and Q are opposite vertices of the trapezoid.

When a trapezoid is diagonal, the relative interior of segment PQ is contained in the interior of the trapezoid. Let us also call such a segment *diagonal*. The set of diagonal segments PQ is a set of segments interior to the polygon $\mathcal{P}$ which do not intersect except at their common endpoints. Once the vertical decomposition of a polygon $\mathcal{P}$ has been computed, the trapezoids in the decomposition may be examined one by one and the diagonal segments may be found in linear time. The diagonal segment decomposes the polygonal region bounded by $\mathcal{P}$ into polygonal sub-regions. We claim that each of these sub-regions is bounded by a polygon which is monotone with respect to the horizontal direction. Indeed, such a polygon cannot have any start or end reflex vertex, because such a vertex in $\mathcal{P}$ is necessarily the endpoint of a diagonal segment, and the claim follows from lemma 12.3.4. It may even be shown that the boundaries of these sub-regions are *unimonotone*, meaning that either the upper monotone chain or the lower monotone chain consists of a single edge. The total complexity of the resulting monotone polygonal sub-regions is $O(n)$, and these regions can be triangulated in linear time, which shows that:

Theorem 12.3.7 *A triangulation of a polygon may be deduced from its vertical decomposition in additional linear time.*

Computing the vertical decomposition of a polygon by sweeping

It remains to show how to build the vertical decomposition of a simple polygon. In fact, a randomized accelerated algorithm that builds the vertical decomposition of a polygon with n vertices was given in section 5.4 and runs in expected time $O(n \log^* n)$. The algorithm we describe here uses the sweep method, and is a variant of the algorithm described in subsection 3.2.2 that computes the intersection points of a set of segments; it runs in time $O(n \log n)$. We then describe a *lazy* version of the same algorithm, whose complexity is lower for a large class of polygons.

To decompose a polygon using the sweep method, we propose to sweep the plane with a vertical line Δ from left to right. The state of the sweep, stored in a structure $\mathcal{Y}$, is the ordered list of *active* edges: these are the edges of the polygon intersected by Δ, ordered according to their intersections along Δ. The structure $\mathcal{Y}$ is implemented using a balanced binary tree, letting us insert, delete, or query active edges in time $O(\log k)$ where k is the number of *active* edges. Moreover, the nodes of the tree store two extra pointers that allow access in constant time to the active edge immediately above or below the active edge E stored in this node. The edge above E is denoted by $above(E)$ and the edge below E by $below(E)$.

The list $\mathcal{Y}$ of active edges changes only when Δ sweeps over a vertex of the polygon. Thus the list of events to be processed is simply the list of vertices of $\mathcal{P}$ sorted by increasing abscissae. Without loss of generality, we may assume that no two vertices have the same abscissa.

The structure $\mathcal{Y}$ initially stores two fictitious edges that intersect the sweep line at $y = +\infty$ and $y = -\infty$ respectively. Processing the event corresponding to a vertex P_i consists of the following operations:

1. Locate P_i in the structure $\mathcal{Y}$ according to its ordinate.

2. Create the walls stemming from P_i.

3. Update the structure $\mathcal{Y}$.

Each of these operations depends on the type of vertex P_i: it can be either a start, an end, or a monotone vertex.

If P_i is a start vertex (see figure 12.8), locating P_i in the list $\mathcal{Y}$ allows us to retrieve the active edges E and E' that lie immediately below or above P_i on Δ. The algorithm builds two walls starting at P_i and butting on E and E'.[1] The edges incident to P_i are inserted in $\mathcal{Y}$ between E and E'.

If P_i is a monotone vertex (see figure 12.9), locating P_i in the structure $\mathcal{Y}$ allows us to retrieve the active edge E_1 that is incident to P_i. The walls stemming from P_i and butting on the edges $below(E_1)$ and $above(E_1)$ are inserted into the decomposition, and edge E_1 is replaced in $\mathcal{Y}$ by the other edge E_2 of $\mathcal{P}$ that is incident to P_i.

---

[1]Details of these operations depend on the particular representation of the vertical decomposition used by the algorithm: simple list of walls, simplified or complete representation of $\mathcal{D}ec(\mathcal{P})$ as described in section 3.3.

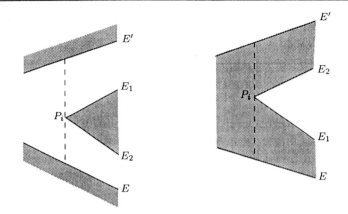

Figure 12.8. Vertical decomposition of a polygon by sweeping: a start vertex.

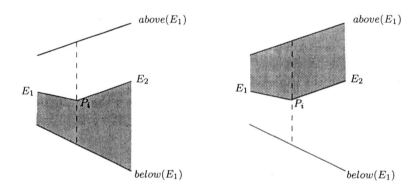

Figure 12.9. Vertical decomposition of a polygon by sweeping: a monotone vertex.

Finally, if P_i is an end vertex (see figure 12.10), locating the vertex P_i in the structure $\mathcal{Y}$ allows us to find both active edges E_1 and E_2 incident to P_i. Say that E_1 is above E_2, then the walls stemming from P_i butt on the edges $below(E_2)$ and $above(E_1)$. The two edges E_1 and E_2 are both removed from the structure $\mathcal{Y}$.

Theorem 12.3.8 *A sweep algorithm builds the vertical decomposition of a polygon with n vertices in time $O(n \log n)$.*

Proof. Sorting the vertices of $\mathcal{P}$ to build the ordered list of events takes time $O(n \log n)$. The number of active edges is always less than n. For each of the n events, locating the current vertex in $\mathcal{Y}$ and updating the structure $\mathcal{Y}$ (which involves at most two insertions or two deletions) require $O(\log n)$ operations, and the remaining operations can be carried out in constant time. $\square$

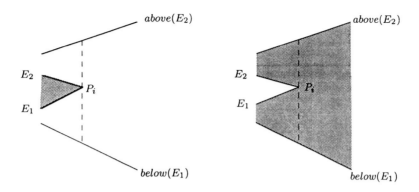

Figure 12.10. Vertical decomposition of a polygon by sweeping: an end vertex.

Remark. The algorithm is based on the fact that the edges of the simple polygon do not intersect except at common vertices. Nowhere do we use explicitly the fact that the edges are connected. The algorithm can therefore be extended straightforwardly to compute the vertical decomposition of a collection of several disjoint polygons, or of polygonal region with holes.

Vertical decomposition of a polygon by lazy sweeping

In its basic version, an algorithm that computes a vertical decomposition using the sweep method requires time $\Omega(n \log n)$ for any kind of polygon, even if it is convex or monotone. In those cases, the vertical decomposition can be easily obtained by other methods in only linear time. The analysis of this algorithm reveals that its complexity is dominated by the cost of initially sorting the vertices of $\mathcal{P}$, the cost of locating the vertices in the structure $\mathcal{Y}$, and the cost of rebalancing the tree after each insertion or deletion of active edges. These operations are not necessary for each vertex of $\mathcal{P}$, however. Indeed, any polygon $\mathcal{P}$ can be considered as the concatenation of monotone polygonal lines, the *monotone chains* whose endpoints are the start and end vertices. The internal vertices of such chains are monotone vertices and are ordered by increasing or decreasing abscissae along the chain. Also, when the algorithm processes the event at a monotone vertex, updating the structure $\mathcal{Y}$ only involves replacing an edge by another one, which requires no rebalancing of the structure. If we manage to successively process the events corresponding to monotone vertices on a chain, we no longer need to locate these vertices in the structure $\mathcal{Y}$.

In its *lazy* version, the algorithm only processes the event when the sweep line Δ sweeps over start or end vertices. So the list of events is now only the list of start or end vertices, sorted by their increasing abscissae, which can be obtained in time $O(n + s \log s)$ where s is the number of these special vertices of $\mathcal{P}$.

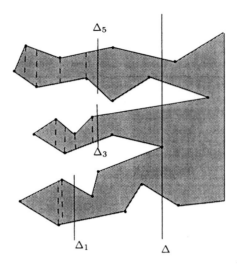

Figure 12.11. Vertical decomposition a polygon using lazy sweeping.

The structure $\mathcal{Y}$ is modified so that it can handle updates in a lazy fashion. This structure is still implemented as a balanced tree, but each node now corresponds to an *active monotone chain*, rather than to an *active* edge: a monotone chain is called *active* if one of the edges on this chain is active, meaning that it intersects the sweep line Δ.

The active edges or monotone chains subdivide the sweep line Δ into an ordered sequence of segments which are alternately *interior* and *exterior*. To simplify the discussion, we only describe how to build the decomposition of the interior of the polygon $\mathcal{P}$, that is, to find the interior walls and trapezoids in the decomposition of $\mathcal{P}$. It thus suffices to consider the interior segments on the sweep line Δ. Each of these segments is bounded by a pair $(\mathcal{C}_i, \mathcal{C}_{i+1})$ of active chains which are consecutive along Δ. To each pair, we dedicate a local sweep line Δ'_i. The local sweep line Δ'_i always lags behind Δ: the edge E'_i of $\mathcal{C}_i$ intersected by Δ'_i precedes (in increasing x-order along the chain $\mathcal{C}_i$) the edge E_i intersected by Δ; similarly, the edge E'_{i+1} of $\mathcal{C}_{i+1}$ intersected by Δ'_i precedes the edge E_{i+1} of $\mathcal{C}_{i+1}$ intersected by Δ. The information stored at the nodes of the tree $\mathcal{Y}$ corresponding to the chains $\mathcal{C}_i$ and $\mathcal{C}_{i+1}$ is relevant only to the edges E'_i and E'_{i+1} intersected by the local sweep line Δ_i.

The local sweep line Δ_i only advances and reaches the global sweep line Δ when a node of $\mathcal{Y}$ corresponding to one of the chains $\mathcal{C}_i$ or $\mathcal{C}_{i+1}$ is visited in order to locate a vertex P in the list of events. When visiting the node of $\mathcal{Y}$ that corresponds to the chain $\mathcal{C}_i$, the algorithm tests whether the active edge E'_i intersected by the local sweep line Δ_i intersects the global sweep line Δ. If not, the local sweep line sweeps over the vertices of $\mathcal{C}_i$ and $\mathcal{C}_{i+1}$ that lie between the two lines Δ_i and Δ. A linear traversal of these two chains (analogous to merging two sorted lists) then builds the wall from each vertex of C_i and C_{i+1}, in constant time for each wall. When the local sweep line has caught up with Δ, the edge E_i' of C_i intersected by the local sweep line Δ_i coincides with the edge E_i intersected by the global sweep line Δ. The vertex P may then be compared with E_i and the location of P in the structure $\mathcal{Y}$ may continue.

The process which advances the local sweep line Δ_i involves merging two sorted lists of vertices and does not need to perform location queries in the structure $\mathcal{Y}$ or to rebalance this structure. Such a process takes time linear in the number of processed monotone vertices. If the number of start or end vertices is s, the number of chains stored in $\mathcal{Y}$ is $O(s)$; thus $\mathcal{Y}$ requires storage $O(s)$, and each location or rebalancing takes time $O(\log s)$. The total complexity of the lazy sweeping is thus $O(n + s \log s)$. As is proved in lemma 12.3.5, for any direction of the sweep line, the number of start and end vertices is at most $r + 2$ if r is the number of reflex vertices in the polygon. The complexity of the lazy sweep algorithm is thus $O(n + r \log r)$.

It is possible to build the portion of the vertical decomposition that lies outside the polygon $\mathcal{P}$ in a similar fashion. It suffices to maintain a local sweep line for each segment of the global sweep line that lies outside the polygon. Finally, we should also mention that the algorithm does not use explicitly the fact that the edges are connected, so that it computes equally well the vertical decomposition of a collection of disjoint polygons, or of a polygonal region with holes.

The following theorem summarizes the results of this paragraph:

Theorem 12.3.9 *The lazy sweep algorithm computes the vertical decomposition of a polygon (or of a collection of disjoint polygons) with a total of n vertices, of which r are reflex vertices, in time $O(n + r \log r)$.*

Note that if a polygon is monotone in some given direction, then the direction of the sweep line may be appropriately chosen as perpendicular to this direction to ensure that the number of start and end vertices for this direction is exactly two, and the lazy sweep algorithm takes time $O(n)$ in this case.

Theorem 12.3.10 *The lazy sweep algorithm computes a vertical decomposition of a monotone polygon in linear time.*

12.4 Exercises

Exercise 12.1 (Decomposition into convex parts) Consider a polygon $\mathcal{P}$ with n vertices and r reflex vertices.

1. Show that any decomposition of the interior of $\mathcal{P}$ into convex parts has at least $\lceil r/2 \rceil + 1$ regions.

2. We present a way to compute, knowing a triangulation of $\mathcal{P}$, a decomposition of the interior of $\mathcal{P}$ into a quasi-minimal number of convex parts. For this, we examine each internal edge in the triangulation in turn, and remove it if the union of the two incident convex regions is also convex. Show that the resulting complex has at most $2r + 1$ convex cells interior to $\mathcal{P}$, and that this is at most four times the minimal size of such a decomposition.

3. Construct a polygon that cannot be decomposed into an optimal number of convex polygonal parts having only vertices on the polygon. Therefore, reaching the optimum may require points to be added inside the polygon.

Exercise 12.2 (Localization in a planar map) This exercise presents a proof of the following result: given a planar map of size $O(n)$, it is possible to build in time $O(n \log n)$ a data structure of size $O(n)$ that allows localization queries in the map to be performed in time $O(\log n)$.

We first consider a planar triangulation $\mathcal{T}$ whose boundary consists of a single triangle. Let n be the number of vertices of $\mathcal{T}$.

1. Show that $\mathcal{T}$ has $3n - 6$ edges and $2n - 5$ triangles.

2. Consider a maximal set of internal vertices of $\mathcal{T}$ such that two vertices are not adjacent and the degree of any of these vertices is at most d. Show that the size n' of any such set is at least

$$n' \geq \frac{(d - 5)}{d(d + 1)} \, n.$$

3. Show that such a set may be found in linear time.

From the triangulation $\mathcal{T}$, it is possible to build a hierarchical structure that allows efficient localization queries in the triangulation $\mathcal{T}$. This structure is analogous to that for 3-polytopes which was described in exercise 9.5 and used in exercise 9.6 to answer several kinds of queries on this polytope. The structure represents a sequence of triangulations

$$\mathcal{T}_0 = \mathcal{T}, \mathcal{T}_1, \ldots, \mathcal{T}_h,$$

such that $\mathcal{T}_h$ has bounded complexity, and $\mathcal{T}_{i+1}$ can be deduced from $\mathcal{T}_i$ by removing a maximal set of non-adjacent internal vertices of $\mathcal{T}_i$ with degree at most d. More precisely, if $\mathcal{S}_i$ is such a subset of vertices of $\mathcal{T}_i$, for each vertex P in $\mathcal{S}_i$, the triangles in the star $\mathcal{T}_i(P)$ of P in $\mathcal{T}_i$ are replaced by a triangulation $\mathcal{T}_i'(P)$ of the boundary of this star. These triangulations are merged to obtain $\mathcal{T}_{i+1}$, so that

$$\mathcal{T}_{i+1} = \mathcal{T}_i - \bigcup_{P \in \mathcal{S}_i} \mathcal{T}_i(P) + \bigcup_{P \in \mathcal{S}_i} \mathcal{T}_i'(P).$$

The underlying data structure is a graph that has a node for each triangle that belongs to one or several successive triangulations, and an edge for each pair (T', T) of triangles such that, for some level $i \in [0, h[$, T belongs to $\mathcal{T}_i$ but not to $\mathcal{T}_{i+1}$, T' belongs to $\mathcal{T}_{i+1}$ but not to $\mathcal{T}_i$, and the intersection $T \cap T'$ has a non-empty relative interior.

4. Show that $h = O(\log n)$, that $\mathcal{T}_{i+1}$ can be computed from $\mathcal{T}_i$ in linear time, and that the graph described above can be built in time $O(n)$ if the triangulation $\mathcal{T}$ is given.

5. Show that the graph can be used to perform localization queries of a point in the triangles of $\mathcal{T}$ in time $O(h) = O(\log n)$.

6. Extend the method to any planar map. If the map has no unbounded edges, it can always be enclosed in a surrounding triangle and the resulting polygonal regions can be triangulated. Otherwise, the regions in the map may be triangulated, by considering points at infinity to triangulate the unbounded regions.

Exercise 12.3 (Triangulating a star-shaped polygon) A polygon $\mathcal{P}$ is called *star-shaped* with respect to a point V if, for any point W belonging to the domain *dom*$(\mathcal{P})$ of the polygon, the relative interior of the segment VW is contained in the interior of the polygon $\mathcal{P}$. Devise a simple algorithm that triangulates a star-shaped polygon with respect to a given point V.

Exercise 12.4 (Triangulating a polygon by sweeping) It is possible to devise a sweep algorithm that triangulates a polygon without building the vertical decomposition into trapezoids. This algorithm merges the techniques of the sweep algorithm that decomposes a polygon as described in subsection 12.3.3 with the algorithm that triangulates a monotone polygon described in subsection 12.3.2.

The plane is swept by a vertical line, and the events are the abscissae of the vertices of the polygon. The algorithm maintains the list of active edges (edges intersected by the sweep line). The active edges subdivide the sweep line into segments alternately inside and outside the polygon $\mathcal{P}$. For each interior segment, the algorithm maintains a chain $V_1, V_2, \ldots, V_k$ of vertices of the polygon such that V_1 and V_k are the right endpoints of the edges that bound the interior segment, and if $k > 2$ the angle between the vectors $V_i V_{i-1}$ and $V_i V_{i+1}$ is greater than π for each $i = 2, \ldots, k - 1$. Finally, the algorithm maintains a pointer towards the vertex of maximal abscissa on each chain.

1. Describe the current step of the algorithm.

2. Show that the complexity of the algorithm is $O(n \log n)$ if the polygon has n vertices.

3. Outline a lazy version of this algorithm whose complexity is $O(n + r \log r)$, where n is the number of vertices and r the number of reflex vertices.

Exercise 12.5 (Shortest paths) Consider a polygon $\mathcal{P}$ with n vertices, and two points P and Q that belong to the boundary or to the interior of the polygon $\mathcal{P}$. We suppose that a triangulation $\mathcal{T}(\mathcal{P})$ of $\mathcal{P}$ has already been computed. Show how to compute in linear time the shortest polygonal line $\pi(P, Q)$ that links P to Q and remains in the interior of the polygon.

Hint: Let T_P be a triangle of $\mathcal{T}(\mathcal{P})$ that contains P and T_Q be a triangle of $\mathcal{T}(\mathcal{P})$ that contains Q. The dual graph of $\mathcal{T}(\mathcal{P})$ is a tree, in which there is a unique path from T_P to T_Q. Thus, there is a sequence of adjacent triangles in $\mathcal{T}(\mathcal{P})$ that links T_P to T_Q. Consider the sequence $E_1, E_2, \ldots, E_l$ of edges adjacent to two consecutive triangles on this path. Let U_i and V_i be the two vertices of E_i. The algorithm computes the shortest paths $\pi(P, U_i)$ and $\pi(P, V_i)$ for increasing i. To compute $\pi(P, U_{i+1})$ and $\pi(P, V_{i+1})$ knowing $\pi(P, U_i)$ and $\pi(P, V_i)$, we use the following observations:

1. Either $U_i = U_{i+1}$, or $V_i = V_{i+1}$. It therefore suffices to compute the shortest path $\pi(P, X)$ for the vertex $X = U_{i+1}$ or $X = V_{i+1}$ that does not belong to $\{U_i, V_i\}$.

2. The shortest paths $\pi(P, U_i)$ and $\pi(P, V_i)$ are polygonal lines; they share an initial polygonal line $\pi(P, A_i)$, and then consist of two concave chains which are the shortest

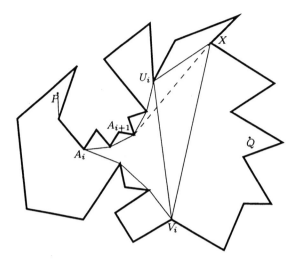

Figure 12.12. The funnel of an edge.

paths $\pi(A_i, U_i)$ and $\pi(A_i, V_i)$ (see figure 12.12). The *funnel* of the edge $U_i V_i$, denoted by $F_P(U_i V_i)$, is the concatenation of the chains $\pi(U_i, A_i)$ and $\pi(A_i, V_i)$. The vertex A_i is the origin of the funnel $F_P(U_i V_i)$. The shortest path $\pi(P, X)$ must also pass through the origin A_i, and is the concatenation of $\pi(P, A_i)$ with the shortest path $\pi(A_i, X)$. This shortest path is the segment $A_i X$ if it does not intersect the funnel $F_P(U_i V_i)$. Otherwise, there is a point A_{i+1} on $F_P(U_i V_i)$ such that $A_{i+1} X$ is tangent to the chain that contains A_{i+1}, and the shortest path $\pi(A_i, X)$ is the concatenation of $\pi(A_i, A_{i+1})$ with $A_{i+1} X$. To compute $\pi(P, X)$ and the funnel $F_P(U_{i+1} V_{i+1})$, it therefore suffices to find A_{i+1}.

3.   To find the point A_{i+1}, we follow $F_P(U_i V_i)$ simultaneously starting at the two endpoints U_i and V_i. The cost of the traversal is proportional to the number of $F_P(U_i V_i)$ visited. But half of the visited vertices of $F_P(U_i V_i)$ do not belong to $F_P(U_{i+1} V_{i+1})$, so they will never be visited again. Show then that the algorithm takes linear time.

Exercise 12.6 (Shortest path tree) Consider a polygon $\mathcal{P}$ with n vertices and P a point that belongs either to the boundary or to the interior of $\mathcal{P}$. The shortest paths that join P to the vertices of $\mathcal{P}$ do not cross, so their union forms a tree whose nodes are vertices of $\mathcal{P}$. Show that it is possible to compute the tree of shortest paths from P in time $O(n \log n)$.

Hint: Let $\mathcal{T}(\mathcal{P})$ be a triangulation of $\mathcal{P}$ and T_P be a triangle of $\mathcal{T}(\mathcal{P})$ that contains P. The dual graph $\mathcal{G}(\mathcal{P})$ of this triangulation can be viewed as a binary tree rooted at the node that corresponds to T_P. Each node of $\mathcal{G}(\mathcal{P})$ has an incoming arc and one or two outgoing arcs corresponding to internal edges in the triangulation $\mathcal{T}(\mathcal{P})$. For simplicity, we make no distinction between the arcs of $\mathcal{G}(\mathcal{P})$ and the edges in $\mathcal{T}(\mathcal{P})$. The algorithm traverses the graph $\mathcal{G}(\mathcal{P})$ in a depth-first fashion. For each node traversed, the algorithm computes the funnel and the shortest path to the endpoints of the outgoing edges, as in the previous exercise.

The cost for each node is proportional to the number of nodes visited in the funnel of the incoming edge. If the node has a single child, the number of visited vertices on the funnel of the incoming edge is proportional to the number of vertices that do not belong to the funnel of the outgoing edge, vertices that will not be visited later. The total cost for these nodes is thus linear. If a node has two children and the funnel of its incoming edge is of size m, the funnels of the outgoing edges have sizes $m_1 + 1$ and $m_2 + 1$, with $m_1 + m_2 = m$. The cost for such a node is proportional to $\min(m_1, m_2)$. For this exercise, say that the *width* of a subtree of $\mathcal{G}(\mathcal{P})$ is the sum of the sizes of all the funnels of the incoming edges of the leaves of this subtree. For each node T of $\mathcal{G}(\mathcal{P})$, denote by $m(T)$ the size of the funnel of the incoming edge, by $e(T)$ the number of arcs of the subtree rooted at this node, and by $m'(T)$ the width of this subtree. It is easily shown that $m'(T) = m(T) + e(T)$, so that in particular $m(T) \leq m'(T)$. From this, show that the total cost $c(m')$ of the binary nodes of a subtree of width m' satisfies the recurrence

$$c(m') = \max_{1 < k' < m'_1} (c(k') + c(m' - k') + O(\min(k', m' - k')))$$

which solves to $c(m') = O(m' \log m')$.

Exercise 12.7 (Shortest path queries) Consider a polygon $\mathcal{P}$ with n vertices and P a point that belongs either to the boundary or to the interior of $\mathcal{P}$. Design a data structure that allows us to find, for any point X on the boundary or in the interior of $\mathcal{P}$, the shortest path $\pi(P, X)$ that links P to X inside $\mathcal{P}$. Each shortest path query must be answered in time $O(\log n + k)$ where n is the number of vertices of $\mathcal{P}$ and k is the number of edges on the shortest path.

Hint: Build the shortest path tree from P as explained in exercise 12.6. The set of regions $\Phi(E)$ bounded by an edge E of $\mathcal{P}$ and the funnel $F_P(E)$ of this edge forms a decomposition of the interior of $\mathcal{P}$. Each region $\Phi(E)$ can be subdivided further by extending the edges of the funnel $F_P(E)$ all the way until they meet E. Each of the sub-regions in the decomposition induced by the tree and these extended edges is in fact triangular, with an edge supported by E, and the opposite vertex is a vertex Q of $F_P(E)$. For each point X in this triangle, the shortest path $\pi(P, X)$ is the concatenation of the path in the shortest path tree that links P to Q with the segment QX. The problem is now replaced by that of locating X in a planar map of size $O(n)$ (see exercise 12.2).

Exercise 12.8 (Visibility polygon) Consider a polygon $\mathcal{P}$ with n vertices and P a point that belongs either to the boundary or to the interior of $\mathcal{P}$. A point X on the boundary of $\mathcal{P}$ or in the interior of $\mathcal{P}$ is *visible* from P if the relative interior of the segment PX is contained in the interior of $\mathcal{P}$. Compute the polygon $V(P, \mathcal{P})$ which encloses the set of points visible from P.

Hint: Build the shortest path tree from P and the planar map formed by the edges of $\mathcal{P}$, the funnel $F_P(E)$ of each edge E of $\mathcal{P}$, and the edges that extend the edges of the funnels as in the previous exercise. This induces a decomposition of $\mathcal{P}$ into sub-edges. Follow the boundary of $\mathcal{P}$, keeping track of whether each sub-edge is visible from P. The visible edges linked in a proper way form the boundary of $V(P, \mathcal{P})$.

Another solution is to sort the vertices of $\mathcal{P}$ by polar angle around P, and to build the visibility polygon $V(P, \mathcal{P})$ using a sweep ray that originates from P and sweeps the plane by rotating with an increasing angle.

Exercise 12.9 (Art gallery) An art gallery is viewed as a polygonal region bounded by a polygon with n vertices. The gallery must be watched over by a set of guards placed at the vertices of the polygon. Each guard keeps an eye on the portion that is visible from its location.

Show that it is always sufficient and sometimes necessary to place $\left\lfloor \frac{n}{3} \right\rfloor$ guards to completely watch over an art gallery with n vertices.

Hint: Let $\mathcal{T}$ be a triangulation of the polygon that encloses the art gallery. It is easy to show that the vertices can be colored with three colors so that each edge is bichromatic, or equivalently that each triangle has a vertex of each color. Simply take the color that is attributed the least number of times, and place a guard at the vertices that have this color.

To prove that $\left\lfloor \frac{n}{3} \right\rfloor$ guards may be necessary, consider a comb-shaped polygon.

Exercise 12.10 (Hierarchical decomposition of a polygon) 1. Show that each binary tree with n nodes has an edge such that its removal creates two trees with at least $\left\lfloor \frac{n+1}{3} \right\rfloor$ nodes each. Show how to compute such an edge in linear time.

2. Apply the result of the previous question to the dual graph $\mathcal{G}(\mathcal{T})$ of a triangulation $\mathcal{T}$ of a polygon $\mathcal{P}$. The edge that splits $\mathcal{G}(\mathcal{T})$ into two balanced subtrees corresponds to an internal edge of $\mathcal{T}$ that splits $\mathcal{P}$ into two sub-polygons P_1 and P_2 each containing at most a fraction $\frac{2}{3}$ of the triangles in $\mathcal{T}$. By recursively splitting these polygons P_1 and P_2, we obtain a hierarchical decomposition with $O(\log n)$ levels.

Exercise 12.11 (Geodesic decomposition) Consider a polygon $\mathcal{P}$ that has n vertices. A *balanced geodesic decomposition* of $\mathcal{P}$ can be obtained using the shortest paths inside $\mathcal{P}$ that join vertices equally spaced on the boundary. More precisely, let P_1, P_2, ..., P_n be the vertices of $\mathcal{P}$. In a first step, compute the shortest paths between P_1 and $P_{\left\lfloor \frac{n}{3} \right\rfloor}$, between $P_{\left\lfloor \frac{n}{3} \right\rfloor}$ and $P_{\left\lfloor \frac{2n}{3} \right\rfloor}$, and between $P_{\left\lfloor \frac{2n}{3} \right\rfloor}$ and P_1. In a second step, connect the pairs $(P_1, P_{\left\lfloor \frac{n}{6} \right\rfloor})$, $(P_{\left\lfloor \frac{n}{6} \right\rfloor} P_{\left\lfloor \frac{n}{3} \right\rfloor})$, etc., and iterate until the pairs consist of adjacent vertices. The interior of $\mathcal{P}$ is then subdivided into regions whose boundaries consist of three shortest paths joining three vertices of the polygon $\mathcal{P}$. Apart from the first region that was bounded by the three shortest paths computed in the first step, a region that appears at step k of the process is bounded by two shortest paths that are computed at step k and a shortest path that was computed at step $k - 1$. These shortest paths may share common edges and the interior of a region (which may occasionally be empty) is a pseudo-triangle bounded by three concave sub-chains of three shortest paths.

1. Show that all the shortest paths built in the process have exactly $n - 3$ distinct edges interior to the polytope $\mathcal{P}$, and that any segment contained inside $\mathcal{P}$ intersects only $O(\log n)$ such edges.

2. Show that the subdivision may be built in overall time $O(n \log n)$.

Hint: To build the shortest path, first compute a triangulation of $\mathcal{P}$ and use the algorithm of exercise 12.5. Note that the time required to compute a shortest path is proportional to the number of triangles in $\mathcal{T}$ traversed by this shortest path, and that the total number of intersections between the computed shortest paths and the edges of $\mathcal{T}$ is $O(n \log n)$.

Exercise 12.12 (Ray shooting) Consider a polygon $\mathcal{P}$ with n vertices. A ray shooting query inside $\mathcal{P}$ consists of identifying the point of the boundary of $\mathcal{P}$ that is hit by a ray originating from a given point Q inside or on the boundary of $\mathcal{P}$ in the direction of a given vector U. Show that a balanced geodesic decomposition can be used to answer such a query in time $O(\log^2 n)$.

Hint: To answer a ray shooting query (Q, U) involves locating the origin Q in the geodesic decomposition, and following the ray (Q, U) in the decomposition. The location can be performed in time $O(\log n)$ if a location structure has been precomputed (see exercise 12.2). The ray (Q, U) intersects at most $O(\log n)$ edges in the geodesic decomposition and the boundary of each region consists of three concave chains, allowing each intersection to be located by binary search.

12.5 Bibliographical notes

The algorithm that triangulates a monotone polygon in linear time, and the idea of using a decomposition of a simple polygon into monotone polygons in order to obtain a triangulation are due to Garey, Johnson, Preparata, and Tarjan [109]. In their article, they obtain the decomposition into monotone polygons using a double sweep whose function is to *regularize* the planar map induced by the edges of the polygon. This operation consists in adding edges to this map so that each vertex of the map, except for the two vertices with minimum and maximum abscissae, is adjacent to at least one vertex of greater abscissa and at least one vertex of smaller abscissa. Later, Fournier and Montuno [107] showed that a decomposition into monotone polygons could be obtained from a vertical decomposition in linear time. Chazelle and Incerpi [56] devised an algorithm that computes the vertical decomposition of a polygon more efficiently than the simple sweep algorithm. The complexity of their algorithm is $O(n \log c)$, where c is the *sinuosity* of the polygon, a parameter that guarantees a certain shape of the polygon and whose value does not exceed three for polygons usually encountered in the applications. The first algorithm that showed that the complexity of triangulating a polygon is $o(n \log n)$ was given by Tarjan and Van Wyk [213], and it computes a vertical decomposition of the polygon in time $O(n \log \log n)$ using the divide-and-conquer method. This algorithm uses Jordan sorting and a few sophisticated data structures such as *finger trees*. Finally, using vertical decompositions in triangulation methods was one of the key ingredients in the linear-time triangulation algorithm designed by Chazelle [44].

Randomized algorithms can compute the vertical decomposition of simple polygons in expected time $O(n \log^* n)$. The algorithm by Clarkson, Tarjan, and Van Wyk is a nice example of a randomized divide-and-conquer, but again it uses Jordan sorting. In contrast, the incremental randomized algorithm described in section 5.4, due to Seidel [204], uses only simple data structures.

There are other triangulation algorithms that do not use vertical decompositions. For instance, Chazelle [41] develops a triangulation algorithm that uses divide-and-conquer and runs in time $O(n \log n)$. This algorithm finds an internal edge that splits the polygon into two balanced sub-polygons, and recursively triangulates them (see exercise 12.10). We must also cite the algorithm by Hertel and Mehlhorn [128] that computes a triangulation directly using the sweep method (see exercise 12.4) and also introduces the lazy sweep method used in subsection 12.3.3. This method is also described in the book by Mehlhorn [162].

The applications of planar triangulations are so many that it is impossible here to give a complete account. From the standpoint of computational geometry, certainly the most important one is to provide a preprocessing step to the localization in a triangular planar map (see exercise 12.2) that was developed by Kirkpatrick [136]. Visibility, shortest paths, and ray shooting problems tackled in exercises 12.5 to 12.12 also provide a fertile application domain for triangulations. The algorithm described in exercise 12.5 that computes the shortest path between two vertices of a polygon is due to Lee and Preparata [148]. In this article, they introduce funnels which are often used by others. For instance, Guibas *et al.* [116] used it to compute shortest paths in a polygon from a vertex or the sub-polygon visible from a given point or segment. Exercises 12.8 to 12.12 are borrowed from this article, but we must point out that the use of *finger trees* to represent funnels allows their algorithms to compute the shortest path tree in linear time. The idea of using the hierarchical decomposition of a polygon (exercise 12.10) to solve ray shooting problems is exploited by Chazelle and Guibas [54] and by Guibas *et al.* [116]. The solution that uses a geodesic decomposition of the polygon (see exercises 12.11 and 12.12) was developed by Chazelle *et al.* [50]. Again, we must point out that the use of sophisticated data structures (*weight-balanced trees* and *fractional cascading*) allows them to answer a ray shooting query in time $O(\log n)$ for a polygon with n vertices. Finally, the art gallery theorem (see exercise 12.9) and its numerous variants are discussed in the book by O'Rourke [183].

Chapter 13

Triangulations in dimension 3

In dimension 3, the possible triangulations of a set of points do not all have the same number of faces. In fact, there are some sets of points which admit triangulations of both linear and quadratic sizes. Moreover, constrained triangulation problems do not always have a solution in dimension 3. For instance, some polyhedra are not triangulable, meaning that the set of faces of the polyhedron cannot be completed into a 3-triangulation so that the vertices of the triangulation are exactly the vertices of the polyhedron. Yet several applications crucially rely on our ability to decompose polyhedral regions into simplices. We must then design a simplicial decomposition scheme. The simplicial decomposition of a polyhedral region is a 3-triangulation whose domain is exactly the polyhedral region (as a closed topological subset of $\mathbb{E}^3$); but this triangulation has additional vertices and edges that are not faces of the polyhedral region, and the edges and 2-faces of the polyhedral region may be split into several faces of the simplicial decomposition. The size of the simplicial decomposition is crucial for subsequent operations, so we aim at minimizing (exactly or approximately) the size of such decompositions. In this chapter, we show how to build a simplicial decomposition from the vertical decomposition. The *vertical decomposition* of a polyhedral region is the three-dimensional analog of the vertical decomposition of a polygonal region introduced in the previous chapter.

Section 13.1 investigates triangulations of a set of points, and presents an algorithm that builds a triangulation of linear size for any set of points such that no three points are collinear. The remainder of the chapter considers constrained triangulation problems. In section 13.2, we present first two unfeasible constrained triangulation problems. Section 13.3 generalizes the notion of a vertical decomposition to polyhedral regions and presents an algorithm that computes a simplicial decomposition for polyhedral regions of genus 0. The resulting simplicial decomposition is not minimal, but its size can be bounded by $O(n+r^2)$ for a polyhedron with n vertices and r reflex edges.

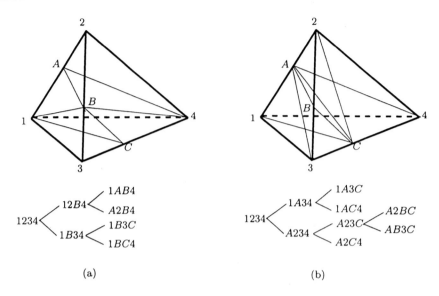

Figure 13.1. Examples of 3-triangulations.

13.1 Triangulation of a set of points

13.1.1 The size of a triangulation

To triangulate a set $\mathcal{A}$ of points in $\mathbb{E}^3$ involves finding a triangulation whose vertices are the points of $\mathcal{A}$ and whose domain is the convex hull $conv(\mathcal{A})$; the relative interiors of the simplices in such a triangulation form a partition of the convex hull $conv(\mathcal{A})$. The main problem arising in dimension 3 is that the *size* of a triangulation of a set $\mathcal{A}$ of points in $\mathbb{E}^3$, defined as the total number of simplices of any dimension, is not entirely determined by the number of points in $\mathcal{A}$ and the number of points of $\mathcal{A}$ on the boundary of the convex hull. There is no relation that binds the numbers of tetrahedra, triangles, and edges of a 3-triangulation of $\mathcal{A}$ as a function of the size of $\mathcal{A}$ and the size of the convex hull $conv(\mathcal{A})$. In fact, the same set of points may be triangulated in several ways into triangulations of different sizes. For instance, a set of seven points is shown in figure 13.1 along with two triangulations, one with four tetrahedra and the other with five.

In section 11.2, it was shown that the size of a 3-triangulation of a set of n points may vary between a linear bound of $n - 3$ and a quadratic bound of $n^2/2 - 5n/2 + 3$, and that both bounds may be achieved.

Linear triangulations

Some sets of points admit a triangulation of linear size, which we call *linear triangulation* for short. In particular, this is true for any set of points in *general position*, meaning that no three points are collinear and no four points are coplanar. Indeed, let $\mathcal{A} = \{A_1, \ldots, A_n\}$ be a set of n points in general position. We may build a triangulation of $\mathcal{A}$ in the following way:

Triangulating the convex hull. Let us first consider the subset formed by the n_e points that are on the boundary of the convex hull $conv(\mathcal{A})$. These points are vertices of the convex hull and the 2-faces of the convex hull are triangles, because of the general position assumption. Choose a vertex A_0 of $conv(\mathcal{A})$, and compute all the tetrahedra that can be obtained as $conv(A_0, F)$ for any 2-face F of the convex hull $conv(\mathcal{A})$ that does not contain A_0. These tetrahedra form a 3-triangulation of $conv(\mathcal{A})$. Theorem 7.2.4 on simplicial polytopes says that $conv(\mathcal{A})$ has $2n_e - 4$ facets, so if g is the number of facets that contain vertex A_0, the number of tetrahedra in our 3-triangulation of $conv(\mathcal{A})$ is $2n_e - 4 - g$.

Adding the internal vertices. Let us now insert the $n_i = n - n_e$ remaining points of $\mathcal{A}$. Those points belong to the domain of the triangulation built above. Since $\mathcal{A}$ is in general position, each point must be contained in the interior of some tetrahedron T in the current triangulation. We decompose this tetrahedron into four smaller tetrahedra that have the newly inserted point as a vertex, by adding the edges that connect this point to the four vertices of T.

Each insertion therefore adds exactly three tetrahedra to the triangulation, so the final triangulation has $2n_e - 4 - g + 3n_i \leq 3n - 11$ tetrahedra. In subsection 13.1.3, we will show how this method leads to an algorithm that computes in time $O(n \log n)$ a linear triangulation of a set of points without three collinear points.

Quadratic triangulations

There are also sets of points that have triangulations with a quadratic size, which we call *quadratic triangulations* for short. Some sets of points even have no triangulation with subquadratic size. Consider for instance the set $\mathcal{A}$ of $2n$ points drawn in figure 13.2. This set of points has n points $A_1, A_2, \ldots, A_n$ situated (in this order) on some given line in $\mathbb{E}^3$, and another n points $B_1, B_2, \ldots, B_n$ on another line that does not lie in the same plane as the first line. A triangulation may be computed as follows: the convex hull of the points is a tetrahedron A_1, A_n, B_1, B_n. Adding the $n - 2$ points $A_2, \ldots, A_{n-1}$ splits this tetrahedron

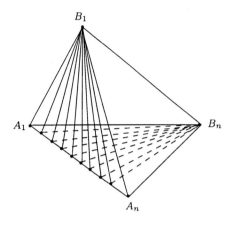

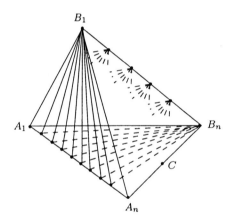

Figure 13.2. A set of points whose only triangulation has quadratic size.

into $n-1$ tetrahedra $A_iA_{i+1}B_1B_n$, $i = 1,\ldots,n-1$. Adding now the $n-1$ points $B_2,\ldots,B_{n-1}$ on the other line splits each tetrahedron $A_iA_{i+1}B_1B_n$ into $n-1$ tetrahedra $A_iA_{i+1}B_jB_{j+1}$, $j = 1,\ldots,n-1$. The resulting triangulation has $(n-1)^2$ tetrahedra, and it is easy to see that this is the only possible triangulation for this set of points.

Linear and quadratic triangulations

Finally, we show that there are sets of points which admit both linear and quadratic triangulations. For instance, consider the set of $2n+1$ points obtained by adding a point C on the line $A_n B_n$ in the preceding example. One way to triangulate it is to consider that C splits the tetrahedron $A_1 B_1 A_n B_n$ into two tetrahedra $A_1 B_1 A_n C$ and $A_1 B_1 C B_n$. Each of these two tetrahedra is then split into $n - 1$ tetrahedra, by adding the points in $\{A_i : i = 2, \ldots, n - 1\}$ and in $\{B_j : j = 2, \ldots, n - 1\}$ respectively into each tetrahedron. The resulting triangulation has $2n - 2$ tetrahedra. Another way is to simply add the point C to the unique triangulation of the set $\{A_1, A_2, \ldots, A_n, B_1, B_2, \ldots, B_n\}$. The addition of C will merely split the tetrahedron $A_{n-1} A_n B_{n-1} B_n$ into two tetrahedra, so that the resulting triangulation has size $(n - 1)^2 + 1$.

Another example of a set of points that has both linear and quadratic triangulations is a set of n points on the moment curve, for which we have shown the existence of a quadratic triangulation in section 11.2. Since this point set is in general position, the discussion above shows that it also admits a triangulation with a linear number of tetrahedra.

Let $\mathcal{A}$ be a set of n points, $\{A_1, \ldots, A_n\}$, in general position in $\mathbb{E}^3$. In order to triangulate this set of points, we may think of extending the incremental method described in subsection 12.1.2 to $\mathbb{E}^3$. According to this method, the points are sorted by lexicographic order of their coordinates x, y, z, then inserted one by one into the triangulation. The triangulation initially consists of a single tetrahedron formed by the first four points. At each incremental step, the algorithm maintains the convex hull $conv(\{A_1, A_2, \ldots, A_{i-1}\})$ of the points already inserted and updates the triangulation by adding the tetrahedra $conv(A_i, F)$ for all the facets F of $conv(\{A_1, A_2, \ldots, A_{i-1}\})$ that are *red* with respect to A_i (recall that these are the facets whose affine hull separates A_i from $\{A_1, A_2, \ldots, A_{i-1}\}$).

We used this method in section 11.2 in order to build a triangulation of the set of points on the moment curve. A major drawback of this method is that, in dimension 3, it can lead to triangulations with a quadratic number of tetrahedra, although the set of points admits a linear triangulation. This is exactly what happens for n points on the moment curve.

In contrast, the algorithm we present here finds a triangulation of a set $\mathcal{A}$ of n points which has linear size if no three points in $\mathcal{A}$ lie on the same line. This algorithm uses the divide-and-conquer method and relies on a theorem proved in the next subsection. Loosely speaking, the theorem shows the existence of a good splitter for the triangulation.

13.1.2   The split theorem

Even though we use the theorem below to triangulate a set of points in dimension 3, the split theorem is true in any dimension d, so in this subsection we assume that the ambient space is $\mathbb{E}^d$. Let $\mathcal{A}$ be a set of n points in general position in $\mathbb{E}^d$, whose convex hull is a d-simplex. We can always rename the points so that the convex hull is the simplex $S = A_1 A_2 \ldots A_{d+1}$. Any point in $\mathcal{A}$ that is not a vertex of S is an internal vertex of any triangulation of $\mathcal{A}$, and is called an *internal* point of $\mathcal{A}$. The set $\mathcal{A}$ therefore has $n' = n - (d+1)$ internal points. Any internal point X of $\mathcal{A}$ splits $S = conv(\mathcal{A})$ into $d+1$ simplices:

$$S_i(X) = A_1 \ldots A_{i-1} X A_{i+1} \ldots A_{d+1}, \quad i = 1, \ldots, d+1.$$

An internal point X in $\mathcal{A}$ is a λ-*splitter* of $\mathcal{A}$ if none of the interiors of the simplices $S_i(X)$ contain more than $\lambda n'$ points of $\mathcal{A}$.

Theorem 13.1.1 (Split theorem) *Any set of n points in general position in $\mathbb{E}^d$ whose convex hull is a simplex contains a $d/(d+1)$-splitter. Such a splitter may be found in linear time, and the ratio $d/(d+1)$ cannot be improved in general.*

Proof. Consider the subset $\mathcal{A}'$ of the n' internal points of $\mathcal{A}$. The proof consists in successively removing the points in $\mathcal{A}'$ that are too close to a vertex of $S = conv(\mathcal{A})$. The remaining points will be good splitters for $\mathcal{A}$.

Let us put $\mathcal{Q}_0 = \mathcal{A}'$. We define a sequence of subsets $\mathcal{Q}_0 \supset \mathcal{Q}_1 \supset \ldots \supset \mathcal{Q}_{d+1}$ in the following way. For each $i = 1, \ldots, d+1$, let N_i be the normal vector to the facet of S that does not contain A_i. By convention, N_i points away from S (see figure 13.3). For each $X \in \mathcal{Q}_{i-1}$, we define an i-ordinate $s_i(X) = (X - A_i) \cdot N_i$. Let Y_i be the $\left\lceil \frac{n'}{d+1} \right\rceil$-th point of $\mathcal{Q}_{i-1}$ with respect to the increasing order of $s_i(X)$. We split $\mathcal{Q}_{i-1}$ into two subsets

$$\begin{aligned}
\mathcal{P}_i &= \{X \in \mathcal{Q}_{i-1} : s_i(X) < s_i(Y_i)\}, \\
\overline{\mathcal{P}_i} &= \{X \in \mathcal{Q}_{i-1} : s_i(X) \leq s_i(Y_i)\}, \\
\mathcal{Q}_i &= \{X \in \mathcal{Q}_{i-1} : s_i(X) \geq s_i(Y_i)\} = \mathcal{Q}_{i-1} \setminus \mathcal{P}_i.
\end{aligned}$$

Let $|\mathcal{P}_i|$ denote the size of $\mathcal{P}_i$ and $|\overline{\mathcal{P}_i}|$ the size of $\overline{\mathcal{P}_i}$. By construction of $\mathcal{Q}_i$, we have

$$\begin{aligned}
|\mathcal{P}_i| &\leq \left\lceil \frac{n'}{d+1} \right\rceil - 1 < \frac{n'}{d+1}, \\
|\overline{\mathcal{P}_i}| &\geq \left\lceil \frac{n'}{d+1} \right\rceil \geq \frac{n'}{d+1}.
\end{aligned}$$

Using these inequalities, we can show that $\mathcal{Q}_{d+1}$ is not empty, and that any point Z in $\mathcal{Q}_{d+1}$ is a $d/(d+1)$-splitter for $\mathcal{A}$. Indeed,

$$\mathcal{Q}_{d+1} = \mathcal{A}' \setminus \bigcup_{i=1}^{d+1} \mathcal{P}_i,$$

$$|\mathcal{Q}_{d+1}| > n' - (d+1)\frac{n'}{d+1} = 0.$$

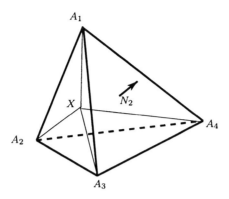

Figure 13.3. A splitter.

Pick a point Z that belongs to $\mathcal{Q}_{d+1}$. For each $i = 1, \ldots, d+1$, we denote by $S_i(Z)$ the simplex $A_1 \ldots A_{i-1} Z A_{i+1} \ldots A_{d+1}$ and by $int(S_i(Z))$ the interior of this simplex. Point Z belongs to all the subsets $\mathcal{Q}_i$, since $\mathcal{Q}_{d+1}$ is contained in $\mathcal{Q}_d$, which is itself contained in $\mathcal{Q}_{d-1}$, etc. Therefore,

$$\forall X \in int(S_i(Z)) \cap \mathcal{A}', \quad (X - A_i) \cdot N_i > (Z - A_i) \cdot N_i \geq (Y_i - A_i) \cdot N_i,$$

and so we can bound the number of points in $S_i(Z)$ by

$$int(S_i(Z)) \cap \mathcal{A}' \quad \subset \quad \mathcal{A}' \setminus \overline{\mathcal{P}_i},$$
$$|int(S_i(Z)) \cap \mathcal{A}'| \quad \leq \quad n' - \frac{n'}{d+1} = \frac{d}{d+1} n'.$$

This proof of existence can be converted into a linear-time algorithm to find a $d/(d+1)$-splitter. Indeed, for $i = 1, \ldots, d+1$, we need only:

1. Find the normal vector N_i.

2. Compute $s_i(X)$ for any point X in $\mathcal{Q}_{i-1}$ and compute the $\left\lceil \frac{n}{d+1} \right\rceil$-th point in $\mathcal{Q}_{i-1}$ with respect to increasing order of $s_i(X)$.

3. Compute $\mathcal{Q}_i$.

For each value of i, step 1 takes constant time, step 3 takes linear time, and step 2 requires the computation of n' dot products, which can be carried out in time $O(n')$. The only delicate point is to select the $\left\lceil \frac{n}{d+1} \right\rceil$-th value of these products. This can nevertheless be performed in linear time (see exercise 3.7), and so step 2 takes linear time as well. The overall cost is thus linear.

Now let us indicate why the ratio $d/(d+1)$ is optimal for a split theorem. It suffices to show that there is a set of points that does not admit a λ-splitter for

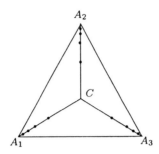

Figure 13.4. A (2/3)-splitter.

$\lambda < d/(d+1)$. Consider the set of points that is a generalization of the planar set of points represented in figure 13.3. To build this set in $\mathbb{E}^d$, we put $d+1$ points at the vertices of a regular simplex $A_1, A_2, \ldots, A_{d+1}$ of center C, and $m(d+1)$ internal points given by

$$B_{i,j} = \left(\frac{1}{2^j} C + (1 - \frac{1}{2^j}) A_i \right), \quad i = 1, \ldots, d+1, \quad j = 1, \ldots, m.$$

Clearly, the best splitters for this set are the points $B_{i,1}$ for $i = 1, \ldots, d+1$, and these splitters are $d/(d+1)$-splitters.   □

13.1.3   An incremental algorithm

The algorithm we present here triangulates a set $\mathcal{A}$ of n points in $\mathbb{E}^3$, and guarantees that the output triangulation has a linear number of vertices if $\mathcal{A}$ does not contain three collinear points. The running time is $O(n \log n + t)$ where t is the size of the output triangulation. The description of the algorithm given below assumes at first that the set of points is in general position and that its convex hull is a simplex. We then show how to remove the second assumption, and finally how to triangulate any set of points.

Triangulating a set of points in general position,
whose convex hull is a simplex

Let us first consider a set $\mathcal{A}$ of n points in general position in $\mathbb{E}^3$ whose convex hull is a tetrahedron. The split theorem 13.1.1 suggests the following algorithm to obtain a linear triangulation of $\mathcal{A}$:

  1. Find a (3/4)-splitter for $\mathcal{A}$, say Z, using the algorithm described in the proof of theorem 13.1.1.

2. Split the tetrahedron $S = conv(\mathcal{A})$ into four tetrahedra $S_i(Z)$ $(i = 1, \ldots, 4)$. For each of these tetrahedra $S_i(Z)$, recursively compute a triangulation for the set $\mathcal{A}_i$ of points in $\mathcal{A}$ contained in $S_i(Z)$, if there are any.

The existence of a $(3/4)$-splitter is shown by the split theorem 13.1.1. Each time a point is chosen as the splitter, four new tetrahedra are added to the triangulation, and they replace one tetrahedron. The final triangulation therefore counts $1 + 3(n - 4) = 3n - 11$ tetrahedra.

The size of the set to be triangulated decreases by a factor of $3/4$ from level to level of the recursion. The depth of the recursion is thus at most $\log_{3/4} n$, and the time needed to process a level of the recursion is linear. So the triangulation is computed in overall time $O(n \log n)$.

Triangulating a set of points in general position

To triangulate a set $\mathcal{A}$ of points in general position whose convex hull is not necessarily a simplex, we preprocess the set as follows:

1. Compute the convex hull $conv(\mathcal{A})$ of $\mathcal{A}$.

2. Triangulate the convex hull. For instance, pick the vertex of maximal degree (the degree of a vertex is the number of incident edges). The facets of the convex hull are triangles because of the general position assumption, hence the collection of simplices of the form $conv(A_0, F)$, where F ranges over all the facets of $conv(\mathcal{A})$ that do not contain A_0, is a triangulation $\mathcal{T}$ of $conv(\mathcal{A})$.

3. Locate the internal points of $\mathcal{A}$ in the triangulation $\mathcal{T}$ obtained in step 2.

Once this preprocessing is over, we are left with a collection of sets of points, each contained in a tetrahedron of $\mathcal{T}$, to which we can apply the algorithm above.

The convex hull $conv(\mathcal{A})$ of a set $\mathcal{A}$ of n points in $\mathbb{E}^3$ can be computed in time $O(n \log n)$ (see chapters 8 and 9). The triangulation in step 2 can be obtained in linear time once the convex hull is known. It has exactly $2n - 4 - g_0$ tetrahedra if the vertex A_0 is incident to g_0 edges of $conv(\mathcal{A})$. To process the location queries, we use the stereographic projection of the triangulation onto a plane Π to transform these queries into location queries in a triangular planar map. Let Π' be a plane supporting $conv(\mathcal{A})$ along the vertex A_0, and Π a plane parallel to Π' that does not intersect $conv(\mathcal{A})$ but such that $conv(\mathcal{A})$ is contained in the slab between Π and Π' (see figure 13.5). The stereographic projection centered at A_0 sends any point P in $\mathbb{E}^3 \setminus \{A_0\}$ onto the point of Π that is the intersection of Π with the line passing through A_0 and P. The set of the projections of facets F of $conv(\mathcal{A})$ that do not contain A_0 forms a

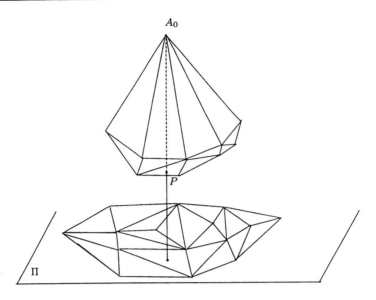

Figure 13.5. Stereographic projection onto Π.

2-triangulation $\mathcal{T}_\Pi$. A point X is contained in the tetrahedron $conv(A_0, F)$ of $\mathcal{T}$ if and only if its projection lies in the projection of F onto Π. The convex hull $conv(\mathcal{A})$ having $O(n)$ facets, there are $O(n)$ tetrahedra in $\mathcal{T}$, so there are also $O(n)$ triangles in $\mathcal{T}_\Pi$, which may be computed straightforwardly in time $O(n)$. The triangulation $\mathcal{T}_\Pi$ may also be preprocessed to support point location queries. The cost of this preprocessing can be made $O(n)$ (see exercise 12.2), and each query takes time $O(\log n)$. The global cost of the location queries is then $O(n \log n)$.

The complexity of preprocessing $\mathcal{A}$ is thus $O(n \log n)$, and the result is a decomposition into subproblems of total size $O(n)$, each of which satisfies the assumptions of the preceding algorithm. Therefore, the set of points can be triangulated in time $O(n \log n)$, and the triangulation has $2n_e - 4 - g_0 + 3n_i$ tetrahedra, if n_e is the number of vertices of $conv(\mathcal{A})$, $n_i = n - n_e$ is the number of internal vertices, and g_0 is the number of edges of the convex hull $conv(\mathcal{A})$ that are incident to A_0.

Remark. The split theorem is true for any dimension d, so the preceding algorithm can easily be extended to compute a triangulation of a set $\mathcal{A}$ of points in general position in $\mathbb{E}^d$. The only difficult step is the location of internal points in the triangulation $\mathcal{T}$ of $conv(\mathcal{A})$, for which the above method does not work in dimension d. For this step we must use another method whose complexity influences the overall complexity of the algorithm. If the size of the convex hull is h, the resulting triangulation still has $h + 3n_i$ tetrahedra.

Triangulating any set of points

It now remains to show how to triangulate a set of points which are not in general position. The general position assumption was used in two ways in the preceding algorithm:

1. To guarantee that the convex hull $conv(\mathcal{A})$ is a simplicial polytope.

2. To guarantee that each internal point in $\mathcal{A}$ that has not yet been chosen as a splitter lies in the interior of a tetrahedron and never falls on a facet or edge of the current triangulation.

If the polytope $conv(\mathcal{A})$ is not simplicial, its 2-faces may always be triangulated, and this triangulation may be used to derive a 3-triangulation of $conv(\mathcal{A})$ as before.

Then, to triangulate a point set in any position, we ignore in a first phase the internal points that, during the location steps, fall on a facet or edge of a tetrahedron in the current triangulation. Unlike the tetrahedra, which may be split, the edges and triangles created by this algorithm will remain in all the triangulations formed in the first phase. Each triangle and edge keeps a pointer to a list that is initially empty, and whenever an internal point is located on a triangle or edge, it is added to the corresponding list and nothing else is done for that point until all the points have been processed. At the end of this phase, the algorithm yields a linear-sized triangulation T' of a subset $\mathcal{A}'$ of $\mathcal{A}$. The tetrahedra in this triangulation do not contain any points of $\mathcal{A}$ in their interiors but triangles and edges may. The points in $\mathcal{A} \setminus \mathcal{A}'$ are stored in the lists of the triangles and edges in whose relative interior they are contained.

In a second phase of the algorithm, the coplanar cases are taken care of. All the triangles in T' which contain points of $\mathcal{A}$ in their relative interior are processed in turn. For such a triangle F, the set of the m_F points that are contained in the relative interior of F is triangulated within F. The points which lie on the incident edges are not taken into account yet. In this way, $2m_F + 1$ new triangles are created and the (at most two) tetrahedra adjacent to F are split into $2m_f+1$ tetrahedra each, by lifting these triangles towards the opposite vertex of the tetrahedron. For each triangle F, the triangulation may be computed in $O(m_F \log m_F)$ time and the number of tetrahedra increases by at most $2(2m_F + 1 - 1) = 4m_F$. This phase can thus be carried out in time $O(\sum_F m_F \log m_F) = O(n \log n)$ and yields a triangulation T'' with a linear number of tetrahedra.

If $\mathcal{A}$ has collinear points, the triangulation T'' may still include edges with points in their relative interiors. In a third phase, the algorithm processes these edges in turn. For each non-empty edge E, the p_E points in $\mathcal{A}$ contained in this edge are sorted along E and each tetrahedron incident to this edge is split into $p_E + 1$ new tetrahedra. As an edge may be incident to a high number of tetrahedra (up to $O(n)$), the triangulation may become quadratic during this phase. The algorithm takes time $O(\sum_E p_E \log p_E) = O(n \log n)$ to sort the points along the edges, and each tetrahedron is created in constant time, so that the overall complexity of the algorithm is $O(n \log n + t)$ if t is the size of the final triangulation.

The following theorem summarizes the characteristics of this algorithm.

Theorem 13.1.2 *A set of n points in $\mathbb{E}^3$ may be triangulated in time $O(n \log n + t)$ where t is the number of tetrahedra in the final triangulation. The algorithm produces a linear triangulation if no three points are collinear.*

13.2   Constrained triangulations

Theorem 12.2.1 does not generalize to dimension 3, and constrained triangulation problems do not always have a solution. A particularly simple example of impossible constrained triangulation is presented in figure 13.6. This example consists of only three orthogonal segments $A_1 B_1$, $A_2 B_2$, and $A_3 B_3$ whose endpoints are situated as follows:

$$\begin{aligned}
A_1 &= (1,0,0), & B_1 &= (-1,0,0), \\
A_2 &= (0,1,-\epsilon), & B_2 &= (0,-1,-\epsilon), \\
A_3 &= (-\epsilon,-\epsilon,1), & B_3 &= (-\epsilon,-\epsilon,-1),
\end{aligned}$$

provided that ϵ is a small enough constant compared to 1. The six points $A_1, B_1, A_2, B_2, A_3, B_3$ define fifteen edges, and if ϵ is small enough the convex hull of the six points $\{A_1, B_1, A_2, B_2, A_3, B_3\}$ is an octahedron which has exactly twelve edges $A_3 A_1$, $A_3 A_2$, $A_3 B_1$, $A_3 B_2$, $B_3 A_1$, $B_3 A_2$, $B_3 B_1$, $B_3 B_2$, $A_1 A_2$, $A_2 B_1$, $B_1 B_2$, and $B_2 A_1$. No triangulation of this point set can include $A_1 B_1$, $A_2 B_2$, and $A_3 B_3$ as edges. Indeed, such a triangulation would have six external vertices ($n = n_e = 6$), three internal edges, and twelve external edges, for a total of fifteen edges ($m = 15$), with eight external facets ($f_e = 8$). Yet the combinatorial relations of subsection 11.2.3 show that the number of tetrahedra is given by

$$t = m - n - n_e + 3 = 6,$$

and that there must be

$$f_i = \frac{1}{2}(4t - f_e) = 8 \tag{13.1}$$

internal facets. Each internal edge of the triangulation must be incident to at least three internal facets. Thus, the presence of the three non-coplanar internal edges $A_1 B_1$, $A_2 B_2$, and $A_3 B_3$ forces the existence of at least nine internal facets. This contradicts equation 13.1 and shows that such a triangulation does not exist.

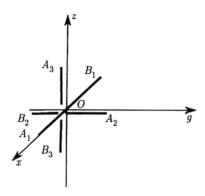

Figure 13.6. An impossible constrained triangulation problem.

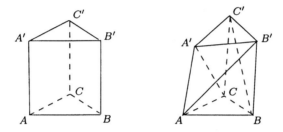

Figure 13.7. Schönhart's polyhedron.

It is possible to build a polyhedron, with edges A_1B_1, A_2B_2, and A_3B_3 as above, that is not triangulable, meaning that no triangulation may have the set of faces of the polyhedron as a boundary (see exercise 13.1). Another famous non-triangulable polyhedron is Schönhart's polyhedron (see figure 13.7) which can be built as follows. Let ABC be an equilateral triangle in the xy-plane. Let $A'B'C'$ be a translated copy of ABC along the z-axis. Schönhart's polyhedron is obtained as a deformation of the prism $ABCA'B'C'$ by slightly rotating the triangle $A'B'C'$ around its axis. The quadrilaterals $ABA'B'$, $BCB'C'$, and $CAC'A'$ cannot remain planar and the facets of the prism are folded towards the interior of the prism along the new edges $A'B$, $B'C$, and $C'A$ so that edges AB', BC', and CA' remain outside the polyhedron. This polyhedron cannot be triangulated, for any tetrahedron that is the convex hull of four of its vertices must include at least one of the edges AB', BC', or CA', and these edges lie outside the polyhedron.

Practical applications, however, rely crucially on the ability to decompose polyhedral regions into simplices. This problem is presented in section 13.3.

13.3   Vertical and simplicial decompositions

As we have just seen, in dimension 3, constrained triangulations problems do not necessarily have a solution and some polyhedra cannot be triangulated. For many applications however, it is very important to deal with a decomposition into tetrahedra or at least into cells of bounded complexity. Thus the problem may be formulated slightly differently to include so-called *Steiner points*, which are not vertices of the polyhedral region but, when added to the constrained triangulation problem, make the decomposition into tetrahedra possible. An optimal decomposition could therefore be introduced as one having either the smallest number of tetrahedra, or the smallest number of Steiner points. Unfortunately, such optimal decompositions are very hard to obtain.

A first method to decompose a polyhedral region is to build a *vertical decomposition* of this region. A vertical decomposition of a polyhedral region consists of a decomposition into vertically cylindrical cells whose horizontal sections are trapezoids. These cells have constant combinatorial complexity, yet they do not form a cell complex. From the vertical decomposition, however, it is easy to construct a 3-triangulation whose domain coincides with the polyhedral region. Subsection 13.3.1 studies vertical decompositions in greater detail and describes an algorithm that computes them.

The major drawback in using vertical decompositions is that the number of simplices in the resulting decomposition may be quadratic even though the polyhedral region may be convex and so may admit a linear-sized triangulation. In the second part of this section, we describe an algorithm which builds a decomposition of a polyhedron with genus 0 into $O(n + r^2)$ tetrahedra, where n is the number of vertices of the polyhedron and r the number of reflex edges: an edge of a polyhedron is *reflex* if the dihedral angle between the two incident facets that lies inside the polyhedron is greater than π. The reflex edges are responsible for the polyhedron being non-convex.

13.3.1   Vertical decomposition of a polyhedral region

The notion of a vertical decomposition in dimension 3 derives from and generalizes the notion of a vertical decomposition introduced in dimension 2 for segments and polygons (see sections 3.3 and 5.4). The definition depends on the choice of a *vertical* direction, here assumed to be that of the z-axis.

In what follows, the polyhedral region bounded by a polyhedron $\mathcal{P}$ is considered as a topologically closed subset of $\mathbb{E}^3$. A vertical decomposition is built by drawing a vertical segment on top of any point P on an edge of $\mathcal{P}$. The segments extend above and beneath P until they touch $\mathcal{P}$ again for the first time. Only the segment lying in the interior of $\mathcal{P}$ will concern us here. A maximal connected

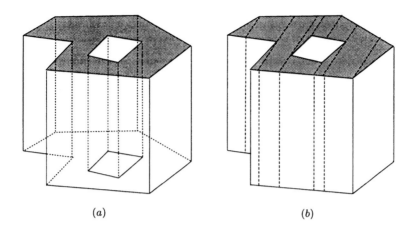

Figure 13.8. Vertical decompositions: (a) a cylindrical cell, (b) walls of type 2.

set of vertical segments stemming from points on the same edge and butting on the same facet of $\mathcal{P}$ forms a vertical trapezoid called a 2-*wall* of type 1. These 2-walls of type 1 decompose the polyhedral region inside $\mathcal{P}$ into cylindrical cells with vertical generator. These cells are called *cylindrical cells*. They each have two non-vertical faces, a lower one called the *floor* and an upper one called the *ceiling*. Each floor or ceiling lies in a unique facet of $\mathcal{P}$ (see figure 13.8a). The floor and ceiling of a cell may have arbitrarily high complexity and may not necessarily be convex or even simply connected. As a consequence, cylindrical cells may be non-elementary (with an arbitrarily high number of vertical facets), non-convex (with reflex vertical edges), or even have genus $g > 0$ (the horizontal cross-sections have holes).

To obtain convex cells of bounded complexities, we decompose each cylindrical cell C in turn. For this, we first consider the floor $Fl(C)$ of the cell, and we decompose it. The decomposition of this polygonal region is described in section 12.3, if we agree on a direction such as the projection of the y-axis on the plane that supports $Fl(C)$. The walls are then segments parallel to this direction stemming from the vertices of $Fl(C)$, contained in $Fl(C)$, and maximal in $Fl(C)$. Call them *1-walls*. On top of these 1-walls, we draw 2-walls as we did before to construct the cylindrical cells. More precisely, from each point on a 1-wall on the floor of C, we draw a maximal segment inside C which extends to the ceiling of C. The set of those maximal vertical segments stemming from a single 1-wall forms a vertical trapezoid called a *2-wall of type 2* (see figure 13.8b). Note that the 2-walls of type 2 of a cylindrical cell C decompose the 2-walls of type 1 of C into 2-walls to which we give type 1′. The 2-walls of type 2 decompose the cylindrical cell C into cylindrical cells which are both convex and elementary: each has a floor and a ceiling which is a trapezoid, occasionally degenerated into a triangle, and with at most four vertical walls which are also trapezoids. These elementary cells are called *prisms*, and the set of all prisms forms the *vertical decomposition* of the polyhedron $\mathcal{P}$.

Decomposition and simplified decomposition

As described above, the vertical decomposition of a polyhedron $\mathcal{P}$ is a *simplified decomposition* in the sense of section 3.3. Indeed, each cell C in the cylindrical decomposition is processed independently, which amounts to considering the 2-walls of type 1 that bounds these cells as infinitely thin but with two distinct *sides*. Each side of a 2-wall of type 1 bounds a unique cylindrical cell, is cut into 2-walls of type $1'$ by the 2-walls of type 2 inside this cell, and has no connection with the 2-walls of type 2 that butt on the other side (see figure 3.2). The simplified decomposition does not describe the infinitely thin cells included inside 2-walls of type 1. As a result, it is a cell complex whose cells are elementary but whose domain does not exactly coincide with the polyhedral region contained within the polyhedron. In this view, two prisms incident to the two different sides of the same 2-wall of type 1 are not adjacent.

To obtain a complete description of the polyhedron $\mathcal{P}$, the two sides of a 2-wall of type 1 are considered one and the same object. The 2-walls of type 2 in one of the two cells C and C' incident to a 2-wall of type 1 decompose this wall into 2-walls of type $1'$ incident to both C and C'. The result is a cell complex whose domain coincides with the polyhedral region inside $\mathcal{P}$ but whose prisms are not elementary any more: each has a floor and a ceiling, and arbitrarily many 2-walls of type $1'$. Nevertheless, these prisms are convex and form the complete vertical decomposition of $\mathcal{P}$. It is easy to derive a simplicial decomposition by triangulating each facet of the decomposition and by lifting the incident triangles of each cell towards one of the vertices of that cell.

Complexity and construction of a vertical decomposition

The complexity of a vertical decomposition (simplified or not) is defined as the total number of its faces (prisms, edges, facets, and vertices). The complexity of the simplified decomposition is simply proportional to the number of prisms since they are elementary. It is also easy to see that the complexity of a complete decomposition is at most twice that of the corresponding simplified decomposition, and that the complexity of the simplicial decomposition is also proportional to the complexity of the simplified decomposition. The following lemma provides a bound for this complexity.

Theorem 13.3.1 *The complexity of the vertical decomposition of a polyhedron with n vertices is $O(n^2)$ and can be obtained in time $O(n^2 \log n)$.*

Proof. The complexity of a simplified decomposition is proportional to the number of prisms of that decomposition which is itself proportional to the number of 2-walls of type 1 or 2. Indeed, each prism is incident to at least one 2-wall of type 2 or to a 2-wall of type 1 which has not been split by a 2-wall of type 2, and either wall is incident to two prisms. The number of 2-walls of type 2 is itself proportional to the number of 2-walls of type 1 in the cylindrical decomposition. Indeed, each 2-wall of type 2 is interior to a cylindrical cell and generated by a 1-wall of the decomposition of the floor of that cell. Each vertex on the floor of a cylindrical cell generates one or two 1-walls in the decomposition of the floor. The number of 2-walls of type 2 in each cylindrical cell is thus at most twice the number of vertices on its floor. The number of vertices on the floor of a cylindrical cell equals the number of edges of the floor and these edges are in one-to-one correspondence with the 2-walls of type 1 that are incident to the cylindrical cell. Each 2-wall of type 1 is incident to two cylindrical cells. As a result, the number of 2-walls of type 2 in the decomposition is at most four times the number of 2-walls of type 1 in the cylindrical decomposition, and it suffices to count these walls only.

To count the number of 2-walls of type 1 in the cylindrical decomposition, we count for each edge E of $\mathcal{P}$ the number of walls of type 1 formed by the vertical segments originating from the points of E. Let H_E be the vertical plane passing through E and let $\mathcal{S}_E$ be the set of segments formed by the intersections of the facets of $\mathcal{P}$ with H_E. The 2-walls of type 1 originating at E are exactly the 2-faces of the 2-dimensional decomposition of $\mathcal{S}_E$ that are incident to an edge contained in E (see figure 13.9).

For each edge E, the size of $\mathcal{S}_E$ is at most equal to the number of facets of $\mathcal{P}$, that is $O(n)$. Two segments in $\mathcal{S}_E$ do not intersect except at a common endpoint, and the vertical decomposition of $\mathcal{S}_E$ has a linear number of 2-faces. The number of 2-walls of type 1 stemming from E is thus $O(n)$. The polyhedron $\mathcal{P}$ having $O(n)$ edges, the total number of 2-walls of type 1 and the overall complexity of the vertical decomposition of $\mathcal{P}$ are thus $O(n^2)$.

This analysis suggests the following algorithm to compute the walls of type 1. For each edge E of $\mathcal{P}$:

1. Compute the set $\mathcal{S}_E$ of segment intersections of the facets of $\mathcal{P}$ with a vertical plane H_E that passes through E.

2. Build the vertical decomposition of $\mathcal{S}_E$ and keep only the 2-faces of this decomposition that are incident to a vertical edge that butt on E. (As we will see in chapter 15, it is also possible to directly compute the 2-faces of the decomposition of $\mathcal{S}_E$ incident to E without computing the entire decomposition of $\mathcal{S}_E$.)

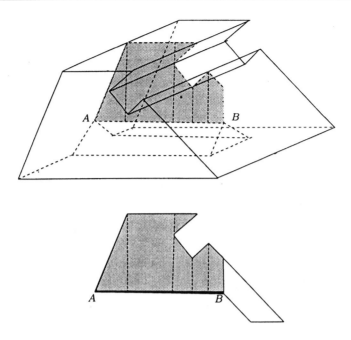

Figure 13.9. Walls of type 1 originating at edge $E = AB$.

The (complete or simplified) vertical decomposition of a polyhedron $\mathcal{P}$ is a cell complex and can be represented by its incidence graph. Starting from the 2-walls of type 1, we first build the incidence graph that corresponds to the decomposition of $\mathcal{P}$ into cylindrical cells. Then each cylindrical cell is decomposed into elementary prisms by decomposing its floor, which also gives the incidence graph of the simplified decomposition. It is easy to compute the incidence graph of the complete decomposition in time linear in the complexity of the vertical decomposition.

For each edge E of $\mathcal{P}$, the set of segments $\mathcal{S}_E$ is obtained in time $O(n)$, and its decomposition computed in time $O(n \log n)$. The 2-walls of type 1 are thus computed in time $O(n^2 \log n)$. Each cylindrical cell C is decomposed into prisms in time $O(n_c \log n_c)$ if n_c is the number of vertices of its floor. The walls of type 2 are thus computed in time $O(\sum_c n_c \log n_c) = O(n^2 \log n)$. $\square$

Remark. By considering the vertical segments whose relative interior is contained in the polyhedral region exterior to $\mathcal{P}$, we can also build the vertical decomposition of the exterior of $\mathcal{P}$. The cylindrical regions formed by the walls of type 1 can then be unbounded, with no floor and a ceiling, or no ceiling and a floor, or neither floor nor ceiling. We can transform this problem into the previous one by adding a large surrounding box: the exterior of the polyhedron now becomes the interior of the box and our polyhedron is now considered as a hole in the box. The decomposition of the exterior of $\mathcal{P}$ can be obtained by removing the floors and ceiling that belong to the boundary of the box, and extending the prisms towards infinity. This shows that there is no real difference between the interior and exterior of a polyhedron as far as vertical decompositions are concerned. One may similarly decompose a polyhedral region whose boundary consists of several polyhedra.

In the rest of this book, we will apply other 3-dimensional decomposition schemes in other settings, such as the decomposition of a cell of an arrangement of triangles (see section 16.2).

13.3.2 Simplicial decomposition of a polyhedron of genus 0

The method described in the previous subsection yields a decomposition into elementary cells (prisms or simplices) of any polyhedral region. The complexity of this decomposition is only guaranteed to be $O(n^2)$ for a polyhedral region with n vertices, however, even though this region may be convex or may be triangulated into $O(n)$ tetrahedra. The method we present in this subsection proceeds by trimming the polyhedral region of its protruding parts before decomposing the resulting polyhedral region. This method can be used for any polyhedral region whose boundary is a polyhedron with genus 0. It yields a decomposition into $O(n + r^2)$ elementary cells if the polyhedron has n edges out of which r are reflex edges.

A few definitions

A few explanations and definitions are useful to describe the algorithm we present below. Given a polyhedron $\mathcal{P}$ of genus 0, we seek a triangulation $\mathcal{T}$ whose domain coincides with the polyhedral region interior to $\mathcal{P}$. The triangulation $\mathcal{T}$ is not necessarily a triangulation of $\mathcal{P}$: it may have additional vertices, called Steiner points, which are not vertices of $\mathcal{P}$. Moreover, the boundary $bd(\mathcal{T})$ is a 2-triangulation whose domain coincides with that of $\mathcal{P}$, but the triangulations $\mathcal{P}$ and $bd(\mathcal{T})$ do not necessarily contain the same triangles.

Consider an edge of $\mathcal{P}$. The planes supporting the two incident facets define a dihedral angle. If this angle is greater than π, the edge is called *reflex*, if it is smaller than π the edge is called *convex*, and we say it is a *flat* edge if this angle equals π. Similarly, a vertex is called *reflex* if it is incident to at least one reflex edge, *flat* if the facets that contain it are contained in at most two planes, and *convex* if it is neither flat nor reflex.

Consider a polyhedron $\mathcal{P}$ without flat vertices and V a convex vertex of $\mathcal{P}$. Let F be a facet of $\mathcal{P}$ that contains V and $H(F)$ the plane that contains this facet, that is, its affine hull. This plane H bounds two half-spaces. The one that locally contains the interior of $\mathcal{P}$, meaning that it contains the intersection of the interior of $\mathcal{P}$ with a small enough ball centered at V, is denoted by $H^+(F)$. The intersection of the closed half-spaces $\overline{H^+(F)}$ for all the facets F of $\mathcal{P}$ that contain V is an unbounded polytope called the *cone*[1] of V.

We denote by $\mathcal{K}(V)$ the set of vertices of the polyhedron $\mathcal{P}$ which are inside the cone of V or on its boundary, and by $\mathcal{K}'(V)$ the set $\mathcal{K}(V) \setminus \{V\}$ of these vertices excluding V.

The *cup* of V can now be described as the difference of convex hulls $conv(\mathcal{K}(V)) \setminus conv(\mathcal{K}'(V))$ (see figure 13.10). The cup of V is a polyhedral region and its boundary is a topological 2-sphere which can be separated into two parts:

- The first part is called the *dome* of V and is formed by the facets of $conv(\mathcal{K}'(V))$ which are not facets of $conv(\mathcal{K}(V))$. These are the facets of $conv(\mathcal{K}'(V))$ which are red[2] with respect to V.

- The second part is called the *lateral boundary* of the cup. It is formed by the facets of $conv(\mathcal{K}(V))$ which are not faces of $conv(\mathcal{K}'(V))$. All the facets on the lateral boundary are incident to V and the lateral boundary is contained in the boundary of the cone of V.

The common boundary between the dome and the lateral boundary is a polygon (although it may not always be contained in a plane), which we call the *crown* of V. The crown of V is formed by the edges of $conv(\mathcal{K}'(V))$ that are incident to a single facet of $conv(\mathcal{K}'(V))$ that is red with respect to V.[3] The vertices and edges of the dome which are not part of the crown are called *internal*. All these definitions are illustrated in figure 13.10.

The following properties are best observed now and will be used later on.

1. The cup of V is star-shaped with respect to V. Its interior is contained in the polyhedral region interior to $\mathcal{P}$, since the segment VW joining V to any point W inside this cup cannot intersect the polyhedron $\mathcal{P}$ in its relative interior.

---

[1]Since V is convex, the cone of V can also be described as the convex hull of the rays cast from V and that contain the edges incident to V, or equivalently as the set of points given by positive combinations $V + \sum_{i=1}^{k} \alpha_i(W_i - V)$ where the W_i's are the vertices of $\mathcal{P}$ adjacent to V and $\alpha_1, \ldots, \alpha_k$ are non-negative reals.

[2]A facet F of a polytope $\mathcal{C}$ is red with respect to a point A if the hyperplane H that supports $\mathcal{C}$ along F separates A from $\mathcal{C}$: A belongs to the open half-space H^- bounded by H which does not intersect the polytope $\mathcal{C}$ (see chapter 8). On the other hand, if A belongs to the open half-space H^+ that contains the interior of $\mathcal{C}$, then this facet F is blue with respect to A.

[3]If the set $\mathcal{K}(V)$ of vertices is in general position, then the facets of $conv(\mathcal{K}'(V))$ are either red or blue, and the edges and vertices of the crown are the *purple* faces of $conv(\mathcal{K}'(V))$ with respect to V, in other words the edges incident both to a blue and a red facet, and the vertices incident to purple edges. If the set $\mathcal{K}(V)$ is not in general position, then V may belong to the plane supporting a facet F of $conv(\mathcal{K}'(V))$ which is neither red nor blue.

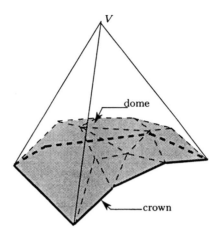

Figure 13.10. The cup of a convex vertex.

2. The lateral boundary of the cup of V is contained in the domain $dom(\mathcal{P})$ of $\mathcal{P}$. Indeed, consider a facet F_i of $\mathcal{P}$ incident to V, and let H_i be the plane that contains this facet and F'_i be the union of all the facets of the lateral boundary that are contained in H_i. Let $\mathcal{K}_i(V)$ be the set of those vertices of $\mathcal{P}$ that belong to H_i, and let $\mathcal{K}'_i(V)$ be the set $\mathcal{K}_i(V) \setminus \{V\}$. Then F'_i can be described as the two-dimensional cup $conv(\mathcal{K}_i(V)) \setminus conv(\mathcal{K}'_i(V))$ (see figure 13.11). As a result, F'_i is star-shaped with respect to V and is contained in the union of the facets of $\mathcal{P}$ contained in H_i.

3. The vertices of the cup of V are vertices of the polyhedron $\mathcal{P}$. The internal vertices of the dome are reflex vertices of $\mathcal{P}$. The vertices of the crown are also reflex vertices of $\mathcal{P}$ except for the vertices V' which are adjacent to V in $\mathcal{P}$ and such that the edge VV' of $\mathcal{P}$ is incident to two non-coplanar facets. From this reasoning it follows that a convex vertex V' may belong to the dome of another convex vertex V only if V and V' are adjacent in $\mathcal{P}$.

4. Unlike the vertices, the edges of the dome (internal edges or edges of the crown) are not necessarily edges of $\mathcal{P}$. The internal edges of the dome of V are reflex edges of the cup of V, whereas the edges of the crown are convex edges of the cup of V. The edges of the lateral boundary are edges of $\mathcal{P}$, however, and they are necessarily convex for the cup when they are incident to two non-coplanar facets of the cup.

A convex vertex of the polyhedron $\mathcal{P}$ is called *free* if its dome does not include any internal vertex, nor any internal edge that is an edge of $\mathcal{P}$. Free vertices are of special interest to us because removing the cup of a free vertex from $\mathcal{P}$ does not create singularities, hence the remaining region is indeed polyhedral.

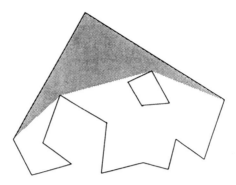

Figure 13.11. A facet of the cup.

The algorithm

Let $\mathcal{P}$ be a polyhedron with n edges of which r are reflex edges. The algorithm first normalizes the facets of this polyhedron so as to obtain a 2-triangulation without flat vertices that has the same domain. For this, it suffices to merge all the connected edges contained in a common line and all the adjacent facets contained in a common plane into a polygonal region, possibly with holes. This region can then be triangulated. The resulting polyhedron is 2-triangulated, still has $O(n)$ edges and the same r reflex edges, but no longer has flat vertices.

The algorithm then proceeds in two phases. In the first phase, or *pull-off* phase, we remove the cups of certain free convex vertices with bounded degree. The cups of these vertices can be triangulated easily and these triangulations can be added back to the triangulation of the remaining polyhedron to yield a triangulation of the original polyhedron. The algorithm keeps on pulling off the cups of those vertices until the size of the remaining polyhedron is $O(r)$. At this point, the second phase of the algorithm computes the vertical decomposition, as explained in the previous subsection, and this decomposition is triangulated into $O(r^2)$ tetrahedra. So the only missing part is the description of the pull-off phase, which we present now.

The pull-off phase is an iterative process. Let $\mathcal{P}_c$ be the current polyhedron. A set of vertices of $\mathcal{P}_c$ is *independent* if its elements are pairwise not adjacent. Recall that the *degree* of a vertex is the number of incident edges. The current step consists of the following operations.

1. Compute a maximal independent set of vertices of $\mathcal{P}_c$ that are convex, free, and of degree smaller than some constant g.

2. Remove those vertices and their attached cup from the polyhedron and its corresponding polyhedral region.

To identify a maximal independent set of convex, free, and bounded-degree vertices, we begin by putting all these convex, free vertices of degree smaller than g into a list $\mathcal{L}$. For this, we examine all the convex vertices of degree smaller than g in turn. Let V be the vertex under examination, and consider the set $\mathcal{H}'(V)$ of vertices adjacent to V in the 2-triangulated polyhedron and the set $\mathcal{H}(V) = \mathcal{H}'(V) \cup \{V\}$. We build the polyhedral region $conv(\mathcal{H}(V)) \setminus conv(\mathcal{H}'(V))$ and we look for reflex edges of $\mathcal{P}_c$ that intersect this region. If there are none, V is free by definition and is put into $\mathcal{L}$. Its cup is $conv(\mathcal{H}(V)) \setminus conv(\mathcal{H}'(V))$ and has been computed already.

To obtain a maximal independent subset of the elements stored in $\mathcal{L}$, we repeat the following procedure until $\mathcal{L}$ is empty: pick any vertex V in $\mathcal{L}$ (for instance the one stored in the first item of the list), put V into some list $\mathcal{L}'$, and remove from $\mathcal{L}$ all the vertices incident to V as well as V.

Pulling off the cups of the selected vertices is done as follows. The cup of V is triangulated using the $O(g)$ tetrahedra $conv(V, F)$ where F is a facet of the dome of V. The remaining polyhedron is obtained by replacing the faces of $\mathcal{P}_c$ that are contained in the lateral boundary of the cup of V, by the faces of the dome of V.

At each step in the pull-off phase, the number of vertices and edges of the current polyhedron decreases, yet the number of reflex edges is unchanged. For reasons that will be made clear (see lemma 13.3.9 below), g must be chosen as an integer greater than 6, so as to allow for another integer t such that $t > 11 + 66/(g - 5)$. Let r be the number of reflex edges of the initial polyhedron (which is also the number of reflex edges of the current polyhedron) and m the total number of edges of the current polyhedron $\mathcal{P}_c$. The pull-off phase is iterated as long as $m > (1 + t)r$.

Analysis

In the following string of lemmas, we prove that if r is the number of reflex edges of the initial polyhedron $\mathcal{P}$, then after the pull-off phase the resulting polyhedron has size $O(r)$ with exactly r reflex edges.

Lemma 13.3.2 *If V and V' are two convex non-adjacent vertices of a polyhedron $\mathcal{P}$, no vertex of $\mathcal{P}$ can be an internal vertex of the domes of both V and V'. Likewise, no edge of $\mathcal{P}$ can be an internal edge of the domes of both V and V'.*

Proof. Let us first show that, if V and V' are two non-adjacent convex vertices of $\mathcal{P}$, the intersection of their cups has an empty interior and is the intersection of their domes. For this, we show that the polyhedral region $\mathcal{I}$ that is the intersection of the cup of V with the cup of V' cannot have a vertex that does not belong to both domes. Since V and V' are not adjacent, neither may qualify as a vertex of $\mathcal{I}$. Owing to the definition of a cup, the interior of a cup does not contain a vertex of $\mathcal{P}$, so that any vertex of $\mathcal{I}$ is the intersection of an edge E of the cup of V with a facet F of the cup of V', or the converse. Since the lateral boundary of a cone is contained in $dom(\mathcal{P})$, however, the facet F and the edge E necessarily belong to the domes of V and V', which proves our assertion.

Suppose now that a vertex P of the polyhedron $\mathcal{P}$ is an internal vertex for both domes of some non-adjacent convex vertices V and V'. Then P must be a reflex vertex for the cup of V and a reflex vertex for the cup of V', hence there must exist a half-ball centered at P contained in the cup of V and a half-ball centered at P contained in the cup of V' respectively. Since the two interiors of the cups are non-intersecting, these half-balls have an empty intersection and so are bounded by the same plane. This shows that the vertex P is a singular face of the polyhedron $\mathcal{P}$, which is not allowed by the definition of a polyhedron.

For the second part of the lemma, it suffices to place a dummy vertex on the edge E of $\mathcal{P}$ that supposedly is internal to the domes of both V and V'. The above discussion brings out the contradiction. $\square$

Lemma 13.3.3 *A vertex internal to the dome of a convex vertex is incident to at least three reflex edges of $\mathcal{P}$.*

Proof. Let V be a convex vertex, and W be a vertex internal to the dome of V. Then W is a reflex vertex of the cup of V and there is a half-ball centered at W entirely contained in the cup of V and thus in the polyhedral region interior to $\mathcal{P}$ as well. The relative interiors of the edges of $\mathcal{P}$ that are incident to W are contained in the open half-space opposite to this half-ball, and because V is not flat, the vertex W is necessarily a vertex of the convex hull of the set of points formed by W and by the vertices of $\mathcal{P}$ adjacent to W. The edges of this convex hull that are incident to W are edges of $\mathcal{P}$ and must also be reflex. $\square$

Lemma 13.3.4 *A reflex vertex of $\mathcal{P}$ is an internal vertex to the domes of at most three convex vertices of $\mathcal{P}$.*

Proof. Let P be a reflex vertex of $\mathcal{P}$, and assume for a contradiction that it is an internal vertex of the domes of four convex vertices X, Y, Z, and T. The four vertices X, Y, Z, and T must be adjacent (lemma 13.3.2) and P must necessarily be included in the tetrahedron $XYZT$. Since P is internal to the domes of X, Y, Z, and T, there is a ball centered at P that is included in each of the four cones of X, Y, Z, and T. Since P is a vertex of $\mathcal{P}$, however, this ball must also intersect the polyhedral region outside $\mathcal{P}$. We may choose a point Q in this ball and exterior to $\mathcal{P}$, such that it belongs to none of the planes passing through P and through an edge of $XYZT$ (see figure 13.12). The point P is

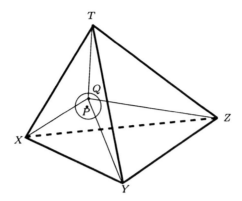

Figure 13.12. For the proof of lemma 13.3.4.

then interior to one of the four tetrahedra defined by Q and three of the points in $\{X, Y, Z, T\}$, say $XYZQ$. Then P cannot belong to the dome of the fourth vertex T, a contradiction. □

Lemma 13.3.5 *For any reflex edge PQ of the polyhedron $\mathcal{P}$, there are at most two convex vertices V and W of $\mathcal{P}$ such that P and Q belong to the crowns of V and W respectively, and such that PQ is an internal edge of the domes of both V and W.*

Proof. Let PQ be a reflex edge of $\mathcal{P}$. Assume for a contradiction that P and Q belong to the crowns of three convex vertices U, V, and W, and that the edge PQ is internal to the domes of U, V, and W.

Each of the segments PU, QU, PV, QV, PW, and QW is contained in $dom(\mathcal{P})$, and the interior of each triangle PQU, PQV, and PQW is contained in the polyhedral region interior to $\mathcal{P}$. As a result, the three affine half-planes bounded by the line PQ that contain the triangles PQU, PQV, and PQW, respectively, are pairwise distinct. Moreover, owing to lemma 13.3.2, U, V, and W are adjacent vertices in $\mathcal{P}$ and each one belongs to the cups of the two others. Since PQ is an internal edge to the dome of U, there exists a plane passing through PQ that separates U from V and W. Likewise, there are two planes containing PQ that separate V from W and U, and W from U and V. Hence the three tetrahedra $PQUV$, $PQVW$, and $PQWU$ have pairwise disjoint interiors and their union contains a small enough ball centered at the midpoint of PQ. Since PQ is an edge of $\mathcal{P}$, this ball must intersect the polyhedral region exterior to $\mathcal{P}$, and we may choose a point Z in this ball, exterior to $\mathcal{P}$, that does not belong to any of the planes containing the triangles PQU, PQV, and PQW respectively (see figure 13.13). Then PQ necessarily intersects one of the triangles UVZ, VWZ,

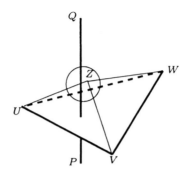

Figure 13.13. For the proof of lemma 13.3.5.

or WUZ, say UVZ. But then PQ cannot be an internal edge of the dome of W because the relative interior of the triangle UVZ lies outside the dome of W. □

Lemma 13.3.6 *A reflex vertex of the polyhedron $\mathcal{P}$ cannot at the same time be internal to the dome of a convex vertex and on the crown of another convex vertex.*

Proof. Let P be a reflex vertex of $\mathcal{P}$. If P is internal to the dome of a vertex of $\mathcal{P}$, then there is a half-ball centered at P contained in the polyhedral region interior to $\mathcal{P}$. If P is on the crown of a vertex V of $\mathcal{P}$, then there is an internal edge E of the dome of V that is incident to P. In the plane $\mathit{aff}(V, E)$ the angle at P internal to the polyhedron is less than π. This contradicts the existence of a half-ball centered at P included in the region interior to $\mathcal{P}$.   □

Lemma 13.3.7 *If $\mathcal{P}$ has r reflex edges, then at most $2r$ convex vertices are not free.*

Proof. The reflex edges of $\mathcal{P}$ may be put into one of three categories:

1. those whose vertices include an internal vertex of a dome,

2. those which are internal edges of a dome with both vertices on the crown of that dome,

3. the others.

These three categories are disjoint by virtue of lemma 13.3.6. If r_1, r_2, r_3 are the respective sizes of these classes, then

$$r_1 + r_2 + r_3 = r.$$

The vertices of the edges in the first category are internal vertices of at most $2r_1$ domes. Indeed, lemma 13.3.4 states that the vertices of an edge in the first category are internal vertices of at most six domes, but lemma 13.3.3 shows that each dome will account for at least three edges in the first category. Finally, lemma 13.3.5 shows that the edges in the second category are internal edges of at most $2r_2$ domes. The total number of convex vertices which are not free is thus at most $2r_1 + 2r_2 \leq 2r$. $\qquad\square$

In what follows, we assume as before that the polyhedron does not have flat vertices. The following two combinatorial lemmas prove the existence of a large enough set of convex free vertices of bounded degree.

Lemma 13.3.8 *Let $\mathcal{P}$ be a polyhedron of genus 0 with n vertices and m edges, r of which are reflex. The number $n'_c(g)$ of convex free vertices with degree at most g of $\mathcal{P}$ is at least*

$$n'_c(g) \geq \frac{m}{3} + 2 - 4r - 2\frac{m-r}{g+1}.$$

Proof. Euler's relation 11.2.4 and its corollary 11.2.5 state that

$$n \geq \frac{m}{3} + 2.$$

If the polyhedron $\mathcal{P}$ has r reflex edges, then at most $2r$ vertices are reflex and the number n_c of its convex vertices satisfies

$$n_c \geq \frac{m}{3} + 2 - 2r.$$

A convex vertex is only incident to convex edges, so the maximum number of incidences between convex vertices and edges is at most $2(m-r)$. The number of convex vertices with degree greater than g is thus at most $\frac{2(m-r)}{g+1}$ and the number $n_c(g)$ of convex vertices with degree at most g satisfies

$$n_c(g) \geq \frac{m}{3} + 2 - 2r - \frac{2(m-r)}{g+1}.$$

Taking lemma 13.3.7 into account, the number $n'_c(g)$ of free convex vertices with degree at most g satisfies

$$n'_c(g) \geq \frac{m}{3} + 2 - 4r - \frac{2(m-r)}{g+1}.$$

$\qquad\square$

Lemma 13.3.9 *Let $\mathcal{P}$ be a polyhedron with n vertices and m edges, r of which are reflex. Consider two integers g and t that satisfy $g \geq 6$ and $t > 11 + \frac{66}{g-5}$. If $m > (1+t)\,r$, then $\mathcal{P}$ has at least $s(m-r)$ free convex vertices of degree at most g, where*

$$s = \frac{t-11}{3t} - \frac{2}{g+1}.$$

Proof. If $m > (1+t)\,r$, then the inequality proved by lemma 13.3.8 becomes

$$
\begin{aligned}
n'_c(g) \;\;&\geq\;\; \frac{m-r}{3} - \frac{11r}{3} - \frac{2(m-r)}{g+1}\\[2mm]
&\geq\;\; (m-r)\left(\frac{t-11}{3t} - \frac{2}{g+1}\right).
\end{aligned}
$$

$\square$

Let us now come back to the analysis of the algorithm. Lemma 13.3.9 serves to prove that each iteration in the pull-off process removes at least a constant fraction of the vertices of the polyhedron. Indeed, if $\mathcal{P}_c$, the current polyhedron, has n vertices, m edges, and r reflex edges, we may assume that $m > (1+t)r$ (otherwise the pull-off phase is over) and that $\mathcal{P}$ has at least $s(m-r)$ free convex vertices of degree at most g. Here g, t, and s are defined as in lemma 13.3.9. The number n'' of pulled-off vertices is thus at least

$$
\begin{aligned}
n'' \;\;&\geq\;\; \frac{s}{g+1}\,(m-r)\\[2mm]
n'' \;\;&\geq\;\; \frac{s}{g+1}\frac{t}{t+1}\,m\\[2mm]
&\geq\;\; \frac{s}{g+1}\frac{t}{t+1}\frac{3}{2}n \;=\; cn.
\end{aligned}
$$

The last inequality follows from the fact that $2m \geq 3n$ since each vertex is incident to at least three edges.

Let V be a vertex of $\mathcal{P}$ of degree at most g. Whether it is convex, and if so the region $conv(\mathcal{H}(V)) \setminus conv(\mathcal{H}'(V))$, may be computed in constant time, and its intersection with the reflex edges of $\mathcal{P}$ may be tested in time $O(r)$. The set of free convex vertices with degree at most g can thus be computed in time $O(nr)$, and a maximal independent subset can be extracted in time $O(n)$. The cup of each pulled-off vertex has complexity $O(g)$, and can be triangulated in constant time into $O(g)$ tetrahedra. The polyhedron can also be patched up in constant time. The number r of reflex edges is constant throughout the algorithm, so an iteration in the pull-off phase has a complexity of $O(r)$ times the number of vertices in the current polyhedron.

The total complexity of the pull-off phase is

$$O(nr + (1 - c)nr + (1 - c)^2 nr + \cdots) = O(nr),$$

and it generates only $O(n)$ tetrahedra.

After this phase is over, the remaining polyhedron is of size $O(r)$. Lemma 13.3.1 shows that this polyhedron may be decomposed into $O(r^2)$ tetrahedra in time $O(r^2 \log r)$. The following theorem summarizes the performances of the algorithm.

Theorem 13.3.10 *A polyhedron of genus 0, with n vertices and r reflex edges, may be decomposed into $O(n + r^2)$ tetrahedra in time $O(nr + r^2 \log r)$.*

Remark. The size of the decomposition into tetrahedra produced by this algorithm is optimal in the worst case up to a constant factor. Indeed, there are polyhedra of size $O(r)$ which cannot be decomposed into fewer than r^2 convex parts (see exercise 13.4).

13.4 Exercises

Exercise 13.1 (A non-triangulable polyhedron) Consider the six points $\{A_1, B_1, A_2, B_2, A_3, B_3\}$ whose coordinates are given in section 13.2. Show that the eight triangles $A_1 B_1 A_2$, $A_1 B_1 A_3$, $A_2 B_2 A_1$, $A_2 B_2 B_3$, $A_3 B_3 B_1$, $A_3 B_3 B_2$, $A_1 B_2 A_3$, and $A_2 B_1 B_3$ define a polyhedron, and that this polyhedron is not triangulable.

Exercise 13.2 (Pull-off) Show that the pull-off phase of the algorithm described in section 13.3 may be implemented in time $O((n + r^2) \log r)$.

Exercise 13.3 (Flat vertices) State explicitly the algorithm that removes the flat vertices of a polyhedron in linear time.

Exercise 13.4 (Chazelle's polyhedron) This polyhedron $\mathcal{P}$ can be described as a rectangular parallelepiped whose top and bottom facets each have $n + 1$ notches. More precisely, the polyhedron $\mathcal{P}$ is bounded by the six planes $x = 0$, $x = 1$, $y = 0$, $y = 1$, $z = 0$, $z = 1$. The top facet of $\mathcal{P}$ in the plane $z = 1$ is split by $n + 1$ notches, each of which is formed by a reflex edge parallel to the x-axis and incident to two quasi-vertical facets (parallel to the xz-plane). The bottom facet of $\mathcal{P}$ in the plane $z = 0$ is split by $n + 1$ notches, each of which is formed by a reflex edge parallel to the y-axis and incident to two quasi-vertical facets (parallel to the yz-plane). The reflex edges of the notches on the bottom facet are contained in the hyperboloid $z = xy$ while the reflex edges of the notches on the top facet are contained in the hyperboloid $z = xy + \epsilon$.

Show that if $\epsilon < \frac{1}{n^2}$, any decomposition of $\mathcal{P}$ into convex parts requires at least $\Omega(n^2)$ convex parts.

Hint: Show that any convex body contained in $\mathcal{P}$ intersects the volume between the two hyperboloids $z = xy$ and $z = xy + \epsilon$ in a region of volume $O(\epsilon)$.

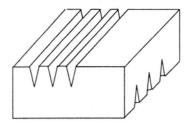

Figure 13.14. Chazelle's polyhedron.

13.5   Bibliographical notes

The algorithm that guarantees a linear triangulation for any set of points in $\mathbb{E}^3$ without collinear points is due to Avis and El-Gindy [19]. In particular, the figures in section 13.1 are borrowed from their article. The split theorem was independently proposed by Avis and El-Gindy [19] and by Edelsbrunner, Preparata, and West [95].

The decomposition into simplices of a polyhedron of genus 0 given in subsection 13.3.2 is borrowed from Chazelle and Palios [58]. They also give the solution to exercise 13.2. The polyhedron described in exercise 13.4 was proposed by Chazelle [42]. This polyhedron with $O(n)$ edges and vertices which cannot be decomposed into fewer than $\Omega(n^2)$ convex parts proves the optimality in the worst case of the decomposition described in subsection 13.3.2. Bajaj and Dey [20] generalized the algorithm by Chazelle and Palios to polyhedra with holes and handles: if the polyhedron has $O(n)$ edges of which r are reflex, their algorithm outputs a decomposition into $O(r^2)$ convex parts and their algorithm runs in time $O(nr + r^{\frac{7}{2}})$.

Part IV

Arrangements

By the *arrangement* of a finite set of curves or arcs in the plane, we mean the decomposition of the plane induced by these curves or arcs. In $\mathbb{E}^d$, we call arrangements the decompositions induced by a finite set of hypersurfaces or portions of hypersurfaces.

Arrangements play a central role and occur in many different applications. This part is divided into three chapters. In chapter 14, we are interested primarily in arrangements of hyperplanes. This interest is spurred mainly by two facts. A set of points transforms into a set of hyperplanes by polarity, and the arrangement of the hyperplanes contains several useful pieces of information on the set of points. Also, arrangements of hyperplanes are particular cases of arrangements of simplices, so the study of the former kind of arrangements provides interesting combinatorial bounds for the latter.

The subsequent chapters investigate a few combinatorial and algorithmic notions related to arrangements of line segments in the plane (chapter 15) and arrangement of triangles in three-dimensional space (chapter 16). The central problem in both chapters is to bound the combinatorial complexity of several parts of these arrangements, and in particular to show that they may be of much smaller complexity than the whole arrangement. Efficient algorithms to compute these portions of arrangements are also sought. These studies are motivated mainly by two applications: computing views and hidden surface removal in computer graphics, and motion planning in robotics.

Chapter 14

Arrangements of hyperplanes

Hyperplane arrangements are the simplest arrangements one may think of. They appear naturally in several applications and the results given below, which are of mostly combinatorial nature, will be useful in the subsequent chapters.

The polarity that maps a set of points to a set of hyperplanes often transforms a problem about points into a problem about hyperplanes. Many examples are provided in the exercises. A preprocessing step for these problems often consists of computing the arrangement of the corresponding set of hyperplanes. This problem is discussed in section 14.4 which takes advantage of a combinatorial result known as the zone theorem, given in section 14.3.

An interesting correspondence between hyperplane arrangements and a certain kind of polytopes, called zonotopes, also sheds more light on problems from crystallography, architecture, or mixture design (see exercises 14.8 and 14.9).

Section 14.5 introduces the notion of levels in hyperplane arrangements, which is central to our analysis of higher-order Voronoi diagrams, studied in chapter 17.

14.1 Definitions

Let $\mathcal{H}$ be a set of n hyperplanes in $\mathbb{E}^d$. The intersection of a finite number of half-spaces is a bounded or unbounded polytope, and so $\mathcal{H}$ induces a decomposition of $\mathbb{E}^d$ into a collection of bounded or unbounded polytopes with pairwise disjoint interiors. These polytopes and their faces form a pure cell complex of dimension d which we call the d-arrangement of $\mathcal{H}$, or more simply the arrangement of $\mathcal{H}$ if d is clearly understood. This cell complex is denoted by $\mathcal{A}(\mathcal{H})$.

From now on, we often use the notions of a set of hyperplanes in general position, or of a simple arrangement. A set $\mathcal{H}$ of n hyperplanes is said to lie in *general position* if the intersection of any $k \leq d$ of them is an affine space of dimension $d - k$, and if moreover the intersection of any $d + 1$ of them is empty. The cells in the arrangement of a set of hyperplanes in general position are simple polytopes. For this reason, such an arrangement is also called a *simple arrangement*.

Using the polarity introduced in subsection 7.1.3, we may map a set of points to the set of its polar hyperplanes. Since polarity preserves adjacency relationships, the set of polar hyperplanes is in general position whenever the set of points is in general position. Recall that a set of points is in general position if any subset of $k + 1 \leq d + 1$ points generates an affine subspace of dimension k.

14.2   Combinatorial properties

This section establishes the combinatorial results related to arrangements that will be used below. The proofs given here are direct; other proofs, however, may be derived from the combinatorial properties of (spherical) polytopes as outlined in exercise 14.1.

Henceforth, $\mathcal{A}(\mathcal{H})$ denotes the d-arrangement of $\mathcal{H}$, $\mathcal{A}(\mathcal{H}\backslash H)$ the d-arrangement of the set $\mathcal{H} \setminus \{H\}$, $\mathcal{A}(\mathcal{H} \cap H)$ the $(d-1)$-arrangement in H of the intersections of the $n - 1$ hyperplanes $(\mathcal{H} \setminus \{H\}) \cap H$. We denote by $n_k(\mathcal{H})$ the number of k-faces in the d-arrangement $\mathcal{A}(\mathcal{H})$, for $k = 0, \ldots, d$, and by $n_k(n, d)$ the maximal number of k-faces in the arrangement of n hyperplanes in $\mathbb{E}^d$.

The following lemma is trivial.

Lemma 14.2.1 *Any k-face F in $\mathcal{A}(\mathcal{H} \setminus H)$ that intersects H gives rise in $\mathcal{A}(\mathcal{H})$ to a $(k-1)$-face, $F \cap H$, and to two k-faces, $F \cap H^+$ and $F \cap H^-$, where H^+ and H^- are the two half-spaces bounded by H. For a given H, all the k-faces of $\mathcal{A}(\mathcal{H})$ that are not faces in $\mathcal{A}(\mathcal{H} \setminus H)$ can be obtained in this way only once.*

By counting separately the faces of $\mathcal{A}(\mathcal{H})$ that do not intersect H, those that are contained in H, and those that intersect H but are not contained in H, the previous lemma yields

$$n_0(\mathcal{H}) = n_0(\mathcal{H} \setminus H) + n_0(\mathcal{H} \cap H) \tag{14.1}$$

$$n_d(\mathcal{H} \cap H) = 0 \tag{14.2}$$

$$n_k(\mathcal{H}) = n_k(\mathcal{H} \setminus H) + n_k(\mathcal{H} \cap H) + n_{k-1}(\mathcal{H} \cap H). \tag{14.3}$$

Theorem 14.2.2 (Euler's relation) *For any d-arrangement of a set $\mathcal{H}$ of n hyperplanes of $\mathbb{E}^d$, we have*

$$\sum_{k=0}^{d} (-1)^k n_k(\mathcal{H}) = (-1)^d.$$

Proof. Let us begin with the following lemma.

Lemma 14.2.3 *The number of k-faces in any simple d-arrangement of d hyperplanes is*

$$n_k(d, d) = 2^k \begin{pmatrix} d \\ k \end{pmatrix}, \ 0 \le k \le d.$$

More generally, $n_k(d, d)$ is also the number of k-faces containing a given vertex in any simple d-arrangement.

Proof of the lemma. Consider the d-arrangement $\mathcal{A}(\mathcal{H})$ of a set $\mathcal{H}$ of d hyperplanes in general position. A k-face of $\mathcal{A}(\mathcal{H})$ is contained in the intersection of $d - k$ hyperplanes in $\mathcal{H}$. Conversely, the intersection of a subset $\mathcal{H}'$ of $d - k$ hyperplanes of $\mathcal{H}$ is an affine space $\mathcal{I}$ of dimension k, and the restrictions to $\mathcal{I}$ of the k hyperplanes in $\mathcal{H} \setminus \mathcal{H}'$ form in $\mathcal{I}$ a k-arrangement, whose k-faces are precisely the k-faces of $\mathcal{A}(\mathcal{H})$ that are contained in $\mathcal{I}$. Thus, we have

$$n_k(d, d) = \begin{pmatrix} d \\ k \end{pmatrix} n_k(k, k).$$

So we are left with the slightly simpler problem of counting the k-faces in a simple k-arrangement of k hyperplanes. Under the general position assumption, such an arrangement has precisely one vertex, S, at the intersection of the k hyperplanes, and every face in the arrangement contains this vertex. This can be shown by induction on the dimension of the faces. If an edge is incident to no vertex, then it must be a line contained in $k - 1$ hyperplanes that has an empty intersection with the k-th hyperplane, and the intersection of the k hyperplanes is empty, contradicting the general position assumption. Suppose that the j-faces in the arrangement all have S as a vertex, for $j = 1, \ldots, i - 1$. Then any i-face F that does not have a vertex cannot have a subface either, since these faces have a vertex, namely S. Then F is an affine $(k - i)$-space contained in i hyperplanes which does not intersect any of the other $k - i$ hyperplanes. This again contradicts the general position assumption and proves our statement.

By induction on k, we can show that $n_k(k, k) = 2^k$. This is obviously true for $k = 1$ and $k = 2$. Assume by induction that it is true for any $k' < k$ and consider a hyperplane H in $\mathcal{H}$. Using the notation as above, the $(k - 1)$-faces of $\mathcal{A}(\mathcal{H} \cap H)$ are contained in exactly two k-faces of $\mathcal{A}(\mathcal{H})$ and each k-face contains only one $(k - 1)$-face of $\mathcal{A}(\mathcal{H} \cap H)$. Therefore, $\mathcal{A}(\mathcal{H})$ has exactly twice as many k-faces as $\mathcal{A}(\mathcal{H} \cap H)$ has $(k - 1)$-faces. The first part of the lemma is thus proved by induction. The second part is immediate if one considers the d hyperplanes containing the given vertex. $\square$

The lemma can be used to verify Euler's relation for d hyperplanes:

$$\sum_{k=0}^{d} (-1)^k 2^k \begin{pmatrix} d \\ k \end{pmatrix} = (1 - 2)^d = (-1)^d.$$

This provides the base case for the inductive proof of Euler's relation, the induction progressing with an increasing number of hyperplanes. Relations 14.1 and 14.2 finish the proof.   □

The following lemma is analogous to lemma 7.1.14 for simple polytopes. It serves in particular to show the Dehn–Sommerville relations (see exercise 14.2), as was the case in chapter 7.

Lemma 14.2.4 *Any i-face of a simple d-arrangement $\mathcal{A}(\mathcal{H})$ is contained in exactly*

$$2^{j-i} \binom{d-i}{d-j}$$

j-faces in the arrangement, for any $0 \leq i \leq j \leq d$.

Proof. The case when $i = 0$ is proved by lemma 14.2.3 and can be used as the basis for a proof by induction on i. Let F be an i-face in the d-arrangement $\mathcal{A}(\mathcal{H})$. Choose any hyperplane H that cuts F and that forms a simple arrangement with the hyperplanes of $\mathcal{H}$. Then $H \cap F$ is an $(i-1)$-face in the simple $(d-1)$-arrangement $\mathcal{A}(\mathcal{H} \cap H)$. The induction hypothesis on $\mathcal{H}$ implies that $H \cap F$ is contained in

$$2^{j-i} \binom{d-i}{d-j}$$

$(j-1)$-faces in $\mathcal{A}(\mathcal{H} \cap H)$. These $(j-1)$-faces are in one-to-one correspondence with the j-faces of $\mathcal{A}(\mathcal{H})$ that contain F: indeed, each of these j-faces contains $F \cap H$, hence also F since $\mathcal{A}(\mathcal{H})$ is a complex. This proves the lemma.   □

To each i-face F of $\mathcal{A}(\mathcal{H})$ $(i = 1, \ldots, d)$ corresponds a *position vector* defined in the following way. For each hyperplane $H \in \mathcal{H}$, we denote by H^+ one of the two open half-spaces bounded by H and denote the other by H^-. The position vector of a point X with respect to $\mathcal{H}$ is the vector $(\varepsilon_1, \ldots, \varepsilon_n)$ where ε_i is $+1$, 0, or -1 according to whether $X \in H_i^+$, $X \in H_i$, or $X \in H_i^-$. The points in a given face have the same position vector, which is naturally called the position vector of the face. If a face F is of dimension i, then it is the intersection of exactly $d-i$ hyperplanes (when the arrangement is simple) and so its position vector has exactly $d-i$ zero components. The position vectors of the cells that contain F can be obtained by replacing all the zero components by $+1$ or -1. Lemma 14.2.4 shows that all possible replacements are realized in the arrangement.

The following theorem gives exact asymptotic bounds for the number of faces and incidences between faces of an arrangement. These bounds are essential for algorithms on arrangements.

Theorem 14.2.5 *The total number of faces in a simple d-arrangement of n hyperplanes is $\Theta(n^d)$, and so is the number of incidences between faces. If the arrangement is not simple, both numbers are still $O(n^d)$.*

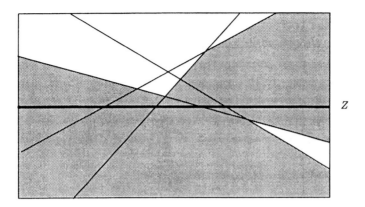

Figure 14.1. A zone in an arrangement of lines.

Proof. For a simple arrangement, the number of k-faces that contain a given vertex is

$$n_k = 2^k \binom{d}{k}, \ 0 \leq k \leq d,$$

as is implied by lemma 14.2.3. But we know that the number of vertices is exactly $\binom{n}{d} = \Theta(n^d)$. Lemma 14.2.4 for $j = i + 1$ states that the number of incidences between an i-face and an $(i + 1)$-face is exactly $2(d - i)$ times the number of those faces, proving the theorem for the case of a simple arrangement. A perturbation argument shows that the number of faces is maximized precisely for simple arrangements. Indeed, if a subset of $k \leq d + 1$ hyperplanes intersect along an affine subspace of dimension $d - k' > d - k$, then these hyperplanes can be perturbed slightly so that the perturbed arrangement is simple. The number of faces only increases, and so does the number of incidences.[1] □

14.3 The zone theorem

Consider the d-arrangement $\mathcal{A}(\mathcal{H})$ of a set $\mathcal{H}$ of n hyperplanes and let Z be a hyperplane that does not belong to $\mathcal{H}$. The complex formed by the d-faces in $\mathcal{A}(\mathcal{H})$ that are intersected by Z and their subfaces is called the *zone* of Z in the arrangement $\mathcal{A}(\mathcal{H})$ (see figure 14.1).

The zone theorem shows that the complexity (the number of faces) of any zone in a d-arrangement is an order of magnitude smaller than the whole arrangement.

---

[1] This is also true when $k = d + 1$, with the convention that an affine subspace of negative dimension is the empty space.

This result is at the core of an optimal algorithm that builds arrangements of hyperplanes, to be described in the next section.

Henceforth, rather than counting the faces in the zone, we will count the pairs (F, C) where C is a d-face in the zone and F is a face of C. Such a pair is called a *side* in the zone, and if F is a k-face we speak of a k-side. Observe that a face F in the arrangement generally belongs to several d-faces in the zone. The number of k-sides of a zone, however, is at most 2^{d-k} times the number of k-faces in the zone.

Let us first prove the zone theorem when $d = 2$.

Lemma 14.3.1 *The zone of a line Z in the arrangement $\mathcal{A}(\mathcal{H})$ of n lines in the plane has complexity $\Theta(n)$ if $\mathcal{H} \cup \{Z\}$ is in general position. If not, it is still $O(n)$.*

Proof. Let Z be the line that defines the zone. Without loss of generality, we may choose a coordinate system such that the x-axis is supported by Z. We may also assume that the lines in the arrangement and Z are in general position, meaning that any two lines in this set intersect in exactly one point and no three lines have a common intersection. A perturbation argument shows that the complexity of the zone is maximized in this case (as is done in the proof of theorem 14.2.5).

Let C be a 2-face in the zone of Z, F an edge of C, and (F, C) the corresponding side of the zone. Let $H(F, C)$ be the half-plane that contains C and is bounded by the line that is the affine hull of F. The side (F, C) is called a *left side* if $H(F, C)$ contains the point $(+\infty, 0)$, a *right side* otherwise. Since no line in $\mathcal{H}$ is parallel to the x-axis, a side may not be simultaneously left and right. We show that the total number of left sides in a zone is at most $3n$, and a symmetric argument shows the desired result.

The proof goes by induction on the number of lines. The result is trivial for $n = 1$. Let H be the line in the arrangement whose intersection with Z has the greatest abscissa (see figure 14.2). By induction, the total number of left sides in the 2-faces of the zone of Z in $\mathcal{A}(\mathcal{H} \setminus H)$ does not exceed $3n - 3$. Let C be the 2-face of this zone that contains the point $(+\infty, 0)$. The intersection F of C and H is a line segment (with endpoints A and B) or a half-line (ending at A). When adding H, (F, C) becomes a new left side and the left sides that contained A, resp. B (if it exists), are both cut into two left sides each. Note that because H is the line in the arrangement whose intersection with Z has the greatest abscissa, (F, C) is the only left side supported by H in the zone of Z in $\mathcal{A}(\mathcal{H})$. The overall number of left sides increases by at most 3, proving the above statement and the lemma. The bound is asymptotically tight whenever $\mathcal{H} \cup \{Z\}$ is in general position.   □

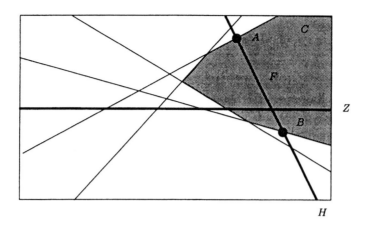

Figure 14.2. Inserting H.

The preceding lemma serves as the basis for an inductive argument that proves the zone theorem in the general case.

Theorem 14.3.2 (Zone theorem) *The complexity of any zone in the d-arrangement $\mathcal{A}(\mathcal{H})$ of n hyperplanes in $\mathbb{E}^d$ is $\Theta(n^{d-1})$ if $\mathcal{H} \cup \{Z\}$ is in general position. If not, this complexity is still $O(n^{d-1})$.*

Proof. The proof is structured as a double induction on the dimension and on the number of hyperplanes. As in section 14.2, we denote by $\mathcal{H}$ the set of n hyperplanes, by H a hyperplane in $\mathcal{H}$, by $\mathcal{A}(\mathcal{H} \setminus H)$ the d-arrangement of the set $\mathcal{H} \setminus H$, and by $\mathcal{A}(\mathcal{H} \cap H)$ the $(d-1)$-arrangement in H of the set formed by the $n-1$ hyperplanes $\{\mathcal{H} \setminus H\} \cap H$. We consider a hyperplane $Z \notin \mathcal{H}$. We denote by $\mathcal{Z}(\mathcal{H})$ the zone of Z in $\mathcal{A}(\mathcal{H})$, by $\mathcal{Z}(\mathcal{H} \setminus H)$ the zone of Z in $\mathcal{A}(\mathcal{H} \setminus H)$, and by $\mathcal{Z}(\mathcal{H} \cap H)$ the zone of $Z \cap H$ in $\mathcal{A}(\mathcal{H} \cap H)$.

We assume again that the set comprising Z and the hyperplanes in the arrangement is in general position. A perturbation argument shows that it is indeed when general position is realized that the complexity of the zone is maximized (see the proof of theorem 14.2.5).

Let (F, C) be a k-side of $\mathcal{Z}(\mathcal{H} \setminus H)$. We distinguish between three cases.

Case 1: $H \cap C = \emptyset$
Then (F, C) gives rise to a single k-side of $\mathcal{Z}(\mathcal{H})$, namely (F, C) itself.

Case 2: $H \cap C \neq \emptyset$ and $H \cap F = \emptyset$
Consider the half-space H_F bounded by H and that contains F, and let $C_F =$

$C \cap H_F$. If Z intersects C_F, then (F, C) gives rise to a k-side of $\mathcal{Z}(\mathcal{H})$, namely (F, C_F). Otherwise (F, C) does not correspond to a side in $\mathcal{Z}(\mathcal{H})$.

Case 3: $H \cap C \neq \emptyset$ and $H \cap F \neq \emptyset$
Consider the two half-spaces H^+ and H^- bounded by H, and let $F^+ = F \cap H^+$, $F^- = F \cap H^-$, $C^+ = C \cap H^+$ and $C^- = C \cap H^-$. If Z intersects C^+ but not C^- (resp. C^- but not C^+), then (F, C) gives rise to a k-side of $\mathcal{Z}(\mathcal{H})$, namely (F^+, C^+) (resp. (F^-, C^-)). Otherwise, Z intersects both C^+ and C^- and (F, C) gives rise to two sides of $\mathcal{Z}(\mathcal{H})$, namely (F^+, C^+) and (F^-, C^-). In the latter case, we note that Z necessarily intersects $C^+ \cap C^- = C \cap H$. Thus $C \cap H$ is a cell of $\mathcal{Z}(\mathcal{H} \cap H)$ and $(F \cap H, C \cap H)$ is a $(k-1)$-side of $\mathcal{Z}(\mathcal{H} \cap H)$.

Let us denote by $z_k(\mathcal{H})$, $z_k(\mathcal{H} \setminus H)$, and $z_k(\mathcal{H} \cap H)$ the numbers of k-sides of $\mathcal{Z}(\mathcal{H})$, $\mathcal{Z}(\mathcal{H} \setminus H)$, and $\mathcal{Z}(\mathcal{H} \cap H)$ respectively, and by $z_k(\mathcal{H}, H)$ the number of k-sides of $\mathcal{Z}(\mathcal{H})$ whose k-face is not contained within H. Any k-side of $\mathcal{Z}(\mathcal{H})$ that is not contained within H is either a k-side of $\mathcal{Z}(\mathcal{H} \setminus H)$ that does not intersect H, or a portion of a k-side of $\mathcal{Z}(\mathcal{H} \setminus H)$ that intersects H. From the preceding discussion, with the convention that $z_l(\mathcal{H} \cap H) = 0$ if $l < 0$, it follows that

$$z_k(\mathcal{H}, H) \leq z_k(\mathcal{H} \setminus H) + z_{k-1}(\mathcal{H} \cap H) \text{ for } 0 \leq k < d. \tag{14.4}$$

This result does not depend on the choice of the hyperplane H, so we may carry out the same analysis for all the hyperplanes H in $\mathcal{H}$, which yields

$$(n - d + k)z_k(\mathcal{H}) \leq \sum_{H \in \mathcal{H}} (z_k(\mathcal{H} \setminus H) + z_{k-1}(\mathcal{H} \cap H)). \tag{14.5}$$

Indeed, a k-side of $\mathcal{Z}(\mathcal{H})$ is counted each time H is not one of the hyperplanes that contains its k-face, and this happens $n - (d - k)$ times.

Let us denote by $z_k(n, d)$ (or by z_k when n and d are clearly understood) the maximum value of $z_k(\mathcal{H})$ over all d-arrangements of n hyperplanes and all choices of Z. Inequality 14.5 can be rewritten as

$$z_k(n, d) \leq \frac{n}{n - d + k}(z_k(n - 1, d) + z_{k-1}(n - 1, d - 1)), \tag{14.6}$$

and holds for any $0 < k \leq d$ and $n > d - k$.

This recurrence can be solved more easily by introducing the quantities $w_k(n, d)$ defined by

$$z_k(n, d) = \binom{n}{d - k} w_k(n, d).$$

Taking upper bounds, inequality 14.6 above can be rewritten as

$$w_k(n, d) \leq w_k(n - 1, d) + w_{k-1}(n - 1, d - 1). \tag{14.7}$$

Solving by induction on n, we obtain

$$w_k(n,d) \leq w_k(d-k+1,d) + \sum_{i=d-k+1}^{n-1} w_{k-1}(i,d-1). \qquad (14.8)$$

An induction on d now yields the asymptotic bound on $w_k(n,d)$ and thus for $z_k(n,d)$. The base case is given by lemma 14.3.1 for the case $d = 2$ which can be stated as

$$z_k(n,2) = O(n).$$

Suppose now that $d > 2$ and that $z_{k'}(n,d') = O(n^{d'-1})$ for all $k' \leq d' < d$, the constant in the big-oh notation depending on d' and k'. Then $w_{k'}(n,d') = O(n^{k'-1})$, and we infer from 14.8 that

$$w_k(n,d) \leq w_k(d-k+1,d) + \sum_{i=d-k+1}^{n-1} O(i^{k-2}).$$

Thus if $k \geq 2$, we have $w_k(n,d) = O(n^{k-1})$ and hence $z_k(n,d) = O(n^{d-1})$.

Unfortunately, for $k = 1$, this only yields $z_1(n,d) = O(n^{d-1} \log n)$. Thus, let us now examine the case of 0-sides and 1-sides. Each d-face in the zone is a convex polytope to which we can apply Euler's relation 7.2.1. For each bounded d-face, the corresponding number n_k of k-faces thus satisfies

$$\sum_{k=0}^{d} (-1)^k n_k = 1,$$

and for unbounded k-faces (see exercise 7.15), we have

$$\sum_{k=0}^{d} (-1)^k n_k = 0.$$

Summing over all the cells in the zone, and assuming that $n \geq d$, we obtain

$$\sum_{k=0}^{d} (-1)^k z_k \geq 0,$$

whether the cell is bounded or not. From this we obtain

$$z_1 - z_0 \leq \sum_{k=2}^{d} (-1)^k z_k = O(n^{d-1}).$$

Each d-face in the zone is a simple polytope, so each vertex of this polytope is incident to d edges. Therefore $z_0 \leq \frac{2}{d} z_1$, and so

$$\left(1 - \frac{2}{d}\right) z_1 \leq z_1 - z_0 = O(n^{d-1}).$$

This provides the desired bound for 0-sides and 1-sides.

The bound is asymptotically tight when $\mathcal{H} \cup \{Z\}$ is in general position. Indeed, the simple $(d-1)$-arrangement of $\mathcal{H} \cap Z$ is of complexity $\Omega(n^{d-1})$, and its k-faces are in one-to-one correspondence with a subset of the $(k+1)$-faces of $\mathcal{Z}(\mathcal{H})$ $(0 \leq k < d)$. $\square$

14.4   Incremental construction of an arrangement

For the incremental construction of an arrangement, we insert the hyperplanes one by one while maintaining the current arrangement. We assume henceforth that the arrangement $\mathcal{A}(\mathcal{H})$ is simple and is represented by the incidence graph of its 2-skeleton (that is, the set of its 0, 1, and 2-faces). These assumptions can be removed (see exercises 14.4 and 14.6).

When inserting a new hyperplane H_i, the set of faces in the current arrangement that are modified or whose incidences are modified is exactly the zone of H_i. The zone theorem is thus at the core of the analysis of the incremental algorithm.

14.4.1   The case of dimension 2

Consider n lines $H_1, \ldots, H_n$ in general position in the plane, whose simple arrangement we want to compute. Let $\mathcal{H}_i = \{H_1, \ldots, H_{i-1}\}$ be the set of lines already inserted and $\mathcal{G}$ the incidence graph of the corresponding arrangement. In order to insert H_i, we first identify an edge E_0 in the current arrangement $\mathcal{A}(\mathcal{H}_i)$ that intersects H_i. Pick any edge E in the arrangement. If this edge intersects H_i, call it E_0. Otherwise, consider the edges in the arrangement $\mathcal{A}(\mathcal{H}_i)$ that are supported by the line that is the affine hull of E. Starting at E, we traverse these edges until we find an edge E_0 that intersects H_i. Let C be a cell in the current arrangement that is incident to E_0; the other such cell is denoted by C'. We find the other edge of C intersected by H_i (if it exists) by traversing all the edges on the boundary of C. Let us call this edge E_1.

Let us assume that E_1 exists (otherwise the description is only simpler). We update $\mathcal{G}$ by creating two new vertices $I_0 = E_0 \cap H_i$ and $I_1 = E_1 \cap H_i$. We insert the new edge $I_0 I_1$ into $\mathcal{G}$, replace E_0 by the two edges formed by the intersections of E_0 with the two half-spaces H_i^+ and H_i^- bounded by H_i, and similarly for E_1. We also replace C by the two cells $C \cap H_i^+$ and $C \cap H_i^-$. Finally, we update the incidence relationships and the sorted lists of edges on the boundary of $C \cap H_i^+$ and $C \cap H_i^-$.

If E_0 is not the only edge of C intersected by H_i, then we consider the cell other than C in the current arrangement that has E_1 as an edge. From neighbor to neighbor, we update all the cells in the current arrangement intersected by

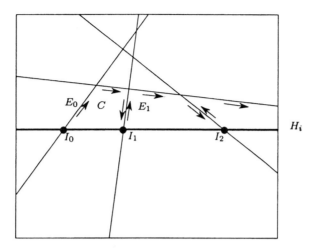

Figure 14.3. Updating $\mathcal{G}$.

the ray supported by H_i originating at I_0. The cells intersected by the opposite ray are obtained similarly, starting with C' instead of C, and this completes the update phase for $\mathcal{G}$. We easily verify that this update operation is carried out in time proportional to the number of edges in the zone of H_i.

Lemma 14.3.1 therefore implies the following theorem.

Theorem 14.4.1 *An arrangement of n lines in the plane may be computed incrementally in time $\Theta(n^2)$.*

14.4.2 The case of dimensions higher than 2

Let $\mathcal{A} = \mathcal{A}(\mathcal{H})$ be the arrangement of a set $\mathcal{H}$ of n hyperplanes in $\mathbb{E}^d$. As mentioned in the introduction to this section, we assume that this arrangement is simple. The algorithm described below shows how to build the 2-skeleton of $\mathcal{A}(\mathcal{H})$ knowing the 2-skeleton of the d-arrangement $\mathcal{A}' = \mathcal{A}(\mathcal{H} \setminus H)$ where H is some hyperplane in $\mathcal{H}$. We might also maintain the complete incidence graph of the faces of any dimension, but it turns out to be simpler to compute the incidence graph of the whole arrangement only after all the 0, 1, and 2-faces have been computed (see exercise 14.4).

Let F be a k-face of $\mathcal{A}'$. Since $\mathcal{A}$ is simple, F cannot be contained in H. F is said to be *active* if $F \cap H \neq \emptyset$. Note that a 0-face cannot be active.

If F is active, $F \cap H$ is a $(k-1)$-polytope whose $(k-2)$-faces correspond to the active $(k-1)$-faces of F. This fact implies the following lemma.

Lemma 14.4.2 *A k-face $(k > 1)$ of $\mathcal{A}'$ is active if and only if it is incident to an active $(k-1)$-face.*

Let $\mathcal{G}_2$ be the 2-skeleton of $\mathcal{A}'$. The active faces of $\mathcal{G}_2$ and the faces of $\mathcal{G}_2 \cap H$ are in one-to-one correspondence. But $\mathcal{G}_2 \cap H$ is the 1-skeleton of the simple $(d-1)$-arrangement formed in H by the intersection of the hyperplanes in $\mathcal{H}$ with H, hence $\mathcal{G}_2 \cap H$ is connected. This implies the following lemma.

Lemma 14.4.3 *The sub-graph of the incidence graph induced on the active 1-faces and 2-faces of $\mathcal{A}'$ is connected.*

The preceding two lemmas allow us to find all the active faces knowing only the active edges. The algorithm can now be described. It proceeds in three phases. We denote by H^+ and H^- the two half-spaces bounded by H.

Phase 1. Find an edge of $\mathcal{A}'$ that cuts H. For this, we start from any edge E in $\mathcal{A}'$. If E intersects H, we are done, otherwise we traverse the edges on the line Δ that is the affine hull of E. Let A be the vertex of E that is the closest to H, and E' the other edge on Δ that contains A as a vertex. Replacing E by E' and iterating eventually leads to the edge on Δ intersected by H.

Phase 2. Mark the 1 and 2-faces of $\mathcal{A}'$ intersected by H as active. This can be achieved by using a list $\mathcal{L}$ of active edges. The list initially contains the edge found in phase 1. While $\mathcal{L}$ is not empty, extract an edge E from it. All its incident 2-faces that have not yet been marked as active are considered in turn. Consider such a face C. Using the incidence graph, the edges of C are traversed until the other edge of C intersected by H is found. If no edge other than E intersects H, then skip to another 2-face. If such an edge is found, however, then it is inserted into $\mathcal{L}$ and marked as active. Phase 2 is over when the list $\mathcal{L}$ is empty and no other 2-face is to be considered. Lemma 14.4.3 shows that this traversal identifies all the active 1 and 2-faces.

Phase 3. Replace the active faces of $\mathcal{A}'$ and update the incidence relationships according to lemma 14.2.1. More precisely, let F be an active 2-face of $\mathcal{A}'$, E and E' the two incident active edges (the case where only one active edge is incident is handled similarly). Denote by H^+ and H^- the two half-spaces bounded by H (see figure 14.4). Create two new vertices $E \cap H$ and $E' \cap H$, five new edges $E_0 = F \cap H$, $E^+ = E \cap H^+$, $E^- = E \cap H^-$, $E'^+ = E' \cap H^+$, $E'^- = E' \cap H^-$, and update their incidence relationships. Create the 2-face $F \cap H^+$ incident to the edges of F contained in H^+, to E^+, E'^+, and to E_0. Similarly, create the 2-face $F \cap H^-$ incident to the edges of F contained in H^-, to E^+, E'^+, and to E_0.

It remains to create the 2-faces, which are intersections of 3-faces of $\mathcal{A}'$ and H. These 3-faces are not represented explicitly in the data structure. It is easy, however, to reconstruct the 2-faces of $\mathcal{A}'$ contained in H, as well as their

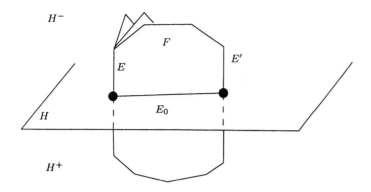

Figure 14.4. Updating the 2-skeleton of an arrangement.

incidence relationships, starting with the vertices and edges just created in H: merely observe that these vertices and edges form the 1-skeleton of a $(d-1)$-arrangement (see exercise 14.4).

The analysis of this algorithm is based on the zone theorem. Phase 1 examines only vertices contained in Δ and their incident edges. There are only $n - d + 1$ such vertices, and any vertex is contained in only $2d$ edges as was proved in lemma 14.2.4. The complexity of phase 1 is thus $O(n)$.

Phases 2 and 3 require time proportional to the number of faces traversed: these are the 0-, 1-, and 2-faces of the zone of a hyperplane. The zone theorem implies that the complexity of these phases is $\Theta(n^{d-1})$. We have thus proved the following theorem.

Theorem 14.4.4 *The simple d-arrangement of n hyperplanes may be computed incrementally in time* $\Theta(n^d)$.

14.5 Levels in hyperplane arrangements

14.5.1 Definitions

Consider the arrangement $\mathcal{A}$ of n hyperplanes $H_1, \ldots, H_n$ in $\mathbb{E}^d$, assumed to be simple, and let O be any point in $\mathbb{E}^d$ that does not belong to any hyperplane in $\mathcal{A}$. This point is called the *reference point*.

Let H_i^- be the open half-space bounded by H_i that does not contain O, and let $\mathcal{H}^-$ be the set of H_i^-, $i = 1, \ldots, n$. A point in $\mathbb{E}^d$ is said to be *at level k* if it belongs to k half-spaces in $\mathcal{H}^-$ (see figure 14.5). All the points in the relative interior of a face are at the same level: this level is by definition called the level of the face. The level of a face is in general different from the levels of its subfaces.

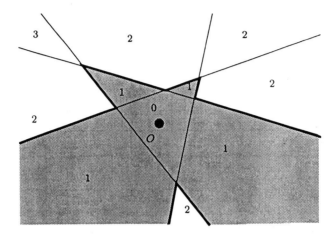

Figure 14.5. Levels of facets in an arrangement of lines. The surface of level 1 is shown in bold.

In particular, an i-face at level k is the intersection of $d - i$ hyperplanes; hence, a cell of $\mathcal{A}$ that contains F may be at any level between k and $k + d - i$.

The subcomplex of $\mathcal{A}$ formed by the cells at level k and by their faces is called the k-*level* of $\mathcal{A}$ and is denoted by $\mathcal{A}_k$. Note that a face on the k-level is not necessarily at level k. The set of faces of $\mathcal{A}$ formed by all the faces whose level is at most k is denoted by $\mathcal{A}_{\leq k}$; it is a sub-complex of $\mathcal{A}$ because the level of a face never exceeds the level of the faces that contain it.

The faces on the boundary of $\mathcal{A}_{\leq k}$ form a $(d-1)$-complex called the *surface of level k*. Note that the faces on the k-level are not necessarily on the surface of level k. In fact, a j-face on the surface of level k has a level l that satisfies

$$k - (d - j) < l \leq k$$

and conversely, a j-face of level l is on the surface of level k for some k that satisfies

$$l \leq k < l + d - j.$$

The polarity as defined in chapter 7 gives a notion equivalent to levels in the dual space. Each hyperplane H_i is mapped to a point H_i^* and a point M is mapped to a hyperplane M^*. Let $\mathcal{H}^*$ be the set of poles $\{H_i^*\}_{i=1,\ldots,n}$. From the equivalence

$$M \in H_i^- \iff H_i^* \in M^{*-}$$

(see lemma 7.1.9), it follows that M belongs to $\mathcal{A}_{\leq k}$ if and only if M^{*-} contains at most k points of $\mathcal{H}^*$. This dual version of the notion of a level is very useful as it leads to the notion of a k-*set* (see exercise 14.18) and allows any result on levels to be translated into a dual statement in terms of k-sets.

14.5.2 Combinatorial properties of levels

A rather direct application of the sampling theorem 4.2.3 bounds the number of faces in the complex $\mathcal{A}_{\leq k}$. Let us recall the context and the notation used in this theorem. Given a set $\mathcal{S}$ of n *objects, regions* each determined by a number (bounded by b) of objects, and a *conflict* relationship between objects and regions, the theorem bounds the number of regions $|\mathcal{F}_{\leq k}(\mathcal{S})|$ determined by the objects in $\mathcal{S}$ that conflict with at most k objects, as a function of the expected number $f_0(\lfloor n/k \rfloor, \mathcal{S})$ of regions defined and without conflict over a random sample in $\mathcal{S}$ of size $\lfloor n/k \rfloor$.

Consider the arrangement $\mathcal{A} = \mathcal{A}(\mathcal{H})$ of a set $\mathcal{H}$ of n hyperplanes in $\mathbb{E}^d$. Without loss of generality, we suppose that the arrangement is simple, since this case maximizes the number of faces in $\mathcal{A}_{\leq k}$. (This can be shown using a perturbation argument analogous to the one given in the proof of theorem 14.2.5). The objects are defined to be the hyperplanes, and the regions are d-tuples of hyperplanes. Thus $b = d$ and the regions defined over $\mathcal{H}$ correspond to vertices of $\mathcal{A}$. We say that a hyperplane H conflicts with a region R if the intersection point of the d hyperplanes in R belongs to the open half-space H^- bounded by H that does not contain O.

Regions defined and without conflict over a set of hyperplanes correspond to vertices of the polytope given as the intersection of the half-spaces that are bounded by those hyperplanes and contain O. If a polytope has r facets, however, the upper bound theorem 7.2.5 states that it has at most $O(r^{\lfloor d/2 \rfloor})$ vertices. Thus $f_0(\lfloor n/k \rfloor, \mathcal{S})$ is bounded above by $O((n/k)^{\lfloor d/2 \rfloor})$. The sampling theorem 4.2.3 shows that the number of vertices of $\mathcal{A}$ at levels at most k is bounded by $O(n^{\lfloor d/2 \rfloor} k^{\lceil d/2 \rceil})$ for $2 \leq k \leq \frac{n}{d+1}$. For other values of k, see remark 1 after theorem 4.2.3. Lemma 14.2.4 can then be used to extend this result to faces of higher dimensions.

Theorem 14.5.1 *The number of faces whose level is at most k ($k \geq 1$) in the d-arrangement of n hyperplanes is bounded by $O(n^{\lfloor d/2 \rfloor} k^{\lceil d/2 \rceil})$.*

This bound is tight as is shown by exercise 14.20.

To bound the complexity of a single level is a much more demanding task. A bound in the planar case is given in exercise 14.19.

14.5.3 Computing the first k levels in an arrangement

In this subsection, we present an on-line algorithm that computes the complex $\mathcal{A}_{\leq k}$ of the first k levels in the d-arrangement $\mathcal{A}$ of a set $\mathcal{H}$ of n hyperplanes in general position.

The algorithm we present now is an incremental on-line algorithm that uses the influence graph method. The words on-line refer to the fact that the algorithm maintains a representation of the arrangement each time a hyperplane is inserted and that it assumes no preliminary knowledge about the hyperplanes to be inserted. This algorithm is deterministic. We give a randomized analysis assuming that the order of insertion of the hyperplanes is random.

We choose to represent the arrangement using the incidence graph of the 0, 1, and 2-faces of $\mathcal{A}_{\leq k}$, and maintain this representation under repeated insertions. Using this description, it is easy to rebuild the total incidence graph and all the faces of $\mathcal{A}_{\leq k}$ (see exercise 14.4).

The framework

We wish to use the framework presented in chapter 4 for the construction of the first k levels. For this, we redefine the problem in terms of objects, regions, and conflicts in a way that differs slightly from that explained in the previous subsection.

The objects are hyperplanes, and regions are line segments. A region is defined over a set $\mathcal{H}$ of hyperplanes if its affine hull is the intersection of $d-1$ hyperplanes of $\mathcal{H}$ and if each endpoint of the line segment is the intersection of d hyperplanes of $\mathcal{H}$. Note that assuming general position implies that each region is defined by a unique subset of $d+1$ hyperplanes. An object and a region conflict if they intersect. Edges in the arrangement $\mathcal{A}$ correspond exactly to regions defined and without conflict over the set $\mathcal{H}$ of hyperplanes.

The algorithm inserts the hyperplanes one by one. At the current step of the construction, the complex $\mathcal{A}'_{\leq k}$ of the first k levels in the arrangement $\mathcal{A}'$ of the current set of hyperplanes is available as the incidence graph of its 0, 1, and 2-faces. Moreover, each edge and each 2-face at level k are marked and a special pointer K gives access to a face at level k.

We now describe what happens when a new hyperplane H is inserted into this structure. The algorithm looks for all the edges of $\mathcal{A}'_{\leq k}$ that conflict with the hyperplane H being inserted, and updates the incidence graph. To quickly find the conflicting edges, the algorithm maintains an influence graph (see section 5.3). Recall that the influence graph is a structure whose goal is to detect the conflicts between any object not necessarily in the current set and the regions defined and without conflict over the current set. It is an oriented graph without cycles that has a node for each region that was defined and without conflict over the current subset of hyperplanes at some previous incremental step. The influence domain of a node is the subset of the objects that conflict with the region corresponding to that node. Arcs in the influence graph connect the nodes so that the influence domain of any node is contained in the union of the influence domains of its parents. In the influence graph we use here, there is a node for each edge at level $\leq k$ in the current arrangement. We observe that, in contrast with other randomized algorithms given in this book, the algorithm is interested in maintaining a subset only of the regions defined and without conflict (the edges of level at most k).

In a first step, the first d hyperplanes are inserted, the incidence graph of the 0, 1, and 2-faces in the first k levels of their arrangement is computed, and the corresponding influence graph is initialized accordingly.

In the current step, a new hyperplane H is inserted. The incremental step consists of a location phase, and of an update phase which can itself be split into three phases.

Locating. The location phase, using a simple traversal of the influence graph, identifies all the nodes in the graph that conflict with H, which correspond to the edges in $\mathcal{A}'_{\leq k}$ intersected by H. If no edge in $\mathcal{A}'_{\leq k}$ is intersected by H, then the hyperplane H does not contain a face in the current complex $\mathcal{A}'_{\leq k}$ of the first k levels, nor in any of the subsequent complexes. So the algorithm may skip to the next incremental step.

Updating. 1. Creating the new faces. In the location phase, all the edges of $\mathcal{A}'_{\leq k}$ intersected by H have been found. Each such edge is cut by H into two parts and generates two new edges $E \cap H^+$ and $E \cap H^-$, where H^+ is the open half-space bounded by H that contains the reference point O, and H^- is the opposite half-space (see figure 14.6). Let us say that these new edges are of *type 1*. Each 2-face F of $\mathcal{A}'_{\leq k}$ intersected by H also generates a new edge, $F \cap H$. Let us say that this new edge is of *type* 2. We compute such an edge in the following fashion. Let E be an edge in $\mathcal{A}'_{\leq k}$ that is intersected by H, and A its intersection point with H. Then all the 2-faces incident to E are intersected by H and generate an edge of type 2. Consider such a 2-face F. Following the part of the boundary of F that is contained in H^-, we can look for the other edge E' incident to F that is intersected by H. If E' does not exist, this means that the edge of type 2 created by F is a semi-infinite ray originating at A. Otherwise, we find an edge E' incident to F and intersecting H at a point B. Then the edge of type 2 created by F is the segment AB. In either case, the incidence relationships between these new edges and vertices are easily taken care of.

We must also update the 2-faces of $\mathcal{A}'_{\leq k}$. We have just seen how to find the 2-faces intersected by H. Such a face F is replaced by the two new faces $F \cap H^+$ and $F \cap H^-$, which we may again call of type 1, and the incidence relationships between these 2-faces and the new edges are updated correspondingly. More tricky is the case of new 2-faces, which we may call of *type 2*, appearing as the intersection of a 3-face with H. Even though the 3-faces are not stored explicitly in the structure, it is not difficult (see exercise 14.4) to reconstruct the 2-faces of type 2 and their incidence relationships, observing that their boundary is made

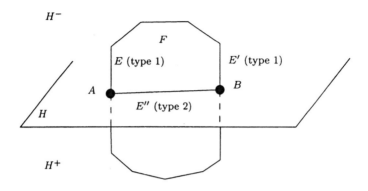

Figure 14.6. Updating.

up of edges that are contained in H and thus of type 2, and these edges have all been created and their incidence relationships have been computed (see exercise 14.4).

Finally, the levels of the faces have changed by at most 1. We update the markers for the edges and for the 2-faces at level k, and collect the created edges that are at level $k+1$ into a list L_{k+1} of edges. More precisely, an edge E at level k in $\mathcal{A}'_{\leq k}$ that is intersected by H generates two new edges of type 1 at levels k and $\overline{k}+1$ respectively. The latter is inserted into L_{k+1}. Similarly, a 2-face at level k in $\mathcal{A}'_{\leq k}$ that is intersected by H generates two new faces of type 1, at levels k and $k+1$ respectively. They share a common incident edge (of type 2) that is at level k and so is marked. Note that the new edges of type 1 obtained as the intersection of H^- and of edges at level $k-1$ in $\mathcal{A}'$ are also at level k in $\mathcal{A}$. So are the edges at level $k-1$ in $\mathcal{A}'$ entirely contained in H^-. Nevertheless, none have been marked yet. These marks will be given in the next phase.

2. Peeling faces at levels greater than k. The faces created in the previous phase do not all belong to $\mathcal{A}_{\leq k}$. The peeling process removes all the faces at levels greater than k from the incidence graph. Of course, this process is not needed if the current subset of hyperplanes has fewer than $k+d-1$ members. The faces to be removed consist of faces that have been created during this incremental step, or of faces of $\mathcal{A}'_{\leq k}$ that are contained in H^-.

As we recall, the algorithm keeps a pointer K to some face at level k in $\mathcal{A}'_{\leq k}$. The peeling process traverses the incidence graph of the set $\mathcal{A}_{k+1}$ of constructed faces at levels greater than k. These faces are necessarily contained in H^-. If L_{k+1} is empty and if the face pointed to by K is contained in H^-, then $\mathcal{A}_{k+1}$ does not intersect H, and its incidence graph is connected. The pointer K may be used as the starting point for a traversal of $\mathcal{A}_{k+1}$. Otherwise, the incidence graph of $\mathcal{A}_{k+1}$ may have several connected components. Nevertheless, each connected component contains one of the new edges of type 1 stored in the list L_{k+1}, which may be used as a starting point for a traversal of this connected component. During the traversal, we can also mark the faces at level k that have not been marked yet, since these faces are incident to at least one face of $\mathcal{A}_{k+1}$.

3. Updating the influence graph. We must also show how to update the influence graph. A new node is created in the graph for each new edge that was not removed during the peeling process. A node corresponding to a new edge of type 1 becomes the child of the node corresponding to the edge of $\mathcal{A}'_{\leq k}$ that contains it. A node corresponding to a new edge of type 2, obtained by intersecting a 2-face F in $\mathcal{A}'_{\leq k}$ with H, has for parents all the nodes corresponding to edges of F either contained in H^- or intersecting H.

A randomized analysis of the algorithm

Let us first estimate the number of nodes in the influence graph. We first bound the number of edges created by the algorithm, by bounding the average number of vertices in the whole arrangement that were created by the algorithm at some point, then using the fact that a vertex is incident to $2d$ edges.

Lemma 14.5.2 *The probability p_j that a vertex at level j in $\mathcal{A}$ is created by the algorithm satisfies*

$$if\ j \leq k,\ \ p_j\ =\ 1$$

$$if\ j > k,\ \ p_j\ =\ \frac{\binom{j}{k}}{\binom{j+d}{k+d}} \leq \left(\frac{k+d}{j}\right)^d.$$

Proof. The proof is trivial if $j \leq k$, since the algorithm always creates all the vertices at levels at most k. Let P be a vertex in the arrangement, let $\mathcal{D}$ be the set of the d hyperplanes that intersect at P, and let $\mathcal{C}$ be the set of the j hyperplanes that determine the level of P, meaning that they separate P from the origin O. If $j > k$, then P is created if and only if, among the hyperplanes in $\mathcal{D} \cup \mathcal{C}$, the first $k+d$ inserted by the algorithm all belong to $\mathcal{D}$. This happens exactly with the probability stated above. $\qquad\square$

Let us denote by $\mathcal{S}_j$, resp. $\mathcal{S}_{\leq j}$, the set of vertices at level j, resp. at most j, in $\mathcal{A}$. The expected number s of vertices created by the algorithm is thus

$$s\ =\ \sum_{j=0}^{n-d} |\mathcal{S}_j| p_j$$

$$= |\mathcal{S}_0| p_0 + \sum_{j=1}^{n-d} (|\mathcal{S}_{\leq j}| - |\mathcal{S}_{\leq j-1}|) \, p_j$$

$$= \sum_{j=0}^{n-d-1} |\mathcal{S}_{\leq j}| \, (p_j - p_{j+1}) + |\mathcal{S}_{\leq n-d}| \, p_{n-d}.$$

Theorem 14.5.1 and lemma 14.5.2 can then be used to bound the expected number of vertices created by the algorithm, yielding

$$s = O(n^{\lfloor d/2 \rfloor}) \sum_{j=k}^{n-d-1} j^{\lceil d/2 \rceil} p_j \left(1 - \frac{j+1}{j+1+d}\right) + O(n^d) \left(\frac{k+d}{n-d}\right)^d$$

from which it follows that

$$s = k^{\lceil d/2 \rceil} O(n^{\lfloor d/2 \rfloor}) \quad \text{for any } k \neq 0.$$

Each time a vertex is created, at most $2d$ incident edges are created. Thus, the expected number of edges created by the algorithm and hence the expected number of nodes in the influence graph are both bounded similarly by $O(n^{\lfloor d/2 \rfloor} k^{\lceil d/2 \rceil})$.

To bound the number of arcs in the influence graph, we observe that whenever the node of a child corresponds to an edge of type 2, the level of the edge corresponding to the parent node increases by 1. The level of a face cannot be greater than $k+1$ otherwise it is removed from the graph, so that a node in the graph cannot have more than k children of type 2, and so not more than $k+1$ children overall. The total space required to store the influence graph is thus $O(n^{\lfloor d/2 \rfloor} k^{\lceil d/2 \rceil + 1})$.

Updating the incidence graph requires as many elementary operations as there are arcs in the created incidence graph. This number is at most k times the number of created edges, since a created edge is traversed at most k times and has at most k children.

The number of faces traversed during the peeling phase is proportional to the number of faces at level $k+1$, which must have been created in a previous incremental step.

Updating the influence graph has a complexity proportional to the number of arcs in the final graph. Hence, the total combined cost of all the update operations is bounded by the overall number of arcs created, which we have shown to be $O(n^{\lfloor d/2 \rfloor} k^{\lceil d/2 \rceil + 1})$.

It remains to estimate the cost incurred by the location phases. Each node in the influence graph having at most $k+1$ children, the number of nodes visited during the location phases is bounded by $k+1$ times the number of conflicts detected during these phases. To estimate the expected number of conflicts detected by the algorithm, we first compute the probability p_{ji} that an edge with endpoints at levels j and $j - i$ is created at some incremental step. Consider a vertex S at level j in the arrangement $\mathcal{A}$, and D one of the d lines supporting an edge incident to S. This line D is the intersection of $d - 1$ hyperplanes among the d hyperplanes that intersect at S. The ray D_S supported by D, that originates from S towards the lower levels in the arrangement, intersects exactly j hyperplanes $H_1, H_2, \ldots, H_j$, ordered according to their intersections $M_l = D_S \cap H_l$ with D_S, $l = 1, \ldots, j$. If the edge SM_i is created during the execution of the algorithm, the corresponding node in the graph will be found to conflict in subsequent location steps at most $i - 1$ times. The condition for the edge SM_i to be created is that at some incremental step in the execution of the algorithm, one of the $d + 1$ hyperplanes that determine this edge is inserted while hyperplanes inserted at previous steps include the d other hyperplanes that determine SM_i but not the hyperplanes H_i' indexed by $i' < i$ nor more than $l \leq k - 1$ of the $j - i$ hyperplanes H_i' indexed by $i' > i$. With this characterization, the probability p_{ji} that SM_i is created can be written as

$$\text{if } j \leq k, \ p_{ji} \ = \ \frac{(d+1)!(i-1)!}{(d+i)!},$$

$$\text{if } j > k, \ p_{ji} \ = \ \sum_{l=0}^{k-1} \frac{\binom{j-i}{l}}{\binom{j+d}{l+d+1}} \frac{d+1}{l+d+1}$$

$$= \ \sum_{l=0}^{k-1} \frac{\binom{j-i}{l}}{\binom{j+d}{l+d}} \frac{d+1}{j-l}.$$

Observing that

$$\frac{(d+1)!(i-1)!}{(d+i)!} = \sum_{l=0}^{j-i} \frac{\binom{j-i}{l}}{\binom{j+d}{l+d}} \frac{d+1}{j-l}$$

and with the convention that $\binom{j-i}{l}$ vanishes for $l > j - i$, we can write in either case

$$p_{ji} = \sum_{l=0}^{k-1} \frac{\binom{j-i}{l}}{\binom{j+d}{l+d}} \frac{d+1}{j-l}.$$

The average number c of conflicts detected by the algorithm is then

$$c = \sum_{j=0}^{n-d} |S_j| d \sum_{i=1}^{j} (i-1) p_{ji}.$$

We are led to compute $g_j = \sum_{i=1}^{j} (i-1) p_{ji}$.

$$g_j = \sum_{i=1}^{j} (i-1) \sum_{l=0}^{k-1} \left[\frac{\binom{j-i}{l}}{\binom{j+d}{l+d}} \frac{d+1}{j-l} \right].$$

An elementary computation shows that

$$\sum_{i=1}^{j} (i-1) \binom{j-i}{l} = \sum_{i=0}^{j-1} (j-1-i) \binom{i}{l}$$

$$= \sum_{m=0}^{j-1} \sum_{i=0}^{m-1} \binom{i}{l} = \sum_{m=0}^{j-1} \binom{m}{l+1}$$

$$= \binom{j}{l+2},$$

so that

$$g_j = \sum_{l=0}^{k-1} \frac{d+1}{j-l} \frac{\binom{j}{l+2}}{\binom{j+d}{l+d}}$$

$$= \sum_{l=0}^{k-1} (d+1) \frac{j-l-1}{j+d} \frac{j!}{(j+d-1)!} \frac{(l+d)!}{(l+2)!}.$$

Put

$$g'_j = \sum_{l=0}^{k-1} (d+1) \frac{j!}{(j+d-1)!} \frac{(l+d)!}{(l+2)!}.$$

Since $g_j \le g'_j$, we can restrict our attention to g'_j, for which we have

$$g'_j = \frac{d+1}{d-1} \frac{1}{\binom{j+d-1}{d-1}} \sum_{l=0}^{k-1} \binom{l+d}{d-2} = \frac{d+1}{d-1} \frac{\binom{k+d}{d-1}}{\binom{j+d-1}{d-1}},$$

and this yields finally

$$g'_j \leq \frac{d+1}{d-1} \left(\frac{k+d}{j}\right)^{d-1}.$$

We can perform an Abel transform analogous to that giving the expected number of vertices created during the execution of the algorithm, to obtain

$$
\begin{aligned}
c &\leq \sum_{j=0}^{n-d} |\mathcal{S}_j| d g'_j \\
&\leq \sum_{j=k}^{n-d-1} d |\mathcal{S}_{\leq j}| \, (g'_j - g'_{j+1}) + d |\mathcal{S}_{\leq n-d}| g'_{n-d},
\end{aligned}
$$

from which we derive, using theorem 14.5.1 once again, that the expected number of conflicts detected by the algorithm is $O(nk \log(n/k))$ if $d = 2$, $O(nk^2 \log(n/k))$ if $d = 3$, and $O(n^{\lfloor d/2 \rfloor} k^{\lceil d/2 \rceil})$ if $d \geq 4$. The overall complexity of all the location sub-phases in all incremental steps is thus $O(nk^2 \log(n/k))$ if $d = 2$, $O(nk^3 \log(n/k))$ if $d = 3$, and $O(n^{\lfloor d/2 \rfloor} k^{\lceil d/2 \rceil+1})$ if $d \geq 4$.

The algorithm that computes the first k levels in the d-arrangement of n hyperplanes, as described in this section, therefore has complexity $O(n^{\lfloor d/2 \rfloor} k^{\lceil d/2 \rceil+1})$.

This result may be somewhat improved as in exercise 14.21. A reference is given in the bibliographical notes. This proves the following theorem.

Theorem 14.5.3 *The first k levels in the simple d-arrangement of n hyperplanes can be computed in time $O(n^{\lfloor d/2 \rfloor} k^{\lceil d/2 \rceil})$ if $d \geq 4$, and in time $O(nk^{\lceil d/2 \rceil} \log \frac{n}{k})$ if $d = 2$ or 3.*

14.6 Exercises

Exercise 14.1 (Projective arrangements and polytopes) Prove all the results in section 14.2 using the combinatorial results on polytopes derived in chapter 11.

Hint: Let $\mathcal{H}$ be a set of hyperplanes in $\mathbb{E}^d$. With each hyperplane in $\mathcal{H}$ we associate an oriented projective hyperplane, and we denote by $\vec{\mathcal{H}}$ the set of these oriented projective hyperplanes. In the spherical model of the oriented projective space, each projective oriented hyperplane is represented by a great $(d-1)$-sphere in S^d. The arrangement of these great $(d-1)$-spheres on S^d is called the oriented projective arrangement of $\vec{\mathcal{H}}$. The projective arrangement of $\mathcal{H}$ is obtained by identifying antipodal points on the sphere. The oriented projective arrangement is a spherical polytope drawn on the sphere S^d to which the same relations as a polytope apply. To each k-face in the projective arrangement of $\mathcal{H}$ there corresponds two k-faces in the oriented projective arrangement, so the formulae for projective arrangements can be obtained by halving the analogous formulae for polytopes. In particular, Euler's relation for a projective arrangement is

$$\sum_{k=0}^{d}(-1)^k p_k(\mathcal{A}) = \frac{1 + (-1)^d}{2}$$

if p_k is the number of k-faces in the arrangement.

The formulae for arrangements in Euclidean space can be obtained by observing that the number of bounded k-faces is the same in Euclidean and in projective arrangements. Unbounded k-faces, however, correspond to two faces in the Euclidean arrangement (one on each side of the hyperplane at infinity), but also to the $(k-1)$-faces on the hyperplanes at infinity. Reversing the argument, two unbounded opposite k-faces in the Euclidean arrangement can be accounted for by the enclosing k-face in the projective arrangement and by the $(k-1)$-face in the projective $(d-1)$-arrangement, so that finally

$$n_k^{(d)} = p_k^{(d)} + p_{k-1}^{(d-1)}.$$

Exercise 14.2 (Dehn–Sommerville relations) Show that the following relations are satisfied for simple d-arrangements:

$$\sum_{i=0}^{k}(-1)^j 2^{k-i} \binom{d-i}{d-k} n_i = n_k.$$

Hint: Use the correspondence described in exercise 14.1 between arrangements and spherical polytopes. The exercise follows from an easy adaptation of the proof of theorem 7.2.2, using lemma 14.2.4 in place of lemma 7.1.14.

Exercise 14.3 (Number of faces) Unlike polytopes, the number of k-faces of a simple d-arrangement of n hyperplanes depends only on n, d, and k, and not on the relative positions of the hyperplanes. Show that the number $p_k(n, d)$ of k-faces in the d-arrangement of n projective hyperplanes (see exercise 14.1) satisfies

$$p_k(n, d) = \binom{n}{d-k} \sum_{i=0}^{k} \binom{n-d-1+k}{i}.$$

Give a similar formula for Euclidean arrangements.

It should not, however, be believed that all arrangements are combinatorially equivalent (have isomorphic incidence graphs). Disprove this by drawing two simple arrangements of lines in the plane or in the projective plane for which no bijection between the two sets of faces preserves the incidence relationships.

Exercise 14.4 (Skeleton of an arrangement) Show that the entire d-arrangement of n hyperplanes in general position may be reconstructed from only the 1-skeleton of this arrangement, that is, only the vertices, edges, and their incidence relationships. Show that this reconstruction can be achieved in optimal $O(n^d)$ time.

Hint: Proceed by rebuilding the faces in order of increasing dimension. Under general position assumptions, a k-face F incident to a $(k-1)$-face G is obtained by relaxing a hyperplane. To know on which side of this hyperplane F is contained, observe that F is contained in the intersection of $d-k$ half-spaces bounded by all but one of the $d-k+1$ hyperplanes whose intersection contains G. When creating a k-face F, we update the incidence relations between F and its sub-faces of dimension $k-1$ (already created). The incidence relations between F and the faces of dimension $k+1$ will be taken care of when processing the $(k+1)$-faces.

Exercise 14.5 (Incidences along a line) Consider a set $\mathcal{H}$ of n hyperplanes, and a line L that is the intersection of $d-1$ hyperplanes in $\mathcal{H}$. Show that the number of edges in the d-arrangement $\mathcal{A}(\mathcal{H})$ that are incident to vertices lying on L is $O(n^{d-1})$.

Exercise 14.6 (Incremental construction) Generalize the algorithm of section 14.4 to compute non-simple arrangements within the same time bounds.

Exercise 14.7 (Canonical triangulation) The canonical triangulation of an arrangement of hyperplanes is defined as follows. We triangulate the faces inductively in order of increasing dimension. A k-face F is triangulated by joining its vertex S_F of smallest d-th coordinate to the $(k-1)$-simplices in the triangulation of the $(k-1)$-faces of F that do not contain S_F. Show that this triangulation has the same complexity as the whole arrangement and that it can be computed in time $O(n^d)$. Show that the number of simplices in this triangulation that are crossed by any given hyperplane is $O(n^{d-1})$.

Exercise 14.8 (Zonotopes) A zonotope is a polytope obtained as the Minkowski sum of a finite set of line segments (see exercise 7.17). Find a bijective transformation that maps the faces of a zonotope $\mathcal{Z}$ in $\mathbb{E}^d$, the sum of n line segments, to the faces of the projective $(d-1)$-arrangement of a set $\mathcal{H}$ of n hyperplanes in $\mathbb{E}^{d-1}$ and which preserves incidence relationships. Using this, give an upper bound on the complexity of a zonotope and an algorithm that computes it, knowing the n segments that define the zonotope. To what does the zone of a hyperplane H in the arrangement correspond in the zonotope?

Hint: Let $S_1, \ldots, S_n$ be the n segments that define $\mathcal{Z}$. Assume that the segments are centrally symmetric through the origin, by translating them if necessary, so that S_i has endpoints A_i and $-A_i$. Denote by $\mathcal{Z}^{\#}$ the polytope polar to $\mathcal{Z}$. From 7.17, deduce that

$$\mathcal{Z}^{\#} = \{X : \sum_{i=1}^{n} |X \cdot A_i| \leq 1\}.$$

Let F be the face defined by

$$F = S_{i_1} \oplus \cdots \oplus S_{i_r} + \varepsilon_{i_r+1} A_{i_r+1} + \cdots + \varepsilon_{i_n} A_{i_n}, \tag{14.9}$$

where $\varepsilon_{i_j} = \pm 1$, $j = r+1, \ldots, n$. Show that the face $F^{\#}$ is

$$F^{\#} = \{X \in \mathcal{Z}^{\#} : X \cdot A_{i_j} = 0 \text{ for } j = 1, \ldots, r, \text{ and } X \cdot (A_{i_r+1} + \cdots + A_{i_n}) = 1\}.$$

From this, it follows that the faces of $\mathcal{Z}^{\#}$ corresponding to faces of $\mathcal{Z}$ that contain a translate of S_i are themselves contained in the hyperplane $H_i = \{X \in \mathbb{E}^d : X \cdot A_i = 0\}$. The correspondence between d-arrangements of hyperplanes passing through the origin and projective $(d-1)$-arrangements finishes the proof.

Exercise 14.9 (The painter's problem) A painter has n buckets at his disposal, each containing a mixture of some basic colors $C_1, \ldots, C_p$. The i-th mixture is characterized by the proportions of each color, which can be modeled as a point S_i in a space $\mathbb{E}^p$. A *product* is obtained by blending some mixtures together. Show that the set of *feasible* products, meaning that they can be derived from the mixtures in the n buckets, can be characterized by the zonotope $\mathcal{Z}$ (see exercise 14.8) built on the segments S_i. Design an algorithm that computes the feasibility of a product using only the mixtures in the n buckets. Note that the solution is not unique in general. In the case of binary blends ($p = 2$), there is an optimal way to compose a product M such that the *residual* set of products, which are still feasible after the necessary quantities of mixtures have been used to make M, is maximal with respect to inclusion. Devise an algorithm that computes this optimal mixture.

Hint: After the product M is created, the residual set of feasible products is contained in $\mathcal{Z}_r = \mathcal{Z} \cap (\mathcal{Z} - M)$. If $p = 2$, then $\mathcal{Z}_r$ is a zonogon (2-zonotope).

Exercise 14.10 (Ray shooting) Consider a set of n disjoint line segments in the plane. A ray shooting query, given a point and a vector, asks for the first segment visible from this point in the direction given by the vector. Equivalently, given a ray, the query asks for the first segment that intersects this ray. Show how to preprocess a data structure in time and space $O(n^2)$, so that ray shooting queries can be answered in time $O(\log n)$.

Hint: Let D be the line that supports the ray. Using polarity, show that finding the segments intersected by D is equivalent to locating the point D^* dual to D in an arrangement $\mathcal{A}$ of $2n$ lines. Show also that, for any point Δ^* in a cell of $\mathcal{A}$, the order in which the segments intersect the dual line Δ is identical. Thus, to a cell A in $\mathcal{A}$ corresponds an ordered list $L(A)$ of segments. If we store this list in an auxiliary structure (such as a dictionary or a balanced binary tree), then once the cell containing D^* is found, the ray shooting query can be answered in additional time $O(\log n)$. For locating that cell, we use the structure described in exercise 12.2. The size and preprocessing time of the entire data structure is thus $O(n^3)$, for a query time of $O(\log n)$. To save a factor n, we must store the lists more compactly. Notice that these lists differ only by one element for two adjacent cells. We may therefore use a persistent dictionary (see exercise 2.7).

Exercise 14.11 (Queries in the plane) Consider a set $\mathcal{P}$ of n points in the plane. Show how to preprocess $\mathcal{P}$ into a structure using space $O(n)$, so as to retrieve a point belonging to any half-plane H^+, bounded by a query line H and containing the origin, in time $O(\log n)$ per query. Show that all these points may then be retrieved in time proportional to their number.

Hint: Peel the consecutive layers $\mathcal{E}_0, \ldots, \mathcal{E}_k$ of $\mathcal{P}$ (as is done in exercise 9.3), and build their vertical decomposition. Polarity with respect to a point inside $\mathcal{E}_k$ transforms these nested polygons into another set of nested polygons $\mathcal{E}'_0, \ldots, \mathcal{E}'_k$ and H into a point H^*. The point-location structure of exercise 12.2 can be used to find in logarithmic time the greatest i such that H^* does not belong to $\mathcal{E}'_i$, which is also the smallest i such that H intersects $\mathcal{E}_i$. A point on $\mathcal{E}_i$ that belongs to H^+ can be found in logarithmic time as well, and the other points can be retrieved by following the boundary chain of a layer or by using the vertical decomposition to go from layer to layer.

Exercise 14.12 (Higher-dimensional queries) Given a set P of n points and a hyperplane H in $\mathbb{E}^d$, explain how to build a data structure of size $O(n^{\lfloor d/2 \rfloor + \varepsilon})$ that finds in logarithmic time a point in P that belongs to a half-space H^+ bounded by a query hyperplane H and containing the origin. Show how to retrieve all these points in time $O(k \log n)$, if k is their number.

Hint: Consider again the dual problem, by using the polarity with respect to the point $(0, \ldots, 0, +\infty)$. The problem is thus to find the hyperplanes dual to the points in P lying above the pole H^* of H. Build a hierarchical structure with $O(\log n)$ layers. The topmost layer represents the canonical triangulation (see exercise 14.7) of the cell at level 0 (with respect to the center of the polarity) in the arrangement of a small sample of the dual hyperplanes. (These cells being unbounded, the d-simplices in the triangulation are more appropriately cylinders based on $(d-1)$-simplices.) Using the tail estimates of exercises 4.5 and 4.6, show that if $\mathcal{H}$ is the set of hyperplanes and $\mathcal{R}$ a random sample of constant size r drawn from $\mathcal{H}$, then with high probability no cylinder in the canonical triangulation of the cell at level 0 in the arrangement of $\mathcal{R}$ intersects more than $O(\frac{n}{r} \log r)$ hyperplanes in $\mathcal{H} \setminus \mathcal{R}$. Taking r big enough, show that the recurrence on the size of the resulting structure solves to $O(n^{\lfloor d/2 \rfloor + \varepsilon})$ (where ε depends on r), and that the structure has $O(\log n)$ layers.

To answer a query, we traverse the structure from the first layer and recursively determine all the hyperplanes lying above H^*. Let $\mathcal{R}$ be the sample attached to the current layer and let S be the cylinder in the triangulation of the arrangement of $\mathcal{R}$ that is intersected by the vertical ray originating at H^*. If H^* belongs to S, then we recursively determine all the hyperplanes crossing S that lie above H^* in S. Otherwise, we systematically test H^* against the n hyperplanes, in time $O(n)$. Let k be the number of hyperplanes lying above H^*. Point H^* lies in no simplex of the canonical triangulation of the cell at level 0 in the arrangement of $\mathcal{R}$ if and only if $\mathcal{R}$ contains at least one of the k hyperplanes lying above H^*. This happens with probability $O(k/n)$ since $\mathcal{R}$ is of size $r = O(1)$. Since the structure has $O(\log n)$ layers, the expected time for a query is $O(k \log n)$ as wanted.

Exercise 14.13 (Computing a zone) Show how to compute the zone of a hyperplane in the arrangement of n hyperplanes in $\mathbb{E}^d$ in time $O(n^{d-1} + n \log n)$ and space $O(n^{d-1})$.

Hint: Adapt the algorithm that computes a cell in the arrangement of segments or triangles as described in section 15.4 and subsection 16.4.3, and use the canonical triangulation of the arrangement defined in exercise 14.7.

Exercise 14.14 (Convex hull of an arrangement) Given n lines, show how to compute the convex hull of all the vertices of their arrangement in $O(n \log n)$ time.

Hint: Only the two extreme vertices on each line may be on the convex hull, so this convex hull has complexity $O(n)$. Computing the zone of the line at infinity (see exercise 14.13) and removing the infinite edges yields a simple polygonal line with $O(n)$ vertices, whose convex hull is exactly the convex hull of the arrangement.

Exercise 14.15 (Sorting by polar angle) Consider a set of n points $M_1, \ldots, M_n$ in the plane. We can define a partial order on the lines $M_i M_j$ passing through any two points M_i, M_j, in the following way: $D = M_i M_j < D' = M_i M_k$ if the slope of D is smaller than that of D'. Show that one can compute a total order on these lines that is compatible with the partial order in optimal time $O(n^2)$.

Hint: We reason in the dual space where an oriented line D of equation $y \cos \theta - x \sin \theta - \delta = 0$ ($\theta \in [-\pi, +\pi[$) is represented by the point (θ, δ). To each point M_i corresponds the pencil of lines passing through this point, and the dual of that pencil is represented by the line dual to M_i. Thus the duals of the points $M_1, \ldots, M_n$ form an arrangement of dual lines. A breadth-first search traversal of the oriented graph of the vertices of this arrangement whose arcs correspond to the edges of the arrangement oriented towards increasing abscissae yields the desired total order.

Exercise 14.16 (Visibility graphs) Consider a set $\mathcal{S}$ of n segments in the plane with disjoint interiors. The *visibility graph* of these segments is the graph whose nodes are endpoints of segments and whose arcs join two nodes that are visible one from another, meaning that the segment joining these two endpoints does not cross any of the other segments. Show that the visibility graph of $\mathcal{S}$ can be computed in time $O((n+k) \log n)$, where k is the size of the visibility graph. When k is high, show how to improve the complexity to $O(n^2)$ which is optimal in the worst-case.

Hint: Compute the downward vertical decomposition of the segments by erecting walls *hanging below* each endpoint. Then rotate the direction of the decomposition from $-\pi/2$ to $3\pi/2$ while maintaining the oriented decomposition. The visibility graph can be computed in the process, since two visible endpoints will share a trapezoid for some orientation. In order to process the events during the rotation in the correct order, sort them in a priority queue (as was done for computing segment intersections by a sweep algorithm in subsection 3.2.2). In order to achieve time $O(n^2)$, sort all the lines that connect two endpoints by polar angle as in exercise 14.15.

Exercise 14.17 (Shortest paths) Let $\mathcal{P}$ be a polygonal region whose boundary is made up of one or several polygons with a total of n edges. Let I and F be two points of $\mathcal{P}$. Show how to compute a shortest path that connects I to F and that is contained inside $\mathcal{P}$ in time $O(n^2)$.

Hint: Use the visibility graph of exercise 14.16.

Exercise 14.18 (k-sets) Let $\mathcal{M}$ be a set of points. A subset of $\mathcal{M}$ is called a k-set if it has exactly k points and if there exists a hyperplane H that separates its points from the other points of $\mathcal{M}$. Show that the number of k-sets of $\mathcal{M}$ equals the number of cells at level k in the arrangement of the hyperplanes dual to the points in $\mathcal{M}$. What does an i-face in this arrangement correspond to in the original space?

Exercise 14.19 (Complexity of a level) Let $\mathcal{A} = \{A_1, \ldots, A_n\}$ be a set of n points in the plane. Show that a set of n points in the plane has at most $O(n\sqrt{k})$ k-sets, and dually, that an arrangement of n lines in the plane has $O(n\sqrt{k})$ faces at level k, for $k > 0$.

Hint: We call a *k-segment* an oriented line segment $A_i A_j$ such that the open half-plane on the right of the line joining A_i to A_j contains exactly k points of $\mathcal{A}$.

1. Let a line Δ sweep the plane parallel to the y-axis while maintaining the number of k-segments intersected by Δ. Show that this number may increase or decrease by at most one, each time Δ sweeps over a point of $\mathcal{A}$.

2. Let Δ be an oriented line containing no points of $\mathcal{A}$, and let Δ_l and Δ_r be the half-planes respectively to the left and right of Δ. Use the previous result to show by induction on k that there are at most $\min(k+1, \lfloor \frac{n}{2} \rfloor)$ k-segments crossing Δ that are segments of the form $A_i A_j$, $A_i \in \Delta_l$ and $A_j \in \Delta_r$.

3. Draw $n-1$ vertical lines $D_1, \ldots, D_{n-1}$ that divide the plane into strips each containing one point of $\mathcal{A}$. Among these lines, pick $p-1$ and call them $\Delta_1, \ldots, \Delta_{p-1}$, such that each of the p vertical strips they define contains at most $\lceil \frac{n}{p} \rceil$ points of $\mathcal{A}$. The number of k-segments that do not cross any of the Δ_i, $i = 1, \ldots, p-1$ is at most

$$p \binom{\lceil \frac{n}{p} \rceil}{2} \leq \frac{n^2 + np}{2p}.$$

The other k-segments must cut a line Δ_i and their number may be bounded using the result in 2. The bound on the number of k-sets is obtained by optimizing the choice of the parameter p.

Summing over all $j \leq k$, we obtain a bound of $O(k^{\frac{3}{2}} n)$ on the number of all j-sets for $0 \leq j \leq k$. This bound is not tight, as shown by theorem 14.5.1.

Exercise 14.20 (A lower bound on the k-level) Show that the k-level of the d-arrangement of n hyperplanes may have as many as $\Omega(n^{\lfloor d/2 \rfloor} k^{\lceil d/2 \rceil - 1})$ faces of all dimensions.

Hint: Consider the dual problem of bounding the number of j-sets for a set $\mathcal{P}$ of n points in $\mathbb{E}^d$ (see exercise 14.18). To establish the lower bound, we place the points in $\mathcal{P}$ on the moment curve $\mathcal{M}$ (see subsection 7.2.4). Any subset $\mathcal{D}$ of d points in $\mathcal{M}$ splits $\mathcal{M}$ into $d+1$ arcs $\mathcal{M}_1, \ldots, \mathcal{M}_{d+1}$ and induces a decomposition of $\mathcal{P}$ into $d+1$ subsets $\mathcal{P}_i = \mathcal{P} \cap \mathcal{M}_i$, $i = 1, \ldots, d+1$. The sets $\mathcal{P}_i$ are alternately on one or on the other side of the hyperplane affine hull of $\mathcal{D}$. The problem is now to count the subsets $\mathcal{D}$ of $\mathcal{P}$ such that

$$\sum_{l=1}^{\lfloor \frac{d+1}{2} \rfloor} |\mathcal{P}_{2l+1}| = j \quad \text{or} \quad \sum_{l=1}^{\lfloor \frac{d+1}{2} \rfloor} |\mathcal{P}_{2l}| = j.$$

For this, show that the number of ways to split an ordered set of s elements into r ordered subsets is

$$\frac{1}{r!}(s+1)^r.$$

Then deduce that the number of j-sets is $\Omega(n^{\lfloor d/2 \rfloor} k^{\lceil d/2 \rceil - 1})$, and that summing over all $j \leq k$ yields a lower bound on the complexity of $\mathcal{A}_{\leq k}$ which is identical to the upper bound shown in theorem 14.5.1.

Exercise 14.21 (Lazy computation of the first k levels) Adapt the randomized algorithm of subsection 15.4.2 to compute the first k levels in the arrangement of n

hyperplanes in expected time $O\left(nk^{\lceil d/2\rceil}\log(n/k)\right)$ if $d=2$ or 3, and $O(n^{\lfloor d/2\rfloor}k^{\lceil d/2\rceil})$ if $d\geq 4$.

Hint: The peeling phase which is systematically processed in the algorithm of subsection 14.5.3 may be replaced by lazy clean-up operations between some well-chosen incremental steps. Use the canonical triangulation (see exercise 14.7).

Exercise 14.22 (Easier computation of the first k levels) Show how to simplify the algorithm that builds the first k levels in the d-arrangement of k hyperplanes when all the hyperplanes contain a face at level 0.

Hint: It suffices to find the first conflict with the polytope at level 0 (a problem analogous to that of finding the conflicts to build a convex hull), and then to use the adjacency graph to detect the other conflicts.

Exercise 14.23 (Optimal separation) Let $\mathcal{B}$ be a set of n blue points and $\mathcal{R}$ be a set of n red points in $\mathbb{E}^d$. Given a hyperplane H, we define its separation defect as the smaller of the two numbers $|\mathcal{B}\cap H^+|+|\mathcal{R}\cap H^-|$ and $|\mathcal{B}\cap H^-|+|\mathcal{R}\cap H^+|$. Show how to compute a hyperplane that best separates $\mathcal{R}$ from $\mathcal{B}$ in time $O(n^d)$.

Hint: Without loss of generality, H^- may account for points of $\mathcal{B}$ and H^+ for $\mathcal{R}$. A blue point B (resp. red point R) will be accounted for by the wrong set if the pole H^* of H belongs to B^{*+} (resp. R^{*-}). The question is now equivalent to finding the sum of the levels of H^* in the dual arrangements of $\mathcal{B}$ and $\mathcal{R}$.

14.7   Bibliographical notes

Combinatorial properties of arrangements and the solutions to exercises 14.1, 14.2, and 14.3 can be found in the book by Grünbaum [114]. Other combinatorial and algorithmic results are presented in the book by Edelsbrunner [89], which is the authoritative reference on arrangements. The proof of the zone theorem given in section 14.3 is adapted from an article by Edelsbrunner, Seidel, and Sharir [97]. Coxeter [73] and McMullen [158] study zonotopes and the correspondence between zonotopes and arrangements of hyperplanes. The application of zonotopes to mixtures is presented in the work by Lacolle, Szafran, and Valentin [143], who give a solution to exercise 14.9.

A theorem analogous to the zone theorem holds for algebraic surfaces of bounded degree in arrangements of hyperplanes. Aronov, Pellegrini, and Sharir [11] showed that the complexity of such a zone is $O(n^{d-1}\log n)$.

The incremental algorithm that constructs the arrangement of n lines in the plane uses $O(n^2)$ working space since it maintains a representation of the current arrangement. Another time-optimal algorithm based on the *topological sweep* method by Edelsbrunner and Guibas uses only $O(n)$ working space [92].

Theorem 14.5.1, which bounds the number of faces in the first k levels in arrangements of hyperplanes, is due to Clarkson and Shor [71]. Lower and upper bounds on the complexity of a single level (see exercise 14.19) are given by Erdős, Lovász, Simmons, and Straus [103] and generalized to higher dimensions in the book by Edelsbrunner [89]. The result in exercise 14.19 was slightly improved by Pach, Steiger, and Szemerédi [186].

The algorithm that computes the first k levels is due to Mulmuley [174] (see also [177]). The algorithm as outlined in exercise 14.21 is fully described (and slightly improved in dimensions 2 and 3) by Agarwal, de Berg, Matoušek, and Schwarzkopf [2]. A deterministic optimal algorithm is given in the planar case by Everett, Robert, and van Kreveld [104], who also give a better solution to exercise 14.23 in the planar case.

Query problems, for a large part unexplored in this book, have spurred a lot of research. A recent account can be found in the books by Agarwal [1] or by Mulmuley [177]. Variants of exercises 14.10 and 14.12 are solved in these books, and the solution to exercise 14.11 is due to Chazelle, Guibas, and Lee [55]. Brönnimann [35] explains how to achieve point location in a polytope with preprocessing time $O(n \log n + n^{\lfloor d/2 \rfloor})$, storage $O(n^{\lfloor d/2 \rfloor})$, and query time $O(\log^2 n)$.

Computing visibility graphs and shortest paths have also motivated a lot of research. Recent developments and references can be found in the articles by Pocchiola and Vegter [188] and by Hershberger and Suri [127].

Chapter 15

Arrangements of line segments in the plane

In an arrangement of n lines in the plane, all the cells are convex and thus have complexity $O(n)$. Moreover, given a point A, the cell in the arrangement that contains A can be computed in time $\Theta(n \log n)$: indeed, the problem reduces to computing the intersection of n half-planes bounded by the lines and containing A (see theorem 7.1.10).

In this chapter, we study arrangements of line segments in the plane. Consider a set S of n line segments in the plane. The arrangement of S includes cells, edges, and vertices of the planar subdivision of the plane induced by S, and their incidence relationships.

Computing the arrangement of S can be achieved in time $O(n \log n + k)$ where k is the number of intersection points (see sections 3.3 and 5.3.2, and theorem 5.2.5). All the pairs of segments may intersect, so in the worst case we have $k = \Omega(n^2)$.

For a few applications, only a cell in this arrangement is needed. This is notably the case in robotics, for a polygonal robot moving amidst polygonal obstacles by translation (see exercise 15.6). The reachable positions are characterized by lying in a single cell of the arrangement of those line segments that correspond to the set of positions of the robot when a vertex of the robot slides along the edge of an obstacle, or when the edge of a robot maintains contact with an obstacle at a point. Since the robot may not cross over an obstacle, it is constrained in always lying inside the same cell of this arrangement. It is therefore important to bound the complexity of such a cell and to avoid computing the whole arrangement. Among the cells of $\mathcal{A}(S)$, a few contain the endpoints of some segments, and the others do not. The latter are naturally convex cells, their complexity is $O(n)$ and each can be computed in time $O(n \log n)$. The complexity of the former cells, however, is more difficult to analyze.

To conduct the combinatorial analysis, we introduce and study a certain class of words over a finite alphabet, the so-called *Davenport–Schinzel sequences* (see section 15.2). These words have a geometric interpretation that is both illuminating and useful: lower envelopes of functions (see section 15.3). Section 15.4 bounds the complexity of a cell and gives an algorithm that computes it. We first show that this complexity is almost linear, in contrast with the entire arrangement which may have $\Omega(n^2)$ edges in the worst case. The complexity of the algorithm is shown to be roughly proportional to the complexity of the cell it computes.

15.1 Faces in an arrangement

Let S be a set of n segments in the plane. To define a cell of their arrangement, we need to distinguish the two sides of a segment, or equivalently to consider that each line segment is a flat rectangle with an infinitesimally small width. The arrangement of S is formed by cells, edges, and vertices of the planar subdivision induced by S, and their incidence relationships. More precisely, the connected components of $\mathbb{E}^2 \setminus S$ are polygonal regions that may have holes (see figure 15.1): the cells of the arrangement are formed by the topological closures of these regions. The edges and vertices of this arrangement are the edges and vertices of the polygons that bound the cell. The arrangement of S will be denoted by $\mathcal{A}(S)$.

15.2 Davenport–Schinzel sequences

Given an alphabet with n symbols, a *word* on this alphabet is an ordered sequence of symbols in this alphabet, and a subsequence of a word $u_1 \ldots u_n$ is a word $u_{i_1} \ldots u_{i_k}$ for some indices $1 \le i_1 < \cdots < i_k \le n$. Given two symbols a and b, an *alternating sequence* of length s is a sequence $u_1 \ldots u_s$ such that $u_i = a$ if i is odd and $u_i = b$ is i is even. An (n, s)-*Davenport–Schinzel sequence* is a word on an alphabet with n symbols such that:

1. Two successive symbols of this word are distinct.

2. For each two symbols a, b in the alphabet, the alternating sequence of length $s + 2$ is not a subsequence of this word.

In other words, no two symbols can alternate more than $s + 1$ times.

So consider the phrase 'A DAVENPORT-SCHINZEL SEQUENCE', considered as a word over the Roman alphabet. The reader will easily verify that the longest alternating subsequence over two symbols is the subsequence

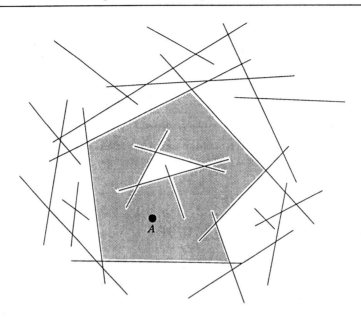

Figure 15.1. A cell in the arrangement of line segments.

'ESESE' (the other subsequences 'ENENE' and 'ECECE' are also suitable). The sequence 'A DAVENPORT-SCHINZEL SEQUENCE' is thus a $(26, 4)$-Davenport–Schinzel sequence!

Denote by $\lambda_s(n)$ the maximal length of an (n, s)-Davenport–Schinzel sequence. First of all, it is not even clear that $\lambda_s(n)$ is finite. In fact, it can be deduced from the connection with lower envelopes (see section 15.3) that $\lambda_s(n) \leq \frac{sn(n-1)}{2} + 1$. The following theorem gives more precise bounds on λ_1, λ_2, and λ_3.

Theorem 15.2.1 *The maximal length $\lambda_s(n)$ of an (n, s)-Davenport–Schinzel sequence is bounded by:*

$$
\begin{aligned}
\lambda_1(n) &= n \\
\lambda_2(n) &= 2n - 1 \\
\lambda_3(n) &= \Theta(n\alpha(n))
\end{aligned}
$$

where $\alpha(n)$ is the very slow-growing inverse of Ackermann's function.[1]

Proof. The proof for $s = 1$ is trivial, since each symbol may appear only once. For $s = 2$, we proceed by induction on n. The result is true for $n = 1$, so we consider an $(n, 2)$-Davenport–Schinzel sequence $(n > 1)$. Let a be its first letter,

---

[1]The definitions and order of magnitude of the inverse Ackermann function are given in subsection 1.1.3.

and put $S = aS'$. If a does not occur in S', then the induction applies for S' and so

$$|S| = 1 + |S'| \leq 1 + 2(n-1) - 1 = 2n - 2.$$

Otherwise we can write $S = aS_1aS_2$, where a does not occur in S_1 and $|S_1| > 0$. If S_2 is empty the length of aS_1a is smaller than $(2n - 2) + 1 = 2n - 1$, as we have just shown. Otherwise, let k be the number of distinct symbols in S_1. By induction, $|S_1| \leq 2k - 1$. Moreover, the definition of a Davenport–Schinzel sequence ensures that no symbol b occurs both in S_1 and S_2, otherwise $abab$ is a subsequence of S. Thus aS_2 may contain at most $n - k$ symbols (note that a may occur in S_2), and by induction we have $|aS_2| \leq 2(n-k) - 1$. Hence

$$|S| = |S_1| + |aS_2| + 1 \leq 2n - 1.$$

To finish the proof for $s = 2$, we must also show that this bound is exact. This can be readily seen by considering the sequence $ab_1ab_2a\ldots ab_{n-1}a$ of length $2n - 1$.

For $s = 3$, the proof goes into very technical details, so we will not prove the announced result here. We can show, however, the simpler result that $\lambda_3(n) = O(n \log n)$. Let S be a $(n, 3)$-Davenport–Schinzel sequence, and $S(a)$ be the subsequence obtained from S by removing all the occurrences of a symbol a. In $S(a)$, there cannot be a subsequence $bcbcb$ and identical consecutive symbols can happen at most twice when the first and the last occurrences of a are surrounded by two b's. Let us call $S'(a)$ the sequence obtained by replacing in $S(a)$ two consecutive symbols b by a single b, whenever this happens. Then $S'(a)$ is an $(n - 1, 3)$-Davenport–Schinzel sequence, and

$$|S| \leq |S'(a)| + 2 + n_a \leq \lambda_3(n-1) + 2 + n_a$$

where n_a stands for the number of occurrences of a in S. Summing over all the symbols a appearing in S, we obtain:

$$n|S| \leq n\lambda_3(n-1) + 2n + |S|.$$

This is true for any sequence S, so that

$$\frac{\lambda_3(n)}{n} \leq \frac{\lambda_3(n-1)}{n-1} + \frac{2}{n-1}$$

whence $\lambda_3(n) = O(n \log n)$. □

15.3 The lower envelope of a set of functions

Consider n continuous functions $f_i(x), i = 1, \ldots, n$ defined over $\mathbb{R}$. The *lower envelope* of the f_i's is the graph of the function defined by

$$f(x) = \min_i f_i(x).$$

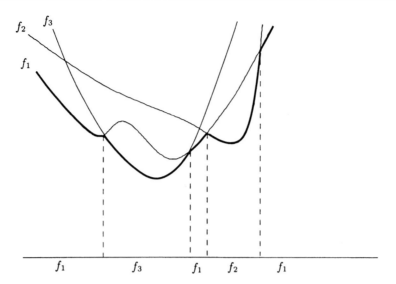

Figure 15.2. The lower envelope of a set of functions, and the corresponding Davenport–
Schinzel sequence.

The lower envelope is formed by a sequence of *curved edges,* where each edge is
a maximal connected subset of the envelope that belongs to the graph of a single
function $f_i(x)$. The endpoints of these edges are located at the intersections of
the graphs of the functions and are called the *vertices* of the envelope.

15.3.1   Complexity

Labeling each edge by the index of the corresponding function, we obtain a sequence of indices by enumerating these labels in the order in which they appear
along the envelope (see figure 15.2). If the graphs of the functions have pairwise
at most s intersection points, then this sequence is an (n, s)-Davenport–Schinzel
sequence. Indeed, let A_i and A_j be two edges appearing in this order along the
envelope, defined over two intervals I and J. The corresponding functions f_i
and f_j being continuous, they must intersect in a point whose abscissa is greater
than the right endpoint of I and smaller than the left endpoint of J. Having
an alternating subsequence of length $s + 2$ for the two symbols i and j implies
the existence of $s + 1$ intersection points between the graphs of f_i and f_j, a
contradiction.

The number of edges on the lower envelope is thus bounded above by the
maximal length $\lambda_s(n)$ of an (n, s)-Davenport–Schinzel sequence.

Consider the case when the functions are defined over closed intervals and not
over the whole of $\mathbb{R}$. The lower envelope is not continuous and the argument used

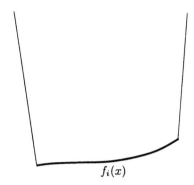

Figure 15.3. Extending the function f_i.

above to bound the number of its edges does not hold any more. This problem may be overcome by extending the domain of definition of the functions f_i to cover the whole of $\mathbb{R}$. More precisely, pick a positive real number μ. If f_i is defined over $[x_{i_{min}}, x_{i_{max}}]$, then we extend the graph of f_i for $x > x_{i_{max}}$ by the semi-infinite ray originating at $(x_{i_{max}}, f_i(x_{i_{max}}))$ whose slope is μ, and symmetrically for $x < x_{i_{min}}$ by the semi-infinite ray originating at $(x_{i_{min}}, f_i(x_{i_{min}}))$ whose slope is $-\mu$ (see figure 15.3). Thus we have a set of functions g_i which extend the functions f_i and are continuous. When μ is large enough, the sequence of labels of the edges on the lower envelope of the g_i's is identical to that of the lower envelope of the f_i's, and this lower envelope can be easily constructed knowing that of the g_i's.

It is readily verified that, for μ large enough, g_i and g_j have at most $s + 2$ intersection points if the corresponding functions f_i and f_j intersect in at most s points. It follows that the sequence of labels of the edges on the lower envelope of $g_1, \ldots, g_n$ is a $(n, s + 2)$-Davenport–Schinzel sequence.

The complexity of the lower envelope of the g_i's is thus bounded above by the maximal length $\lambda_{s+2}(n)$ of an $(n, s + 2)$-Davenport–Schinzel sequence. The complexity of the lower envelope of the f_i's is also bounded by $\lambda_{s+2}(n)$.

Example. Consider the case of line segments. Two line segments intersect in at most one point, so the sequence of labels on the lower envelope of a set of segments is an $(n, 3)$-Davenport–Schinzel sequence. The complexity of this lower envelope is thus $O(n\alpha(n))$. In fact, this bound is achievable and one may actually construct line segments whose lower envelope has super-linear complexity $\Theta(n\alpha(n))$ (see the bibliographical notes at the end of this chapter).

Let us now consider the case when the functions f_i are only defined over semi-infinite intervals. We first consider the functions f_i whose domains of definition are intervals defined by $x \geq x_{i_{min}}$. If we extend these functions by a half-line starting at $(x_{i_{min}}, f_i(x_{i_{min}}))$ of slope $-\mu$ for μ big enough, then we obtain functions g_i, defined over $\mathbb{R}$, whose graphs have pairwise at most $s + 1$ intersection points if the graphs of the f_i's had pairwise at most s intersection points. The sequence of labels on the lower envelope $\mathcal{L}_r$ of the g_i's is an $(n, s+1)$-Davenport–Schinzel sequence. The complexity of $\mathcal{L}_r$ is thus $\lambda_{s+1}(n)$.

A similar result obviously holds for the lower envelope $\mathcal{L}_l$ of the functions f_j whose domains of definition are defined by $x < x_{j_{max}}$. The lower envelope of the n functions f_i is the lower envelope of the union of $\mathcal{L}_r$ and $\mathcal{L}_l$. Its complexity is $O(n_r + n_l) = O(\lambda_{s+1}(n))$ since both $\mathcal{L}_r$ and $\mathcal{L}_l$ are monotone chains.

Example.   The lower envelope of n half-lines has complexity $O(n)$.

15.3.2   Computing the lower envelope

We now present an algorithm that computes the lower envelope of n functions f_i, $i = 1, ..., n$, defined over $\mathbb{R}$ such that no two graphs of these functions have more than s intersection points. The algorithm recursively computes the lower envelope $\mathcal{I}_1$ of $f_1(x), \ldots, f_{\lfloor n/2 \rfloor}$, and the lower envelope $\mathcal{I}_2$ of $f_{\lfloor n/2 \rfloor + 1}, \ldots, f_n(x)$. Both envelopes are monotone chains of complexity $\lambda_s(\frac{n}{2}) \leq \lambda_s(n)$, as was shown in the previous subsection. Monotonicity implies that we can compute the lower envelope of the union of $\mathcal{I}_1$ and $\mathcal{I}_2$ by sweeping the plane with a line parallel to the y-axis, in a manner that is similar to merging two sorted lists (see section 3.1.2). Let us call the current edges the two edges of $\mathcal{I}_1$ and $\mathcal{I}_2$ intersecting the sweep line. When the sweep line passes over a vertex of a current edge $\mathcal{I}_1$ or $\mathcal{I}_2$, this current edge is replaced by the edge that follows on the corresponding lower envelope. If this edge is part of the constructed lower envelope, a new edge is created for the lower envelope. When the sweep line encounters an intersection point between the two current edges, a new edge is created on the lower envelope. In either case, the next intersection point between the two current edges is computed. Merging the lower envelopes in this fashion takes time proportional to the total number of edges on $\mathcal{I}_1$ and $\mathcal{I}_2$, and to the number of intersection points between $\mathcal{I}_1$ and $\mathcal{I}_2$ which is $O(\lambda_s(n))$ as was shown in the previous section. We have thus proved that:

Theorem 15.3.1 *The lower envelope of n functions f_i, $i = 1, ..., n$, defined over $\mathbb{R}$ and whose graphs have pairwise at most s intersection points, has complexity $O(\lambda_s(n))$ and can be computed in time $O(\lambda_s(n) \log n)$.*

15.4   A cell in an arrangement of line segments

Let $\mathcal{S}$ be a set of n line segments in the plane. In the arrangement of $\mathcal{S}$, we may distinguish between cells whose boundaries contain at least one endpoint of a segment (the *non-trivial* cells) and the cells whose boundaries contain no endpoints (the *trivial* cells). Trivial cells are convex and their complexity is $O(n)$. In section 15.3, we have seen that the complexity of the lower envelope of a set of n line segments in the plane can be $\Theta(n\alpha(n))$. So we can conclude that $\Omega(n\alpha(n))$ is a lower bound on the worst-case complexity of a non-trivial cell in the arrangement of n line segments. To show this, consider a set of n line segments whose lower envelope has complexity $\Theta(n\alpha(n))$. To $\mathcal{S}$, we add $2n$ segments, almost vertical, and long enough so that each of them stands above an endpoint of a segment in $\mathcal{S}$ (see figure 15.3). We also add a horizontal segment lying above all the segments in $\mathcal{S}$ while cutting all the almost vertical segments that we added. The new set of segments $\mathcal{S}'$ has $3n+1$ segments, and the edges on the boundary of the unbounded cell lying below all the segments are in one-to-one correspondence with the edges of the lower envelope of $\mathcal{S}'$. But the $\Omega(n\alpha(n))$ edges on the lower envelope of $\mathcal{S}$ also correspond to a subset of the edges on the lower envelope of $\mathcal{S}'$. It follows that the unbounded cell is at least as complex as the lower envelope of $\mathcal{S}$, so that it also has complexity $\Omega(n\alpha(n))$.

As we will see, this bound is also an upper bound, which shows that the complexity of cells in the arrangement of line segments depends almost linearly on the number of segments, while the total arrangement may have up to $\Omega(n^2)$ edges in the worst case. We will then explain how to efficiently compute such a cell.

15.4.1 Complexity

Consider a set $\mathcal{S}$ of n line segments in the plane. We will assume that these segments are in general position, meaning that no three segments have a common intersection and that any two segments intersect in at most one point. A standard perturbation argument shows that the complexity of a cell is maximal in this case. Indeed, if the segments are not in general position, one may perturb them slightly so that they are in general position, without decreasing the number of edges or vertices of the cell under consideration.

From now on, and as was done in section 15.1, we consider that each line segment S is a rectangle of infinitely small width whose boundary is formed by two copies of the segment S called the *sides* of S, and two infinitely short perpendicular segments at the vertices. Under the general position assumption, the boundary of the union of these rectangles is homeomorphic to the union of all the segments. Henceforth, we will thus make a distinction between a segment, considered as a infinitely thin rectangle, and a segment side. The number of sides is $2n$.

We orient the rectangles counter-clockwise, which induces a clockwise orientation for the connected components of the boundaries of each cell.

Let Γ be a connected component of the boundary of some cell C in the arrangement $\mathcal{A}(\mathcal{S})$ of $\mathcal{S}$. Note that a segment may contain several edges of Γ.

Lemma 15.4.1 *Consider a segment S that contains at least one edge of Γ. The edges of Γ contained in S are traversed on the boundary of Γ in the same order as they are traversed on the boundary of S.*

Proof. Consider the infinitely thin rectangle S and the region R bounded by Γ that does not contain C. Then S is contained in R, and the result follows from a slight adaptation of the proof of theorem 9.4.1. $\qquad\qquad\square$

We label each edge of Γ by the index of the side of the segment of $\mathcal{S}$ to which it belongs. The sequence Σ_Γ of these labels forms a circular sequence which we break into a linear sequence by choosing some origin O on Γ. The number of distinct labels in Σ_Γ is at most the number of sides, which is $2n$. Two successive labels are distinct. Since two segments have only one intersection point, it is tempting to conjecture that the sequence Σ_Γ is a $(2n,3)$-Davenport–Schinzel sequence. The choice of O may induce some additional repeats, however. Indeed, if *ababab* is not a subsequence of the circular sequence, it may not always be possible to choose O so that the same is true for the linear sequence. For instance, consider figure 15.4: the linear sequence

$$\Sigma_\Gamma = a_1\ c_2\ c_1\ a_1\ a_2\ c_1\ b_1\ b_2\ c_1\ c_2\ b_2\ a_2\ a_1\ b_2\ b_1$$

does contain the subsequence $a_1 c_1 a_1 c_1 a_1$. We solve this technical problem by constructing another sequence Σ_Γ^* on at most $3n$ symbols which is at least as long as Σ_Γ. Let L be a side of a segment that supports several edges along Γ. These edges are naturally ordered by the orientation of L, so we let I be the first point of L that belongs to Γ and F the last point of L that belongs to Γ. The points I and F subdivide Γ into two chains ending at I and F. Denote by γ the oriented chain that contains the origin O. The idea is to give a different label to the edges on Γ that belong to $L \cap \gamma$ according to whether they are before O or after O. Then the new sequence Σ_Γ^* is merely the linear sequence of these new labels along the edges Γ. For instance, on figure 15.4, we now have

$$\Sigma_\Gamma^* = a_1''\ c_2\ c_1\ a_1''\ a_2\ c_1\ b_1''\ b_2\ c_1\ c_2\ b_2\ a_2\ a_1'\ b_2\ b_1' \ .$$

Σ_Γ^* has at most $3n$ distinct labels, since only one side of each segment needs to be relabeled.

Lemma 15.4.2 Σ_Γ^* *is a $(3n,3)$-Davenport–Schinzel sequence.*

Proof. We already know that Σ_Γ^* has at most $3n$ distinct labels and does not contain two identical consecutive elements. It remains to see that *ababa* is not a subsequence of Σ_Γ^* for any two symbols $a \neq b$.

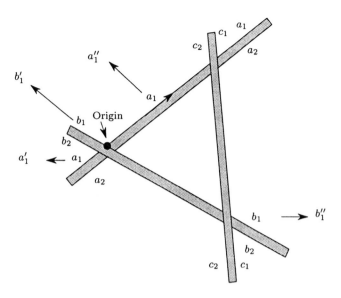

Figure 15.4. Circular and linear sequences.

We first show that, if $abab$ is a subsequence of Σ_Γ^*, the sides labeled a and b must intersect. For this, let the subsequence $abab$ correspond to the edges E_a', E_b', E_a'', E_b'' on Γ. Let S_a be the side labeled a that contains E_a' and E_a''. Pick a point A_1 in the relative interior of E_a' and a point A_2 in the relative interior of E_a'' (see figure 15.5). We define S_b, B_1 and B_2 similarly.

Let Λ be the union of the subchain Γ_{12} of Γ that joins A_1 to A_2 and of the simple polygonal chain contained in the interior[2] of S_a. Then Λ is a simple closed polygonal chain. The bounded polygonal region Δ enclosed by Λ contains, in a neighborhood of B_1, a portion of the segment B_1B_2. Indeed, if Λ is oriented by the orientation induced by Γ, then in a neighborhood of A_1 the side S_a is on the right of Λ and the cell lies to the left, and a similar statement holds for S_a in a neighborhood of A_2 and for S_b in a neighborhood of B_1. Moreover, Λ cannot cross the portion of Γ that joins A_2 to B_2, so that Δ cannot contain B_2. The segment B_1B_2 must therefore cross Λ. It cannot cross Γ_{12}, however, hence it must cross $\Lambda \setminus \Gamma_{12}$, and therefore also A_1A_2.

Assume now for a contradiction that $ababa$ is a subsequence of Σ_Γ^*. In addition to the notation above, let us pick a point A_3 in the relative interior of E_a'' that is after A_2 on E_a'', and another point A_4 in the relative interior of the third edge E_a''' labeled a, and so supported by S_a. From the preceding argument, we know that A_1A_2 and B_1B_2 intersect, and similarly for B_1B_2 and A_3A_4 (simply consider

---

[2]We assume that the segments are in general position, and that they are infinitely thin rectangles.

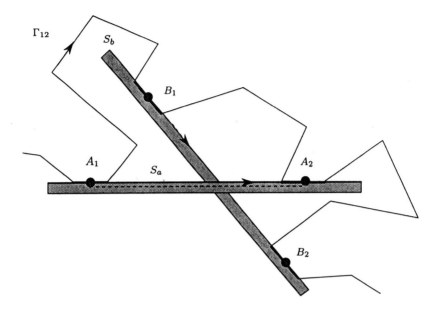

Figure 15.5. For the proof of lemma 15.4.2.

the subsequence *baba* of Σ_Γ). The two intersection points must be distinct since, owing to the relabeling and to lemma 15.4.1, the points A_i are all distinct and necessarily appear in the order A_1, A_2, A_3, A_4 on S_a. But this latter condition implies that $A_1 A_2$ and $A_3 A_4$ cannot intersect, so that S_b must cut S_a twice. This is impossible as two segments may only cross once.   □

An immediate consequence of this lemma is:

Theorem 15.4.3 *The complexity of a cell in the arrangement of n line segments in the plane is $O(n\alpha(n))$.*

As we mentioned in section 15.3, it is possible to place segments in the plane so that the cell containing, say, the origin has complexity $\Omega(n\alpha(n))$, so the bound in the theorem above is tight.

15.4.2 Computing a cell

The algorithm

Again let S be a set of n line segments in the plane, assumed to be in general position, and pick a point A that does not belong to any of the segments in S. Our goal is to compute the cell $C(S)$ in the arrangement of S that contains A.

The algorithm we present here is a variant of the randomized on-line algorithm that computes the vertical decomposition of S. The reader unfamiliar with that algorithm is invited to refer to subsection 5.3.2 for more details. We will only recall here the main definitions. The vertical decomposition is obtained by casting a ray upwards and downwards from any endpoint of the segments. The ray stops as soon as it encounters a segment in S (see figure 5.4a,d). The vertical segments (sometimes half-lines) traced by the rays are called walls, and together with the segments in S they decompose the plane into trapezoids that may degenerate into triangles or unbounded trapezoids. The algorithm also computes the vertical adjacencies of the trapezoids.[3]

To apply the formalism of chapter 4, we defined the problem in subsection 5.3.2 in terms of objects, regions, and conflicts between objects and regions. For this problem, an object is a segment. A region is a trapezoid in the decomposition of a subset of the segments. Each region is determined by at most four segments. There is a conflict between an object and a region if and only if the segment intersects the trapezoid. Computing the vertical decomposition is thus the same as computing the set of regions defined and without conflicts over S.

The algorithm we present to compute a cell $C(S)$ in fact computes a vertical decomposition of that cell (see figure 15.6). To generalize the algorithm of subsection 5.3.2 to compute only a single cell is not straightforward, however: the regions that interest us are not all the trapezoids defined and without conflict over the set S of segments, but only those contained in the cell $C(S)$. Unfortunately, whether a trapezoid is contained in the cell $C(S)$ cannot be decided locally by examining only that trapezoid and the segments that define it. This forbids *verbatim* use of the formalism and results of chapters 4 and 5.

To avoid this difficulty, we proceed as follows. Let R be the subset of segments already inserted into the data structure, and let $C(R)$ be the cell in the arrangement of R that contains A. We allow the algorithm to compute, in addition to the trapezoids in the decomposition of $C(R)$, other trapezoids in the arrangement of R that are not trapezoids of $C(R)$. In order not to degrade the performances of the algorithm, at certain incremental steps we perform a *clean-up* step, during which we remove the trapezoids that do not belong to the cell $C(R)$. Only the trapezoids that belong to $C(R)$ will be subdivided during subsequent incremental insertions. To distinguish between these trapezoids, we traverse the connected component in the vertical adjacency graph G of the current vertical decomposition that contains the trapezoid containing A. This latter trapezoid is maintained throughout the incremental steps. The other leaves of the graph that are not traversed are *deactivated*: they correspond to trapezoids in the current decomposition that are not contained in the cell $C(R)$. These trapezoids will not be subdivided, and the corresponding leaves in the graph will not have children in subsequent insertions. Figure 15.7 shows an intermediate situation in the algorithm.

---

[3]Recall that two trapezoids are vertically adjacent if they share a common vertical wall.

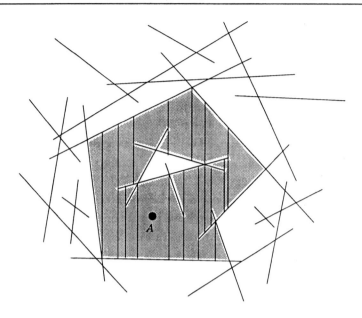

Figure 15.6. Vertical decomposition of the cell that contains A.

Between two clean-up steps, the algorithm is similar to the one described in subsection 5.3.2, apart from a few details which will be noted below. For each insertion of a new segment S, we locate S using the influence graph, then update the decomposition by subdividing the active trapezoids intersected by S. In the influence graph, this corresponds to creating new children for the active nodes that conflict with S.

Between two clean-up steps, and inside each trapezoid which has not been deactivated, we build the decomposition of the arrangement of the segments which conflict with this trapezoid and are inserted between the two clean-up steps. Let $\mathcal{T}_p$ be the set of nodes in the influence graph which were not deactivated during the previous clean-up step p. To each node in $\mathcal{T}_p$ we assign a secondary influence graph. This *secondary graph* is rooted at T and its nodes are the descendants of T created between step p and the next clean-up step. The secondary graph computed just as in subsection 5.3.2 under the incremental insertions of the segments inserted between step p and the next clean-up step. Its construction differs from that of a usual influence graph in a minor detail: the removal of superfluous walls. When inserting a segment S, if it intersects a wall, then only one of the two parts of that wall intersected by S is a wall in the new arrangement, and the other part must be removed and the two adjacent trapezoids must be merged. This procedure is detailed in subsection 5.3.2, and we apply it here to adjacent trapezoids that belong to the same secondary influence graph and also to trapezoids in different secondary influence graphs. A merge of the latter kind is called an *ex-*

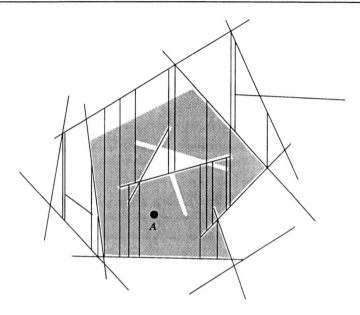

Figure 15.7. Intermediate situation in the computation of the cell that contains A. The shaded zone represents the final cell. The trapezoids which are neither entirely nor partially shaded are deactivated.

ternal merge. The vertical adjacencies are updated accordingly. External merges therefore introduce certain links between nodes in distinct secondary influence graphs.

The clean-up steps ensure that not too many trapezoids are created. Nevertheless, they must not be so frequent that the algorithm becomes inefficient. We perform clean-up steps after the insertion of the 2^i-th segment, for $i = 1, \ldots, \lfloor \log n \rfloor - 1$. Note that the last clean-up step was performed at step p_f where p_f is the greatest power of 2 such that $2p_f \leq n$.

Analysis of the algorithm

Suppose for now that n is a power of 2. We will analyze the complexity of the algorithm between two clean-up steps p and $2p$.

Denote by $\mathcal{S}_i$ the set of segments inserted during steps $1, \ldots, 2^i$. Each trapezoid T of $C(\mathcal{S}_p)$ is subdivided into trapezoids by the segments with which it conflicts. Let $\mathcal{S}_{2p}^T$ stand for the set of segments of $\mathcal{S}_{2p}$ that conflict with T, and let Σ_{2p}^T be the corresponding chronological sequence. The portion of the decomposition of the segments in $\mathcal{S}_{2p}^T$ that lies inside T has complexity $O(|\mathcal{S}_{2p}^T|^2)$.

To make things simpler, we assume that the algorithm does not perform the external merges. The number of nodes is only greater, so the location phase is always more complex. The cost of the external merges is proportional to the number of nodes killed (and hence visited) during the steps, so the external merges are accounted for by the location phase. The bounds we obtain on the complexity of the algorithm that does not perform the external merges will thus still be valid for the algorithm that performs the external merges.

Subsection 5.3.2 provides us with a bound on the number of nodes and the storage needed by the secondary influence graphs. For a trapezoid T the bound is $O(|\mathcal{S}_{2p}^T|^2)$. To bound the storage required by all the secondary influence graphs computed between steps p and $2p$ (and ignoring the external merges), we sum this quantity over all the trapezoids of $C(\mathcal{S}_p)$. Here we need a moment theorem analogous to theorem 4.2.6, but usable in a context where regions are not defined locally. Such a theorem is stated in exercise 4.4 and bounds this sum by a function of the expected complexity $g_0(r, \mathcal{Z})$ of a cell in the arrangement of a random sample of r segments in a set $\mathcal{Z}$. (Note that this complexity is linearly equivalent to the complexity of its vertical decomposition.) Therefore, the number of nodes in all the secondary influence graphs computed between steps p and $2p$ is

$$\sum_T O(|\mathcal{S}_{2p}^T|^2) = O\left(\left(\frac{2p}{p} \right)^2 g_0(p, \mathcal{S}_{2p}) \right) = O(p\alpha(p)).$$

The storage needed by the whole influence graph is thus

$$\sum_{i=1}^{\lfloor \log n \rfloor - 1} O(2^i \alpha(2^i)) = O(n\alpha(n)).$$

The complexity of the algorithm can be accounted for by three terms that correspond to the location phase, the update phase, and the clean-up steps. The location phase is analyzed in much the same way as the storage. We first evaluate the average number of nodes visited during step p, in the secondary influence graph rooted at a node that corresponds to a trapezoid T in $C(\mathcal{S}_p)$. The node is also denoted by T for simplicity. As before, we denote by $\mathcal{S}_{2p}^T$ (resp. Σ_{2p}^T) the subset (or the chronological sequence) of the segments that conflict with T and that are inserted before step $2p$. Denote by $\mathcal{S}_T$ (resp. Σ_T) the subset (or the chronological sequence) of all the segments that conflict with T. Note that $\mathcal{S}_{2p}^T$ is a random subset of $\mathcal{S}_T$. Under the assumption above, it all happens as if we were locating the segments in the sequence $\mathcal{S}_T$ in the secondary graph rooted at T. This graph is the influence graph corresponding to the decomposition of $\mathcal{S}_{2p}^T$ inside the interior of T. A slight adaptation of the proof of theorem 5.3.4 yields an upper bound on the expected number of nodes visited in the secondary influence graph. Let $f_0(r, \mathcal{Z})$ be the expected number of trapezoids in the decomposition of a random sample of r segments in a set $\mathcal{Z}$. Then $f_0(r, \mathcal{Z}) = O(r^2)$. If we assume that $\mathcal{S}_{2p}^T$ is given, then we may use theorem 5.3.4 to bound the cost of inserting the last object by

$$O\left(\sum_{r=1}^{|\mathcal{S}_{2p}^T|} \frac{f_0(\lfloor r/2 \rfloor, \mathcal{S}_{2p}^T)}{r^2}\right).$$

This expression bounds the cost of inserting all the segments in $\mathcal{S}_{2p}^T$ as well as the cost of locating in the secondary graph rooted at T all the segments in $\mathcal{S}_T \setminus \mathcal{S}_{2p}^T$ that are inserted after step $2p$. The number of nodes visited in the secondary influence graph during the successive insertions, averaged over all random samples $\mathcal{S}_{2p}^T$ in $\mathcal{S}_T$, is thus

$$O\left(E\left(|\mathcal{S}_T| \sum_{r=1}^{|\mathcal{S}_{2p}^T|} \frac{f_0(\lfloor r/2 \rfloor, \mathcal{S}_{2p}^T)}{r^2}\right)\right) = O\left(|\mathcal{S}_T| \, E(|\mathcal{S}_{2p}^T|)\right).$$

Hence, the expected number of segments inserted before step $2p$ that conflict with T is

$$E(|\mathcal{S}_{2p}^T|) = \frac{p}{n-p}|\mathcal{S}_T| = O\left(\frac{p}{n}|\mathcal{S}_T|\right),$$

since $p \leq \frac{n}{2}$. The average number of nodes visited in the secondary influence graph rooted at T is finally $O\left(\frac{p}{n}|\mathcal{S}_T|^2\right)$.

Summing over all the nodes of $C(\mathcal{S}_p)$, we then obtain a bound on the expected number m of nodes visited in all the secondary influence graphs rooted at these nodes:

$$m = O\left(\frac{p}{n} \sum_T |\mathcal{S}_T|^2\right).$$

Once again, we can use the adapted moments theorem in exercise 4.4, to obtain

$$m = O\left(\frac{p}{n}\left(\frac{n}{p}\right)^2 p\alpha(p)\right) = O(n\alpha(p)). \tag{15.1}$$

Summing over all the clean-up steps, we get

$$\sum_{i=1}^{\lfloor \log n \rfloor - 1} O(n\alpha(2^i)) = O(n\alpha(n)\log n).$$

The update phases and clean-up steps are easily analyzed. Indeed, the update phases require time proportional to the number of nodes created, which is $O(n\alpha(n))$. Identifying the trapezoids of $C(\mathcal{S}_p)$ during the clean-up step p requires time proportional to the number of trapezoids in $C(\mathcal{S}_p)$, which is $O(p\alpha(p))$.

To deactivate the trapezoids during the different clean-up steps takes time proportional to the number of created nodes, which is again $O(n\alpha(n))$. The total cost of the clean-up steps is thus

$$\sum_{i=1}^{\lfloor \log n \rfloor - 1} O(2^i \alpha(2^i)) = O(n\alpha(n)).$$

This finishes the proof of the theorem stated below when n is a power of 2. To analyze the general case, we must also analyze the cost of inserting the segments at steps $2p_f + 1, \ldots, n$. But this is word for word the same as the analysis above and produces the same results (and notably equation 15.1) if we note that $p_f > \frac{n}{4}$.

Theorem 15.4.4 *A single cell in the arrangement of n line segments in the plane can be computed in expected time $O(n\alpha(n) \log n)$ and storage $O(n\alpha(n))$.*

15.5   Exercises

Exercise 15.1 (Optimal computation of lower envelopes) Show that the lower envelope of n line segments in the plane can be computed in optimal time $O(n \log n)$.

Hint: The lower bound $\Omega(n \log n)$ is proved by reduction to sorting. As for the upper bound, first project the endpoints of the segments on the x-axis. They define $2n - 1$ consecutive non-overlapping intervals. Build a balanced binary tree whose leaves are the intervals in the appropriate order. To a node corresponds an interval which is the union of the intervals at the leaves in the subtree. A segment S is assigned to the node whose interval is the smallest that still contains the projected endpoints of the segments. (This node is the first common ancestor of all the leaves covered by S.) Show that the lower envelope of the segments assigned to a single node has complexity $O(m)$ and not $O(m\alpha(m))$, using that there exists a vertical line that intersects all these segments and using also the result on half-lines mentioned in section 15.3. Observing that the projections of two segments assigned to different nodes at the same level in the tree do not overlap, show that the lower envelopes of the segments assigned to the nodes on a given level of the tree also have linear complexity, and can be computed in time $O(n \log n)$. These $O(\log n)$ lower envelopes can be merged in time $O(n\alpha(n) \log \log n)$, which is $O(n \log n)$.

Exercise 15.2 (Airport scheduling) Consider a set $\mathcal{M}$ of n points in $\mathbb{E}^d$ that move along algebraic curves of bounded degree at given constant speeds. At each moment t, we want to know the point $M(t)$ in $\mathcal{M}$ that is the closest to the origin. Show that the sequence Σ of points $M(t)$ for $t \in [0, +\infty[$ is almost linear, and that it may be computed in time $O((n + |\Sigma|) \log n)$.

Exercise 15.3 (Computing a view) Consider a scene formed by n line segments in the plane (not necessarily disjoint). Show that a view from a given point (defined as the portions of segments visible from that point) may be computed in optimal time $O(n \log n)$. Similar question for more general objects.

Hint: A projective transformation sends the origin to $(-\infty, 0)$, and the corresponding problem is exactly that of computing the lower envelope of n segments, see exercise 15.1.

Exercise 15.4 (Convex hull of objects) Show that computing the convex hull of n objects in the plane reduces to computing the lower envelope of n functions. If the objects are convex and disjoint, the graphs of these functions have at most two intersection points. Give bounds on the combinatorial and computational complexities of convex hulls of curved objects, in particular circles, ellipses, etc.

Hint: For each object, consider the set of its tangent lines, and use polarity to work in the dual plane (where points correspond to lines in the original plane).

Exercise 15.5 (Stabbing lines) Given n objects in the plane, compute the set of lines that simultaneously stab all of them.

Hint: Use the same polarity as in the preceding exercise.

Exercise 15.6 (Motion planning of a polygon under translation) Consider a polygon $\mathcal{M}$ with m sides that moves under translation within a polygonal region $\mathcal{E}$ with n sides.

1. Show that the set of translations that bring a vertex of $\mathcal{M}$ (resp. of $\mathcal{E}$) in contact with an edge of $\mathcal{E}$ (resp. of $\mathcal{M}$) is a set $\mathcal{C}$ of mn line segments (identifying the vector OM of the translation with the point M). Conclude that the set of feasible translations of $\mathcal{M}$ in $\mathcal{E}$ consists of one or several polygonal regions with total complexity $O(m^2 n^2)$. If $\mathcal{M}$ is convex, the complexity is only $O(mn)$ (see exercise 19.8).

2. Show that the set of positions of $\mathcal{M}$ in $\mathcal{E}$ that are accessible from a given position I is the cell in the arrangement of $\mathcal{C}$ that contains I. Conclude that it is possible to determine whether two positions I and J are accessible one from the other, and if so, compute a feasible path for $\mathcal{M}$ from I to J in time $O(mn\alpha(mn)\log(mn))$.

3. Show that the complexity of the arrangement of $\mathcal{C}$ may be as bad as $\Omega(m^2 n^2)$ (see figure 15.8).

4. Show that, in some cases, any path from I to J may have complexity $\Omega(mn)$. For instance, consider a carpenter's folding rule with m segments, but in a semi-folded rigid configuration, that tries to pass through n consecutive doors.

Exercise 15.7 (Non-trivial boundary) Consider two connected polygonal regions $\mathcal{B}$ and $\mathcal{R}$, which may have holes. A connected component of the intersection $\mathcal{B} \cap \mathcal{R}$ is called *non-trivial* if its boundary includes at least one vertex of $\mathcal{B}$ or $\mathcal{R}$. The *non-trivial boundary* of the intersection $\mathcal{B} \cap \mathcal{R}$ is the union of the polygons that bound all the non-trivial connected components of this intersection. Show that the complexity of the non-trivial boundary of the intersection $\mathcal{B} \cap \mathcal{R}$ is $O(|\mathcal{B}| + |\mathcal{R}|)$, where $|\mathcal{B}|$ and $|\mathcal{R}|$ respectively stand for the number of sides of $\mathcal{B}$ and $\mathcal{R}$. Hence any cell of the intersection of two polygonal regions $\mathcal{B}$ and $\mathcal{R}$ has complexity $O(|\mathcal{B}| + |\mathcal{R}|)$.

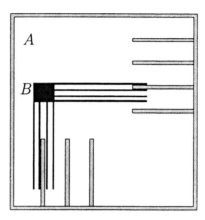

Figure 15.8. There are $\Omega(m^2 n^2)$ feasible positions of $\mathcal{M}$ that belong to the same number of distinct cells of $\mathcal{C}$.

Hint: Put $\mathcal{V} = \mathcal{B} \cap \mathcal{R}$ for the intersection of $\mathcal{B}$ and $\mathcal{R}$. Since $\mathcal{R}$ and $\mathcal{B}$ play entirely symmetric roles, it suffices to look at the contribution of the boundary of $\mathcal{B}$ to the non-trivial boundary of $\mathcal{V}$. For this, follow the edges on the boundary of $\mathcal{B}$, and count the number of edges of $\mathcal{V}$ contained in each edge of $\mathcal{B}$.

For each edge E on the boundary of $\mathcal{B}$, count the edges of $\mathcal{V}$ contained in E. We distinguish the first one along E from the others. Among the others, count separately those that belong to the same connected component of $\mathcal{V}$, those that do not belong to the same connected component of the boundary of $\mathcal{V}$, and the remaining edges.

Exercise 15.8 (Computing the non-trivial boundary) Show that the non-trivial boundary of two polygonal regions $\mathcal{B}$ and $\mathcal{R}$ (see exercise 15.7) can be computed in time $O(m \log m)$, if m is the total number of edges of $\mathcal{B}$ and $\mathcal{R}$.

Hint: We sweep the plane with a line going in two directions, first going from left to right and then from right to left. During the sweep, we maintain three structures which respectively represent the segments of $\mathcal{B}$, of $\mathcal{R}$, and of the resulting non-trivial boundary that intersect the sweep line. During the left-to-right sweep, we only create a new interval for the result when the current event is a vertex of $\mathcal{B}$ contained in $\mathcal{R}$, or a vertex of $\mathcal{R}$ contained in $\mathcal{B}$. We call such a vertex a *remarkable* vertex. Then we are assured that this interval is contained in a non-trivial cell. We do not discover the entire non-trivial cell, however, rather we only know the portion of this cell that can join a remarkable vertex by a decreasing x-monotone path. This is why we need to sweep the plane in the other direction, from right to left.

Exercise 15.9 (Computing a cell) Devise a deterministic algorithm that computes a single cell in the arrangement of n line segments in the plane, in time $O(n \log^2 n)$. The cell is characterized by a point A that belongs to it.

Hint: Use the divide-and-conquer method. Split the set of n segments into two subsets of roughly the same size to obtain two cells C_1^A and C_2^A in the sub-arrangements that contain A. Merging these two cells can be done using a variant of the sweep method of exercise 15.8. This variant in fact computes the non-trivial boundary of the intersection $C_1^A \cap C_2^A$ as well as the boundary of the cell in this intersection that contains A, even if it does not belong to the non-trivial boundary. It remains to extract the description of the cell C_A in the current divide-and-conquer step that contains A.

Exercise 15.10 (Half-lines) Show that the complexity of a cell in the arrangement of n half-lines is $O(n)$. Devise a deterministic algorithm that computes it in optimal time $O(n \log n)$.

Hint: Applying a rotation if necessary, we may assume that no half-line is vertical. Suppose that the cell is characterized by a point that belongs to it. Distinguish between the subset $\mathcal{E}^+$ of the half-lines that intersect $y = +\infty$ and the subset $\mathcal{E}^-$ of the half-lines that intersect $y = -\infty$. For each of these subsets, we explain how to compute the unbounded cell that contains the origins of some half-lines. (The other cells can be computed in a similar way.) For $\mathcal{E}^-$, we compute a left tree and a right tree by sweeping the plane from top to bottom with a line parallel to the x-axis. At each intersection I between two half-lines, we keep only the portion of the half-lines which lies to the left (for the left tree) or to the right (for the right tree). The boundary of the unbounded cell of $\mathcal{E}^-$ is obtained by computing the boundary of the unbounded cell of the union of both trees. Finally, exercise 15.8 can be used to compute the intersection of the cells of $\mathcal{E}^+$ and $\mathcal{E}^-$ that contain A.

Exercise 15.11 (Curved arcs) Bound the complexity of a cell of an arrangement of curved arcs in the plane and devise an algorithm that computes the cell that contains some given point.

Exercise 15.12 (Manipulator) A planar manipulator is formed by two rigid bodies articulated in a point A. One body is fixed to the origin O. The manipulator has two degrees of freedom: a rotation around O and a rotation around A. The configuration of the manipulator is parameterized by the corresponding two angles. Given some obstacles, devise an algorithm that computes the set of configurations for which the manipulator does not collide with an obstacle. Devise also an algorithm that determines whether two positions are reachable one from the other and, if so, outputs a path that realizes this change of configuration.

Hint: Express the constraints that limit the motion of the manipulator in the configuration space (which has dimension 2) and use exercise 15.11.

15.6 Bibliographical notes

The connection between lower envelopes of functions and Davenport–Schinzel sequences was established in a paper by Davenport and Schinzel [74]. Atallah [13], then Sharir and collaborators [1, 122, 206] proved bounds on the length of Davenport–Schinzel sequences. Wiernik and Sharir [219] showed how to realize a lower envelope of n segments in the plane that has complexity $\Omega(n\alpha(n))$. The solution to exercise 15.1 is due to Hershberger [124].

The analyses of the complexity and of the computation of the unbounded cell in the arrangement of line segments are given by Pollack, Sharir, and Sifrony [189]. Their result is extended by Guibas, Sharir, and Sifrony [118] to the case of a cell in the arrangement of curved arcs. Other results on curved arcs are given in [64, 90]. The complexity and computation of m cells is studied by Edelsbrunner, Guibas, and Sharir in [91, 93]. Solutions to exercises 15.7, 15.8, and 15.9 can be found in their papers.

Alevizos, Boissonnat, and Preparata [7] study the arrangements of half-lines and give a solution to exercise 15.10. These arrangements find applications in pattern recognition [8, 33].

The randomized algorithm that computes a single cell described in this chapter is due to de Berg, Dobrindt, and Schwarzkopf [76]. This algorithm can be generalized to dimension 3, which is not the case for a previous algorithm due to Chazelle *et al.* [52].

A comprehensive survey of Davenport–Schinzel sequences and their geometric applications can be found in the book by Sharir and Agarwal [207].

Chapter 16

Arrangements of triangles

The questions of chapter 15 can also be asked in spaces of dimensions greater than 2. Unfortunately, there are very few efficient algorithms for dimensions greater than 3 and so this chapter studies problems in dimension 3 only. Known extensions to higher dimensions and their corresponding references are given in the exercises and in the bibliographical notes.

The notion of an arrangement of segments in the plane can be extended to that of an arrangement of triangles in $\mathbb{E}^3$. Given a set $\mathcal{T}$ of n triangles in $\mathbb{E}^3$, its arrangement is formed by the faces of dimensions 0, 1, 2, and 3 (respectively called vertices, edges, facets, and cells) of the spatial subdivision of $\mathbb{E}^3$ induced by $\mathcal{T}$, and their incidence relationships.

It is easy to see that the complexity of an arrangement $\mathcal{A}$ of n triangles in $\mathbb{E}^3$ is $\Theta(n^3)$ in the worst case, which is the same as that of an arrangement of n planes. In this chapter, we are mostly interested in the complexity analysis and computation of only parts of $\mathcal{A}$: the lower envelope of a set of triangles and the cell of the arrangement that contains a given point.

The algorithm that computes the lower envelope of a set of triangles has an important application in computer graphics since it allows us to compute an orthographical view of a polyhedral scene. It is also useful in robotics (see exercise 16.12) and for computing Voronoi diagrams for L_1 and L_∞ norms as is shown in section 18.4. Computing a cell is essentially motivated by the computation of the free space of a polyhedral robot moving by translation among polyhedral obstacles.

The main problem that arises when trying to extend the results of the preceding chapter to a space of dimension higher than 2 is the difficulty of defining a decomposition scheme into elementary cells that has a complexity roughly comparable with that of the arrangement. This question is tackled in subsection 16.2 and again in subsection 16.4.2. Section 16.3 presents the known results concerning envelopes of triangles. Finally, the complexity of a cell in the arrangements of triangles is bounded in section 16.4, and we describe an algorithm that computes it which is similar to that presented in the previous chapter for the case of dimension 2.

16.1   Faces in an arrangement

Given a set $\mathcal{T}$ of n triangles in $\mathbb{E}^3$, the arrangement of $\mathcal{T}$ is formed by faces of dimension 0, 1, 2, and 3 (respectively called vertices, edges, facets, and cells) of the subdivision of $\mathbb{E}^3$ induced by $\mathcal{T}$, and by the adjacency relationships between these faces. To define a cell in an arrangement of triangles, we need to distinguish the two sides of a triangle, in other words to consider each triangle as an infinitely thin flat prism whose boundary contains two parallel copies of the triangles at an infinitesimal distance. We still speak of the interior of a triangle or face, and we mean its relative interior.

More precisely, a cell in the arrangement is the topological closure of a connected component of $\mathbb{E}^3 \setminus \mathcal{T}$; a facet of a cell is the topological closure of a connected component of the intersection of the boundary of the cell with a triangle of $\mathcal{T}$; the vertices and edges of the arrangement are the vertices and edges of the polygons that bound the facets. An edge in the arrangement is contained in the edge of a triangle of $\mathcal{T}$ or in the intersection of two triangles. A vertex of the arrangement is either a vertex of a triangle in $\mathcal{T}$, the intersection of a triangle in $\mathcal{T}$ and the edge of another triangle in $\mathcal{T}$, or the intersection of three triangles in $\mathcal{T}$.

16.2   Decomposing an arrangement of triangles

In several applications, it is useful to decompose the cells of an arrangement into elementary cells that are convex with bounded complexity. In chapter 12, we showed how to triangulate or decompose a polygonal region. For the case of line segments in the plane, each cell in the arrangement can be considered as a polygonal region, so the problem can be solved. The resulting decomposition has the same complexity as the arrangement itself, up to some multiplicative factor.

In $\mathbb{E}^3$, the situation is somewhat complicated by the fact that triangulating a polyhedron with n vertices may require up to $\Omega(n^2)$ tetrahedra in the worst case (see exercise 13.4). The number of vertices in the arrangement of triangles can be cubic, leading to a very onerous decomposition.

Subsection 16.2.1 presents a better solution obtained by generalizing the idea in the decomposition of line segments and by using Davenport–Schinzel sequences. The result is a vertical decomposition whose complexity is very close to that of the arrangement itself. We have already used a similar construction to triangulate a polyhedron (see section 13.3). Subsection 16.2.2 then shows how to adapt the vertical decomposition to subdivide all the non-convex cells in the arrangement of n triangles into $O(n^2)$ convex cells. The vertical decompositions are very useful, as will be shown in several algorithms described in this chapter. They also allow arrangements of triangles to be triangulated, a very easy operation when the vertical decomposition is known.

Henceforth, $\mathcal{T}$ will stand for a set of n triangles in $\mathbb{E}^3$. We assume that no triangle in $\mathcal{T}$ is parallel to the z-axis.

16.2.1 Vertical decomposition

In this subsection, as before, each triangle has two sides: the upper side and the lower side. This is equivalent to considering a triangle as an infinitely thin flat prism.

The decomposition that we describe below is a very close relative of the decomposition of polyhedral regions described in subsection 13.3.1. From each point P on an edge of the arrangement, we draw two vertical segments (parallel to the z-axis) upwards and downwards until they hit another triangle in $\mathcal{T}$.

A connected set of vertical segments whose endpoints belong to a given edge in the arrangement and to a given triangle in the arrangement constitutes a vertical trapezoid which we call a *2-wall of type 1*. These vertical walls decompose the cells of the arrangement into cylindrical subcells with at most two non-vertical faces: an upper face called the *ceiling* of the cell, and a lower face called its *floor*. The decomposition process is not yet over, because the cylindrical subcells may have many vertical faces and may not even be convex or may have holes (see figure 16.1). To get convex cells of bounded complexity, we subdivide each cylindrical subcell C again in the following manner.

1. We decompose the floor $Fl(C)$ of C.[1] This decomposition is obtained by drawing, from every vertex of $Fl(C)$ towards its interior, a line segment of maximal length that is parallel to the projection of the y axis on the non-vertical plane containing $Fl(C)$. These segments form the 1-walls of the floor and have both endpoints on the boundary of $Fl(C)$. This decomposition is shown in bold dotted lines in figure 16.1.

2. From each point P on a 1-wall of the floor of C, draw a vertical line segment towards the interior of C until it meets the ceiling. The set of vertical segments that pass through a given 1-wall of $Fl(C)$ constitutes a trapezoid which we call a *2-wall of type 2*.

Each cylindrical cell is processed separately, so we must distinguish the two sides of a 2-wall of type 1 (just as we did for the simplified decomposition of a set of

---

[1]The cell C is cylindrical, so decomposing the ceiling would yield exactly the same result.

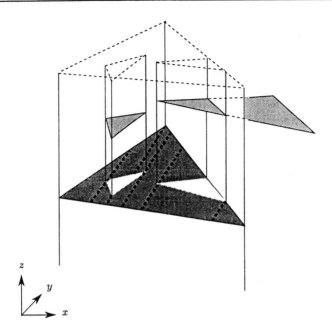

Figure 16.1. Decomposing an arrangement of triangles. The walls of type 1 are represented
by light lines, and the traces of walls of type 2 on the floor of the cell are
represented in dashed bold.

segments in section 3.3). Each side of a 2-wall of type 1 is subdivided by 2-walls
of type 2 into new walls which we call 2-walls of type 1′. All cells in the resulting
decomposition are convex and have bounded complexity: they have at most six
vertical sides and an upper and lower face, and all these faces are trapezoidal.
We call such cells *prisms*.

To estimate the number of prisms in this decomposition, we compute the number of 2-walls, and observe that any prism has at least one 2-wall and that a
2-wall belongs to at least two prisms. The number of walls of type 1′ is at most
the sum of the numbers of walls of types 1 and 2. The number of 2-walls of
type 2 is at most twice the number of 2-walls of type 1. Indeed, each vertex of
the floor $Fl(C)$ of a cylindrical cell C is the endpoint of at most two 1-walls of
$Fl(C)$, and the number of vertices of $Fl(C)$ also equals the number of edges of
the vertical faces (2-walls of type 1) of C. Hence it is enough to count all the
vertical walls of type 1. Let E be an edge of some triangle or the intersection of
two triangles, and let $\mathcal{R}$ be the vertical strip generated by all the vertical lines
passing through a point of E. We denote by $\mathcal{R}_h$ the part of $\mathcal{R}$ that lies above
E and by $\mathcal{R}_l$ the part that lies below. Let $\mathcal{S}_h$ (resp. $\mathcal{S}_l$) be the set of segments
obtained as the intersections of triangles in $\mathcal{T}$ and $\mathcal{R}_h$ (resp. $\mathcal{R}_l$). The vertical
edges of the walls of type 1 that have an endpoint on E are precisely the vertical

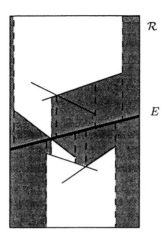

Figure 16.2. Vertical faces that intersect the edge E.

line segments that join a point of E and a vertex on the lower envelope (in $\mathcal{R}$) of $\mathcal{S}_h$ or on the upper envelope of $\mathcal{S}_l$. From theorem 15.3.1, we derive that the number of 2-walls of type 1 that are incident to E is $O(n\alpha(n))$. If these triangles do not intersect, this number is $O(n)$.

Summing over all the segments E that are either edges of a triangle in $\mathcal{T}$ or the intersection of two triangles in $\mathcal{T}$, we find that the total number of walls of type 1 (and hence also the number of walls of types 2 and 1′) is $O((n+p)n\alpha(n))$, if the number p of intersecting pairs of triangles is not zero. If it is, then the number of walls of type 1 is $O(n^2)$. The theorem below summarizes this result.

Theorem 16.2.1 *For any arrangement of n triangles in $\mathbb{E}^3$, with $p > 0$ intersecting pairs of triangles, there exists a decomposition of complexity $O(n(n+p)\alpha(n))$. If $p = 0$, the complexity of the decomposition is $O(n^2)$.*

In the worst case, we have $p = O(n^2)$ and the decomposition has complexity $O(n^3\alpha(n))$, which is very close to the optimum, since the arrangement itself has complexity $\Omega(n^3)$ in the worst case.

One may also bound the complexity of the decomposition by $O(n^2\alpha(n)\log n + t)$, where t is the complexity of the arrangement (see exercise 16.4). This bound is better than that of the previous theorem when $p = \Omega(n\log n)$.

16.2.2 Convex decomposition

In this section, we show how to decompose the non-convex cells in the arrangement $\mathcal{A}$ of a set $\mathcal{T}$ of n triangles into $O(n^2)$ convex parts. Overall and in the worst case, we obtain a decomposition into $O(n^3)$ convex parts for the whole arrangement. The proof is constructive and is based on building a portion of the vertical decomposition of the non-convex cells.

A face in the arrangement $\mathcal{A}$ is said to be *outer* if it is contained in a face on the boundary of a triangle in $\mathcal{T}$. Otherwise, it is an *inner* face. Note that the boundary of a cell contains an outer face if and only if the cell is non-convex.

Theorem 16.2.2 *In the arrangement of n triangles in $\mathbb{E}^3$, the union of the non-convex cells can be decomposed into $O(n^2)$ convex parts. The union of all the cells can therefore be decomposed into $O(n^3)$ convex parts.*

Proof. As indicated above, only the first statement needs a proof. For the second statement, simply note that the cells that do not contain outer faces are convex, and that they form a subset of the cells of the arrangement of the planes that support the triangles. They must therefore have overall complexity $O(n^3)$. For the non-convex cells, we first show how to decompose the non-convex cells of the arrangement of n line segments into $O(n)$ convex parts. This is achieved as for the usual vertical decomposition of a set of line segments, except that we draw only the walls from the endpoints of the segments and not from their intersection points. The cells in the decomposition induced by the walls and the segments are convex, since the only non-convex vertices of a cell in the arrangement of line segments are the endpoints of the segments. There are at least one and at most two walls incident to each non-convex vertex, so there are only twice as many convex parts as there are endpoints, hence $O(n)$ convex parts in this decomposition.

Consider now the case of dimension 3. We decompose the arrangement in a way similar to the case $d = 2$. More specifically, we consider each edge on the boundary of a triangle in $\mathcal{T}$. Through each point P of such an edge we draw a vertical line segment (parallel to the z-axis) upwards and downwards until it hits a triangle in $\mathcal{T}$ other than that which contains P. In this way, we build 2-walls of *type 1*. Note that, contrary to the decomposition described in subsection 16.2.1, we do not build walls of type 1 on top of the edges of the arrangement that are contained in the intersection of two triangles. Each cell in the resulting decomposition is a vertical cylinder, bounded above and below by convex domes. The cylinders are not necessarily convex since their horizontal projections are generally not convex either. Yet the reflex edges must be vertical, and they correspond to segments drawn on top of the vertex of some triangle, or on top of the intersection of a triangle and an edge. We obtain convex parts by adding more 2-walls of *type 2*. More precisely, if $\Pi(C)$ is the projection on the xy-plane a non-convex cell C, then $\Pi(C)$ may be decomposed into convex parts as we did for the two-dimensional case, by building 1-walls parallel to the y-axis for each non-convex vertex of $\Pi(C)$. These vertices are vertices of the triangles, or the intersection of a triangle and an edge. We then build vertical 2-walls of type 2

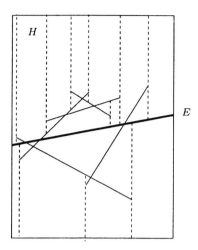

Figure 16.3: Walls in the convex decomposition induced by the triangles in a vertical plane H.

passing through these 1-walls. Each side of a 2-wall of type 1 is subdivided by the 2-walls of type 2 into new 2-walls which we call of type $1'$.

Each cell now reduces to a convex truncated cylinder whose generators are vertical lines. The number of cells is proportional to the number of 2-walls. We first estimate the number of 2-walls of type 1. In the vertical plane H that contains an edge E of a triangle in $\mathcal{T}$, the line segments formed by the intersections of the triangles in $\mathcal{T}$ with H induce an arrangement $\mathcal{A}_H$. The 2-walls of type 1 which intersect H decompose the non-convex cells of $\mathcal{A}_H$ incident to E into convex parts (see figure 16.3). The 2-walls of type 1 contained in H are either cells of this decomposition, or convex cells of $\mathcal{A}_H$ incident to E. The former are in number $O(n)$ as we saw above, and the latter correspond to a subset of the intervals in the decomposition of E by the triangles in $\mathcal{T}$. The number of 2-walls of type 1 in a plane H is thus $O(n)$, accounting for $O(n^2)$ walls of type 1 overall. (There are $O(n)$ possible planes H.) There can also exist only $O(n^2)$ walls of type 2, because the number of 2-walls of type 2 erected in a cell C is proportional to the number of vertices of $\Pi(C)$, which is itself proportional to the number of vertical edges in C and thus to the number of walls of type 1 in C. Finally, the number of walls of type $1'$ equals the sum of the numbers of walls of types 1 and 2. This finishes the proof of the theorem. $\square$

16.3   The lower envelope of a set of triangles

We now consider a set $\mathcal{T} = T_1, \ldots, T_n$ of n triangles in $\mathbb{E}^3$. For the sake of simplicity, we will assume that the triangles are not vertical, so that any vertical

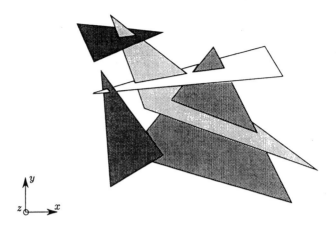

Figure 16.4. The lower envelope of five triangles, as seen from below projected onto the xy-plane.

line intersects any triangle in at most one point.

Each triangle can be considered as the graph of a linear function $z = T_i(x, y)$ defined over a triangle in $\mathbb{R}^2$. We define the *lower envelope* $\mathcal{E}$ of the triangles in $\mathcal{T}$ as the graph of the function $\min_{1 \leq i \leq n} T_i(x, y)$ (see figure 16.4). The connected components of the intersection of the envelope with a triangle of $\mathcal{T}$ are called the facets of the envelope. These facets are polygonal regions whose vertices and edges are vertices and edges of the lower envelope. The complexity of the lower envelope is the total number of its faces.

The faces of the lower envelope can be projected into the xy-plane, and this yields a planar map $\underline{\mathcal{E}}$ whose cells correspond to the facets of the lower envelope. We will assume that the triangles are in *general position*, by which we mean that:

1. two triangles do not overlap (their intersection cannot be two-dimensional),

2. an edge of a triangle intersects other triangles only in their relative interiors (not on an edge or at a vertex),

3. a vertex of a triangle does not belong to another triangle,

4. the projections of any two edges of the triangles do not overlap, and

5. the projections of any three edges of triangles do not intersect in a common point.

The general position assumption is not necessarily satisfied, especially when the triangles are facets of a polyhedron. By a standard perturbation argument, however, the triangles can be slightly perturbed into a general position while only augmenting the number of faces on the lower envelope. Therefore, the upper bounds we give below on the complexity of lower envelopes of triangles in general position apply to degenerate configurations as well.

We note that, under this general position assumption, the number of vertices (edges) on the envelope is at most twice the number of vertices (edges) of the planar map $\mathcal{E}$ obtained by projecting the faces of the lower envelope on the xy-plane. This is because a vertex (edge) in this planar map is the projection of at most two vertices (edges) of the lower envelope.

16.3.1 Complexity

This subsection is devoted to proving the following theorem, which bounds the complexity of the lower envelope of triangles:

Theorem 16.3.1 *The lower envelope of n triangles in $\mathbb{E}^3$ has worst-case complexity $\Theta(n^2\alpha(n))$. When the triangles do not intersect, this worst-case complexity drops to $\Theta(n^2)$.*

Proof of the upper bound. We first count the number $s(\mathcal{E})$ of vertices in the planar map $\mathcal{E}$ obtained by projecting the faces of the lower envelope $\mathcal{E}$ on the xy-plane. For this, we consider the lower envelope $\mathcal{E}'$ of $\mathcal{T} \setminus \{T\}$, $T \in \mathcal{T}$, and we estimate the increase $s(\mathcal{E}) - s(\mathcal{E}')$ in the number of vertices of the map when T is reinserted.

The new vertices of the envelope (which are vertices of $\mathcal{E}$ but not of $\mathcal{E}'$) are either on T or vertically below an edge of T. The triangles being in general position, the new vertices in the map $\mathcal{E}$ are the projections of points of the following kinds:

1. a vertex of T,

2. the intersection of an edge of T with the interior of a triangle in $\mathcal{T}\setminus\{T\}$,

3. the intersection of the interior of T and an edge of a triangle in $\mathcal{T}\setminus\{T\}$,

4. a point on an edge of T that lies vertically above an edge of a triangle in $\mathcal{T} \setminus \{T\}$,

5. the intersection of the interior of T and an edge of $\mathcal{E}'$ that is contained in the intersection of two triangles in $\mathcal{T}\setminus\{T\}$,

6. a point on an edge of T that lies vertically above an edge of $\mathcal{E}'$ that is contained in the intersection of two triangles in $\mathcal{T}\setminus\{T\}$.

The vertices of types 1–4 do not create problems: there are at most $O(n)$ of them. When we insert T, each edge of $\underline{\mathcal{E}'}$ is contained in the intersection of two

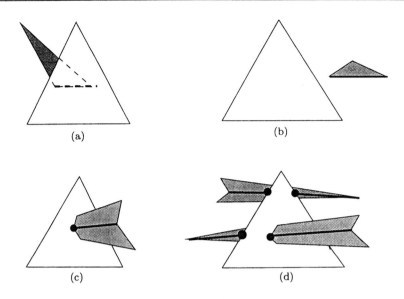

Figure 16.5. Inserting a triangle: an edge can be (a) hidden, (b) unchanged, (c) shortened, or (d) split into two edges.

triangles, and can be (a) hidden by T, (b) unchanged, (c) shortened, or (d) split into two edges of $\mathcal{E}$ (see figure 16.5). Only in case (d) does the number of vertices in the projected map increase. More precisely, there are two new vertices and at least one of them is of type 6. The number of vertices increases by at most twice the number of vertices of type 6. Their number is $O(n\alpha(n))$. Indeed, if E is an edge of T and H is the vertical plane strip formed by the vertical rays originating from E and going upwards, the triangles in $\mathcal{T} \setminus T$ intersect H along segments, and the vertices of type 6 contained in E are vertices of the lower envelope of these segments in H. Theorem 15.3.1 shows that there are $O(n\alpha(n))$ of them. This is true for all three edges of T, so there are at most $O(n\alpha(n))$ new vertices. Therefore

$$s(\mathcal{E}) = s(\mathcal{E}') + O(n\alpha(n)).$$

By inserting all the triangles in $\mathcal{T}$ successively, and denoting by $s(n)$ the maximum number of vertices in the planar map projection of the lower envelope of n triangles, we get the recurrence

$$s(n) \leq s(n-1) + O(n\alpha(n)),$$

which solves to $s(n) = O(n^2\alpha(n))$. Under the general position assumption, each vertex of the projected map $\mathcal{E}$ has degree 2 or 3, so that the map has $O(n^2\alpha(n))$ edges as well, and Euler's relation yields the same bound for the number of its cells. As we saw above, the same bounds apply for edges and facets of $\mathcal{E}$.

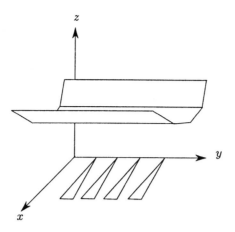

Figure 16.6. Configuration for the lower bound.

If the segments do not intersect, only cases (a) and (d) above are allowed, and they account for $O(n)$ new vertices. Hence a similar recurrence shows that the complexity of the lower envelope is simply $O(n^2)$.

Proof of the lower bounds.   To construct the example, we use the fact mentioned above that the lower envelope of n segments may have $\Omega(n\alpha(n))$ edges (see section 15.6). Now take $\frac{n}{2}$ segments in the half-plane $y = 0, z > 0$, so that their lower envelope has $\Theta(n\alpha(n))$ faces. Picking a point A far enough away on the y-axis, we can construct $\frac{n}{2}$ triangles almost parallel to the y-axis by taking the convex hull of these segments with the point A. The idea is to duplicate this complexity $\Theta(n\alpha(n))$ a number of times: place $\frac{n}{2}$ disjoint, thin triangles in the plane $z = 0$, long enough so that each intersects from side to side the vertical projections of the $\frac{n}{2}$ triangles constructed above. Then the lower envelope of these n triangles has complexity $\Theta(n^2\alpha(n))$.

A similar construction when the triangles do not intersect leads to the $\Omega(n^2)$ bound.

16.3.2   Vertical decomposition

We define the *vertical decomposition* $\mathcal{D}ec(\mathcal{E})$ of the lower envelope $\mathcal{E}$ of a set $\mathcal{T}$ of triangles as the portion of the entire vertical decomposition of $\mathcal{T}$ that lies below the lower envelope.

The decomposition $\mathcal{D}ec(\mathcal{E})$ projects vertically onto the xy-plane along the decomposition of the planar map $\underline{\mathcal{E}}$ whose faces are the projections of the faces of the lower envelope $\mathcal{E}$: the cells in this decomposition are trapezoids, sometimes degenerate (unbounded or triangles, see section 3.3 and subsection 12.3.3). To each trapezoid in the planar map corresponds a vertical cylinder whose horizontal section is the trapezoid: such a cylinder is called a *prism*. Each prism is unbounded in the direction of the negative z-axis, and is bounded in the opposite direction by a facet contained in a single triangle of $\mathcal{T}$. This facet is called the *ceiling* of the prism. The collection of these prisms and of their faces constitutes the *vertical decomposition* of the lower envelope of the triangles. If $\mathcal{T}$ consists of n triangles, the complexity of this decomposition is $O(n^2 \alpha(n))$, and so is the complexity of the lower envelope. Note that the ceiling of a prism C is a trapezoid of the decomposition of a set $\mathcal{S}_C$ of segments contained in the triangle T_C that contains the ceiling of the prism. This set $\mathcal{S}_C$ is formed by

1. the edges of T_C,

2. the intersections of T_C with other triangles of $\mathcal{T}$,

3. the vertical projections onto T_C of edges or portions of edges of triangles that lie below T_C.

Among the 2-walls in the decomposition of the lower envelope, we make a distinction between the 2-walls *of type 1* hanging below the edges of $\mathcal{E}$ and the 2-walls *of type 2* hanging below the 1-walls of the planar map $\underline{\mathcal{E}}$. By construction, the non-vertical edges of the walls of type 2 (which are edges of the ceilings of the prisms) all contain a vertex of $\mathcal{E}$.

We note that a prism may be adjacent to many prisms since both sides of a wall of type 1 may be subdivided differently by the abutting walls of type 2 (see also the discussion in subsection 16.2.1). Nevertheless, the adjacency graph is planar and its complexity is linear with respect to the number of prisms.

16.3.3   Computing the lower envelope

The algorithm we present here is incremental and uses the influence graph method described in chapter 5. The analysis is randomized. To simplify the description of the algorithm, we suppose that the plane $z = +\infty$ is a triangle in $\mathcal{T}$ and that it is inserted first.

The triangles are inserted one after the other, and the algorithm maintains the *vertical decomposition* of the current lower envelope as well as the adjacency relationships of the prisms in the decomposition.

To fit the framework of randomized algorithms, we redefine the problem in terms of objects, regions, and conflicts. Here, the objects are triangles and the regions are prisms (vertical cylindrical cells with horizontal trapezoidal sections). A region is defined by a set of triangles if it is a prism in the decomposition of the lower envelope of a subset of triangles. A prism C is determined by the triangle T_C that contains its ceiling, and by the at most four edges of $\mathcal{S}_C$ contained in

T_C that determine the edges on its ceiling. Each segment being either an edge of T_C, the intersection of T_C and a triangle in $\mathcal{T} \setminus \{T_C\}$, or the projection on T_C of an edge of a triangle in $\mathcal{T} \setminus \{T_C\}$, it follows that a region is determined by at most five triangles.

An object conflicts with a region whenever they have a non-empty intersection. Observe that the regions that are defined by the triangles of $\mathcal{T}$ and do not conflict with these triangles are exactly the prisms in the decomposition of the lower envelope of $\mathcal{T}$.

Let T be a new triangle inserted into $\mathcal{T}$, let $\mathcal{E}'$ be the current lower envelope before inserting T, and let $\mathcal{Dec}(\mathcal{E}')$ be the current decomposition. We first identify the prisms in $\mathcal{Dec}(\mathcal{E}')$ that intersect T, then we update the lower envelope and its decomposition. Let us call $\mathcal{E}$ the new lower envelope after updating and let $\mathcal{Dec}(\mathcal{E})$ be its decomposition. To efficiently find the regions that conflict with T, the algorithm also maintains an influence graph. We may recall that the influence graph is a structure whose goal is to detect rapidly the conflicts between a new object and the regions defined and without conflict over the current set of triangles. The influence graph is an oriented acyclic graph that has a node for each region that was a region defined and without conflict over a subset of the set of triangles; this subset was the current set of triangles during a previous incremental step. The arcs in the influence graph connect the nodes in such a way that the influence domain of a node (the subset of objects that conflict with the region that corresponds to this node) is contained in the union of the influence domains of its parents. At each step of the algorithm, the regions defined and without conflict over the current subset are stored in the leaves of the influence graph.

We now describe how to perform the current incremental step by performing a location phase and an update phase.

Locating. The location phase is used to retrieve all the leaves in the influence graph that conflict with T. These leaves correspond to prisms in $\mathcal{Dec}(\mathcal{E}')$ intersected by T. A simple traversal of the influence graph that backtracks each time it encounters a node that either was already traversed or that does not conflict with T identifies all the nodes that conflict with T. If no leaf in the graph intersects T, then T does not appear on the lower envelope $\mathcal{E}$ or on any lower envelope that will subsequently be computed. The algorithm may skip updating the structure and directly insert the next segment.

Updating. Among the prisms in $\mathcal{Dec}(\mathcal{E}')$ that intersect T, we make a distinction between the prisms that are split by T into two distinct connected components, and the others of which we say that they are *pierced* by T.

We first construct all the prisms in the decomposition $\mathcal{Dec}(\mathcal{E})$ of the new lower

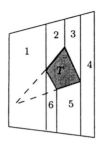

Figure 16.7. The six sub-prisms whose ceiling is not supported by T, projected onto a
horizontal plane.

envelope $\mathcal{E}$ whose ceiling is not supported by T. Consider a prism C pierced by
T. The portion of C that appears on $\mathcal{E}$ is subdivided into at most six sub-prisms
with a trapezoidal section (see figure 16.7). A node in the influence graph is
created for each of the nodes corresponding to these six sub-prisms, and these
nodes become children of the node corresponding to C.

Some of these nodes are only temporary, as we will see shortly. Indeed, some
sub-prisms may share a wall of type 2 that has to be removed, and the prisms have
to be merged to obtain the decomposition of $\mathcal{E}$. This merge process is analogous
to the merges performed by the incremental algorithm that builds the planar
vertical decomposition of segments (see section 3.3), and can be described as
follows. Each wall of type 2 in $\mathcal{D}ec(\mathcal{E}')$ that is intersected by T is subdivided into
at most four sub-walls: a sub-wall above T, another below T, and occasionally a
sub-wall on each side of T (see figure 16.8). The first sub-wall must disappear, the
second will be processed later. As for the occasional last two sub-walls, neither
of their upper edges is contained in T and the one that does not contain a vertex
of the lower envelope must be removed, because it does not induce a wall in the
new subdivision. The two incident sub-prisms are merged and the adjacency
graph is updated accordingly. In the influence graph, the temporary nodes that
correspond to the sub-prisms are merged into a single node that inherits the
parents of all these sub-prisms.

In this way, we have computed all the prisms whose upper face is not contained
in T. It remains to compute the other prisms, whose union $\mathcal{U}$ is a portion of the
union of the prisms that conflict with T and whose ceilings are supported by T.

This union $\mathcal{U}$ is computed using the adjacency graph of the prisms. $\mathcal{U}$ is not
connected in general, as is shown for instance in figure 16.9.

Let $\mathcal{F}$ be a connected component of this union $\mathcal{U}$. Then $\mathcal{F}$ corresponds to a
face F of the lower envelope $\mathcal{E}$ that is supported by T. The prisms in the decomposition of $\mathcal{E}$ whose ceilings are supported by F result from the 2-decomposition
of F along the y'-axis in F that projects onto the y-axis in a horizontal plane. To

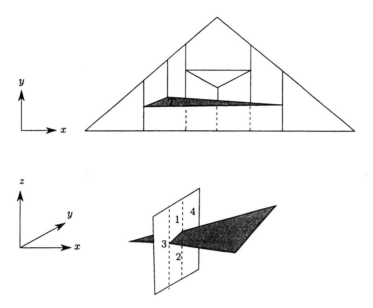

Figure 16.8. Merging the adjacent temporary prisms: the shaded triangle is inserted and a wall is subdivided into at most four sub-walls. The projected decomposition of the lower envelope is represented above, and the walls to be removed are shown in dashed lines.

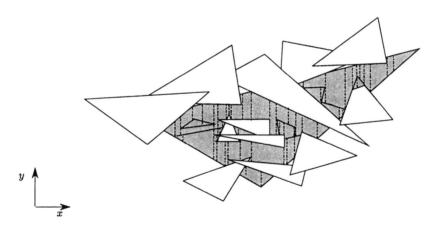

Figure 16.9. The union $\mathcal{U}$ of the prisms intersected by T, shaded and shown in projection in the xy-plane. The vertical decomposition in the facets supported by T and induced by the triangles is shown in dashed lines.

maintain the property that the influence domain of a child is contained within the union of the influence domains of its parents, each prism should be attached to all the prisms that conflict with T and intersect $\mathcal{F}$; this would result in an unbounded number of children for a node, and the usual randomized analysis does not apply in this case. We remedy this drawback by creating a new node N in the influence graph that corresponds to a prism whose ceiling is F. This node is of a special kind and, unlike the other nodes, does not correspond to a prism in the decomposition of the lower envelope. We attach N to the nodes in the influence graph that correspond to prisms that conflict with T (either pierced or split) and that intersect $\mathcal{F}$. Face F (and hence the prism corresponding to N) may have a large number of edges and may be not simply connected. It can be decomposed using the planar randomized incremental algorithm (in the affine hull of F) and we hang 2-walls vertically below the 1-walls of F. In this way, we obtain a secondary influence graph that represents the planar decomposition in the facet F, and hence a decomposition of the space lying below F. This secondary influence graph is rooted at N.

Analysis of the algorithm

We make a distinction between the *primary* nodes in the influence graph of the triangles, which we also refer to as the *primary influence graph*, and the *secondary* nodes in the secondary influence graphs. Primary nodes correspond to a region defined and without conflict over a subset of the current set of triangles; this subset was the current subset during some previous incremental step.

Recall that $\mathcal{U}$ is the union of the sub-prisms obtained by subdividing the prisms that conflict with T and whose ceilings are supported by T. At the i-th step, the number u of edges on the boundary of $\mathcal{U}$ is bounded by the number of primary nodes that conflict with T. Moreover, lemma 5.2.4 implies that the average number of secondary nodes created at step i is proportional to u. The expected number of secondary nodes is thus proportional to the number of primary nodes.

Therefore it suffices to bound the number of primary nodes. These nodes are in one-to-one correspondence with the prisms defined and without conflict over the current subset at some previous incremental step.

Even though the situation does not exactly fit the framework of theorem 5.3.4, because of the existence of secondary nodes, we can still apply its proof *verbatim* to count the number of primary nodes created, and the number of primary nodes visited during a location phase. So we conclude that the number of primary nodes in the influence graph is

$$O\left(\sum_{r=1}^{n} \frac{f_0(r, \mathcal{T})}{r}\right),$$

where $f_0(r, \mathcal{T})$ is the number of prisms defined and without conflict over a random sample of r triangles in $\mathcal{T}$. Each node in the primary influence graph has a bounded number of children, so this bound is also valid for the number of arcs in the primary influence graph. Theorem 16.3.1 states that $f_0(r) = O(r^2\alpha(r))$ and so the total number of primary nodes created during the incremental insertions of n triangles is $O(n^2\alpha(n))$.

We now have to estimate the costs incurred by maintaining the secondary influence graphs. The results of subsection 5.3.2 that refer to the construction of the decomposition of a set of segments in the plane show that the location in the secondary graph takes expected time $O(\log n)$. Likewise, the secondary graphs can be computed in expected time $O(n_T \log n_T)$ where n_T is the number of primary nodes that conflict with T. Therefore the costs of updating and locating in the secondary graphs are proportional by a factor of $\log n$ to the number of primary nodes visited during an incremental step.

The number of primary nodes visited upon inserting the i-th triangle is given by theorem 5.3.4:

$$O\left(\sum_{r=1}^{n} \frac{f_0(\lfloor r/2 \rfloor), \mathcal{T}}{r^2}\right) = O(n\alpha(n)).$$

We conclude that the costs of locating and updating the structures when inserting the n-th triangle are both $O(n\alpha(n)\log n)$.

If the triangles do not intersect, we have $f_0(r) = O(r^2)$ and the cost of inserting the n-th triangle becomes $O(n\log n)$.

We have not yet analyzed the cost of updating the adjacency graph. This cost is not bounded easily since the algorithm keeps track of all the adjacencies between prisms and yet a prism may be adjacent to several others. Nevertheless, an argument similar to that presented in exercise 5.3 shows that the cost of these updates is also $O(n\alpha(n))$.

This finishes the analysis of the algorithm. Our findings are summarized in the following theorem.

Theorem 16.3.2 *The lower envelope of a set of n triangles in E^3 can be computed using an on-line algorithm that maintains an influence graph. The expected time required for inserting the n-th triangle is $O(n\alpha(n)\log n)$, or $O(n\log n)$ if the triangles do not intersect.*

In several cases, $f_0(r, \mathcal{T}) = O(r)$ holds, for instance when studying certain kinds of Voronoi diagrams (see section 18.4). Using this better bound in the above discussion implies:

Corollary 16.3.3 *Let $\mathcal{T}$ be a set of n triangles in E^3, such that the expected number of faces of the lower envelope of a random sample of r triangles in $\mathcal{T}$ has complexity $O(r)$. Then the lower envelope of $\mathcal{T}$ can be computed using an on-line algorithm that maintains an influence graph. The expected time required for inserting the n-th triangle is $O(\log^2 n)$.*

16.4   A cell in an arrangement of triangles

16.4.1   Complexity

As we saw above, the complexity of a lower envelope of a set $\mathcal{T}$ of n triangles in $\mathbb{E}^3$ is bounded below by $\Omega(n^2\alpha(n))$ (see theorem 16.3.1). This bound is also a lower bound on the complexity of a cell in the arrangement of n triangles in $\mathbb{E}^3$. Indeed, it suffices to add to each triangle T in $\mathcal{T}$ three planar facets obtained by casting vertical rays in the direction of the increasing z-coordinates from each point of the edges of T. The lower envelope of $\mathcal{T}$ corresponds to the non-vertical faces of the unbounded cell that lies below all the triangles in $\mathcal{T}$ in the arrangement of $\mathcal{T}$ and all the added triangles.

An upper bound that is very close to this lower bound is given by the following theorem.

Theorem 16.4.1 *The complexity of a single cell in an arrangement of n triangles in $\mathbb{E}^3$ is $O(n^2 \log n)$.*

The remainder of this subsection is devoted to proving a slightly worse bound $O(n^2 \log^2 n)$. Refining this analysis yields the bound stated in the theorem (see also the bibliographical notes at the end of this chapter.)

The proof we present here is similar in spirit to the proof of the zone theorem 14.3.2. It is split into several lemmas that prove that the number of edges of a cell in the arrangement of n triangles in $\mathbb{E}^3$ is $O(n^2 \log^2 n)$. Euler's relation then finishes the proof.

Let $\mathcal{T}$ be a set of n triangles in $\mathbb{E}^3$, and pick a point O in $\mathbb{E}^3$ that does not belong to any triangle in $\mathcal{T}$. Our goal is to bound the complexity of the cell C in the arrangement of $\mathcal{T}$ that contains O. We assume that the triangles are in general position (see subsection 16.3.1), since a standard perturbation argument shows that the complexity is maximized in that case.

As in subsection 16.2.2, we say that a face of C is *outer* if it is contained in an edge on the boundary of a triangle in $\mathcal{T}$. Otherwise, we say that the edge is *inner*. We note that the outer faces are of dimension 0 or 1, and that an inner k-face ($k \le 3$) is contained in the intersection of the interiors of $3 - k$ triangles.

Lemma 16.4.2 *The number of outer vertices and edges contained in a single cell is $O(n^2)$.*

Proof. Each edge of a triangle is subdivided by the other triangles of $\mathcal{T}$ into at most n segments. Therefore, there can be no more than $O(n^2)$ edges. But there are at most two outer vertices per outer edge.   □

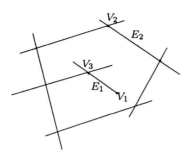

Figure 16.10. A cell in the arrangement of line segments. The definitions of *outer*, *inner*, and *popular* are analogous to the three-dimensional case. The vertex V_1 is outer, while V_2 and V_3 are inner. V_2 is not popular while V_3 is popular. The edge E_1 is popular, while E_2 is not.

Henceforth, we will therefore be only interested in inner faces. They include the facets, the edges contained in the intersection of the interiors of two triangles, and the vertices contained in the intersection of the interiors of three triangles.

By the *side* of an inner k-face F, we mean one of the regions bounded by the $3 - k$ planes that support the triangles containing F. Lemma 14.2.4 states that an inner k-face in the arrangement of $\mathcal{T}$ has 2^{3-k} sides.

We say that an inner face F of C is *popular* if all its sides are contained in C in a neighborhood of F. The situation is depicted in two dimensions in figure 16.10.

The concept of popularity plays a central role in the proof. The following three lemmas successively bound the numbers of popular edges, popular facets, and edges on the boundary of non-popular facets.

Lemma 16.4.3 *There are $O(n^2)$ popular vertices and popular edges on the boundary of a cell.*

Proof. We count the number of lower endpoints of popular edges. The lower endpoint A of a popular edge E has the smallest z-coordinate, and is either an inner or an outer vertex. The number of outer vertices is $O(n^2)$, by lemma 16.4.2. If A is an inner vertex, then it lies at the intersection of E with the interior of a triangle T, and so it is the intersection of three planes: the affine hull of T and the affine hulls of the two triangles that contain E. Among the sides of A, there must be one, say R, for which A is the lower endpoint. Let us decompose C into convex sub-cells as was done in subsection 16.2.2, and let us call C' the sub-cell which intersects R in a neighborhood of A. Since C' is convex, it must be contained in R and so A is the lower endpoint to C'. By theorem 16.2.2, the number of convex sub-cells in this decomposition is $O(n^2)$. Since each sub-cell has a unique lower endpoint, which must be incident to three popular edges since the triangles are in general position, the result follows for popular edges.

Now a popular vertex is always incident to six popular edges when T is in general position, so the number of popular vertices is proportional to the number of popular edges.   □

Lemma 16.4.4 *The number of inner edges contained in popular facets that are on the boundary of a cell is $O(n^2 \log n)$.*

Proof. Consider a triangle T in $\mathcal{T}$ and let C' be the cell in the arrangement of $\mathcal{T} \setminus T$ that contains O. When inserting T in this arrangement, we estimate the increase in the number of inner edges in the cell that contains O and that belong to popular facets. We let E be an inner edge in C' that is contained in a popular facet F in C'.

Case 1: $T \cap E = \emptyset$
Whether E is an edge of C or not, the number of edges that belong to E does not increase when inserting T.

Case 2: $T \cap E \neq \emptyset$
Let P be the intersection point of E and T. Then E is cut into two edges E_1 and E_2. If both E_1 and E_2 are edges of C, and only in this case, the number of edges that belong to E increases by one when inserting T. The edge E' of C that is contained in $F \cap T$ and incident to P must be popular when the number of edges that belong to E increases. Therefore P must be the endpoint of a popular edge of C.

Denote by q and $q'(T)$ the number of inner edges of C and C' contained in popular facets, and let $q''(T)$ be the number of inner edges of C contained in T. Moreover, denote by $r(T)$ the number of vertices of popular edges of C contained in T.

From what was said before, we know that

$$q - q''(T) \leq q'(T) + r(T). \tag{16.1}$$

This result is independent of the choice of T. So we may carry out a similar analysis for all the triangles in $\mathcal{T}$, and by summing the resulting equations, we count each inner edge E of a popular facet of C $(n - 2)$ times (once for each triangle that does not contain E), and we count each vertex S of C three times (once for each triangle that contains S). Overall, we obtain

$$(n - 2)q \leq \sum_{T \in \mathcal{T}} q'(T) + 3r. \tag{16.2}$$

Finally, if we denote by $q(n)$ the maximum value of q over all possible arrangements of n triangles and all possible choices for the origin O, inequality 16.2 becomes

$$q(n) \leq \frac{n}{n-2} q(n-1) + O(n), \qquad (16.3)$$

by using the fact that the number of popular edges of C is $O(n^2)$ (see lemma 16.4.3). We solve this recurrence by putting

$$q(n) = \binom{n}{2} w(n).$$

Inequality 16.3 becomes

$$w(n) \leq w(n-1) + O\left(\frac{1}{n}\right). \qquad (16.4)$$

This implies that $w(n) = O(\log n)$, which in turn shows that $q(n) = O(n^2 \log n)$.
$\square$

We can now finish the proof of the theorem, if we know a bound on the number of inner edges in non-popular facets of C. This bound is provided by our last lemma.

Lemma 16.4.5 *The number of inner edges in the non-popular facets of C is $O(n^2 \log^2 n)$.*

Proof. The proof is similar to that of the previous lemma. Let T be a triangle of $\mathcal{T}$, C' be the cell of the arrangement of $\mathcal{T} \setminus T$ that contains the origin O, and E be an inner edge of C' that does not belong to any popular facet of C'. We insert T into the arrangement and estimate the increase in the number of inner edges in the non-popular facets of the new cell C that contains O.

As in the previous lemma, the number of these inner edges that belong to E increases (by one) only when T cuts E into two edges E_1 and E_2 which are both edges of F. Assuming this is the case, let P be the intersection point of E and T, and F_T be the facet of C contained in T and incident to P. Then F_T is popular and contains vertex P. Note that F_T is the only popular facet of C that has P as a vertex.

We now denote by q and $q'(T)$ the number of inner edges in C and C' that are contained in the non-popular facets of C that belong to T. Moreover we denote by $r(T)$ the number of vertices on the popular facets of C that belong to T, and by r the number of vertices on the popular facets of C. Equations analogous to equations 16.1 and 16.2 hold as well:

$$\begin{aligned}
q - q''(T) &\leq q'(T) + r(T), \\
(n-2)q &\leq \sum_{T \in \mathcal{T}} q'(T) + r.
\end{aligned}$$

If we denote by $q(n)$ the maximum of q over all possible arrangements of n triangles in $\mathbb{E}^2$ and all the possible choices for the origin O, using the fact that $r = O(n^2 \log n)$ as was proved in lemma 16.4.4, we derive the recurrence

$$q(n) \leq \frac{n}{n-2} q(n-1) + O(n \log n), \tag{16.5}$$

which solves to $q(n) = O(n^2 \log^2 n)$. $\square$

16.4.2   Vertical decomposition

Consider a set $\mathcal{T}$ of n triangles in $\mathbb{E}^3$ and its arrangement $\mathcal{A}$, and pick a point O that does not belong to any triangle. In this subsection we bound the complexity of the vertical decomposition of the cell $C(\mathcal{T})$ that contains O.

We first show that the complexity of the vertical decomposition of $C(\mathcal{T})$ is proportional to the sum of the complexity of $C(\mathcal{T})$ and of the number of pairs of edges of $C(\mathcal{T})$ that are mutually visible, meaning that they can be connected by a vertical line segment whose relative interior does not intersect any triangle in $\mathcal{T}$. For this, consider a line segment E, either an edge of a triangle or the intersection of two triangles, such that E contains at least one edge of $C(\mathcal{T})$. Let $\mathcal{R}$ be the vertical strip described by a vertical line passing through the edge E. By $\mathcal{R}_a$, resp. $\mathcal{R}_b$, we denote the portion of $\mathcal{R}$ that lies above E, resp. below E. We obtain a set of segments $\mathcal{S}_a$ (resp. $\mathcal{S}_b$) by intersecting the triangles in $\mathcal{T}$ with the half-strip $\mathcal{R}_a$ (resp. $\mathcal{R}_b$). For reasons similar to those given in subsection 16.2.1 for bounding the complexity of the entire arrangement, the complexity of the decomposition of a single cell is bounded by the total number of vertices on the lower envelope (in the plane of $\mathcal{R}$) of $\mathcal{S}_a$ and on the upper envelope of $\mathcal{S}_b$ (see figure 16.2). Between the vertices of these envelopes that are contained in $C(\mathcal{T})$, we make a distinction between the intersection points of the envelopes and E, and the others. This shows that the number of vertices on the envelope is accounted for by the vertices of $C(\mathcal{T})$ that belong to E and the edges of $C(\mathcal{T})$ that are visible from E. Summing over all the segments E that are edges of triangles or intersections of two triangles, we see that our first statement is true.

We obtained a bound on the complexity of a cell in the previous section, so we are only interested in pairs of mutually visible edges of $C(\mathcal{T})$.

Consider two edges E_1 and E_2 of $\mathcal{A}$. We say that E_1 and E_2 are *k-visible* if there exists a vertical line segment that connects them and whose relative interior intersects $C(\mathcal{T})$ and exactly k triangles in $\mathcal{T}$. These triangles are said to *obscure* the pair. A 0-visible pair, or visible pair for short, always consists of two edges of $C(\mathcal{T})$.

As in the previous subsection, we make a distinction between the *inner* pairs of edges for which each edge is contained in the intersection of two triangles, and the other pairs called *outer* for which at least one edge is contained in the boundary of a triangle.

Lemma 16.4.6 *The number of 0- and 1-visible outer pairs is $O(n^2\alpha(n))$.*

Proof. We use an argument very similar to that used in subsection 16.2.1, which was recalled above. Since we are only interested in outer pairs, we need only consider the edges of triangles and not the intersections of two triangles. The number of outer visible pairs that involve some edge E of a triangle equals the number of vertices on the upper and lower envelopes of two sets of segments, which is $O(n\alpha(n))$. Since there are at most $3n$ such edges, the number of outer visible pairs is $O(n^2\alpha(n))$ The random sampling theorem 4.2.3 and its corollary 4.2.4 imply similar bounds for 1-visible pairs: here the objects are triangles, the regions are vertical segments joining an edge of a triangle and either an edge of another triangle or the intersection of two triangles, and a triangle conflicts with a segment if they intersect.[2] $\qquad\square$

We now count the number $q_0(\mathcal{T})$ of inner visible pairs. Let us consider two inner edges $E_1 = T_1 \cap T_2$ and $E_2 = T_3 \cap T_4$ of $C(\mathcal{T})$ that are mutually visible, where the T_i's are triangles in $\mathcal{T}$. Assume that E_1 is above E_2 and denote by S the vertical segment that connects E_1 to E_2. By hypothesis, the interior of S does not intersect any triangle of $\mathcal{T}$.

Consider an extensible vertical segment S', that occupies the same position as S initially. We slide S' successively in four different directions. For the first move, we constrain the upper endpoint of S' to belong to E_1, while the lower endpoint belongs to T_3 and S' intersects T_4. There is a single degree of freedom, and we move S' along this direction until one of the following situations occurs (see figure 16.11):

1. S' reaches the end of E_1; this endpoint is a vertex of $C(\mathcal{T})$,

2. S' encounters an edge E of T_3 or T_4; then (E_1, E) is an outer visible pair if E is an edge of T_4, or a 1-visible outer pair if E is an edge of T_3,

3. S' meets the edge E of a triangle of $\mathcal{T} \setminus \{T_1, T_2, T_3, T_4\}$, between E_1 and T_3; then S' contains a vertical segment S'' that connects the outer pair (E_1, E), which is 1-visible if S'' intersects T_4, or visible otherwise,

4. S' reaches an edge that is the intersection of T_3 with some triangle T in $\mathcal{T} \setminus \{T_1, T_2, T_3, T_4\}$.

---

[2]Note that the proof also applies *verbatim* to pairs of edges in the arrangement that can be connected by a vertical line segment intersecting zero or one triangle in $\mathcal{T}$ (but not necessarily in $C(\mathcal{T})$).

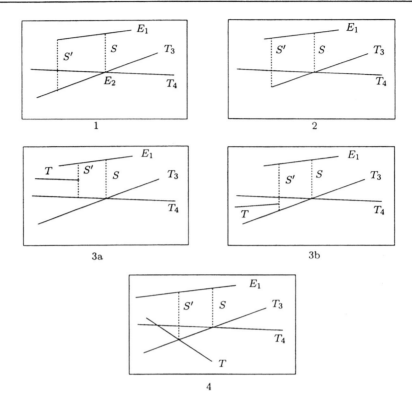

Figure 16.11. The four different kinds of events and their variants, represented in the vertical
plane that contains E_1. E_2 intersects this plane in a single point, and T_3 and
T_4 in two segments.

Likewise, for the second move, we slide S' while keeping its upper endpoint on
E_1 and its lower endpoint on T_4, in the direction where S' intersects T_3. We stop
S' in one of the analogous four cases (exchanging T_3 and T_4). Switching the roles
of E_1 and E_2 gives the four different moves.

It is important to notice that each event above is encountered only once in all
the moves along a given direction. Indeed, when moving a vertical segment in one
direction, an inner visible pair is met again only after one of the above events.

We count the total number of events encountered during the moves corresponding to each inner visible pair of edges of $C(\mathcal{T})$.

1. A vertex of $C(\mathcal{T})$ is encountered during at most six events of type 1. Indeed,
the guiding edge must be one of the at most six edges incident to this vertex.
The total number of events of type 1 is at most six times the number of vertices
of $C(\mathcal{T})$, which is $O(n^2 \log n)$ because of theorem 16.4.1.

2. The outer pair (E_1, E) is encountered during a single event of type 2 when
sliding along E_1. It follows that the total number of events of type 2 equals the

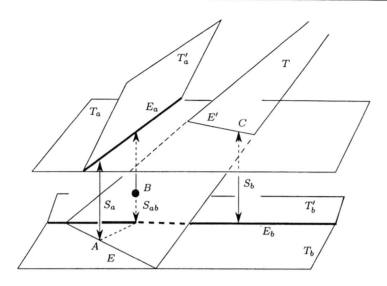

Figure 16.12. The 1-visible pair (E_a, E_b) is reached by two motions from the 0-visible pairs (E_a, E) and (E_b, E').

number of outer pairs that are visible or 1-visible. This number is $O(n^2 \alpha(n))$ because of lemma 16.4.6.

3. The same reasoning holds for the events of type 3 whose number is also $O(n^2 \alpha(n))$.

4. The events of type 4 can be encountered at most four times. Yet the analysis only works with a more precise account: the following lemma bounds the number of events of type 4 that are encountered more than twice.

Lemma 16.4.7 *There are at most $O(n^2 \log n)$ events of type 4 that are encountered more than twice.*

Proof. Consider a 1-visible inner pair (E_a, E_b), and assume that E_a is contained in the intersection of the two triangles T_a and T'_a, that E_b is contained in the intersection of the two triangles T_b and T'_b, and that E_a lies above E_b (see figure 16.12). We denote by T the triangle that obscures the pair (E_a, E_b) and by F the facet of the arrangement contained in T that is stabbed by the vertical segment S_{ab} that connects E_a and E_b.

If the pair is encountered more than twice during events of type 4, then it must have been so after at least one motion along E_a and at least one motion along E_b. So let us consider a visible pair from which we started one of the motions, and assume it is a pair (E_a, E). (For pairs (E_b, E'), the situation is entirely symmetrical.) Let S_a be the vertical segment that connects E_a and E (see for instance figure 16.12 where we also show the vertical segment S_b that connects E_b and E'.) The edge E is contained in the intersection of T with T_b or T'_b: otherwise, E would involve a triangle T' not in $\{T, T_b, T'_b\}$ and then either T' would obscure (E_a, E_b) which would not be 1-visible, a contradiction, or the motion of S along E_a from (E_a, E) would stop on an edge of the boundary of T' before it reached (E_a, E_b), which is also impossible. So we may assume that E is contained in $T \cap T_b$.

We note that E is an edge of F and that F is a popular facet of $C(T)$. Connect the points $A = S_a \cap F$ and $B = S_{ab} \cap F$ by a line segment (contained in F) called the *trace on F* of the motion from (E_a, E) to (E_a, E_b).

We connect B to the at most three other visible pairs from which we started motions that ended on (E_a, E_b), and proceed similarly for all the inner 1-visible pairs obscured by F that were encountered more than twice during events of type 4. We thus obtain a set S of segments contained in F (see figure 16.13) which are the traces on F of all these different motions. A segment in S cannot intersect a triangle in T. Moreover, two segments in S cannot intersect (except at common endpoints), otherwise the intersection point would correspond to a 1-visible pair obscured by F that would have stopped one of the motions ending at (E_a, E_b) before it reached its endpoint.

By construction, any segment in S has exactly one endpoint on the boundary of F. The endpoints that are not on the boundary of F correspond to events encountered more than twice: these endpoints are thus incident to three or four segments in S. Let us identify the endpoints of segments of S that belong to a given edge on the boundary of F: we obtain in this way a planar graph G whose external nodes are the edges of F, whose internal nodes are the endpoints of the segments of S that are not on the boundary of F, and whose arcs correspond to the segments in S.

Lemma 16.4.8 *The number of internal nodes in the graph G, hence the number of events of type 4 corresponding to F that are encountered more than twice, is $O(|F|)$.*

Proof. We first show that the graph G cannot contain cycles of length 2. Indeed, each arc of G connects an internal node to an external node. An internal node, that corresponds to the 1-visible pair (E_a, E_b), is connected to three or four external nodes that represent the four distinct edges $F \cap T_a$, $F \cap T'_a$, $F \cap T_b$, $F \cap T'_b$. As before, we denote by T_a and T'_a the triangles whose intersection contains E_a, and by T_b and T'_b those whose intersection contains E_b.

Therefore G is a planar graph without cycles of length 2. It has $n_e = O(|F|)$ external nodes and internal nodes of degree 3 or 4, which are incident to external nodes only. To prove the lemma, we construct from G another graph G' that does not have internal nodes as follows. We distinguish the two sides of an arc

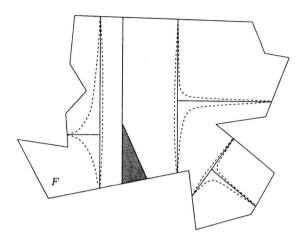

Figure 16.13. The segments in S are shown in solid lines. The shaded face corresponds to a cycle of length 2 in the graph $\mathcal{G}$. Replacing these solid lines by the dashed lines and identifying the vertices on a common edge of F yields the graph $\mathcal{G}'$.

of $\mathcal{G}$, and the three or four sides of a node of $\mathcal{G}$. We then replace two arcs that are incident to a common internal node by a single arc in $\mathcal{G}'$. It is easy to see that $\mathcal{G}'$ is planar and does not have a cycle of length 2. It has the same number n_e of nodes, and the same number of arcs, as $\mathcal{G}$. Euler's relation shows that the number of arcs of $\mathcal{G}'$, and hence of $\mathcal{G}$, is proportional to the number of its nodes, and hence is $O(|F|)$.

The total number of internal vertices of $\mathcal{G}$ is thus $O(|F|)$. □

Summing over all the popular facets of $C(\mathcal{T})$, we conclude that the total number of events of type 4 that are encountered more than twice is $O(n^2 \log n)$ because of lemma 16.4.4. □

The preceding discussion shows that the total number of events of type 4 is bounded by

$$4q_0(\mathcal{T}) \leq 2q_1(\mathcal{T}) + O(n^2 \log n),$$

if we denote by $q_k(\mathcal{T})$ the number of k-visible inner pairs. It follows that

$$\frac{n-2}{n} q_0(\mathcal{T}) \leq \frac{n-4}{n} q_0(\mathcal{T}) + \frac{1}{n} q_1(\mathcal{T}) + O(n \log n). \qquad (16.6)$$

We claim that $\frac{n-4}{n} q_0(\mathcal{T}) + \frac{1}{n} q_1(\mathcal{T})$ is bounded by the expected number of visible inner pairs in the arrangement of a random sample of $n-1$ triangles in $\mathcal{T}$. Indeed, an inner visible pair in $\mathcal{T}$ is an inner pair visible in the sample if and only if the removed triangle is not one of the four triangles that define the pair, which happens with probability $\frac{n-4}{n}$. An inner 1-visible pair of $\mathcal{T}$ is an inner visible pair in the sample if and only if the removed triangle is crossed by the segment that connects the pair, which happens with probability $\frac{1}{n}$. This provides the desired upper bound. Observe that it is an upper bound but not an equality: indeed, it does not account for the visible pairs in the sample that were visible in another cell of the arrangement of $\mathcal{T}$ adjacent to $C(\mathcal{T})$ through the removed triangle.

Denoting by $q_k(n)$ the maximum number of k-visible inner pairs in the arrangement of n triangles in $\mathbb{E}^3$, we have

$$\frac{n-4}{n}\, q_0(n)\, +\, q_1(n) \leq q_0(n-1),$$

from which and from inequality 16.6 we derive the recurrence equation

$$\frac{n-2}{n}\, q_0(n) \leq q_0(n-1) + O(n\log n),$$

whose solution is $q_0(n) = O(n^2 \log^2 n)$.

This finishes the proof of the following theorem:

Theorem 16.4.9 *The complexity of the vertical decomposition of a single cell in the arrangement of n triangles in $\mathbb{E}^3$ is $O(n^2 \log^2 n)$.*

16.4.3   Computing a cell

Let $\mathcal{T}$ be a set of n triangles in $\mathbb{E}^3$, and pick a point O that does not belong to any triangle. In this section, we give an algorithm that computes the cell $C(\mathcal{T})$ in the arrangement of $\mathcal{T}$ that contains O.

The algorithm we present here is an extension of the incremental randomized algorithm described in subsection 15.4.2 that computes a single cell in the arrangement of line segments in the plane. It consists of inserting the triangles in turn. Let $\mathcal{R}$ be the current subset of triangles, introduced in the previous incremental steps, and $C(\mathcal{R})$ the cell in the arrangement of $\mathcal{R}$ that contains O. After inserting a triangle, we update the decomposition of the triangles (see subsection 16.2.1) without worrying that some prisms in this decomposition may lie entirely outside $C(\mathcal{R})$. We thus allow the algorithm to compute, in addition to the prisms in the decomposition of $\mathcal{R}$, some other prisms that are not prisms of $C(\mathcal{R})$. To keep the complexity within reasonable limits, it is necessary to stop the construction of the decomposition outside $C(\mathcal{R})$ at certain *clean-up* steps. During those clean-up steps, we *deactivate* the prisms that do not belong to the cell $C(\mathcal{R})$. Only the active prisms are subdivided in the subsequent incremental steps.

Between two clean-up steps, the algorithm is similar to the algorithm presented in subsection 16.3.3 to compute the lower envelope of triangles.

The algorithm maintains an influence graph similar to that which is used to compute the lower envelope of triangles, and an adjacency graph. It also maintains a pointer to the prism in the current decomposition that contains the origin O. The nodes in the influence graph represent the prisms computed by the algorithm. If $\mathcal{R}$ is the current subset of triangles, the prisms in the decomposition of the cell $C(\mathcal{R})$ correspond to the active leaves of the influence graph. Only those leaves will have children in the subsequent steps. The nodes in the adjacency graph $\mathcal{G}$ represent the active prisms and an arc in $\mathcal{G}$ connects two active prisms that share a common vertical wall. We also maintain pointers between the nodes of the influence graph and the node of the adjacency graph.

After an initial step which consists simply of creating a node that represents the whole of $\mathbb{E}^3$, the triangles are inserted each in turn in a random order. The current incremental step can be subdivided into three phases: locating the prisms intersected by the triangle T, updating, and occasionally cleaning up.

Locating. During the location phase, we retrieve the leaves of the influence graph that intersect T by traversing the influence graph and backtracking whenever the next node to visit is already visited or does not conflict with T. If no leaf in the influence graph conflicts with T, then T does not appear on the boundary of $C(\mathcal{R})$ or in any of the cells that are computed subsequently. So the algorithm skips directly to the insertion of the next triangle.

Updating. Each active prism P is subdivided into a constant number of new sub-prisms, which become children of P. (A case analysis shows that this number is at most nineteen, to be precise.) As happens when decomposing line segments in the plane or when computing the lower envelope of triangles, some of these sub-prisms are not prisms in the new decomposition and they must be merged or removed.

Among the new prisms, we make a distinction between those whose floor or ceiling is contained in T and the others. Merging the latter kind of prisms does not create particular problems and can be performed in a way that is very similar to what was done in subsection 16.3.3.

As for the new prisms whose ceiling is contained in T, we build a secondary influence graph as in subsection 16.3.3. We proceed similarly for the prisms whose floor is contained in T.

Cleaning up. At special steps 2^i, for $i = 1, \ldots, \lfloor \log n \rfloor - 1$, we visit all the prisms contained in the current cell $C(\mathcal{R})$, by traversing the adjacency graph of the current decomposition, starting from the cell that contains O. The nodes that are not visited in this process are deactivated.

The complexity analysis of the algorithm combines the analyses of subsections 15.4.2 and 16.3.3. We leave to the reader the careful examination of the complexity, and the proof that it is $O(|\mathcal{D}ec(n)| \log n)$ where $|\mathcal{D}ec(n)|$ denotes the maximum complexity of a cell in the arrangement of n triangles in $\mathbb{E}^3$. Theorem 16.4.9 finishes the proof.

Theorem 16.4.10 *It is possible to compute a single cell in the arrangement of n triangles in $\mathbb{E}^3$ in time $O(n^2 \log^3 n)$ using storage $O(n^2 \log^2 n)$.*

16.5   Exercises

Exercise 16.1 (The complexity of a lower envelope of simplices) Show that the complexity of the lower envelope of n $(d-1)$-simplices in $\mathbb{E}^d$ is $\Theta(n^{d-1}\alpha(n))$.

Exercise 16.2 (Deterministic computation of lower envelopes) Given a set T of n triangles in $\mathbb{E}^3$, show how to deterministically compute the lower envelope of T in time $O(n^2 \log n)$ if the triangles are disjoint, and $O(n^2\alpha(n) \log n)$ otherwise.

Hint: For a set S of triangles, we denote by $\mathcal{A}(S)$ the arrangement of the affine hulls of the projected edges of the triangles in T in the xy-plane. We denote by $\underline{\mathcal{E}}(S)$ the planar map obtained by projecting the lower envelope of S, and by $\underline{\mathcal{E}}^*(S)$ the refinement of $\underline{\mathcal{E}}(S)$ obtained by superimposing $\underline{\mathcal{E}}(S)$ and $\mathcal{A}(S)$. To use the divide-and-conquer method, consider two subsets T_1 and T_2 of T of roughly the same size. Compute $\underline{\mathcal{E}}^*(T_1)$ and $\underline{\mathcal{E}}^*(T_2)$ recursively. Superimposing $\underline{\mathcal{E}}^*(T_2)$ and $\mathcal{A}(T_1)$ yields a planar map $\underline{\mathcal{E}}_{12}^{\#}$ and, likewise, superimposing $\underline{\mathcal{E}}^*(T_2)$ and $\mathcal{A}(T_1)$ yields a planar map $\underline{\mathcal{E}}_{21}^{\#}$. Show that $|\underline{\mathcal{E}}_{12}^{\#}| = O(n^2\alpha(n))$ and that $|\underline{\mathcal{E}}_{21}^{\#}| = O(n^2\alpha(n))$ (think of inserting the lines in the arrangements $\mathcal{A}(T_i)$ one by one). Compute $\mathcal{A}(T)$ and, for each cell in $\mathcal{A}(T)$, compute the portion of the lower envelope whose projection onto $z = 0$ is this cell. This portion is a convex dome, the intersection of two convex domes corresponding to a cell of $\underline{\mathcal{E}}_{12}^{\#}$ and a cell of $\underline{\mathcal{E}}_{21}^{\#}$. The factor $\log n$ can be removed; see the references in the bibliographical notes.

Exercise 16.3 (Lower envelopes of surfaces) Devise a generalization of the algorithm described in subsection 16.3.3, which computes the lower envelope of a set of triangles, to the case of algebraic surface patches in $\mathbb{E}^3$.

Exercise 16.4 (Decomposition of an arrangement of triangles) Show that the complexity of the vertical decomposition of the arrangement of n triangles in $\mathbb{E}^3$ is $O(n^2\alpha(n) \log n + t)$, if t is the complexity of the arrangement. Devise a randomized algorithm that computes such a decomposition in expected time $O((n^2\alpha(n) \log n + t) \log n)$.

Hint: Adapt the proof of the theorem which bounds the complexity of the vertical decomposition of a single cell.

Exercise 16.5 (Vertical decomposition of surfaces) Generalize the notion of a vertical decomposition of an arrangement of triangles to the case of algebraic surfaces in $\mathbb{E}^3$. Bound the complexity of such a decomposition and devise an algorithm to compute it that is similar to the one presented in subsection 16.2.1.

Exercise 16.6 (Computing a view) Consider a polyhedral scene with a total of n triangular facets (which may intersect). Show that the view from a given point, which is the portion of the facets visible from this point, has complexity $O(n^2)$ if the relative interiors of the triangles are disjoint, $O(n^2\alpha(n))$ otherwise. Show that these bounds are best possible in the worst case. Give an algorithm that computes the view from a given point with a cost proportional to these bounds by a factor $\log n$.

Exercise 16.7 (Stabbing) Given n polyhedral objects in $\mathbb{E}^3$, compute the set of planes that simultaneously stab them all.

Hint: Use polarity as in the planar analogue, exercise 15.5.

Exercise 16.8 (Levels in arrangements of simplices) Let S be a set of n $(d-1)$-simplices in $\mathbb{E}^d$ which we assume are in general position, and let $\mathcal{A}(S)$ be their arrangement. To each $(d-2)$-face of a simplex in S corresponds a vertical $(d-1)$-face generated by the vertical rays cast upwards from the $(d-2)$-face. A simplex and these vertical $(d-1)$-faces generated by its $(d-2)$-faces bound a truncated cylinder with vertical generators. Let $\mathcal{A}'(S)$ be the arrangement of the corresponding cylinders. We note that the vertical faces of $\mathcal{A}'(S)$ further subdivide the faces of $\mathcal{A}(S)$.

To maintain an analogy with what was said about levels in hyperplane arrangements, we say that a point in $\mathbb{E}^3$ is at level k if it belongs to the interior of exactly k cylinders. All the points in the relative interior of a face of $\mathcal{A}'(S)$ are at the same level. By definition, this level is the level of the face. We denote by $\mathcal{A}'_{\leq k}(S)$ the sub-complex of $\mathcal{A}'(S)$ formed by all the faces of $\mathcal{A}'(S)$ at level at most k.

Show that for any integer k, the number of faces at any level $0 \leq j \leq k$ is bounded by $O(kn^{d-1}\alpha(\frac{n}{k}))$.

Conclude that if $\mathcal{T}$ is such that any vertical line intersects at most k simplices, then the complexity of the arrangement of $\mathcal{T}$ is $O(kn^{d-1}\alpha(\frac{n}{k}))$.

Hint: Use theorem 16.3.1, exercise 16.1, and the sampling theorem 4.2.3.

Exercise 16.9 (Computing the first k levels) Devise and analyze an algorithm that computes the first k levels in the arrangement of n triangles in $\mathbb{E}^3$.

Hint: Adapt the algorithm that computes the lower envelope of a set of triangles, and the algorithm that computes the first k levels in an arrangement of hyperplanes (see section 14.4).

Exercise 16.10 (A cell in higher dimensions) Show that the complexity of a single cell in the arrangement of n $(d-1)$-simplices in $\mathbb{E}^d$ is $O(n^{d-1}\log^{d-1} n)$.

Exercise 16.11 (Motion planning for a polyhedron) Consider a polyhedron $\mathcal{M}$ with m facets that moves inside a polyhedral region $\mathcal{E}$ with n facets. Without loss of generality, we assume that the facets are triangular.

1. Show that the set of translations that bring a vertex of $\mathcal{M}$ into contact with a facet of $\mathcal{E}$, or a vertex of $\mathcal{E}$ into contact with a facet of $\mathcal{M}$, or an edge of $\mathcal{M}$ into contact with an edge of $\mathcal{E}$, is a set $\mathcal{C}$ of at most mn triangles and parallelograms (when we identify the vector OM of the translation with its endpoint M).

2. Show that the set of positions of $\mathcal{M}$ inside $\mathcal{E}$ that can be accessed from a given position I corresponds to the cell that contains I in the arrangement of the triangles and parallelograms of $\mathcal{C}$. Conclude that it may be determined whether two positions I and F in $\mathcal{E}$ may be connected by a path along which $\mathcal{M}$ remains entirely inside $\mathcal{E}$, and, if so, compute such a path in time $O(m^2 n^2 \log^3(mn))$.

Exercise 16.12 (Flying saucers) Consider a polyhedral flying saucer that flies above a terrain modeled by a function $z(x, y)$ that is piecewise linear. Show that the set of translations of $\mathbb{E}^3$ for which the flying saucer is strictly above the ground is characterized as the region above the upper envelope of certain triangles in $\mathbb{E}^3$. Give a lower bound on the complexity of such a set.

16.6   Bibliographical notes

The bound on the complexity of the lower envelope of triangles was shown by Pach and Sharir in [185]. The proof presented in subsection 16.3.1 is shaped after the proof given by Halperin and Sharir in [120] for the case of algebraic surfaces. They show that the lower envelope of n algebraic surface patches of bounded degree in $\mathbb{E}^d$ is $O(n^{d-1+\varepsilon})$ for any ε (but the constant increases when ε goes to 0). The algorithm that computes the envelope of triangles in $\mathbb{E}^3$ described in subsection 16.3.3 and its extension to algebraic surfaces (see exercise 16.3) are due to Boissonnat and Dobrindt [30].

The bound on the complexity of a single cell in the arrangement of triangles is due to Aronov and Sharir [9, 10], who also generalize it to simplices in any dimension (see exercise 16.10). This result was extended to algebraic surfaces by Halperin and Sharir [119, 121] who showed that a single cell in the arrangement of n algebraic surface patches of bounded degree in $\mathbb{E}^3$ has almost quadratic complexity. How to compute a cell in dimension 3 is explained by de Berg, Dobrindt, and Schwarzkopf in [76].

The bound on the complexity of the vertical decomposition of a single cell in the arrangement of triangles in dimension 3 given in subsection 16.4.2 and the bound on the complexity of the entire arrangement (see exercise 16.4) are due to Tagansky [212]. A slightly worse bound and a deterministic algorithm that computes the vertical decomposition of the arrangement of triangles are given by de Berg, Guibas, and Halperin in [77]. A bound of $O(n^4 \log n)$ on the vertical decomposition of triangles in dimension 4 is given by Guibas, Halperin, Matoušek, and Sharir in [115]. Decomposing an arrangement of algebraic surfaces is a problem studied by Clarkson *et al.* in [64] and by Chazelle *et al.* in [51]. In the latter, they bound the complexity of their cylindrical decomposition by $O(n^{2d-3}\beta(n))$, where $\beta(n) = 2^{\alpha(n)^c}$ for a constant c that does not depend on n.

Many references and results concerning hidden surface removal in computer graphics can be found in the book by de Berg [75]. Schwartz and Sharir [197] give a wide survey of combinatorial and algorithmic techniques motivated by motion planning and robotics.

Part V

Voronoi diagrams

Voronoi diagrams are very useful structures, frequently encountered in several disciplines because they model growth processes and distance relationships between objects: it is not surprising to see them appear in the study of crystal growth or in studies on the great structures of the universe. In nature, they can be observed in crystalline structures, on the shell of a turtle, or on the neck of a reticulate giraffe.

Voronoi diagrams are very closely related to the geometric structures encountered so far: polytopes, triangulations, and arrangements. Their mathematical properties are particularly numerous and interesting. Chapter 17 is entirely devoted to Voronoi structures with a Euclidean metric, whereas other metrics are studied in chapter 18. Chapter 19 presents results specific to dimension 2 that have no analogue in higher dimensions.

Voronoi diagrams can also be used as data structures to solve numerous problems: nearest neighbors and motion planning are two outstanding instances. Several examples are given in the exercises and throughout chapter 19.

Chapter 17

Euclidean metric

This chapter is concerned with the simplest case of Voronoi diagrams, where the objects are points and the distance is given by the usual Euclidean metric in $\mathbb{E}^d$. The cells in the Voronoi diagram of a set $\mathcal{M}$ of points are then the equivalence classes of the equivalence relation "to have the same nearest neighbor in $\mathcal{M}$". It is possible to show (see section 17.2) that such cells can be obtained by projecting the facets of a polytope in $\mathbb{E}^{d+1}$ onto $\mathbb{E}^d$, which enables us to use several results concerning polytopes for Voronoi diagrams as well. Bounds can be obtained in this way for the complexity of Voronoi diagrams and of their computation. In section 17.3, we define a dual of the Voronoi diagram, the Delaunay complex, that enjoys several properties which make it desirable in applications such as numerical analysis in connection with finite-element methods. The last section of this chapter introduces a first generalization of Voronoi diagrams (see section 17.4): the higher-order Voronoi diagrams. The cells in the diagram of order k are the equivalence classes of the equivalence relation "to have the same k nearest neighbors in $\mathcal{M}$", a notion that is often very helpful in data analysis.

17.1   Definition

Let $\mathcal{M}$ be a set of n points in $\mathbb{E}^d$, $M_1, \ldots, M_n$, which we call the *sites* to avoid confusion with the other points in $\mathbb{E}^d$. To each site M_i we attach the region $V(M_i)$ in $\mathbb{E}^d$ that contains the points in $\mathbb{E}^d$ closer to M_i than to any other point in $\mathcal{M}$:

$$V(M_i) = \{X \in \mathbb{E}^d \ : \ \delta(X, M_i) \leq \delta(X, M_j) \text{ for any } j \neq i\}.$$

In this chapter, δ denotes the Euclidean distance in $\mathbb{E}^d$. Other distances will be considered in chapter 18.

The set of points closer to M_i than to another site M_j is the half-space that contains M_i and that is bounded by the *perpendicular bisector* of the segment

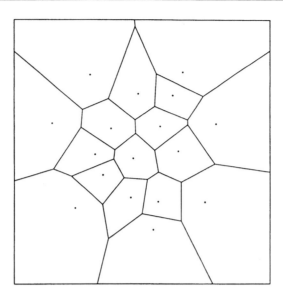

Figure 17.1. The Voronoi diagram of a set of points in the plane.

$M_i M_j$: this is the hyperplane perpendicular to $M_i M_j$ that intersects $M_i M_j$ at the midpoint of M_i and M_j. The region $V(M_i)$ is thus the intersection of a finite number of closed half-spaces, bounded by the perpendicular bisectors of $M_i M_j$, $j = 1, \ldots, n$, $j \neq i$. This shows that $V(M_i)$ is a convex polytope, which may or may not be bounded. As we will see later, the $V(M_i)$'s and their faces form a cell complex whose domain is the whole of $\mathbb{E}^d$. This complex is called the *Voronoi diagram* of $\mathcal{M}$ and is denoted by $Vor(\mathcal{M})$ (see figure 17.1).

A first and useful interpretation of the Voronoi diagram (another interpretation is given in the next chapter) views the cell $V(M_i)$ as the set of centers of balls such that the boundary of such a ball contains M_i and its interior does not contain another site M_j, $j \neq i$. In particular, this point of view leads to the interpretation of a Voronoi diagram in $\mathbb{E}^d$ as a polytope in $\mathbb{E}^{d+1}$, which also enables it to be computed efficiently. This interpretation is developed in section 17.2 where we represent spheres of $\mathbb{E}^d$ as points in $\mathbb{E}^{d+1}$.

From now on, we say that the sites are in L_2-*general* position if no sphere can contain $d + 2$ sites on its boundary.

17.2 Voronoi diagrams and polytopes

17.2.1 Power of a point with respect to a sphere

Consider the Euclidean space of dimension d, $\mathbb{E}^d$, and let O be its origin, and Σ be a sphere of $\mathbb{E}^d$ centered at C with radius r. Its equation is given by $\Sigma(X) = 0$

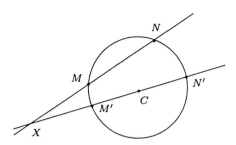

Figure 17.2. Power of a point with respect to a circle.

where

$$\Sigma(X) = XC^2 - r^2. \tag{17.1}$$

By the *interior* of a sphere Σ, we mean the set of points X such that $\Sigma(X)$ is negative. The *exterior* is the set of points X such that $\Sigma(X)$ is negative. A point X is said to be *on*, *inside* or *outside* a sphere if it belongs to the sphere, respectively to its interior, to its exterior. For any point X in $\mathbb{E}^d$, $\Sigma(X)$ is called the *power* of X with respect to Σ. The power of the origin with respect to Σ is also denoted by σ and we have

$$\sigma = \Sigma(O) = C^2 - r^2. \tag{17.2}$$

If D is any line that contains X, and if M and N are the intersection points of D with Σ, then

$$\Sigma(X) = XM \cdot XN. \tag{17.3}$$

This is obvious when D is the line connecting X and C. Otherwise let D' be the line that contains X and C, and let M' and N' be its intersection points with Σ (see figure 17.2). The triangles XMM' and $XN'N$ are similar (the angles $\widehat{M'MN}$ and $\widehat{M'N'N}$ are supplementary), which proves equation 17.3. In the case where X belongs to the exterior of Σ and D is tangent to Σ at T, then $M = N = T$ and the previous equation can be rewritten

$$\Sigma(X) = XT^2. \tag{17.4}$$

17.2.2 Representation of spheres

Let ϕ be the mapping that takes a sphere Σ in $\mathbb{E}^d$, of center C and whose power with respect to O is σ, to the point $\phi(\Sigma) = (C, \sigma)$ in $\mathbb{E}^{d+1}$. Using ϕ enables us to treat spheres in $\mathbb{E}^d$ just as points in $\mathbb{E}^{d+1}$.

We embed $\mathbb{E}^d$ as the hyperplane in $\mathbb{E}^{d+1}$ whose equation is $x_{d+1} = 0$. As usual, the direction of the x_{d+1}-axis is called the vertical direction and we use the words *above* and *below* in connection with this vertical ordering. We denote by X a point of $\mathbb{E}^d$ or its coordinate vector $(x_1, \ldots, x_d)$ indifferently, and by $\underline{X}$ a point in $\mathbb{E}^{d+1}$ or its coordinate vector $(x_1, \ldots, x_{d+1})$. By the above embedding, $\phi(\Sigma)$ projects vertically onto C. Later on, we will often use vertical projections and, unless mentioned otherwise, the word *projection* refers to the vertical projection from $\mathbb{E}^{d+1}$ onto $\mathbb{E}^d$.

We also use homogeneous coordinates and the matrix notation. We denote by $\boldsymbol{X} = (x_1, \ldots, x_d, t)$ (resp. $\underline{\boldsymbol{X}} = (x_1, \ldots, x_{d+1}, t)$) the homogeneous coordinate vector of a point X in $\mathbb{E}^d$ (resp. a point $\underline{X}$ in $\mathbb{E}^{d+1}$). The equation of the sphere Σ can then be rewritten with homogeneous coordinates as

$$\boldsymbol{X} \, \boldsymbol{\Sigma} \, \boldsymbol{X}^t = 0 \quad \text{with} \quad \boldsymbol{\Sigma} = \begin{pmatrix} \mathbb{I}_d & -C^t \\ -C & \sigma \end{pmatrix},$$

where $\mathbb{I}_d$ denotes the $d \times d$ identity matrix.

17.2.3   The paraboloid $\mathcal{P}$

From equation 17.2, it follows that the images under ϕ of points in $\mathbb{E}^d$, considered as spheres of radius 0, belong to the paraboloid of revolution $\mathcal{P}$ with vertical axis and equation

$$x_{d+1} = \sum_{i=1}^{d} x_i^2 = X \cdot X \quad \text{with} \quad X = (x_1, \ldots, x_d).$$

In a homogeneous system of coordinates $\mathcal{P}$ is given by

$$\underline{\boldsymbol{X}} \, \Delta_{\mathcal{P}} \, \underline{\boldsymbol{X}}^t = 0 \quad \text{where} \quad \Delta_{\mathcal{P}} = \begin{pmatrix} \mathbb{I}_d & 0 & 0 \\ 0 & 0 & -1/2 \\ 0 & -1/2 & 0 \end{pmatrix}.$$

Identifying a point X and the sphere centered at X with radius 0 shows that ϕ maps any point X in $\mathbb{E}^d$ to the point $\phi(X)$ in $\mathbb{E}^{d+1}$ obtained by lifting X onto $\mathcal{P}$.

The set of concentric spheres in $\mathbb{E}^d$, centered at C, is mapped by ϕ onto the vertical line in $\mathbb{E}^{d+1}$ that contains C (and hence $\phi(C)$). Let Σ be such a sphere. Equation 17.2 implies that the signed vertical distance from $\phi(\Sigma)$ to $\phi(C)$ equals r^2 (see figure 17.3). Thus, the *real* spheres, whose squared radii are non-negative, are mapped by ϕ to the points lying on or below the paraboloid, while the points lying above the paraboloid are the images under ϕ of the *imaginary* spheres, whose squared radii are negative.

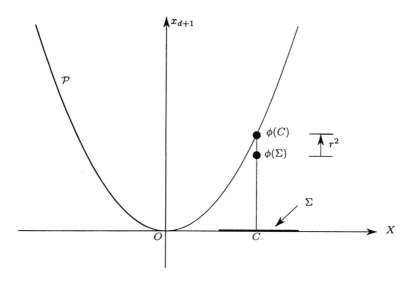

Figure 17.3. The paraboloid $\mathcal{P}$ (the coordinate system is not normed, so as to simplify the representation).

17.2.4 Polarity

Consider a quadric $\mathcal{Q}$ in $\mathbb{E}^{d+1}$ defined by its homogeneous equation

$$\mathcal{Q}(\underline{X}) = \underline{X} \Delta_{\mathcal{Q}} \underline{X}^t = 0.$$

Henceforth, $\mathcal{Q}$ will be the paraboloid $\mathcal{P}$, but we treat the case of any quadric for generality as it introduces no additional difficulty. Two points X and Y are *conjugate* with respect to $\mathcal{Q}$ if

$$\mathcal{Q}(\underline{X}, \underline{Y}) = \underline{X} \Delta_{\mathcal{Q}} \underline{Y}^t = 0.$$

The polarity with respect to $\mathcal{Q}$ described in section 7.3 is an involution between points and hyperplanes in $\mathbb{E}^{d+1}$ which maps any point $\bar{A}$ to its *polar hyperplane* $\bar{A}^*$ of equation

$$\underline{A} \Delta_{\mathcal{Q}} \underline{X}^t = 0,$$

and maps any hyperplane $\underline{H}$ to a point $\underline{H}^*$ whose polar hyperplane is $\underline{H}$. The point $\underline{H}^*$ is called the *pole* of $\underline{H}$.

Note that if $\mathcal{Q}$ is the paraboloid $\mathcal{P}$ and if we put $\phi(\Sigma) = (C, \sigma)$, the equation of the polar hyperplane $\phi(\Sigma)^*$ of $\phi(\Sigma)$ can be rewritten as

$$x_{d+1} = 2C \cdot X - \sigma.$$

An essential property of polarity is that it preserves incidences (see section 7.3): a point $\underline{X}$ belongs to a hyperplane $\underline{H}$ if and only if its polar hyperplane $\underline{X}^*$

contains the pole $\underline{H}^*$ of $\underline{H}$. Moreover (see exercise 7.14), when the quadric is the paraboloid $\mathcal{P}$, we have

$$\underline{X} \in \underline{H}^+ \iff \underline{H}^* \in \underline{X}^{*+}$$
$$\underline{X} \in \underline{H}^- \iff \underline{H}^* \in \underline{X}^{*-},$$

if we denote by $\underline{H}^+$ and $\underline{H}^-$ the half-spaces bounded by $\underline{H}$ and that lie respectively above and below $\underline{H}$.

17.2.5   Orthogonal spheres

Two spheres Σ_1 and Σ_2 centered at C_1 and C_2 and with radii r_1 and r_2 are *orthogonal* if

$$\Sigma_1(C_2) = r_2^2, \tag{17.5}$$

or equivalently if

$$\Sigma_2(C_1) = r_1^2.$$

A simple verification shows that, if the spheres are real, then they are orthogonal if and only if the angle (IC_1, IC_2) at any intersection point I of $\Sigma_1 \cap \Sigma_2$ is a right angle, or equivalently, if and only if the dihedral angle of the tangent hyperplanes at I is a right angle.

Expression 17.5 may be rewritten as

$$C_1 \cdot C_2 - \frac{1}{2}(\sigma_1 + \sigma_2) = 0, \quad \text{with} \quad \sigma_i = C_i^2 - r_i^2 \ (i = 1, 2),$$

which shows that two spheres Σ_1 and Σ_2 are orthogonal if the two points $\phi(\Sigma_1)$ and $\phi(\Sigma_2)$ are conjugate with respect to the paraboloid $\mathcal{P}$. This implies that:

Lemma 17.2.1 *The set of spheres in $\mathbb{E}^d$ that are orthogonal to a given sphere is mapped by ϕ to the polar hyperplane $\phi(\Sigma)^*$ of $\phi(\Sigma)$.*

Let us now consider the points in $\mathbb{E}^d$ as spheres of radius 0. The set of spheres in $\mathbb{E}^d$ that pass through a given point $X \in \mathbb{E}^d$ is also the set of spheres orthogonal to the sphere centered at X with radius 0. Therefore its image under ϕ is the hyperplane $\phi(X)^*$ polar to $\phi(X) \in \mathcal{P}$. This hyperplane must be tangent to $\mathcal{P}$ and to $\phi(X)$: indeed, the only sphere of radius 0 which is orthogonal to X is X itself, and hence $\phi(X)^*$ intersects $\mathcal{P}$ in a single point $\phi(X)$.

Let Σ be a sphere in $\mathbb{E}^d$. The intersection of $\phi(\Sigma)^*$ with $\mathcal{P}$ is the image under ϕ of the set of spheres with radius 0 that are orthogonal to Σ, namely Σ itself (considered as a set of points, or equivalently as a set of spheres of radius 0). Consequently, $\phi(\Sigma)^* \cap \mathcal{P}$ in $\mathbb{E}^{d+1}$ projects onto Σ in $\mathbb{E}^d$. More generally, we have the following result.

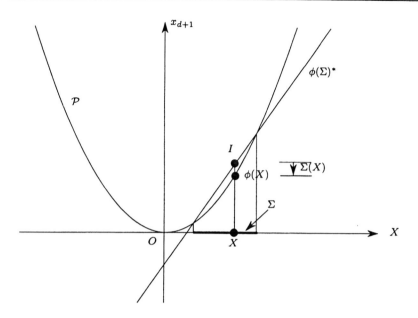

Figure 17.4. Interpretation in $\Sigma(X)$ in $\mathbb{E}^{d+1}$.

Lemma 17.2.2 *The intersection of the paraboloid $\mathcal{P}$ with a hyperplane $\underline{H}$ projects onto $\mathbb{E}^d$ as a sphere $\phi^{-1}(\underline{H}^*)$ whose center is the vertical projection of $\underline{H}^*$. Conversely, the points of a sphere Σ of $\mathbb{E}^d$ lifted on the paraboloid $\mathcal{P}$ in $\mathbb{E}^{d+1}$ belong to a unique hyperplane that intersects $\mathcal{P}$ exactly at these points. This hyperplane is the polar hyperplane $\phi(\Sigma)^*$ of $\phi(\Sigma)$.*

It follows from this lemma that the power of a point X with respect to a sphere Σ equals the square of the radius of the sphere Σ_X orthogonal to Σ and centered at X (Σ_X is imaginary if X is inside Σ). The power $\Sigma(X)$ can be easily computed in the space $\mathbb{E}^{d+1}$ that represents the spheres of $\mathbb{E}^d$. Indeed (see figure 17.4), Σ_X is mapped by ϕ to a point $\underline{I}$ in $\mathbb{E}^{d+1}$ that is the intersection of the vertical line that passes through X (which corresponds to the spheres centered at X) with the polar hyperplane $\phi(\Sigma)^*$ of $\phi(\Sigma)$ (which corresponds to the spheres orthogonal to Σ). The x_{d+1}-coordinates of $\phi(x)$ and $\underline{I}$ are respectively X^2 and $\Sigma_X(O) = X^2 - \Sigma(X)$ since the square of the radius of Σ_X equals the power of X with respect to Σ. The difference of these x_{d+1}-coordinates is called the *signed vertical distance*. This proves the following lemma.

Lemma 17.2.3 *The power of X with respect to a sphere Σ equals the signed vertical distance from the point $\phi(X)$ to the hyperplane $\phi(\Sigma)^*$.*

We thus have the following lemma:

Lemma 17.2.4 *Let X and Σ be respectively a point and a sphere in $\mathbb{E}^d$. If $\underline{H}$ is a hyperplane in $\mathbb{E}^{d+1}$, we denote by $\underline{H}^-$ the half-space lying below $\underline{H}$. Then:*

$$
\begin{aligned}
X \in \Sigma &\iff \phi(X) \in \phi(\Sigma)^* \iff \phi(\Sigma) \in \phi(X)^* \\
X \in int(\Sigma) &\iff \phi(X) \in \phi(\Sigma)^{*-} \iff \phi(\Sigma) \in \phi(X)^{*-} \\
X \in ext(\Sigma) &\iff \phi(X) \in \phi(\Sigma)^{*+} \iff \phi(\Sigma) \in \phi(X)^{*+}
\end{aligned}
$$

The equivalences on the left are consequences of the two preceding lemmas, and the ones on the right are proved by the special properties of polarity (see subsection 17.2.4 and exercise 7.13).

Any point in the half-space that lies below $\phi(X)^*$ in $\mathbb{E}^{d+1}$ is thus the image under ϕ of a sphere whose interior contains X. Likewise, any point in the half-space that lies above $\phi(X)^*$ in $\mathbb{E}^{d+1}$ is the image under ϕ of a sphere whose exterior contains X, and the points on $\phi(X)^*$ are the images of the spheres passing through X.

Remark. Lemma 17.2.3 shows that the squared distance $\|XA\|^2$ separating points X and A, which is also the power of X with respect to the sphere centered at A with radius 0, equals the absolute value of the vertical distance between $\phi(A)^*$ and $\phi(X)$. Points X and A play symmetric roles, so $\|XA\|^2$ also equals the absolute value of the vertical distance between $\phi(X)^*$ and $\phi(A)$.

17.2.6   Radical hyperplane

Let Σ_1 and Σ_2 be two spheres in $\mathbb{E}^d$. The set of points in $\mathbb{E}^d$ that have the same power with respect to these two spheres is a hyperplane, called the *radical hyperplane* and denoted by H_{12}, whose equation is given by

$$
H_{12}: \quad \Sigma_1(X) - \Sigma_2(X) = 0.
$$

As we observed in subsection 17.2.5, the power of a point X with respect to a sphere Σ equals the square of the radius of the sphere orthogonal to Σ centered at X. A point has the same power with respect to Σ_1 as with respect to Σ_2 if it is the center of a sphere orthogonal to both Σ_1 and Σ_2. Lemma 17.2.1 shows that the set of spheres in $\mathbb{E}^d$ that are orthogonal to a given sphere Σ is mapped by ϕ onto the polar hyperplane $\phi(\Sigma)^*$. The spheres orthogonal to Σ_1 and Σ_2 are thus mapped by ϕ to the affine subspace of dimension $d-1$ that is the intersection of $\phi(\Sigma_1)^*$ and $\phi(\Sigma_2)^*$. The projection onto $\mathbb{E}^d$ of this affine subspace is exactly the set of points that have the same power with respect to Σ_1 and Σ_2.

17.2.7   Voronoi diagrams

Let $\mathcal{M} = \{M_1, \ldots, M_n\}$ be a set of n points in $\mathbb{E}^d$. As before, we embed the Euclidean space $\mathbb{E}^d$ of dimension d into $\mathbb{E}^{d+1}$ as the hyperplane $x_{d+1} = 0$, and

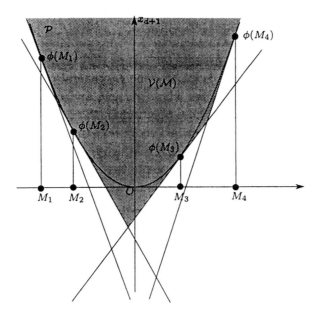

Figure 17.5. The Voronoi polytope $\mathcal{V}(\mathcal{M})$.

we let $\phi(M_i)^*$ denote the hyperplane in $\mathbb{E}^{d+1}$ that is tangent to the paraboloid $\mathcal{P}$ at the point $\phi(M_i)$ obtained by lifting M vertically onto the paraboloid $\mathcal{P}$, for each $i = 1, \ldots, n$. The preceding discussion shows that the set of spheres (real or imaginary) whose interiors contain no point of $\mathcal{M}$ is mapped by ϕ to the intersection of the n half-spaces lying above the hyperplanes $\phi(M_1)^*, \ldots, \phi(M_n)^*$. This intersection is an unbounded polytope which contains $\mathcal{P}$. We call it the *Voronoi polytope* and denote it by $\mathcal{V}(\mathcal{M})$ (see figure 17.5).

Theorem 17.2.5 *The Voronoi diagram of $\mathcal{M}$, denoted by $\mathrm{Vor}(\mathcal{M})$, is a cell complex of dimension d in $\mathbb{E}^d$ whose faces are obtained by projecting onto $\mathbb{E}^d$ the proper faces of the Voronoi polytope $\mathcal{V}(\mathcal{M})$.*

Proof. The boundary of $\mathcal{V}(\mathcal{M})$ is a pure cell complex of dimension d, hence so is $\mathrm{Vor}(\mathcal{M})$. Let $\overline{A}$ be a point on a facet of $\mathcal{V}(\mathcal{M})$ that is contained in the hyperplane tangent to $\mathcal{P}$ at $\phi(M_i)$. Then $\overline{A}$ is the image under ϕ of a sphere Σ_A that passes through M_i and whose interior contains no other point of $\mathcal{M}$ (see lemma 17.2.4). There cannot be a site in $\mathcal{M}$ closer to the center of Σ_A than M_i. But this center is exactly the projection A of $\overline{A}$ onto $\mathbb{E}^d$. In other words, A belongs to the cell $V(M_i)$ of the Voronoi diagram. $\qquad\square$

This theorem implies that the combinatorial properties of Voronoi diagrams follow directly from those of polytopes as studied in chapter 7. In particular, if the points $\phi(M_i)^*$ are in general position in $\mathbb{E}^{d+1}$, then $\mathcal{V}(\mathcal{M})$ is a simple $(d+1)$-polytope. Each vertex is thus incident to $d+1$ hyperplanes. Expressed in terms of M_i's, the *general condition assumption* means that no $d+2$ points in $\mathcal{M}$ lie on the boundary of a sphere: this is exactly the L_2-general position assumption. If it is satisfied, $Vor(\mathcal{M})$ is a complex whose vertices are all equidistant from some $d+1$ points in $\mathcal{M}$ and closer to these points than to any other point in $\mathcal{M}$: they are the centers of spheres circumscribed to $(d+1)$-tuples whose interiors do not contain any point in $\mathcal{M}$. More generally, a k-face of $Vor(\mathcal{M})$ is the projection of a k-face of $\mathcal{V}(\mathcal{M})$. It is thus the set of points that are equidistant from $d+1-k$ points in $\mathcal{M}$ and closer to these points than to any other point in $\mathcal{M}$.

Theorem 17.2.5 reduces the problem of computing the Voronoi diagram of n points in $\mathbb{E}^d$ to the computation of the intersection of n half-spaces of $\mathbb{E}^d$. The algorithms described in this book that compute half-space intersections, be they deterministic, randomized, static or dynamic, output-sensitive or not, can all be used to compute Voronoi diagrams.

Corollary 17.2.6 *The complexity (namely, the number of faces) of the Voronoi diagrams of n points in $\mathbb{E}^d$ is $\Theta(n^{\lceil d/2 \rceil})$. We may compute such a diagram in time $O(n \log n + n^{\lceil d/2 \rceil})$, which is optimal in the worst case.*

Proof. The upper bounds on the complexity and running time of the algorithm are immediate consequences of the upper bound theorem 7.2.5 and of results of the previous sections.

That $\Omega(n^{\lceil d/2 \rceil})$ is a lower bound on the complexity of the Voronoi diagram of n points in $\mathbb{E}^d$ is a consequence of exercise 7.11, where it is shown how to construct a maximal polytope whose vertices lie on the paraboloid, and of theorem 17.3.1 below.

That $\Omega(n \log n)$ is a lower bound on computing the Voronoi diagram in the plane is a consequence of the fact that the unbounded edges of the Voronoi diagram of a set $\mathcal{M}$ of points correspond to projections of the edges of the convex hull of $\mathcal{M}$. We also comment on this below.   □

17.3   Delaunay complexes

17.3.1   Definition and connection with Voronoi diagrams

Given a set of n points $\mathcal{M} = \{M_1, \ldots, M_n\}$ in $\mathbb{E}^d$, we lift the points onto the paraboloid $\mathcal{P}$ to $\{\phi(M_1), \ldots, \phi(M_n)\}$, and consider the unbounded polytope $\mathcal{V}(\mathcal{M})$ that is the intersection of the n half-spaces that lie above the hyperplanes $\phi(M_1)^*, \ldots, \phi(M_n)^*$, where $\phi(M_i)^*$ is tangent to $\mathcal{P}$ at $\phi(M_i)$.

We denote by $\mathcal{D}(\mathcal{M})$ the convex hull of the points $\phi(M_1), \ldots, \phi(M_n)$ and a point O' on the x_{d+1}-axis, with $x_{d+1} > 0$ large enough so that the facial structure of

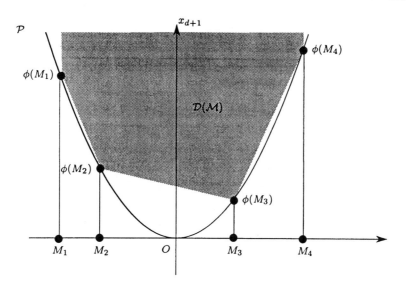

Figure 17.6. The polytope $\mathcal{D}(\mathcal{M})$.

that convex hull is stable as O' vanishes to infinity (see figure 17.6). The faces of $\mathcal{D}(\mathcal{M})$ that do not contain O' form the lower envelope of $conv(\phi(M_1), \ldots, \phi(M_n))$ (see also exercise 7.14). Their projections onto $\mathbb{E}^d$ form a complex whose vertices are exactly the M_i's. The domain of this complex is the projection of the convex hull of the $\phi(M_i)$'s: it is therefore the convex hull $conv(\mathcal{M})$ of the M_i's. This complex is called the Delaunay complex of $\mathcal{M}$ and is denoted by $\mathcal{D}el(\mathcal{M})$. For $k = 0, \ldots, d$, the k-faces of $\mathcal{D}el(\mathcal{M})$ are thus in one-to-one correspondence with the k-faces of $\mathcal{D}(\mathcal{M})$ that do not contain O'.

As shown in exercise 7.14, there exists a bijection between the faces of $\mathcal{V}(\mathcal{M})$ and the faces of $\mathcal{D}(\mathcal{M})$ that do not contain O'. This bijection maps the facet of $\mathcal{V}(\mathcal{M})$ containing $\phi(M_i)^*$ to the point $\phi(M_i)$. More generally, the k-faces of $\mathcal{V}(\mathcal{M})$ are in one-to-one correspondence with the $(d - k)$-faces of $\mathcal{D}(\mathcal{M})$ that do not contain O'. Moreover, this bijection reverses inclusion relationships.

Owing to theorem 17.2.5, the k-faces of $\mathcal{V}or(\mathcal{M})$ are also in bijection with the k-faces of the unbounded polytope $\mathcal{V}(\mathcal{M})$. So we have a bijection between the k-faces of $\mathcal{V}or(\mathcal{M})$ and the $(d - k)$-faces of $\mathcal{D}el(\mathcal{M})$ that reverses inclusion relationships. The Delaunay complex $\mathcal{D}el(\mathcal{M})$ is therefore dual to the Voronoi diagram $\mathcal{V}or(\mathcal{M})$.

Notice that the duality above maps a face of $\mathcal{V}or(\mathcal{M})$, formed by the points equidistant from m sites in $\mathcal{M}$, to the face of $\mathcal{D}el(\mathcal{M})$ that is the convex hull of these sites.

The preceding discussion leads to the following theorem:

Theorem 17.3.1 *The Delaunay complex of n points $M_1, \ldots, M_n$ in $\mathbb{E}^d$ is a complex dual to the Voronoi diagram. Its faces are obtained by projecting the faces of the lower envelope of the convex hull of the n points $\phi(M_1), \ldots, \phi(M_n)$, obtained by lifting the M_i's onto the paraboloid $\mathcal{P}$.*

The preceding theorem reduces the computation of the Delaunay complex of n points in $\mathbb{E}^d$ to the computation of the convex hull of n points in $\mathbb{E}^{d+1}$. All the convex hull algorithms described in this book, be they deterministic or randomized, static or dynamic, output-sensitive or not, therefore provide algorithms of the same kind that compute Delaunay complexes.

Theorem 17.3.1 also gives a lower bound on the complexity of the Delaunay complex. Indeed, exercise 7.11 exhibits polytopes in $\mathbb{E}^{d+1}$ with n vertices on the paraboloid, whose complexity is $\Theta(n^{\lceil d/2 \rceil})$. The same bound therefore applies to Delaunay complexes, and dually to Voronoi diagrams.

Corollary 17.3.2 *The Delaunay complex of n points in $\mathbb{E}^d$ can be computed in time $O(n \log n + n^{\lceil d/2 \rceil})$, and this is optimal in the worst case.*

17.3.2   Delaunay triangulations

Under L_2-general position assumptions, $\mathcal{V}(\mathcal{M})$ is a simple polytope, $\mathcal{D}(\mathcal{M})$ is a simplicial polytope, and $Del(\mathcal{M})$ is a simplicial complex which we call in this case the *Delaunay triangulation* (see figure 17.7). If there is a subset $\mathcal{M}' \subset \mathcal{M}$ of $l > d + 1$ co-spherical points and if the interior of the sphere circumscribed to $\mathcal{M}'$ (namely the sphere that passes through all the points in $\mathcal{M}'$) does not contain points in $\mathcal{M} \setminus \mathcal{M}'$, the Delaunay complex $\mathcal{D}(\mathcal{M})$ is not simplicial any more since $conv(\mathcal{M}')$ is a d-face of the Delaunay complex and it is not a simplex. Note however that this face may always be triangulated, and other non-simplicial faces of the complex may be triangulated as well. There are many ways to triangulate these faces, and any such triangulation is called a *Delaunay triangulation*. Henceforth, we denote by $Det(\mathcal{M})$ any such triangulation.

17.3.3   Characteristic properties

The Delaunay complex has remarkable properties, all due to the fact that it is dual to the Voronoi diagram.

Theorem 17.3.3 *Let $\mathcal{M}$ be a set of n points $M_1, \ldots, M_n$ in $\mathbb{E}^d$. Any d-face in the Delaunay complex can be circumscribed by a sphere that passes through all its vertices, and whose interior contains no point in $\mathcal{M}$.*

Proof. Let us pick a d-face T of the Delaunay complex. Then T is the convex hull $T = conv(M_{i_0}, \ldots, M_{i_l})$ of l co-spherical points $M_{i_0}, \ldots, M_{i_l}$. (If the points are in

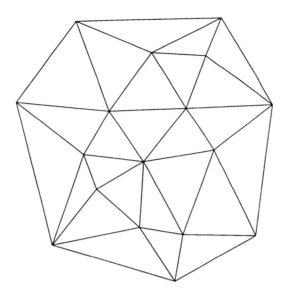

Figure 17.7. The Delaunay triangulation that corresponds to the Voronoi diagram shown in figure 17.1.

L_2-general position we have $l = d$.) The convex hull $conv(\phi(M_{i_0}), \dots, \phi(M_{i_l}))$ is a d-face F of the convex hull $conv(\phi(M_1), \dots, \phi(M_n))$, because of theorem 17.3.1. The intersection of the hyperplane $\underline{H}_F$ that supports F and of the paraboloid projects onto $\mathbb{E}^d$ as a sphere Σ circumscribed to $conv(M_{i_0}, \dots, M_{i_l})$, and its center is the projection on $\mathbb{E}^d$ of the pole $\underline{H}_F^*$ of $\underline{H}_F$ (see lemma 17.2.2). $\underline{H}_F^*$ is a vertex of $\mathcal{V}(\mathcal{M})$, and more precisely is the intersection of the polar hyperplanes $\phi(M_{i_0}^*), \dots, \phi(M_{i_l}^*)$. C is the vertex of the Voronoi diagram that is incident to the cells that correspond to the sites $M_{i_0}, \dots, M_{i_l}$, and the interior of Σ cannot contain any other point in $\mathcal{M}$. $\qquad\square$

Our next theorem extends this result into a necessary and sufficient condition for the convex hull of some points in $\mathcal{M}$ to be a face of the Delaunay complex of $\mathcal{M}$.

Theorem 17.3.4 *Let $\mathcal{M}$ be a set of points in $\mathbb{E}^d$, and $\mathcal{M}_k = \{M_{i_0}, \dots, M_{i_k}\}$ be a subset of k points in $\mathcal{M}$. The convex hull of $\mathcal{M}_k$ is a face of the Delaunay complex if and only if there exists a $(d-1)$-sphere passing through $M_{i_0}, \dots, M_{i_k}$ and such that no point in $\mathcal{M}$ belongs to its interior.*

Proof. The necessary condition immediately results from the preceding theorem and from the fact that a sphere circumscribed to a face is also circumscribed to its subfaces. Assume that there exists a $(d-1)$-sphere Σ that passes through $M_{i_0}, \dots, M_{i_k}$ and whose interior contains no point in $\mathcal{M}$. Let $\underline{H}$ be the hyperplane $\phi(\Sigma)^*$ in $\mathbb{E}^{d+1}$. This hyperplane contains the points $\phi(M_{i_0}), \dots, \phi(M_{i_k})$

and the half-space $\underline{H}^-$ lying below $\underline{H}$ does not contain points in $\phi(\mathcal{M})$ (according to lemma 17.2.4). Thus $\underline{H}$ is a hyperplane supporting $\mathcal{D}(\mathcal{M})$ along $conv(\phi(M_{i_0}), \ldots, \phi(M_{i_k}))$. Hence $conv(\phi(M_{i_0}) \ldots \phi(M_{i_k})) = \underline{H} \cap \mathcal{D}(\mathcal{M})$ is a face of $\mathcal{D}(\mathcal{M})$. It follows from theorem 17.3.1 that $M_{i_0}, \ldots, M_{i_k}$ is a face of the Delaunay complex of $\mathcal{M}$. $\square$

Corollary 17.3.5 *Any Delaunay triangulation of a set $\mathcal{M}$ of points in $\mathbb{E}^d$ is such that the sphere circumscribed to any d-simplex in the triangulation contains no point of $\mathcal{M}$ in its interior. Conversely, any triangulation satisfying this property is a Delaunay triangulation.*

The next theorem now exhibits a local characterization of Delaunay triangulations that will be put to good use later on. Let us consider any Delaunay triangulation $\mathcal{T}(\mathcal{M})$ of a set $\mathcal{M}$ of points in $\mathbb{E}^d$ and let $S_1 = M_1 \ldots M_d M_{d+1}$ and $S_2 = M_1 \ldots M_d M_{d+2}$ be a pair of adjacent d-simplices in $\mathcal{T}(\mathcal{M})$ that share a common face $F = M_1 \ldots M_d$. The pair (S_1, S_2) is called *regular* if M_{d+1} does not belong to the interior of the sphere Σ_2 circumscribed to S_2. If the sphere Σ_1 circumscribed to S_1 differs from Σ_2, the regularity condition is equivalent to the property that M_{d+2} does not belong to the interior of Σ_1. Indeed, M_{d+1} does not belong to the interior of Σ_2 if and only if $\Sigma_2(M_{d+1}) > 0$. But the hyperplane H_F that supports F is the radical hyperplane of Σ_1 and Σ_2. Since $\Sigma_1(M_{d+1}) = 0$, the half-space bounded by H_F that contains M_{d+1} (resp. M_{d+2}) consists of the points whose power with respect to Σ_1 is smaller (resp. greater) than their power with respect to Σ_2, and therefore

$$\Sigma_1(M_{d+2}) > \Sigma_2(M_{d+2}) = 0,$$

which proves that M_{d+2} does not belong to the interior of Σ_1.

Theorem 17.3.6 *Consider a triangulation $\mathcal{T}(\mathcal{M})$ of a set $\mathcal{M}$ of points in $\mathbb{E}^d$. Then $\mathcal{T}(\mathcal{M})$ is a Delaunay triangulation if and only if all the pairs of adjacent d-simplices in $\mathcal{T}(\mathcal{M})$ are regular.*

Proof. That the condition is necessary is a consequence of theorem 17.3.3. We must now show that it suffices. To alleviate the notation, we denote by $\phi(S)$ the k-simplex in $\mathbb{E}^{d+1}$ whose vertices are the images of the vertices of a k-simplex S in $\mathbb{E}^d$, and by $\mathcal{C}$ the union of the $\phi(S)$'s for all the faces S of the Delaunay triangulation $\mathcal{T}(\mathcal{M})$. The proof consists of proving that $\mathcal{C}$ is the graph of a convex real-valued function over the convex hull $conv(\mathcal{M})$.

As above, we consider two adjacent d-simplices $S_1 = M_1 \ldots M_d M_{d+1}$ and $S_2 = M_1 \ldots M_d M_{d+2}$ in $\mathcal{T}(\mathcal{M})$ that share a common face $F = M_1 \ldots M_d$. We denote by Σ_1 and Σ_2 the spheres circumscribed to S_1 and S_2. Owing to lemma 17.2.4, the regularity condition is equivalent to $\phi(M_{d+1}) \in \phi(\Sigma_2)^{*+}$ and also to $\phi(M_{d+2}) \in \phi(\Sigma_1)^{*+}$ because of the discussion above. Therefore, if the pair (S_1, S_2) is regular, then the $(d-1)$-face $\phi(F)$ is locally convex, meaning that there is a hyperplane that contains $\phi(F)$ such that $\phi(S_1)$ and $\phi(S_2)$ belong to the half-space lying above this hyperplane. This is true for any $(d-1)$-face of $\mathcal{C}$ incident to two d-faces, and so $\mathcal{C}$ is locally convex at any point. Moreover, $\mathcal{C}$ is defined over a convex subset of $\mathbb{E}^d$, namely the convex hull of $\mathcal{M}$. Therefore, $\mathcal{C}$ is convex and is the lower envelope of the polytope $\mathcal{D}(\mathcal{M})$, which proves that $\mathcal{T}(\mathcal{M})$ is a Delaunay triangulation of $\mathcal{M}$. □

17.3.4 Optimality of Delaunay triangulations

As we have seen in chapter 11, there exist several ways to triangulate a set of points. Some are not very interesting in practice, and in many applications certain criteria must be optimized, and an optimal triangulation is desirable. There are several ways to define optimality. In this section, we show that Delaunay triangulations maximize two criteria, compactness and equiangularity.

Compactness

The preceding theorem was concerned with spheres circumscribed to simplices in the triangulation. The next theorem considers the *smallest enclosing sphere* for each simplex S: this sphere is the circumscribed sphere of S if the center of the latter belongs to S, or otherwise is a sphere centered on some k-face ($k < d$) of S and passes through the $k+1$ centers of this face.

As before, we consider a set $\mathcal{M}$ of points in $\mathbb{E}^d$ and $\mathcal{T}(\mathcal{M})$ a triangulation of $\mathcal{M}$. To $\mathcal{T}(\mathcal{M})$ corresponds a function $\Sigma_T(X)$ defined over $conv(\mathcal{M})$ as the power of a point X with respect to the sphere Σ circumscribing any d-simplex of $\mathcal{T}(\mathcal{M})$ that contains X. By the results of subsection 17.2.6, $\Sigma_T(X)$ is well-defined when X belongs to several cells.

Lemma 17.3.7 *Let* $\mathcal{D}et(\mathcal{M})$ *be a Delaunay triangulation of* $\mathcal{M}$ *and* $\mathcal{T}(\mathcal{M})$ *be any other triangulation of* $\mathcal{M}$. *Then*

$$\forall X \in conv(\mathcal{M}), \ \Sigma_{\mathcal{D}et}(X) \geq \Sigma_T(X).$$

Proof. Consider a d-simplex T in $\mathcal{T}(\mathcal{M})$ that contains X, Σ its circumscribed sphere, and $\phi(T)$ the d-simplex of $\mathbb{E}^{d+1}$ whose vertices are the images under ϕ of the vertices of T. (Recall that these vertices are obtained by lifting the vertices of T onto the paraboloid $\mathcal{P}$.) Lemma 17.2.3 shows that $\Sigma_T(X)$ is the signed vertical distance (here negative) from $\phi(\Sigma)^*$ to $\phi(X)$. Notice that $\phi(\Sigma)^*$ is the affine hull of $\phi(T)$. For a given X, this signed vertical distance is maximized when $\phi(T)$ is a face of the convex hull of $\phi(\mathcal{M})$ in $\mathbb{E}^{d+1}$: in other words, when T is a simplex of a Delaunay triangulation of $\mathcal{M}$. $\square$

Lemma 17.3.8 *If T is a d-simplex and if Σ_T is its circumscribed sphere, then*

$$\min_{X \in T} \Sigma_T(X) = \Sigma_T(C'_T) = -r'^{\,2}_T$$

where C'_T and r'_T are respectively the center and the radius of the smallest sphere enclosing T.

Proof. Let Σ_T be the sphere circumscribed to T, C_T its center, and r_T its radius. Then

$$\Sigma_T(X) = XC_T^2 - r_T^2$$

is minimized when $X = C_T$ and is therefore greater than $-r_T^2$. If C_T is contained in T, the smallest enclosing sphere of T is Σ_T, hence $r'_T = r_T$ and the lemma is trivial. Otherwise, the smallest enclosing sphere of T is centered on a k-face $(k < d)$, namely the face F such that the orthogonal projection of C_T onto the plane that supports F falls inside F. The radius r'_T of this sphere is that of the $(k-1)$-sphere circumscribed to F. Its center C'_T minimizes the value of $XC_T \cdot XC_T$ when $X \in T$. Pythagoras' theorem then shows that

$$C_T C'^{\,2}_T + r'^{\,2}_T = r_T^2,$$

which finishes the proof. $\square$

Let $\mathcal{T}(\mathcal{M})$ be any triangulation of a set $\mathcal{M}$ of points in $\mathbb{E}^d$. For each simplex T in $\mathcal{T}(\mathcal{M})$, we let r'_T denote the smallest radius of a sphere that encloses T, and the *maximum min-containment radius* of $\mathcal{T}(\mathcal{M})$ is defined by

$$C(\mathcal{T}(\mathcal{M})) = \max_{T \in \mathcal{T}(\mathcal{M})} r'_T.$$

The *most compact* triangulations are then defined as the triangulations that minimize the maximum min-containment radius.

Theorem 17.3.9 *Delaunay triangulations are the most compact among all the triangulations of $\mathcal{M}$.*

Note that since the maximum min-containment radius $C(\mathcal{T}(\mathcal{M}))$ is defined only by the simplices T of $\mathcal{T}(\mathcal{M})$ such that $C(\mathcal{T}(\mathcal{M})) = r'_T$, triangulations other than Delaunay triangulations might also be most compact among the triangulations of $\mathcal{M}$.

Proof. Let $\mathcal{T}(\mathcal{M})$ be any triangulation of $\mathcal{M}$ and let $\mathcal{D}et(\mathcal{M})$ be any Delaunay triangulation of $\mathcal{M}$. We denote by $X_{\mathcal{T}}$ the point X that minimizes $\Sigma_{\mathcal{T}}(X)$ and by $X_{\mathcal{D}et}$ the point X that minimizes $\Sigma_{\mathcal{D}et}(X)$. From lemma 17.3.8, we know that $X_{\mathcal{T}}$ is the center of the smallest sphere that encloses the simplex in $\mathcal{T}(\mathcal{M})$ which contains $X_{\mathcal{T}}$. We denote its radius by $r'_{\mathcal{T}}$. Likewise, $X_{\mathcal{D}et}$ is the center of the smallest sphere that encloses the simplex in $\mathcal{D}et(\mathcal{M})$ that contains $X_{\mathcal{T}}$ and its radius is denoted by $r'_{\mathcal{D}et}$. The maximum min-containment radius of $\mathcal{T}(\mathcal{M})$ equals $r'_{\mathcal{T}}$ and that of $\mathcal{D}et$ equals $r'_{\mathcal{D}et}$. Using lemmas 17.3.7 and 17.3.8, we obtain

$$\Sigma_{\mathcal{T}}(X_{\mathcal{T}}) = -{r'_{\mathcal{T}}}^2 \le \Sigma_{\mathcal{T}}(X_{\mathcal{D}et}) \le \Sigma_{\mathcal{D}et}(X_{\mathcal{D}et}) = -{r'_{\mathcal{D}et}}^2.$$

$\square$

Equiangularity ($d = 2$)

We now restrict the discussion to triangulations of a set of points in the plane. Given a triangulation $\mathcal{T}(\mathcal{M})$ of a set $\mathcal{M}$ of n points in the plane, we define its *angle vector* as the vector $Q(\mathcal{T}(\mathcal{M})) = (\alpha_1, \ldots, \alpha_{3t})$ where the α_i's are the angles of the t triangles of $\mathcal{T}(\mathcal{M})$ sorted by increasing value. We know that $\sum_{i=1}^{3t} \alpha_i = t\pi$. Note that a triangulation that maximizes the angle vector for the lexicographic order also maximizes the smallest of its angles. Such a triangulation is called *globally equiangular*.

Theorem 17.3.10 *A globally equiangular triangulation of a set $\mathcal{M}$ of points in the plane is always a Delaunay triangulation.*

Proof. We must prove that, among all the triangulations of $\mathcal{M}$, the ones that maximize the angle vector for the lexicographic order are always Delaunay triangulations. Let us thus consider two triangles $T_1 = ABC$ and $T_2 = BCD$ in some triangulation $\mathcal{T}(\mathcal{M})$, such that the union of T_1 and T_2 is a strictly convex quadrilateral Q. (This means that A, B, C, and D are all vertices of the convex hull $conv(A, B, C, D)$.) In order to increase the equiangularity, we can flip the diagonal as follows (shown in figure 17.8). If the triangles $T'_1 = ABD$ and $T'_2 = ACD$ are such that $Q(T'_1, T'_2) > Q(T_1, T_2)$, then replace $\mathcal{T}(\mathcal{M})$ by a triangulation $\mathcal{T}^1(\mathcal{M})$ which contains T'_1 and T'_2 instead of T_1 and T_2.

The previous rule may be dubbed a *regularization* rule since it transforms a pair of adjacent triangles into a regular pair of triangles: if the two triangles do not form a convex quadrilateral, then the pair is obviously regular, and the rule does not apply; otherwise, $T_1 \cup T_2$ is convex and the pair is transformed into a regular pair. Indeed, let Σ_1 and Σ_2 be the circles circumscribed to T_1 and T_2. We will show that the diagonal AD is flipped if and only if D is contained inside the circle Σ_1. Let α, β, γ, and δ be the angles at the vertices of the quadrilateral $ABCD$, α, β_1, and γ_1 the angles at the vertices of T_1, and β_2, γ_2, and δ the angles

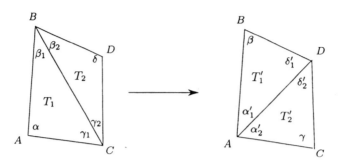

Figure 17.8. Flipping a diagonal to increase the equiangularity.

at the vertices of T_2. Moreover, we denote by α'_1, β, and δ'_1 the angles at the vertices of T'_1, and α'_2, γ, and δ'_2 the angles at the vertices of T'_2. The situation is depicted in figure 17.8. If α is the smallest angle in T_1 and T_2, then the diagonal is not flipped. But then

$$\delta = \pi - \beta_2 - \gamma_2 < \pi - \alpha,$$

so that $\alpha + \delta < \pi$, which shows that A is not contained inside Σ_2, and this implies that D is not contained inside Σ_1. The situation is entirely symmetric when the smallest angle is δ. When the smallest angle is β_1, then we flip the diagonal only if δ'_2 is greater than β_1, which only happens when D is contained inside Σ_1. Of course, the cases when the smallest angle is γ_1, β_2, or γ_2 are entirely similar, so we have shown that the diagonal is flipped if and only if it transforms the irregular pair (T_1, T_2) into a regular pair (T'_1, T'_2).

Clearly, after a flip we have $Q(\mathcal{T}^1(\mathcal{M})) > Q(\mathcal{T}(\mathcal{M}))$. Flipping the edges whenever possible progressively increases the angle vector of the triangulation. Since there are only a finite number of triangulations, this process eventually reaches a triangulation that has only regular pairs of adjacent triangles. This triangulation is a Delaunay triangulation as is shown by theorem 17.3.6.   □

Note that this local regularization always leads to a Delaunay triangulation. When the points are in L_2-general position, there is only one Delaunay triangulation: the result of the procedure described above therefore does not depend on the starting configuration, nor on the order chosen to flip the diagonals.

When the points are not in L_2-general position, however, the theorem above shows that flipping diagonals only reaches a Delaunay triangulation. Yet there are several Delaunay triangulations, which may not all have the same angle vectors. Still, there is an algorithm that can reach a globally equiangular triangulation (see the bibliographical notes).

17.4 Higher-order Voronoi diagrams

In this section, we define Voronoi diagrams of order k and show the connection between these diagrams and the faces at level k in a hyperplane arrangement in $\mathbb{E}^{d+1}$. As usual, the Euclidean space $\mathbb{E}^d$ of dimension d is embedded in $\mathbb{E}^{d+1}$ as the hyperplane $x_{d+1} = 0$, and $\phi(M)^*$ denotes the hyperplane in $\mathbb{E}^{d+1}$ that is tangent to the paraboloid $\mathcal{P}$ at the point $\phi(M)$ obtained by lifting M vertically onto the paraboloid.

In section 17.2, we established the connection between the Voronoi diagram of a set $\mathcal{M}$ of n points $M_1, \ldots, M_n$ in $\mathbb{E}^d$ and the polytope in $\mathbb{E}^{d+1}$ that is the intersection of the n half-spaces $\phi(M_i)^{*+}$ that lie above the hyperplanes $\phi(M_i)$. Equivalently, $\mathcal{V}(\mathcal{M})$ is the cell at level 0 in the arrangement $\mathcal{A}$ of the hyperplanes $\phi(M_1)^*, \ldots, \phi(M_n)^*$, if the reference point is on the x_{d+1}-axis, sufficiently high so that it is above all the hyperplanes. Let us recall that a point is at level k in $\mathcal{A}$ if it belongs to exactly k open half-spaces $\phi(M_{i_1})^{*-}, \ldots, \phi(M_{i_k})^{*-}$, such that each $\phi(M_{i_j})^{*-}$ is bounded by $\phi(M_{i_j})^*$ and does not contain the reference point (see section 14.5).

It is tempting to consider the cells at levels $k > 0$. We define below a cell complex that spans $\mathbb{E}^d$, called the *Voronoi diagram of order k* of $\mathcal{M}$, and show in theorem 17.4.1 that the cells of this complex are the non-overlapping projections onto $\mathbb{E}^d$ of the cells at level k in the arrangement $\mathcal{A}$.

Let $\mathcal{M}_k$ be a subset of size k of $\mathcal{M}$. The Voronoi region of $\mathcal{M}_k$ is the polytope $V_k(\mathcal{M}_k)$ of the points in $\mathbb{E}^d$ that are closer to all the sites in $\mathcal{M}_k$ than to any other site in $\mathcal{M} \setminus \mathcal{M}_k$. Formally,

$$V_k(\mathcal{M}_k) = \{X : \forall M_i \in \mathcal{M}_k, \forall M_j \in \mathcal{M} \setminus \mathcal{M}_k, \|XM_i\| \leq \|XM_j\|\}.$$

Let us consider all the subsets of size k of $\mathcal{M}$ whose Voronoi regions are not empty. As proved in the theorem below, these polytopes and their faces form a d-complex whose domain is $\mathbb{E}^d$. This complex is called the *Voronoi diagram of order k* of $\mathcal{M}$ (see figures 17.9 and 17.10). It is denoted by $\mathcal{V}or_k(\mathcal{M})$. When $k = 1$, we recognize the definition of the usual Voronoi diagram.

Theorem 17.4.1 *The Voronoi diagram $\mathcal{V}or_k(\mathcal{M})$ of order k of a set $\mathcal{M} = \{M_1, \ldots, M_n\}$ of n points in $\mathbb{E}^d$ is a cell complex of dimension d in $\mathbb{E}^d$. The cells of this complex correspond to the cells at level k in the arrangement $\mathcal{A}$ of the hyperplanes $\phi(M_1)^*, \ldots, \phi(M_n)^*$ in $\mathbb{E}^{d+1}$, when the reference point is on the x_{d+1}-axis above all the hyperplanes. A cell of $\mathcal{V}or_k(\mathcal{M})$ is obtained by projecting vertically onto $\mathbb{E}^d$ the corresponding cell in $\mathcal{A}$. The l-faces of $\mathcal{V}or_k(\mathcal{M})$ are obtained by projecting the l-faces common to cells at level k in $\mathcal{A}$.*

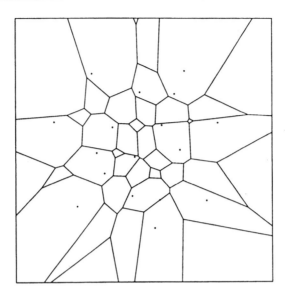

Figure 17.9. The Voronoi diagram of order 2 of the points in figure 17.1.

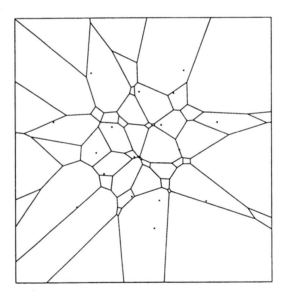

Figure 17.10. The Voronoi diagram of order 3 of the points in figure 17.1.

Proof. The proof relies on lemma 17.2.4. A sphere in $\mathbb{E}^d$ whose interior contains k points is mapped by ϕ to a point at level k in the arrangement $\mathcal{A}$ of the hyperplanes $\phi(M_1)^*, \ldots, \phi(M_n)^*$.

More precisely, X belongs to the cell $V_k(\mathcal{M}_k)$ in the Voronoi diagram of order k, if and only if X is the center of a sphere Σ whose interior contains the points

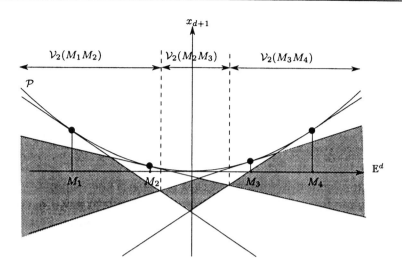

Figure 17.11. The Voronoi diagram of order 2 is obtained by projecting the cells at level 2 in $\mathcal{V}(\mathcal{M})$.

of $\mathcal{M}_k$ and only those. Then $\phi(\Sigma)$ belongs to the k closed half-spaces below the hyperplanes $\phi(M_j)^*$ for the M_j's in $\mathcal{M}_k$, and only to those half-spaces. The cells of the Voronoi diagram of order k are obtained by projecting vertically the cells at level k of $\mathcal{A}$ (see figure 17.11).

It is easily verified that any vertical line intersects at least one cell at level k in $\mathcal{A}$ and does not intersect the interior of more than one cell at level k. It follows that the l-faces, for $l < d$, of the Voronoi diagram of order k are obtained by vertically projecting the l-faces common to several cells at level k. If the M_i's are in L_2-general position, then the hyperplanes $\phi(M_1)^*, \ldots, \phi(M_n)^*$ are in general position. In that case, it was shown in section 14.5 that the cells of $\mathcal{A}$ that contain an l-face F at level k have levels that vary between k and $k + d + 1 - l$. Among those, there is only one cell at level k and one cell at level $k + d + 1 - l$, and several cells at levels $k < j < k + d + 1 - l$. It follows that the vertical projection of F is an l-face of the Voronoi diagrams of orders $k + 1, k + 2, \ldots, k + d - l$. $\square$

Having computed the Voronoi diagram of order k of the sites, looking for the k nearest sites of any point X in $\mathbb{E}^d$ can be performed by finding the cell of the diagram that contains X (see exercises 17.2 and 17.4).

It follows from the construction that the total complexity of the Voronoi diagrams of all orders k, $1 \le k \le n - 1$, is $O(n^{d+1})$: indeed it is exactly the complexity of the arrangement of the hyperplanes in $\mathbb{E}^{d+1}$. Moreover, these diagrams can be computed in time $O(n^{d+1})$ (see theorem 14.4.4). An upper bound on the complexity of the Voronoi diagrams $Vor_1(\mathcal{M}), \ldots, Vor_k(\mathcal{M})$ of orders between 1 and k is provided by theorem 14.5.1, which bounds the complexity of the first k levels in the arrangement of n hyperplanes in $\mathbb{E}^{d+1}$. Moreover, $Vor_{\leq k}(\mathcal{M}) = \{Vor_1(\mathcal{M}), \dots, Vor_k(\mathcal{M})\}$ can be computed by computing the first k levels in this arrangement (see theorem 14.5.3 and exercise 14.22) and projecting back onto $\mathbb{E}^d$.

Theorem 17.4.2 *The overall complexity of the first k Voronoi diagrams of a set of n points in $\mathbb{E}^d$ is $O(n^{\lfloor (d+1)/2 \rfloor} k^{\lceil (d+1)/2 \rceil})$. These k diagrams may be computed in time $O(n^{\lfloor (d+1)/2 \rfloor} k^{\lceil (d+1)/2 \rceil})$ if $d \geq 3$, or $O(nk^2 \log \frac{n}{k})$ if $d = 2$.*

Let us close this section by observing that Vor_{n-1} is the complex whose cells consist of the points further from a particular site than from any other site. This is why this diagram is sometimes called the *furthest-point Voronoi diagram*. The vertices of this diagram are the centers of spheres circumscribed to $d + 1$ sites and whose interiors contain all the other sites. Its cells are all unbounded. The furthest-point diagram can be obtained by computing the intersection of the n *lower* half-spaces bounded by the hyperplanes $\phi(M_i)^*$, $i = 1, \dots, n$.

17.5   Exercises

Exercise 17.1 (Shortest edge) Denote by $\mathcal{F}$ and $\mathcal{G}$ two finite sets of points in $\mathbb{E}^d$. Show that the shortest edge that connects a point in $\mathcal{F}$ to a point in $\mathcal{G}$ is an edge of the Delaunay triangulation of $\mathcal{F} \cup \mathcal{G}$. From this, conclude that each point is adjacent to its nearest neighbor in the Delaunay triangulation.

Exercise 17.2 (Nearest neighbor in the plane) Show that, given the Voronoi diagram of a set $\mathcal{M}$ of points in the plane, it may be preprocessed in linear time to answer nearest neighbor queries (that is, find the nearest site to a query point) in logarithmic time. Same question for the set of k nearest neighbors (k fixed).

Hint: One may use the result of exercise 12.2.

Exercise 17.3 (On-line nearest neighbor) We place point sites in the plane and we want to maintain a data-structure on-line so as to answer nearest neighbor queries on the current set of sites (that is, find the nearest site to a query point). Devise a structure that stores n sites using storage $O(n)$, such that under the assumption that the points are inserted in a random order, the expected time needed to insert a new site is $O(\log n)$, and that answers any query in expected time $O(\log^2 n)$.

Hint: One may build a two-level data-structure in the following way. The first level corresponds to a triangulation of the Voronoi diagram, obtained by connecting a site to all the vertices of its Voronoi region. Build an influence graph for the regions defined as the triangles in this triangulation (it is a variant of the influence graph used in exercise 17.10). Each triangle points to the site that kills it, and all the triangles created after the insertion of a site are sorted in polar angle around this site and stored into an array: this is the second level of the data structure. Show that the query point belongs to $O(\log n)$ triangles on the average, and hence that only $O(\log n)$ binary searches are performed in the arrays of the second level.

Exercise 17.4 (Nearest neighbor) Consider a set of n point sites in $\mathbb{E}^d$. Explain how to design a data structure of size $O(n^{\lceil d/2 \rceil + \varepsilon})$ that allows the nearest site to any query point P to be found in logarithmic time.

Hint: Use the solution of exercise 14.12 in $\mathbb{E}^{d+1}$. The exponent ε can be removed, at the cost of increasing the query time to $O(\log^2 n)$ (see the bibliographical notes at the end of chapter 14).

Exercise 17.5 (Union of balls) Use lemma 17.2.4 to reduce the problem of computing the union of n balls in $\mathbb{E}^d$ to that of computing the intersection of the paraboloid $\mathcal{P}$ with a polytope in $\mathbb{E}^{d+1}$. Conclude that the complexity of the union of n balls is $O(n^{\lceil \frac{d}{2} \rceil})$. Devise an algorithm that computes the union of n balls in expected time $\Theta(n \log n + n^{\lceil \frac{d}{2} \rceil})$.

Exercise 17.6 (Intersection of balls) The results of exercise 17.5 are also valid for the intersection of n balls in $\mathbb{E}^d$. In $\mathbb{E}^3$, show that if the balls have same radius, the complexity of the intersection is only $O(n)$ and propose an algorithm that computes this intersection in expected time $\Theta(n \log n)$.

Hint: Show that each face of the intersection is "convex", meaning that given any two points in any face, there is an arc of a great circle joining these points which is entirely contained in that face; then use Euler's relation. For the algorithm, use a variant of the randomized incremental algorithm of section 8.3.

Exercise 17.7 (Minimum enclosing ball) Show that the center of the smallest ball whose interior contains a set $\mathcal{M}$ of points in $\mathbb{E}^2$ is either a vertex of the furthest-point Voronoi diagram (of order $n-1$) of $\mathcal{M}$, or else the intersection of an edge of this diagram (on the perpendicular bisector of two sites A and B) with the edge AB.

Exercise 17.8 (Centered triangulation) Consider a triangulation $\mathcal{T}(\mathcal{M})$ of a set $\mathcal{M}$ of points in $\mathbb{E}^d$. Show that, if each simplex in $\mathcal{T}(\mathcal{M})$ contains the center of its circumscribed sphere, then $\mathcal{T}(\mathcal{M})$ is a Delaunay triangulation of $\mathcal{M}$. Construct a counterexample for the converse.

Hint: Show that any adjacent pair of d-simplices is regular.

Exercise 17.9 (Non-optimality of the Delaunay triangulation) Construct a set of points in the plane whose Delaunay triangulation does not minimize the greatest angle among all the possible triangulations. Same question to show that the Delaunay triangulation does not necessarily minimize the length of the longest edge, nor the total length of the triangulation (sum of the lengths of the edges).

Exercise 17.10 (Incremental algorithm) Let $Del(\mathcal{M})$ be the Delaunay triangulation of a set $\mathcal{M}$ of points in L_2-general position in $\mathbb{E}^d$. Let A be a point of $\mathbb{E}^d$ distinct from the points in $\mathcal{M}$, let $\mathcal{S}$ be the sub-complex formed by the d simplices in $Del(\mathcal{M})$ whose circumscribed spheres contain A in their interior, and let $\mathcal{F}$ be the set of $(d-1)$-faces on the boundary of $\mathcal{S}$. Show that if A belongs to the convex hull of $\mathcal{M}$, then the d-simplices in $Del(\mathcal{M} \cup \{A\})$ are exactly the simplices in $Del(\mathcal{M})$ that do not belong to $\mathcal{S}$ and the simplices $conv(A, F)$, $F \in \mathcal{F}$. Generalize this result to the case when A is outside the convex hull of $\mathcal{M}$, and derive an incremental algorithm that computes the Delaunay triangulation. Show that this algorithm runs in time $\Theta(n^{\lceil \frac{d}{2} \rceil + 1})$ in the worst case. Show that if the points are inserted in random order, then the algorithm runs in expected time $O(n \log n + n^{\lceil \frac{d}{2} \rceil})$, which is optimal. Devise a dynamic algorithm that also allows points to be removed.

Hint: Use a randomized algorithm with an influence graph. Objects are sites, regions are the balls circumscribed to $d+1$ sites, and an object conflicts with a region if it belongs to that region. Show that the ball circumscribed to any new simplex $S = conv(A, F)$ is contained in the union of the two balls circumscribed to T and V, the two d-simplices that share the common facet F. Build an influence graph in which each node is the child of only two nodes, namely the node corresponding to F is the child of the nodes corresponding to T and V. The number of children of a node is not bounded, but the analysis can be carried out using biregions (see exercise 5.7).

Exercise 17.11 (Flipping the diagonals) Devise an incremental algorithm to compute the Delaunay triangulation of points in the plane which, at each step, connects the new point to the edges of the triangle that contains it, and then regularizes the triangulation as in the proof of theorem 17.3.10. Show that, if the points are inserted in random order, the algorithm can be made to run in expected time $O(n \log n)$, which is optimal.

Exercise 17.12 (Flipping in higher dimensions) Generalize the local regularization rule introduced in the proof of theorem 17.3.10 to the triangulation of point sites in $\mathbb{E}^3$ and in higher dimensional spaces. Show that this does not always result in the Delaunay triangulation of the points, in contrast with the planar case.

Hint: As was done for planar triangulations (proof of theorem 17.3.10), local regularization in $\mathbb{E}^3$ corresponds to replacing the upper facets of a simplex in $\mathbb{E}^4$ by its lower facets. A simplex in $\mathbb{E}^4$ having five facets, local regularization in $\mathbb{E}^3$ leads to replacing two adjacent tetrahedra T_1 and T_2 by three tetrahedra T_3, T_4, and T_5 that are pairwise adjacent (and have the same vertices as T_1 and T_2), or the converse. Show that the local regularization rule cannot always be applied even though the triangulation is not regular everywhere.

Exercise 17.13 (Flipping in higher dimensions) Show that if one adds a new point P to a Delaunay triangulation $Det(\mathcal{M})$ of a set $\mathcal{M}$ of points in $\mathbb{E}^d$, the Delaunay triangulation $Det(\mathcal{M} \cup \{P\})$ can be obtained by splitting the simplex of $Det(\mathcal{M})$ that contains P into $d + 1$ new simplices, and then applying the generalized local regularization of exercise 17.12. Show that if the n points in $\mathcal{M}$ are inserted in a random order, this incremental algorithm computes $Det(\mathcal{M})$ in expected time $O(n \log n + n^{\lceil \frac{d}{2} \rceil})$, which is optimal.

Hint: As in exercise 17.12, each regularization in $\mathbb{E}^d$ corresponds to replacing the upper facets of a $(d+1)$-simplex in $\mathbb{E}^{d+1}$ by its lower facets. Show that any $(d+1)$-simplex S involved in any step of the local regularization has P as a vertex and that the convex hull of its vertices other than P is a d-simplex of $\mathcal{D}et(\mathcal{M})$ that is destroyed by the local regularization.

Exercise 17.14 (Complexity of the Voronoi diagram of order k) Show that the complexity of the Voronoi diagram of order k of n points in $\mathbb{E}^2$ is always $O(k(n-k))$ and can be $\Omega(k(n-k))$ in the worst case.

Exercise 17.15 (Higher-order Voronoi diagrams and polytopes) Let $\mathcal{M}$ be a set of n points $M_1, \ldots, M_n$ in $\mathbb{E}^d$. With each subset $\mathcal{M}_k = \{M_{i_1}, \ldots, M_{i_k}\}$ of size k of $\mathcal{M}$, we associate its center of gravity $G(\mathcal{M}_k) = \frac{1}{k}\sum_{j=1}^k M_{i_j}$, and the real number $\sigma(\mathcal{M}_k) = \frac{1}{k}\sum_{j=1}^k M_{i_j}^2$. Show that the Voronoi diagram of order k of $\mathcal{M}$ is the projection of the polytope in $\mathbb{E}^d$ defined as the intersection of the half-spaces lying above the hyperplanes polar to the points $(G(\mathcal{M}_k), \sigma(\mathcal{M}_k))$, for all subsets $\mathcal{M}_k$ of size k of $\mathcal{M}$.

Hint: From the fact that

$$\frac{1}{k}\sum_{j=1}^k (X - M_{i_j}) \cdot (X - M_{i_j}) = X^2 - \frac{2}{k}\sum_{j=1}^k M_{i_j} \cdot X + \frac{1}{k}\sum_{j=1}^k M_{i_j}^2,$$

we infer that the k nearest neighbors of X are the points in $\mathcal{M}_k$ if and only $\mathcal{M}_k$ is the subset for which X has the smallest power with respect to the sphere centered at $G(\mathcal{M}_k)$ and whose power with respect to the origin is $\sigma(\mathcal{M}_k)$.

Exercise 17.16 (Euclidean minimum spanning tree) Consider a set $\mathcal{M}$ of n points in L_2-general position in $\mathbb{E}^d$. A *Euclidean minimum spanning tree*, or EMST for short, is a tree whose nodes are the points in $\mathcal{M}$ and whose total edge length is minimal. Show that such a tree is a subgraph of the Delaunay triangulation of $\mathcal{M}$. For the planar case, show that an EMST can be computed in time $O(n \log n)$. Consider the case where the set of points is not in L_2-general position any more.

Hint: Show that the following greedy algorithm produces a minimum spanning tree. Denote by $\mathcal{A}$ the set of points of $\mathcal{M}$ that are already connected to the current tree. The greedy algorithm picks the shortest segment that does not induce a cycle in the current subtree. This edge connects a point of $\mathcal{A}$ to a point of $\mathcal{M} \setminus \mathcal{A}$. The latter point is added to $\mathcal{A}$, the edge is added to the tree, and so on until the tree spans $\mathcal{M}$. Show that this yields an EMST, even if the points are not in L_2-general position. Exercise 17.1 shows that it can be completed into a Delaunay triangulation of $\mathcal{M}$. Explain how to make the algorithm run in time $O(n \log n)$.

17.6 Bibliographical notes

Voronoi diagrams have been used for a long time and in various disguises. Voronoi, a Russian mathematician of the early twentieth century, was the first to give them a precise definition and study them for their own sake, but they had already been used by

Wigner and Steiz in crystallography, by Thiessen in geography, and even by Descartes in astronomy. Delaunay [78] established most of the fundamental properties about the triangulation that bears his name.

More recently, the connection between Voronoi diagrams and polytopes was discovered by Brown [38] and by Edelsbrunner and Seidel [96]. The presentation in this book is based on the works of Boissonnat, Cérézo, Devillers, and Teillaud [26] and of Devillers, Meiser, and Teillaud [82]. The connection with polytopes answers exercise 17.4.

The optimality of the Delaunay triangulation was established by Rajan [193] for the compactness, and by Lawson [144] for the equiangularity. Mount and Saalfeld [170] have proposed an algorithm to compute a globally equiangular triangulation when the points are not in L_2-general position. In the context of approximating surfaces by piecewise linear patches, controlling the equiangularity serves to control the quality of the approximation even though it is more profitable to minimize the greatest angle as was shown by Nielson and Franke [179]. The Delaunay triangulation does not generally minimize the greatest angle, nor the total edge length, even though it often works well for practical instances. Recent references on these topics can be found in the works of Edelsbrunner and Tan [101], Edelsbrunner, Tan, and Waupotitsch [102], and Dickerson, McElfresh, and Montague [83]. Rippa [195], and also Rippa and Schiff [194], gave other useful criteria in the context of approximating surfaces for which the Delaunay triangulation is optimal. Desnogués [79] provides a good survey of polyhedral approximation.

An incremental algorithm that computes the Voronoi diagram of a set of points (see exercise 17.10) was given by Green and Sibson [113]. The Delaunay tree introduced by Boissonnat and Teillaud [31, 32] improves on the average performance when the sites are inserted in random order. This algorithm was made fully dynamic by Devillers, Meiser, and Teillaud [81]. An algorithm that proceeds by flipping diagonals was proposed by Lawson [144], then dynamized in the plane by Guibas, Knuth, and Sharir [117] who present a randomized analysis and also solutions to exercises 17.3 and 17.11. Its generalization to higher dimensions was studied by Joe [131, 132], Rajan [193], and Edelsbrunner and Shah [98], who provide a solution to exercise 17.13.

Lee proposed the first algorithm that computes Voronoi diagrams of higher orders in the plane [145]. He also gave a solution to exercise 17.14. The connection between Voronoi diagrams of order k and polytopes (exercise 17.15) was established by Aurenhammer [15]. Boissonnat, Devillers, and Teillaud [29] and also Mulmuley [174] proposed semi-dynamic or even fully dynamic algorithms that compute the Voronoi diagrams of all orders up to k in any dimension. Clarkson [67], Aurenhammer and Schwarzkopf [18], and also Agarwal, de Berg, and Matoušek [2] gave randomized algorithms that compute Voronoi diagrams of a single order k, rather than all the diagrams of orders $\leq k$.

The connection between Delaunay triangulations and Euclidean minimum spanning trees (see exercise 17.16) is discussed in the book by Preparata and Shamos [192], where one can also find a linear time algorithm that computes the Euclidean minimum spanning tree knowing the Delaunay triangulation. Conversely, Devillers [80] gave a randomized algorithm that computes the Delaunay triangulation of a set of n points in the plane knowing its Euclidean minimum spanning tree in expected time $O(n \log^* n)$.

For other references on Voronoi diagrams, the reader is referred to the book by Okabe, Boots, and Sugihara [182] or to the survey articles by Aurenhammer [16] and Fortune [105].

Chapter 18

Non-Euclidean metrics

In the previous chapter, we established a correspondence between the points in $\mathbb{E}^d$ and certain hyperplanes in $\mathbb{E}^{d+1}$, namely the hyperplanes tangent to the paraboloid $\mathcal{P}$. It is tempting to define the analogue of $\mathcal{V}(\mathcal{M})$ for a more general set of hyperplanes that may not necessarily be tangent to $\mathcal{P}$. In that case, the intersection of the n half-spaces lying above the hyperplanes is again a polytope whose proper faces, projected onto $\mathbb{E}^d$, form a cell complex that covers $\mathbb{E}^d$ entirely. This complex generalizes the Voronoi diagram and can be considered as the Voronoi diagram of a family of spheres, when the distance is defined as the power of a point with respect to one of the spheres. This interpretation, to be detailed in section 18.1, justifies the appellation *power diagrams* for such diagrams. These diagrams play a central role in several generalizations of Voronoi diagrams: in particular, we explore affine diagrams, which are Voronoi diagrams of point sites for a general quadratic distance (see section 18.2), and diagrams for weighted distances (see section 18.3).

Not all Voronoi diagrams for different metrics can be cast into power diagrams. For instance, polyhedral distances (and especially L_1 and L_∞) have important applications and are studied in section 18.4, and an application of hyperbolic Voronoi diagrams (see section 18.5) is given in the next chapter.

The representation of spheres introduced and used in the previous chapter is again very useful for computing power diagrams and hyperbolic Voronoi diagrams. In addition to this representation of spheres, we introduce in this chapter a new way of looking at Voronoi diagrams that is helpful for studying weighted diagrams, L_1 and L_∞ diagrams, and for the algorithms in the next chapter. Intuitively, the Voronoi diagram of a set $\mathcal{M}$ of points can be interpreted as the result of a growth process starting with the points in $\mathcal{M}$. Indeed, imagine crystals growing from each point of $\mathcal{M}$ at the same rate in all directions. The growth of a crystal stops at the points where it encounters another crystal, because of the constraint that the crystals may not interpenetrate. The crystal originating at a point M_i in $\mathcal{M}$ covers the region that is reached by that crystal first, or in other words the points that are closer to M_i than to any other point in $\mathcal{M}$: this is exactly the Voronoi cell of M_i.

This growth process in $\mathbb{E}^d$ can be visualized in $\mathbb{E}^{d+1}$ by adding another coordinate, considered as the *time* elapsed since the start of the growth process. Thus $\mathbb{E}^d$ corresponds to the hyperplane $x_{d+1} = 0$ in $\mathbb{E}^{d+1}$, and the isotropic growth of a point M_i is a cone of revolution with vertex M_i and vertical (that is, parallel to the x_{d+1}-axis) axis. The faces of the Voronoi diagram appear as the projections onto $\mathbb{E}^d$ of faces on the lower envelope of the cones. If the sites do not start growing at the same time, the cones are translated vertically: this leads to Voronoi diagrams with additive weights. If the sites do not grow at the same rate, then the angles of the cones are different: the resulting Voronoi diagrams have multiplicative weights. If the sites do not grow isotropically (namely at the same rate in all directions), the cones are no longer cones of revolution: in this way we can generate Voronoi diagrams for the L_1 and L_∞ distances, and more generally for polyhedral distances (see exercise 19.3).

Throughout this chapter, the "distances" we consider are not exactly distance functions in the mathematical sense. In fact, we will only require that the distance function is increasing.

18.1   Power diagrams

18.1.1   Definition and computation

Let $\mathcal{S} = \{\Sigma_1, \ldots, \Sigma_n\}$ be a set of n spheres in $\mathbb{E}^d$. To each Σ_i corresponds a region $P(\Sigma_i)$ of $\mathbb{E}^d$, consisting of the points whose power with respect to Σ_i is smaller than their powers with respect to the other spheres:

$$P(\Sigma_i) = \{X \in \mathbb{E}^d \ : \ \forall j \neq i, \ \Sigma_i(X) \leq \Sigma_j(X)\}.$$

The region $P(\Sigma_i)$ is the intersection of a finite number of half-spaces (bounded by the radical hyperplanes H_{ij}, $j = 1, \ldots, n$, $j \neq i$). It is therefore a convex polytope, occasionally empty or unbounded. The $P(\Sigma_i)$'s and their faces form a cell complex which covers $\mathbb{E}^d$: this complex is called the *power diagram* of $\mathcal{S}$ and we denote it by $\mathcal{P}ow(\mathcal{S})$ (see figure 18.1).

As in the previous chapter, we map a sphere Σ in $\mathbb{E}^d$, centered at C and of equation $\Sigma(X) = 0$, to the point $\phi(\Sigma) = (C, \Sigma(0))$ in $\mathbb{E}^{d+1}$. The hyperplane polar to $\phi(\Sigma)$ with respect to the paraboloid $\mathcal{P}$ is denoted by $\phi(\Sigma)^*$: if $\mathbb{E}^d$ is embedded in $\mathbb{E}^{d+1}$ as the hyperplane $x_{d+1} = 0$, then $\phi(\Sigma)^*$ is the hyperplane that intersects $\mathcal{P}$ along the quadric obtained by lifting Σ onto $\mathcal{P}$ (see figure 17.4).

Let $\mathcal{P}(\mathcal{S})$ be the intersection of the half-spaces bounded below by the polar hyperplanes $\phi(\Sigma_1)^*, \ldots, \phi(\Sigma_n)^*$ to the points $\phi(\Sigma_1), \ldots, \phi(\Sigma_n)$.

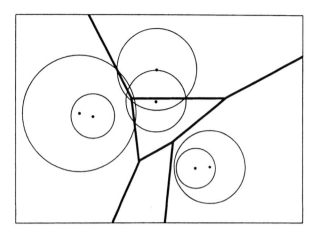

Figure 18.1. A power diagram.

Theorem 18.1.1 *The power diagram $Pow(\mathcal{S})$ of a set $\mathcal{S} = \{\Sigma_1, \ldots, \Sigma_n\}$ of n spheres in $\mathbb{E}^d$ is a cell complex in $\mathbb{E}^{d+1}$. Its faces are obtained by projecting the proper faces of the unbounded $(d+1)$-polytope $\mathcal{P}(\mathcal{S})$, the intersection of n half-spaces in $\mathbb{E}^{d+1}$ bounded below by the polar hyperplanes $\phi(\Sigma_1)^*, \ldots, \phi(\Sigma_n)^*$.*

Proof. Let $\underline{A}$ be a point on a facet of $\mathcal{P}(\mathcal{S})$ which is supported by the polar hyperplane $\phi(\Sigma_i)^*$, and let A be its projection onto $\mathbb{E}^d$. The power of A with respect to Σ_i is the signed vertical distance from $\underline{A}$ to $\phi(A)$ (lemma 17.2.3). Since $\underline{A}$ belongs to a facet of $\mathcal{P}(\mathcal{S})$, the power of A with respect to Σ_i is less than or equal to the power of A with respect to any other sphere in $\mathcal{S}$. In other words, A belongs to the cell $P(\Sigma_i)$ that corresponds to Σ_i in the power diagram. $\qquad\square$

Note that when the hyperplanes $\phi(\Sigma_i)^*$ are in general position, $\mathcal{P}(\mathcal{S})$ is a simple polytope in $\mathbb{E}^{d+1}$, so each vertex is incident to $d+1$ hyperplanes. In terms of the spheres Σ_i, this general position assumption means that no subset of $d+2$ spheres in $\mathcal{S}$ are orthogonal to a common sphere in $\mathbb{E}^d$, or equivalently that no point in $\mathbb{E}^d$ has the same power with respect to $d+2$ spheres in $\mathcal{S}$. In this case, we say that the spheres are in *general position*. The power diagram $Pow(\mathcal{S})$ is a cell complex whose vertices have the same power with respect to $d+1$ spheres in $\mathcal{S}$ (and are therefore the centers of spheres orthogonal to $d+1$ spheres in $\mathcal{S}$), and have a greater power with respect to the other spheres in $\mathcal{S}$. More generally, a k-face in $Pow(\mathcal{S})$ is formed by the points that have the same power with respect to $d+1-k$ given spheres in $\mathcal{S}$, and a greater power with respect to the other spheres in $\mathcal{S}$.

Corollary 18.1.2 *The complexity of the power diagram of n spheres in $\mathbb{E}^d$ is $\Theta(n^{\lceil d/2 \rceil})$. The diagram can be computed in time $O(n \log n + n^{\lceil d/2 \rceil})$, which is* *optimal in the worst case.*

Remark 1. In the case of Voronoi diagrams, all the hyperplanes $\phi(M_i)^*$ are tangent to the paraboloid $\mathcal{P}$, and each of them contributes a facet to $\mathcal{V}(\mathcal{M})$. For power diagrams, however, a polar hyperplane $\phi(\Sigma_i)^*$ does not necessarily contribute a face to $\mathcal{P}(\mathcal{S})$. Such a hyperplane is called *redundant*. In the power diagram, it means that the cell $P(\Sigma_i)$ is empty: Σ_i does not contribute a cell to $\mathcal{P}ow(\mathcal{S})$.

Remark 2. There is no particular difficulty if some, even all, the spheres in $\mathcal{S}$, are imaginary. This fact is used in section 18.3.2.

Remark 3. Any polytope in $\mathbb{E}^{d+1}$ that is the intersection of upper half-spaces corresponds to a power diagram: if $H_1, \ldots, H_n$ are the hyperplanes that bound these half-spaces, then their upper envelope projects onto the power diagram of the spheres $\phi^{-1}(H_1^*), \ldots, \phi^{-1}(H_n^*)$.

18.1.2   Higher-order power diagrams

As was done for Voronoi diagrams in section 17.4, we may define power diagrams of higher orders.

Let $\mathcal{S}_k$ be a subset of $\mathcal{S}$ of size k. We call the power cell of $\mathcal{S}_k$ the set $P(\mathcal{S}_k)$ of points in $\mathbb{E}^d$ that have a smaller power with respect to any sphere in $\mathcal{S}_k$ than to any sphere in $\mathcal{S} \setminus \mathcal{S}_k$:

$$P(\mathcal{S}_k) = \{ X \in \mathbb{E}^d \ : \ \forall \Sigma_i \in \mathcal{S}_k, \ \forall \Sigma_j \in \mathcal{S} \setminus \mathcal{S}_k, \ \Sigma_i(X) \le \Sigma_j(X) \}.$$

Consider all the subsets of size k of $\mathcal{S}$ whose corresponding power cell is not empty. These regions and their faces form a cell complex that covers $\mathbb{E}^d$ entirely, and that is called the *power diagram of order* k of $\mathcal{S}$. We denote it by $\mathcal{P}ow_k(\mathcal{S})$.

This fact is a consequence of the theorem below, whose proof closely resembles that of theorem 17.4.1. This theorem clarifies the links between power diagrams of order k in $\mathbb{E}^d$ and faces at level k in the arrangement of n hyperplanes in $\mathbb{E}^{d+1}$. As usual, the Euclidean space of dimension d is identified with the hyperplane $x_{d+1} = 0$ in the space $\mathbb{E}^{d+1}$ of dimension $d+1$, and $\phi(\Sigma)^*$ stands for the polar hyperplane of $\phi(\Sigma)$.

Theorem 18.1.3 *Consider a set $\mathcal{S} = \{\Sigma_1, \ldots, \Sigma_n\}$ of spheres in $\mathbb{E}^d$, and let $\mathcal{A}$ be the arrangement of their polar hyperplanes $\phi(\Sigma_1)^*, \ldots, \phi(\Sigma_n)^*$. The power diagram of order k, $\mathcal{P}ow_k(\mathcal{S})$, is a cell d-complex in $\mathbb{E}^d$. Its cells are the vertical projections of the cells at level k in the arrangement $\mathcal{A}$, the reference point being on the x_{d+1}-axis above all the hyperplanes $\phi(\Sigma_i)^*$, $i = 1, \ldots, n$. The l-faces of $\mathcal{P}ow_k(\mathcal{S})$, $l < d$, are obtained by projecting the l-faces common to at least two cells of $\mathcal{A}$ at level k.*

From theorems 14.5.1 and 14.5.3, we derive the following result.

Theorem 18.1.4 *The complexity of the first k power diagrams of a set of n spheres in $\mathbb{E}^d$ is $O(n^{\lfloor(d+1)/2\rfloor}k^{\lceil(d+1)/2\rceil})$. These k diagrams can be computed in time $O(n^{\lfloor(d+1)/2\rfloor}k^{\lceil(d+1)/2\rceil})$ if $d \geq 3$, and in time $O(nk^2 \log\frac{n}{k})$ if $d = 2$.*

18.2 Affine diagrams

The notion of a Voronoi diagram can be extended to more general sites or to non-Euclidean distances. A particularly interesting extension occurs when the locus of points equidistant from two sites is a hyperplane: in this case, the diagram is called an *affine diagram*. Voronoi diagrams and power diagrams are affine diagrams, and we will show that any affine diagram is a power diagram. Moreover, certain non-affine diagrams can be derived from an affine diagram and therefore from a power diagram: this is notably the case of diagrams with additive or multiplicative weights studied in section 18.3.

18.2.1 Affine diagrams and power diagrams

An affine diagram is a diagram defined for object sites and for a distance such that the set of points equidistant from two objects is a hyperplane. The cells of such diagrams are thus convex polytopes and affine diagrams can be identified with cell complexes.

To any affine diagram of n objects corresponds a set of $\binom{n}{2}$ perpendicular bisectors H_{ij}, $1 \leq i < j \leq n$. These hyperplanes must satisfy the relations

$$H_{ij} \cap H_{jk} = H_{ij} \cap H_{ik} = H_{ik} \cap H_{jk} \overset{\text{def}}{=} I_{ijk}$$

for any $1 \leq i < j < k \leq n$.

We say that the diagram is *simple* if the I_{ijk} are disjoint and not empty.

Theorem 18.2.1 *Any simple affine diagram in $\mathbb{E}^d$ is the power diagram of a set of spheres in $\mathbb{E}^d$.*

Proof. We embed $\mathbb{E}^d$ in $\mathbb{E}^{d+1}$ as the hyperplane $x_{d+1} = 0$. The proof consists of constructing a set of n hyperplanes $P_1, \ldots, P_n$ in $\mathbb{E}^{d+1}$ such that the vertical projection of $P_i \cap P_j$ for $i < j$ is exactly H_{ij}. Assuming these hyperplanes are known, to each P_i corresponds a sphere $\Sigma_i = \phi^{-1}(P_i^*)$ whose polar hyperplane is exactly P_i: Σ_i is also the projection on $\mathbb{E}^d$ of the intersection of P_i with the paraboloid $\mathcal{P}$. Hence H_{ij} is the radical hyperplane of Σ_i and Σ_j for all i and j

such that $1 \leq i < j \leq n$. It follows that the affine diagram is exactly the power diagram of the spheres Σ_i, $i = 1, \ldots, n$.

We now show how to build the P_i's. Denote by h_{ij} the vertical projection of H_{ij} onto P_i (note that $i < j$ by the definition of H_{ij}).

Let us take for P_1 any non-vertical hyperplane, and for P_2 any non-vertical hyperplane that intersects P_1 along h_{12}. For $k \geq 3$, we must take for P_k the hyperplane that intersects P_1 along h_{1k} and P_2 along h_{2k}: such a hyperplane exists because h_{1k} and h_{2k} intersect along the affine subspace of dimension $d - 2$ that is the projection of I_{12k} onto P_1, P_2, or P_k.

It remains to see that the vertical projection of $P_i \cap P_j$ is exactly H_{ij}. By construction, this is true for $P_1 \cap P_2$, $P_1 \cap P_j$, and $P_2 \cap P_j$, $j \geq 3$. For $3 \leq i < j \leq n$, we know that $P_i \cap P_j \cap P_1$ projects onto $\mathbb{E}^d$ along I_{1ij}, and that $P_i \cap P_j \cap P_2$ projects onto $\mathbb{E}^d$ along I_{2ij}. The diagram being simple, I_{1ij} and I_{2ij} must be distinct. The projection of $P_i \cap P_j$ must therefore contain I_{1ij} and I_{2ij}, and hence also their affine hull which is nothing other than H_{ij}. $\square$

Below, we rather use

Theorem 18.2.2 *The affine diagram whose hyperplanes H_{ij} have equations*

$$-2(C_i - C_j) \cdot X + \sigma_i - \sigma_j = 0$$

is the power diagram of the spheres Σ_i, $i = 1, \ldots, n$ centered at C_i and with respect to which the origin has power σ_i.

Proof. We may simply check that the equation of H_{ij} can be written as $\Sigma_i(X) - \Sigma_j(X) = 0$, which is exactly that of the radical hyperplane of Σ_i and Σ_j (see subsection 17.2.6). $\square$

18.2.2   Diagrams for a general quadratic distance

Consider two points X and A in $\mathbb{E}^d$. By the *general quadratic distance* from A to X, we mean the quantity

$$\delta_Q(X, A) = (X - A)\Delta(X - A)^t + p(A),$$

where Δ is a real symmetric $d \times d$ matrix, and where $p(A)$ is a real number.

The diagrams encountered so far are all particular cases of Voronoi diagrams for a quadratic distance:

- Standard Voronoi diagrams are obtained for $\Delta = \mathbb{I}_d$ and $p(X) = 0$.

- Furthest-point Voronoi diagrams (introduced as diagrams of n points of order $n - 1$) are obtained for $\Delta = -\mathbb{I}_d$ and $p(X) = 0$.

- Power diagrams correspond to $\Delta = \mathbb{I}_d$ and $p(X) \neq 0$.

For any pair of points A and B, the set of points X that are equidistant from A and B is the hyperplane H_{AB} equation

$$H_{AB} \;:\; 2(B-A)\Delta X^t + A\Delta A^t - B\Delta B^t + p(A) - p(B) = 0.$$

The Voronoi diagram of a finite set of points for a general quadratic distance is thus an affine diagram by theorem 18.2.2.

Theorem 18.2.3 *The Voronoi diagram of n points for an arbitrary general quadratic distance in $\mathbb{E}^d$ has complexity $\Theta(n^{\lceil d/2 \rceil})$. It can be computed in time $\Theta(n \log n + n^{\lceil d/2 \rceil})$.*

18.3 Weighted diagrams

This section introduces two kinds of diagrams which are not affine. They are defined for finite sets of point sites and for a Euclidean distance that is weighted additively or multiplicatively. Each distance is appropriately defined in the subsection below.

These diagrams are not cell complexes those we have been studying so far. Nevertheless, they can be given a facial structure that is similar to that of cell complexes. Consider the equivalence relation shared by the points in $\mathbb{E}^d$ that have the same nearest neighbors. The equivalence classes subdivide $\mathbb{E}^d$ into (open) regions whose closures we call the *faces* of the diagram. The cells of the diagram span $\mathbb{E}^d$ entirely and the intersection of two faces is a (possibly empty) collection of lower-dimensional faces.

As we see below, the faces of these diagrams are not polytopes, and may not even be connected. Nevertheless, these weighted diagrams can be derived simply from power diagrams.

18.3.1 Weighted diagrams with additive weights

Let $\mathcal{M} = \{M_1, \ldots, M_n\}$ be a set of n points in $\mathbb{E}^d$. To each M_i corresponds a real r_i called the *weight* of M_i. The *additive weighted distance*, or *additive distance* for short, from a point X in $\mathbb{E}^d$ to M_i is the quantity

$$\delta_+(X, M_i) = \|XM_i\| - r_i.$$

The *diagram of $\mathcal{M}$ with additive weights* is defined like the Voronoi diagram except that the distance used is not the Euclidean distance but the additive distance defined above. This diagram is denoted by $\mathcal{V}or_+(\mathcal{M})$. An instance is

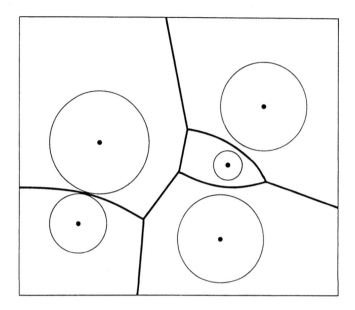

Figure 18.2. A diagram with additive weights. Sites are the centers and their corresponding weights are the radii of the circles. In this example, the diagram of the
points with additive weights is also the Voronoi diagram of the circles for the
Euclidean metric.

shown in figure 18.2. We observe that it is the first example of a non-affine
diagram shown in this book.

Note that adding the same constant to all the points does not modify the
diagram. This lets us assume that all the r_i's are non-negative.

The representation of spheres introduced in section 17.2 is not very helpful here
and we use another which shows a very natural correspondence between weighted
Voronoi diagrams and affine diagrams in dimension $d + 1$.

Consider the sphere Σ_i in $\mathbb{E}^d$ centered at M_i with radius r_i, and let ψ be the
bijection that maps Σ_i to the point $\psi(\Sigma_i) = (M_i, r_i) \in \mathbb{E}^{d+1}$.

The spheres of zero radius correspond to the points in the hyperplane of equation $x_{d+1} = 0$ in $\mathbb{E}^{d+1}$.

The points at additive distance r from M_i can be considered as the centers of
the spheres of radius $|r|$ tangent to Σ_i, that are inside or outside Σ_i according to
whether r is negative or positive. The images under ψ of these spheres generate
a cone of revolution $\mathcal{C}(\Sigma)$ of equation

$$\mathcal{C}(\Sigma) \;:\; x_{d+1} = \|XC\| - r$$

which has apex $(C, -r)$, is symmetrical to $\psi(\Sigma)$ with respect to the hyperplane
$x_{d+1} = 0$, and has an aperture angle of $\frac{\pi}{4}$. The vertical projection I_X of a point

X in $\mathbb{E}^d$ on the cone $\mathcal{C}(\Sigma)$ is the image under ψ of the sphere centered at X and tangent to Σ. The signed vertical distance from X to I_X equals the additive distance from X to C weighted by r.

To each sphere Σ_i, $i = 1, \ldots, n$, corresponds the cone $\mathcal{C}(\Sigma_i)$, also denoted by $\mathcal{C}_i$. It follows from the discussion above that the projection of the lower envelope of the cones $\mathcal{C}_i$ onto $\mathbb{E}^d$ is exactly $\mathcal{V}or_+(\mathcal{M})$.

The set of points in $\mathbb{E}^d$ that are equidistant (with respect to the additive distance) from two points of $\mathcal{M}$ is thus the projection of the intersection of two cones. This intersection is a quadric contained in a hyperplane. Indeed, we have

$$
\begin{aligned}
\mathcal{C}_1 &: (x_{d+1} + r_1)^2 = XM_1^2, \quad x_{d+1} + r_1 > 0, \\
\mathcal{C}_2 &: (x_{d+1} + r_2)^2 = XM_2^2, \quad x_{d+1} + r_2 > 0.
\end{aligned}
$$

The intersection of the two cones is contained in the hyperplane H_{12} whose equation is obtained by subtracting the two sides of the above equations:

$$
H_{12} : -2(M_1 - M_2) \cdot X - 2(r_1 - r_2)x_{d+1} + M_1^2 - r_1^2 - M_2^2 + r_2^2 = 0.
$$

This and theorem 18.2.2 show that there exists a correspondence between the diagram $\mathcal{V}or_+(\mathcal{M})$ and the power diagram of the spheres Σ_i' in $\mathbb{E}^{d+1}$ ($i = 1, \ldots, n$), where Σ_i' is centered at $\psi(\Sigma_i)$ and has radius $r_i\sqrt{2}$ (see figure 18.3). More precisely, the cell of $\mathcal{V}or_+(\mathcal{M})$ that corresponds to M_i is the projection of the intersection of the cone $\mathcal{C}_i$ with the cell of the power diagram corresponding to the sphere Σ_i'. Indeed, X is in $\mathcal{V}or_+(M_i)$ if and only if the projection X_i of X onto $\mathcal{C}_i$ has a smaller x_{d+1}-coordinate than the projections of X onto the other cones $\mathcal{C}_j$, $j \neq i$. In other words, the coordinates (X, x_{d+1}) of X_i must obey

$$
\begin{aligned}
(x_{d+1} + r_i)^2 &= XM_i^2 \\
(x_{d+1} + r_j)^2 &\leq XM_j^2 \quad \text{for any } j \neq i,
\end{aligned}
$$

and by subtracting both sides, it follows that $\Sigma_i'(X_i) \leq \Sigma_j'(X_i)$ for all j.

The additive diagram can be computed using the following algorithm:

1. Compute Σ_i', for $i = 1, \ldots, n$.

2. Compute the power diagram of the Σ_i''s.

3. For all $i = 1, \ldots, n$, project onto $\mathbb{E}^d$ the intersection with the cone $\mathcal{C}_i$ of the cell of the power diagram that corresponds to Σ_i'.

The power diagram of the Σ_i' can be computed in time $O(n^{\lfloor d/2 \rfloor + 1})$. The intersection involved in step 3 can be computed in time proportional to the number of faces of the power diagram of the Σ_i''s, which is $O(n^{\lfloor d/2 \rfloor + 1})$. We have thus proved that:

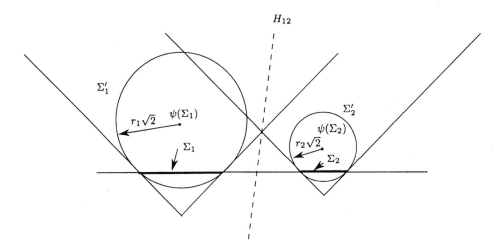

Figure 18.3. Any Voronoi diagram for the additive distance can be derived from a power
diagram in $\mathbb{E}^{d+1}$.

Theorem 18.3.1 *The Voronoi diagram of a set of n points in $\mathbb{E}^d$ with additive
weights has complexity $O(n^{\lfloor d/2 \rfloor + 1})$ and can be computed in time $O(n^{\lfloor d/2 \rfloor + 1})$.*

This result is optimal in odd dimensions, since the bounds above coincide with
the corresponding bounds for the Voronoi diagram of points under the Euclidean
distance. It is not optimal in dimension 2, however, as we now show. We also
conjecture that it is not optimal in any even dimension.

In the plane, we have seen that additive diagrams can be thought of as the
projection onto $\mathbb{E}^2$ of the lower envelope of cones with vertical axis and aperture
angle $\frac{\pi}{4}$. Therefore, each cell is connected. Moreover, the vertices of the diagram
are incident to exactly three edges, under the general position assumption, and
these edges are arcs of hyperbolas, each of which is the projection of the intersection of two cones. Euler's relation shows that the diagram has complexity
$O(n)$. A perturbation argument shows that the general position assumption is
not restrictive, since allowing degeneracies only merges some vertices and makes
some edges disappear. In section 19.1, it is shown that such a diagram can be
computed in optimal time $O(n \log n)$.

18.3.2   Weighted diagrams with multiplicative weights

Let $\mathcal{M} = \{M_1, \ldots, M_n\}$ be a set of n point sites in $\mathbb{E}^d$. To each M_i corresponds a
positive real number $p(M_i)$ called the *weight* of M_i. To simplify the presentation,
we suppose that the $p(M_i)$'s are all distinct, but the extension to the more general
case presents no additional difficulties.

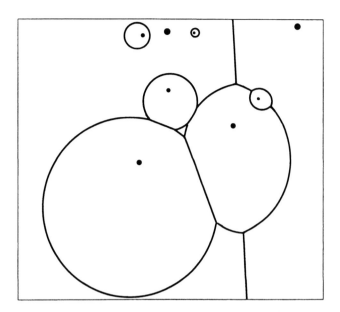

Figure 18.4. A diagram with multiplicative weights. Sites are represented by small disks, and the weight of a site is inversely proportional to the diameter of the disk.

By the *distance with multiplicative weights*, or *multiplicative distance* for short, from a point X to a point M_i, we mean the quantity

$$\delta_*(X, M_i) = p(M_i) \, \|X M_i\|.$$

The *Voronoi diagram of* $\mathcal{M}$ *for the multiplicative distance* is defined like the Voronoi diagram, except that the distance is not the Euclidean distance but rather the multiplicative distance. We denote this diagram by $Vor_*(\mathcal{M})$ (see figure 18.4). Observe that a cell of the diagram need not be connected.

The set of points at equal multiplicative distance from two sites M_i and M_j is a sphere Σ_{ij} of equation

$$p_i \, (X - M_i)^2 = p_j \, (X - M_j)^2$$

with $p_i = p(M_i)^2$. In normalized form, we obtain

$$X^2 - 2 \frac{p_i M_i - p_j M_j}{p_i - p_j} \cdot X + \frac{p_i M_i^2 - p_j M_j^2}{p_i - p_j} = 0.$$

Using the representation of section 17.2, this sphere is represented in $\mathbb{E}^{d+1}$ as the point

$$\phi(\Sigma_{ij}) = \left(\frac{p_i M_i - p_j M_j}{p_i - p_j}, \frac{p_i M_i^2 - p_j M_j^2}{p_i - p_j} \right).$$

Its polar hyperplane H_{ij} with respect to the paraboloid $\mathcal{P}$ has equation

$$H_{ij}(X, x_{d+1}) = (p_i - p_j)x_{d+1} - 2p_i M_i \cdot X + 2p_j M_j \cdot X + p_i M_i^2 - p_j M_j^2 = 0.$$

The hyperplanes H_{ij} are the radical hyperplanes of spheres Σ_i in $\mathbb{E}^{d+1}$ ($i = 1, \ldots, n$). The sphere Σ_i, possibly imaginary, is centered at $(p_i M_i, -\frac{p_i}{2})$, and with respect to it the origin has power $p_i M_i^2$. This establishes a correspondence between the diagram $Vor_*(\mathcal{M})$ and the power diagram of the spheres Σ_i. More precisely, the cell $V_*(M_i)$ in $Vor_*(\mathcal{M})$ that corresponds to M_i is the projection of the intersection of the paraboloid $\mathcal{P}$ with the cell $P(\Sigma_i)$ that corresponds to Σ_i in the power diagram of the Σ_i's. Indeed, if X is a point in $\mathbb{E}^d$ and $\phi(X)$ is its vertical projection onto the paraboloid $\mathcal{P}$ of equation $x_{d+1} = X^2$, then we have

$$
\begin{aligned}
X \in V_*(M_i) &\iff p_i(X - M_i)^2 \leq p_j(X - M_j)^2 && \forall j \neq i \\
&\iff H_{ij}(X, X^2) \leq 0 && \forall j \neq i \\
&\iff \Sigma_i(\phi(X)) \leq \Sigma_j(\phi(X)) && \forall j \neq i \\
&\iff \phi(X) \in P(\Sigma_i).
\end{aligned}
$$

An algorithm that computes the diagram of $\mathcal{M}$ with multiplicative weights is:

1. Compute Σ_i, for $i = 1, \ldots, n$.

2. Compute the power diagram of the Σ_i's.

3. For $i = 1, \ldots, n$, project the intersection of the cell that corresponds to Σ_i in the power diagrams of the Σ_i's with the paraboloid $\mathcal{P}$.

This proves the following theorem.

Theorem 18.3.2 *The Voronoi diagram of a set of n points in $\mathbb{E}^d$ with multiplicative weights has complexity $O(n^{\lfloor d/2 \rfloor + 1})$ and can be computed in time $O(n^{\lfloor d/2 \rfloor + 1})$.*

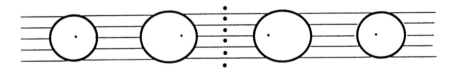

Figure 18.5. An instance of a quadratic multiplicative diagram in dimension 2: $\frac{n}{2}$ points are put on a given vertical line and are given the same weight, while $\frac{n}{2}$ other points are aligned on a horizontal line and have the same weight, which is much larger than the weight given to the points in the first half.

This result is optimal in odd dimensions, since in that case these bounds match those of the Voronoi diagram of n points in $\mathbb{E}^d$ for the Euclidean distance. It is also optimal in even dimensions (see exercise 18.4). Figure 18.5 shows a quadratic multiplicative diagram in dimension 2.

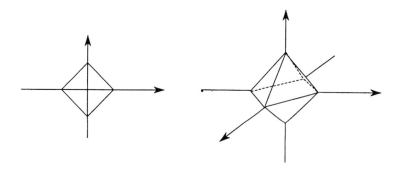

Figure 18.6. Co-cubes in dimensions 2 and 3.

18.4 L_1 and L_∞ metrics

The L_1 distance of a point $X = (x_1, \ldots, x_d)$ in $\mathbb{E}^d$ to a point $M = (m_1, \ldots, m_d)$ in $\mathbb{E}^d$ is defined as

$$\delta_1(X, M) = \sum_{i=1}^{d} |x_i - m_i|.$$

The points at a given distance r from M are thus on a polytope whose vertices are given by their coordinates $x_i = m_i \pm r$ and $x_j = m_j$ if $i \neq j$, for $j = 1, \ldots, d$. In dimension 2 this polytope is a tilted square, and in dimension 3 it is a regular octahedron (see figure 18.6). This polytope is dual to the cube and we call it a *co-cube*. Henceforth, a co-cube always means a polytope dual to a cube whose edges are parallel coordinate axes.

Let $\mathcal{M} = \{M_1, \ldots, M_n\}$ be a set of n point sites in $\mathbb{E}^d$. The *Voronoi diagram of $\mathcal{M}$ for the L_1 distance* is defined similarly to the Voronoi diagram, except that the distance used in the definition of the cells is not the Euclidean distance but the L_1 distance. It is denoted by $Vor_{L_1}(\mathcal{M})$ (see figure 18.7).

We can define a facial structure for this diagram by using the equivalence relation R shared by the points in $\mathbb{E}^d$ that have the same subset of nearest neighbors. The equivalence classes of R subdivide the space $\mathbb{E}^d$ in open regions whose closures are called the *faces* of the diagram. The faces of the diagram are piecewise affine.

If the points in $\mathbb{E}^d$ are identified with the hyperplane $x_{d+1} = 0$ in $\mathbb{E}^{d+1}$, then, in a way similar to what was explained in subsection 18.3.1, to each point M_i there corresponds a pyramid $\mathcal{P}_i$ of equation

$$x_{d+1} = \delta_1(X, M_i).$$

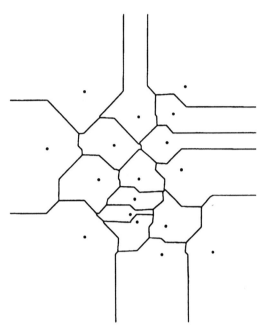

Figure 18.7. Diagram for the L_1 metric.

Let us consider the lower envelope of the $\mathcal{P}_i$'s, that is, the graph of the function $\min_{1\leq i\leq n}\delta_1(X,M_i)$. The portion of the lower envelope that belongs to any $\mathcal{P}_i$ projects onto the hyperplane $x_{d+1}=0$ as the cell of the diagram $\mathcal{V}or_{L_1}(\mathcal{M})$ that corresponds to M_i. The facets of the $\mathcal{P}_i$'s form a collection of d-pyramids. The lower envelope of these pyramids is a collection of d-faces, and their lower-dimensional faces include all the lower-dimensional faces of the lower envelope of the $\mathcal{P}_i$'s. The vertical projections onto $x_{d+1}=0$ of the d-faces of the lower envelope of the pyramids form a refinement of the faces of the diagram $\mathcal{V}or_{L_1}(\mathcal{M})$. The complexity of the diagram $\mathcal{V}or_{L_1}(\mathcal{M})$ can thus be bounded by combining theorem 16.3.2 and exercise 16.1, which bound the complexity of the lower envelope of n d-simplices in $\mathbb{E}^{d+1}$. This yields

$$|\mathcal{V}or_{L_1}(\mathcal{M})| = O(n^d\alpha(n)).$$

This bound is almost tight for certain sets of points that are not in general position (see exercise 18.9). We conjecture, however, that for points in general enough position, this bound is not attained and that these diagrams have the same complexity as their Euclidean counterparts. Later on, we show that this is indeed the case in dimension 2, for which we give a linear bound. It is also the case in dimension 3 (see exercise 18.10). If $d=2$, the bisector for the L_1-distance of two points is, in general, a polygonal line formed by three linear pieces; if the line connecting the two points is parallel to one of the main bisectors, however,

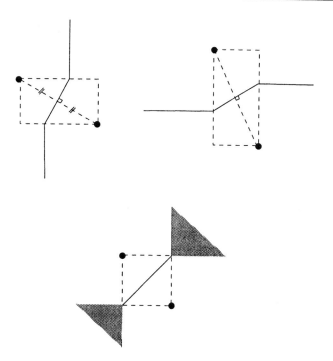

Figure 18.8. Bisectors for the L_1 distance. If the line connecting the two points is parallel to one of the main bisectors, the bisector is not a polygonal line.

the L_1-bisector is no longer a polygonal line and contains two faces of dimension 2 (see figure 18.8).

We say that two points are in L_1-*general position* if no two points are connected by a line parallel to one of the main bisectors, and if no four points belong to a common co-cube. In this case, the bisectors are polygonal lines formed of three line segments, and $Vor_{L_1}(\mathcal{M})$ contains n connected cells: indeed, for any $i \in \{1, \ldots, n\}$, the cell $V_1(M_i)$ that corresponds to a point M_i is star-shaped with respect to M_i (meaning that if $X \in V_1(M_i)$, then the segment XM_i is contained in $V_1(M_i)$), and is therefore connected. Moreover, each vertex in the diagram is incident to two or three edges because of the L_1-general position assumption. The diagram $Vor_{L_1}(\mathcal{M})$ is therefore a planar map with n cells whose vertices have degree two or three and whose edges consist of at most three segments. Euler's relation then shows that the complexity of the diagram is $O(n)$.

If the points are not in L_1-general position, then some regions may correspond to pairs of points (see figure 18.8) and some vertices may be of degree higher than 3. This second complication can be straightened out by simply perturbing the diagram so as to replace each vertex of degree $k > 3$ by a small polygonal chain with $k - 2$ vertices of degree 3 and $k - 3$ edges. The number of faces does not

increase in the process, and the number of vertices increases by the same amount
as the number of edges; hence Euler's relation still guarantees that the complexity
of the diagram is $O(n)$. The first complication, however, is more serious and may
allow the size to grow up to quadratic: exercise 18.9 presents such an example
and a way to avoid this problem. The example generalizes to higher dimensions
and the lower bound $\Omega(n^d)$ may be shown to hold for the complexity of Voronoi
diagrams of n points in $\mathbb{E}^d$ for the L_1 distance.

If the points are in L_1-general position, the complexity of the diagram is thus
$O(n)$ in dimension 2 and the algorithm that computes the lower envelope of n
triangles in space (see subsection 16.3.3) can be used to compute this diagram
in time $O(n \log^2 n)$ (see corollary 16.3.3). An optimal algorithm exists that computes such a diagram in time $O(n \log n)$ (see exercise 19.2).

The situation for the L_∞ distance is very similar to the one just described for
the L_1 distance. Its complexity in dimensions higher than 3 is easier to analyze,
however. The L_∞ *distance* of a point $X = (x_1, \ldots, x_d)$ in $\mathbb{E}^d$ from a point
$M = (m_1, \ldots, m_d)$ in $\mathbb{E}^d$ is given by

$$\delta_\infty(X, M) = \max_{i=1,\ldots,d} |x_i - m_i|.$$

The points at a distance r from M are thus on a cube centered at M whose facets
are parallel to the coordinate axes, and whose side is $2r$.

The *Voronoi diagram of $\mathcal{M}$ for the L_∞ distance* is denoted by $\mathcal{V}or_{L_\infty}(\mathcal{M})$. An
instance is shown in figure 18.9.

The cells of this diagram can be obtained by projecting onto the hyperplane
$x_{d+1} = 0$ in $\mathbb{E}^{d+1}$ the cells on the lower envelope of the n pyramids $\mathcal{Q}_i$ of equation

$$x_{d+1} = \delta_\infty(X, M_i).$$

The facets of the $\mathcal{Q}_i$'s form a collection of d-pyramids. The faces on the lower
envelope of these pyramids form a refinement of the faces on the lower envelope
of the $\mathcal{Q}_i$'s. Hence, the vertical projections onto the hyperplane $x_{d+1} = 0$ of the
faces on the lower envelope of these pyramids form a refinement of the faces of
the diagram $\mathcal{V}or_{L_\infty}(\mathcal{M})$. The complexity of the Voronoi diagram $\mathcal{V}or_{L_\infty}(\mathcal{M})$ is
thus bounded by the complexity of a lower envelope of n simplices in $\mathbb{E}^{d+1}$:

$$|\mathcal{V}or_{L_\infty}(\mathcal{M})| = O(n^d \alpha(n)).$$

This bound is almost tight for certain sets of points that are not in general
position (see exercise 18.9). If the points are in so-called L_∞-general position,
then it is possible to show that the complexity of Voronoi diagrams for the L_∞
metric is the same as that for Euclidean Voronoi diagrams, namely $O(n^{\lceil d/2 \rceil})$ (see
exercise 18.10). We show this for the case $d = 2$. When $d = 2$, $\mathcal{V}or_{L_\infty}(\mathcal{M})$ can

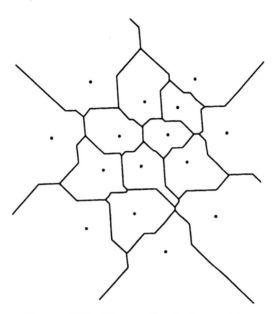

Figure 18.9. Diagram for the L_∞ metric.

be identified with the diagram $Vor_{L_1}(\mathcal{M})$ studied previously by simply rotating the coordinate system by an angle of $\frac{\pi}{4}$. The points are in L_∞-*general position*, if no two points are connected by a line parallel to the coordinate axes, and no four points belong to a common cube whose facets are parallel to the coordinate axes. If so, then the complexity of $Vor_{L_\infty}(\mathcal{M})$ is $O(n)$ in dimension 2, and the algorithm described in subsection 16.3.3 that computes the lower envelope of triangles can be used to compute this diagram in time $O(n \log^2 n)$ (see exercise 19.2 for a better algorithm).

The distances considered here, L_1 and L_∞, are particular cases of polyhedral distances, so-called because their unit ball is a polytope. Voronoi diagrams for polyhedral distances are studied in exercise 19.3.

18.5 Voronoi diagrams in hyperbolic spaces

18.5.1 Pencils of spheres

A *pencil of spheres* in $\mathbb{E}^d$ is a set $\mathcal{S}$ of spheres that are affine combinations of two given spheres Σ_1 and Σ_2:

$$\mathcal{F} = \{\Sigma : \exists \lambda \in \mathbb{R}, \forall X \in \mathbb{E}^d, \Sigma(X) = \lambda \Sigma_1(X) + (1 - \lambda)\Sigma_2(X)\}.$$

If we apply to spheres the mapping ϕ introduced in section 17.2, we map the spheres in $\mathbb{E}^d$ to points in $\mathbb{E}^{d+1}$. From the results of section 17.2, it follows that the image under ϕ of a pencil $\mathcal{F}$ is the line $\phi(\mathcal{F})$ in $\mathbb{E}^{d+1}$ that connects the points $\phi(\Sigma_1)$ and $\phi(\Sigma_2)$.

We may distinguish between four kinds of pencils, according to whether the line that is the image under ϕ of the pencil intersects the paraboloid in one point (transversally), in two points, is tangent to $\mathcal{P}$, or does not intersect $\mathcal{P}$ (see figure 18.10).

- If $\phi(\mathcal{F})$ intersects $\mathcal{P}$ transversally in only one point, then $\mathcal{F}$ contains a single sphere of zero radius, and $\phi(\mathcal{F})$ is a *pencil of concentric spheres*.

- If the line $\phi(\mathcal{F})$ intersects $\mathcal{P}$ in two points, $\mathcal{F}$ contains two spheres of radius zero, called the *limit points* of the pencil.

- If the line $\phi(\mathcal{F})$ is tangent to $\mathcal{P}$, then $\mathcal{F}$ may be considered as a pencil whose two limit points are identical, or as a pencil supported by a sphere that reduces to a point. Such a pencil is called a *tangent pencil*.

- If the line $\phi(\mathcal{F})$ does not intersect $\mathcal{P}$, there exists a family of hyperplanes tangent to $\mathcal{P}$ that contain $\phi(\mathcal{F})$. Let $\phi(\Sigma_{\mathcal{F}})$ be the set of points of $\mathcal{P}$ at which these hyperplanes are tangent to $\mathcal{P}$. Then $\phi(\Sigma_{\mathcal{F}})$ is the image under ϕ of the set $\Sigma_{\mathcal{F}}$ of points that belong to all the spheres in the pencil $\mathcal{F}$. Coming back to the definition of a pencil, we have $\Sigma(X) = 0$ for all values of λ, and this implies that $\Sigma_1(X) = \Sigma_2(X) = 0$ and that $\Sigma_{\mathcal{F}}$ can be identified with the $(d-1)$-sphere $\Sigma_1 \cap \Sigma_2$. All the d-spheres in the pencil $\mathcal{F}$ intersect along the $(d-1)$-sphere obtained as the intersection of any two spheres in the pencil. For this reason, $\Sigma_{\mathcal{F}}$ is called the *supporting sphere* of the pencil.

The very definition of a pencil of spheres implies that any point in the radical hyperplane H_{12} of two spheres Σ_1 and Σ_2 in the pencil has same power with respect to any sphere in the pencil. We may therefore define the *radical hyperplane of a pencil of spheres* as the radical hyperplane of any two spheres in the pencil.

The radical hyperplane of a pencil supported by a sphere is the affine hull of the supporting sphere. A concentric pencil has no radical hyperplane. The radical hyperplane of a pencil with limit points is the perpendicular bisector of these two points. The radical hyperplane of a tangent pencil is the hyperplane tangent to all the spheres in the pencil.

18.5.2   Voronoi diagrams in hyperbolic spaces

The Poincaré model of the hyperbolic space of dimension d is the half-space $\mathbb{H}^d = \{X \in \mathbb{E}^d \ : \ x_d > 0\}$. We will not define the hyperbolic distance precisely.

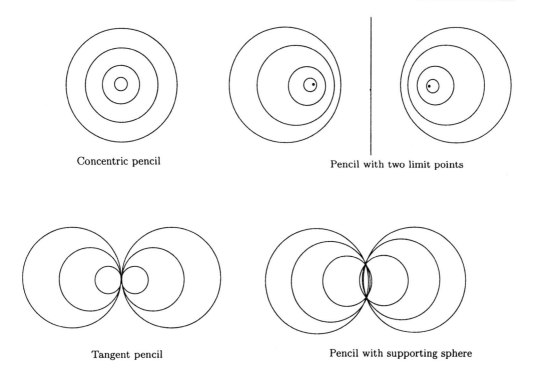

Concentric pencil

Pencil with two limit points

Tangent pencil

Pencil with supporting sphere

Figure 18.10. The four kinds of pencils.

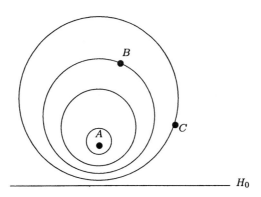

Figure 18.11. B is closer to A for the hyperbolic distance than C is.

The interested reader will find a more precise account in the classical references on the topic ([22] for instance). To define the hyperbolic diagram, it suffices to decide, given three points A, B, and C in $\mathbb{H}^d$, whether B or C is closer to A. For this, we consider the pencil $\mathcal{F}_A$ of spheres with limit points A and A', where A' denotes the symmetric of A with respect to the hyperplane H_0 of equation

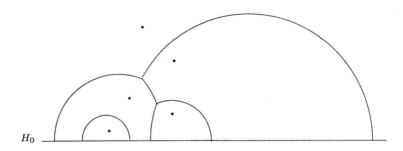

Figure 18.12. A hyperbolic Voronoi diagram in the Poincaré half-space.

$x_d = 0$. (Note that H_0 is the radical hyperplane of $\mathcal{F}_A$.) We say that B is closer to A for the hyperbolic distance if the sphere in $\mathcal{F}_A$ that passes through B has a smaller radius than the sphere in $\mathcal{F}_A$ that passes though C (see figure 18.11).[1]

Given a set $\mathcal{M} = \{M_1, \ldots, M_n\}$ of n points in the Poincaré half-space $\mathbb{H}^d$, there corresponds a region $V_h(M_i)$ in $\mathbb{H}^d$ to each point M_i in $\mathcal{M}$. This region consists of the points in $\mathbb{H}^d$ that are closer to M_i than to any other point in $\mathcal{M}$:

$$V_h(M_i) = \{X \in \mathbb{H}^d, \delta_h(X, M_i) \leq \delta_h(X, M_j) \quad \text{for any} \quad j \neq i\},$$

The *Voronoi diagram for the hyperbolic distance* of $\mathcal{M}$, also called the hyperbolic diagram of $\mathcal{M}$, is the subdivision of the Poincaré half-space induced by the equivalence relation shared by the points that have the same nearest neighbors for the hyperbolic distance. The faces of the diagram are the closures of the equivalence classes. The $V_h(M_i)$'s form the cells of the diagram (see figure 18.12).

$V_h(M_i)$ is the set of points $X \in \mathbb{H}^d$ that have M_i as a nearest neighbor. Since the locus of points in $\mathbb{H}^d$ at a given hyperbolic distance from a given point $A \in \mathbb{H}^d$ is a sphere of the pencil $\mathcal{F}_A$, it follows that, for any point X in $V_h(M_i)$, the interior of the sphere in the pencil $\mathcal{F}_X$ that passes through M_i contains no point of $\mathcal{M}$.

We can also embed $\mathbb{H}^d$ into $\mathbb{E}^{d+1}$ by identifying it with the half-hyperplane $x_{d+1} = 0$, $x_d > 0$. The hyperplane H_0 is therefore identified with the subspace $\{x_{d+1} = x_d = 0\}$. The pencil $\mathcal{F}_X$ is mapped by ϕ into a line in $\mathbb{E}^{d+1}$ parallel to the x_d-axis. Indeed, if X' is the symmetric of X with respect to H_0, the pencil $\mathcal{F}_X$ has limit points at X and X' that are mapped by ϕ to $\phi(X)$ and to $\phi(X')$, and both these images are symmetric with respect to the hyperplane $x_d = 0$ in $\mathbb{E}^{d+1}$. This implies that a point X belongs to $V_h(M_i)$ if and only if the ray parallel to the

---

[1]It is tempting to define the hyperbolic distance from A to a point B as the radius of the sphere in $\mathcal{F}_A$ that passes through B. This "distance" is not symmetric, however, and is not the true hyperbolic distance defined for instance in [22]. Nevertheless, in what follows, taking the pseudo-distance to be this radius or indeed any other increasing function of this radius leads to the same diagram.

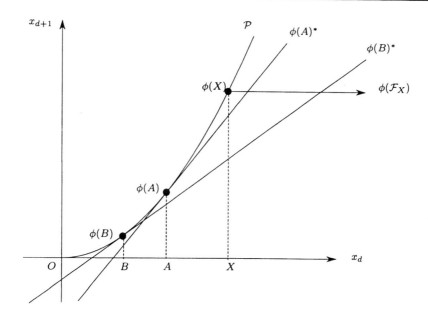

Figure 18.13. X belongs to $V_h(A)$ $(d{=}1)$.

x_d-axis in $\mathbb{E}^{d+1}$ originating at $\phi(X)$ (which is entirely contained in the paraboloid $\mathcal{P}$) and directed towards $x_d > 0$ intersects the hyperplane $\phi(M_i)^*$ polar to $\phi(M_i)$ before any of the other polar hyperplanes $\phi(M_j)^*$ (see figure 18.13).

This observation has important consequences:

1. The *bisecting surface* of two points for the hyperbolic distance is a half-sphere: indeed, a point X is equidistant from A and B if and only if $\mathcal{F}_X$ contains a sphere that passes through A and B, that is, if and only if $\phi(\mathcal{F}_X)$ intersects $\phi(A)^* \cap \phi(B)^*$. In other words, $\phi(X)$ belongs to Γ, the projection of $\phi(A)^* \cap \phi(B)^*$ parallel to the x_d-axis onto the paraboloid $\mathcal{P}$ (more exactly, the half of the paraboloid that is in the half-space $x_d > 0$). But Γ is the intersection of $\mathcal{P}$ with a hyperplane H in $\mathbb{E}^{d+1}$ parallel to the x_d-axis. Its vertical projection onto $x_{d+1} = 0$ is a sphere Σ_{AB} (lemma 17.2.2), centered on H_0 by symmetry. Moreover, Σ_{AB} belongs to the pencil with limit points A and B. Indeed, the spheres in this pencil are mapped by ϕ to the points

$$\phi(\Sigma) = \lambda\phi(A) + (1 - \lambda)\phi(B).$$

The corresponding polar hyperplanes have equations

$$\phi(\Sigma)^* = \lambda\phi(A)^* + (1 - \lambda)\phi(B)^*,$$

and they are all the hyperplanes that contain $\phi(A)^* \cap \phi(B)^*$. H is thus a hyperplane polar to a sphere in the pencil $\mathcal{F}_{AB}$ that has two limit points A and B.

But H is the hyperplane polar to $\phi(\Sigma_{AB})$ (see lemma 17.2.2). As a result, Σ_{AB} belongs to $\mathcal{F}_{AB}$. Finally, Σ_{AB} is the unique sphere in $\mathcal{F}_{AB}$ that is centered on H_0.

2. A point X is equidistant from $d+1$ points $A_0, \ldots, A_d$ if and only if $\phi(X)$ is the projection of $\bigcap_{i=0}^{d} \phi(A_i)^*$ parallel to the x_d-axis, onto the half-paraboloid. The point at equal hyperbolic distance from $d+1$ points is the limit point of the pencil that contains the sphere circumscribed to the $d+1$ points of radical hyperplane H_0.

3. The hyperbolic Voronoi diagram can be obtained by projecting the polytope $\mathcal{V}(\mathcal{M}) = \bigcap_{i=1}^{n} \phi(A_i)^{*+}$ parallel to the x_d-axis onto the half-paraboloid, then projecting the result vertically onto the hyperplane $x_{d+1} = 0$. Note that the projection parallel to the x_d-axis does not map all the points of $\mathcal{V}(\mathcal{M})$ onto the half-paraboloid. This double projection establishes an injective correspondence between the Euclidean and the hyperbolic Voronoi diagrams of $\mathcal{M}$. More directly, these two projections can be avoided by performing the single following transformation. Replace the planar $(d-1)$-faces of the Euclidean diagram that are (at least partly) contained in the half-space $x_d > 0$, by the corresponding portions of spheres (hyperbolic bisectors limited to $x_d > 0$); a k-face ($k < d-1$) of the Euclidean diagram is the intersection of $d-k+1$ planar $(d-1)$-faces, and is replaced by the portion of surface that is the intersection of the $d-k+1$ corresponding spherical faces. From the injective correspondence between Euclidean and hyperbolic diagrams, we deduce the following theorem.

Theorem 18.5.1 *The complexity of the hyperbolic Voronoi diagram of n points in the hyperbolic Poincaré half-space $\mathbb{H}^d$ is $\Theta(n^{\lceil d/2 \rceil})$. Such a diagram can be computed in time $\Theta(n \log n + n^{\lceil d/2 \rceil})$.*

18.6   Exercises

Exercise 18.1 (Greatest empty rectangle) Let X and A be two points in $\mathbb{E}^2$. The quadratic distance $\delta_{\mathcal{Q}}(X, A)$ is defined as

$$\delta_{\mathcal{Q}}(X, A) = (X - A)\Delta(X - A) \quad \text{with} \quad \Delta = \begin{pmatrix} 0 & 1 \\ 1 & 0 \end{pmatrix}.$$

Show that $\delta_{\mathcal{Q}}(X, A)$ is the area of the rectangle whose sides are parallel to the coordinate axes and of which A and X are two opposite vertices. Given a set $\mathcal{S}$ of points in the plane, show that its diagram for this quadratic distance function can be used to compute the rectangle of greatest area whose sides are parallel to the coordinate axes, whose sides each contain at least one point of $\mathcal{S}$, and whose interior does not contain any point of $\mathcal{S}$.

Hint: To find a greatest empty rectangle, use a divide-and-conquer algorithm. The merge step consists in finding the greatest empty rectangle intersected by the separating line. The greatest empty rectangle for which three points of contact lie on one side of the separating line (and the fourth on the other side) can be found easily. Two points of contact on one side of the separating line define a corner of the rectangle, and the corners of empty rectangles are the so called maxima and can be found in $O(n \log n)$ time. The greatest empty rectangle with two points of contact on either side of the separating line is defined by an opposite pair (A, B) of maxima such that the segment connecting them is an edge of the affine diagram defined for the generalized quadratic distance $\delta_Q(A, B)$. The complexity of the merge step is $O(n \log n)$, hence the total algorithm runs in time $O(n \log^2 n)$.

Exercise 18.2 (Lower envelope of cones) Show that the lower envelope of n vertical cones of revolution in $\mathbb{E}^d$ has complexity $O(n^{\lfloor d/2 \rfloor + 1})$ and can be computed in time $O(n^{\lfloor d/2 \rfloor + 1})$. If the vertices of the cones are all contained in a given horizontal hyperplane, and if their angles are all identical, then the complexity of the lower envelope drops to $O(n^{\lfloor \frac{d+1}{2} \rfloor})$ and it can be computed in time $O(n \log n + n^{\lfloor \frac{d+1}{2} \rfloor})$.

Exercise 18.3 (Spheres and disks) According to the general definition of Voronoi diagrams, we may define the Voronoi diagram of a set of disks $D_1, \ldots, D_n$ as usual, where the distance of a point X from a disk D_i centered at C_i and of radius r_i is defined by

$$d_S(X, D_i) = \max(0, \|XC_i\| - r_i).$$

Show that the Voronoi diagram of n disks in $\mathbb{E}^d$, where $d \geq 3$, has complexity $O(n^{\lfloor d/2 \rfloor + 1})$ and that it can be computed in time $O(n^{\lfloor d/2 \rfloor + 1})$. If $d = 2$, show that these bounds are $O(n)$ and $O(n \log n)$ respectively.

Hint: To each C_i, give a weight r_i and compute the diagram of the disks knowing the additive diagram of their centers. In the discussion of subsection 18.3.1, the cone C_i, $i = 1, \ldots, n$ must be replaced by the same cone truncated by the halfspace $x_{d+1} \geq 0$.

Exercise 18.4 (Diagrams with multiplicative weights) Show that $\Omega(n^{\lfloor d/2 \rfloor + 1})$ is a lower bound on the complexity of the Voronoi diagram of n points in $\mathbb{E}^d$ with multiplicative weights.

Hint: Generalize the example of figure 18.5.

Exercise 18.5 (Regular complex) Let $\mathcal{C}$ be a d-complex in $\mathbb{E}^d$. We say that $\mathcal{C}$ is *regular* if it can be obtained as the vertical projection of a polytope in $\mathbb{E}^{d+1}$. Show that $\mathcal{C}$ is regular is and only if it is a power diagram. Show that any simple complex (meaning that its cells all consist of simple polytopes) is regular (the best-known examples are arrangements of hyperplanes in general position). Devise an algorithm that determines whether a complex is regular and, if so, computes the corresponding polyhedron.

Hint: Use theorem 18.2.1. For hyperplane arrangements, the connection with zonotopes is particularly helpful (see exercise 14.8).

Exercise 18.6 (The inverse problem) Show that it is possible to determine whether a complex is a Voronoi diagram and, if so, to compute the corresponding sites in time linear in the total complexity of its cells.

Exercise 18.7 (Spider webs) By a *spider web*, we mean the 1-skeleton of a 2-complex that covers $\mathbb{E}^2$. Show that if the spider web is the skeleton of a power diagram, then we can assign a tension to each edge such that each vertex is in an equilibrium state.

Hint: For the tension of an edge, take the length of the dual edge. An edge and its dual edge are perpendicular, and the dual edges of the edges incident to a vertex S form a cycle that we orient counter-clockwise. At a vertex S, the sum of the tensions equals the sum of the vectors of the dual edges, so that the total tension vanishes at the vertices.

Exercise 18.8 (Cubes and co-cubes) Show that in $\mathbb{E}^3$, several homothetic cubes or co-cubes may pass through four points even though these points are in L_∞-general position.

Exercise 18.9 (Degenerate positions for L_1 and L_∞ distances) Show that in $\mathbb{E}^2$, the Voronoi diagram for the L_1 metric of points that are along one of the main bisectors is quadratic. Show that if the bisector of two points on a line parallel to one of the main bisectors is redefined as the Euclidean perpendicular bisector, then the complexity of the diagram becomes linear, and a cell is formed by the set of points that share exactly one common nearest neighbor for the L_1 distance (but do not necessarily have the same subset of nearest neighbors). Generalize the example above to show that $\Omega(n^d)$ is a lower bound on the complexity of a Voronoi diagram of n points in $\mathbb{E}^d$ for the L_1 metric. Also give similar results for the L_∞ metric.

Exercise 18.10 (Complexity of $\mathcal{V}or_{L_\infty}$) Show that the complexity of a Voronoi diagram for the L_∞ metric of a set $\mathcal{M}$ of n points in $\mathbb{E}^d$ in L_∞-general position is $O(n^{\lceil d/2 \rceil})$.

Hint: It suffices to bound the number of so-called maximal placements of a maximal cube whose facets are perpendicular to the coordinate axes, and whose interior contains no point of $\mathcal{M}$. A contact is a pair formed by a facet of such a cube and by a point in $\mathcal{M}$. A placement realizes a contact of multiplicity k at a point if this point belongs to k facets of the corresponding cube. If $k = 1$, the contact is said to be simple. For a given maximal placement, the sum of the multiplicities of the points of contact is $d + 1$. First show that any maximal placement realizes two contacts, called parallel contacts, whose facets are parallel. We say that a maximal placement is reducible if at least one of its parallel contacts is simple and if the other does not have multiplicity d. Show then that it suffices to bound the number of irreducible maximal placements. For this, charge a reducible placement to an irreducible one by applying the following procedure as many times as needed: Scale up the cube by a homothety centered at one of its vertices that lies on a facet involved in some parallel contact. In this way we obtain a smaller cube contained in the preceding one but whose multiplicity is increased for at least one of the contacts. Show that an irreducible placement is charged by at most $O(1)$ reducible placements. Finally show that the number of irreducible placements is $O(n^{\lceil d/2 \rceil})$. For this, notice that the centers of such placements belong to some affine subspace of dimension at most $d - 3$. In this subspace, the centers of maximal placements correspond to the vertices of a union of n cubes of same size, so we may use the result of exercise 4.8.

Exercise 18.11 (Simplicial distance) Let S be a $(d+1)$-simplex in $\mathbb{E}^d$ that contains the origin O. We denote by λS the image of this polytope under the homothety centered at O and of ratio λ. The simplicial distance $\delta_S(X, A)$ from point X to point A is defined as the smallest real $\lambda \geq 0$ such that $X - A$ belongs to λS. Show that the complexity of a Voronoi diagram for a simplicial distance of a set $\mathcal{M}$ of n points in $\mathbb{E}^d$ is $O(n^{\lceil d/2 \rceil})$.

Hint: We define a reducible placement of S as in exercise 18.10: it is a placement that has several simple contacts. The number of irreducible placements is $O(n^{\lceil d/2 \rceil})$ and we can also show that the same bound holds for the reducible placements.

Exercise 18.12 (Hyperbolic bisector) Show that the equation of the hyperbolic bisector Σ_{AB} of two points $A = (a_1, \ldots, a_d)$ and $B = (b_1, \ldots, b_d)$ is

$$(a_d - b_d) X \cdot X + 2(b_d A - a_d B) \cdot X - b_d A \cdot A + a_d B \cdot B = 0.$$

Exercise 18.13 (The Poincaré disk) Rather than using the Poincaré half-space $\mathbb{H}^2$ as a model of the hyperbolic space, we introduce the Poincaré disk $\mathbb{D}$ which can be derived from $\mathbb{H}^2$ by a homographic transformation. More precisely, if the Poincaré half-space is identified with the complex half-plane $\{z \in \mathbb{C}, \operatorname{Im} z > 0\}$, the homographic map defined by

$$h(z) = \frac{z - i}{z + i}$$

is a bijection from $\mathbb{H}^2$ into $\mathbb{D}$. Show that the edge that joins two points remains a circle centered on the boundary of $\mathbb{D}$, and also that the points at equal distance from A are on a circle that belongs to the pencil that has A as a limit point and that contains the boundary of $\mathbb{D}$. From this, explain how to compute the Voronoi diagram of a set of points in $\mathbb{H}^2$.

Exercise 18.14 (Dual of a hyperbolic diagram) Show that we may dualize the hyperbolic Voronoi diagram of a set of points $\mathcal{M}$ in $\mathbb{H}^d$ by projecting the convex hull of $\phi(\mathcal{M})$ parallel to the x_d-axis onto the half-paraboloid, and then projecting the result of this first projection onto the hyperplane $x_{d+1} = 0$. Show that this dual is in bijection with a sub-complex of the Delaunay complex.

18.7 Bibliographical notes

Power diagrams were studied by Aurenhammer [14] and by Imai, Iri, and Murota [129]. Affine diagrams are defined in [17] by Aurenhammer and Imai, who also show their connection with power diagrams and diagrams with additive and multiplicative weights. Solutions to exercises 18.5 and 18.6 are due to Aurenhammer. The solution to exercise 18.1 is adapted from that given by Chazelle, Drysdale, and Lee [47] and Aurenhammer. Spider webs were already analyzed by Maxwell in the nineteenth century. Recent references can be found in the article by Ash, Bolker, Crapo, and Whiteley [12].

Diagrams for the L_1 and L_∞ metrics in the plane were studied by Lee and Wong [149] then by Lee and Drysdale [147]. The generalization to general convex distances is tackled by Chew and Drysdale [61]. Voronoi diagrams for the L_1 and L_∞ metrics in dimensions 3 and higher (see exercise 18.10) and also simplicial distances (see exercise 18.11) are treated by Boissonnat, Sharir, Tagansky, and Yvinec [34]. In the plane, Klein proposes a notion of abstract Voronoi diagram [139] and Klein, Mehlhorn, and Meiser describe a randomized algorithm that computes such diagrams [141].

Diagrams for the hyperbolic distance are studied by Boissonnat, Cérézo, Devillers, and Teillaud [26], who present an application to shape reconstruction from plane sections.

Chapter 19

Diagrams in the plane

In the two preceding chapters, we have shown how to compute several types of Voronoi diagrams in $\mathbb{E}^d$ by computing the upper envelope of hyperplanes in $\mathbb{E}^{d+1}$ or $\mathbb{E}^{d+2}$. This often leads to optimal algorithms: this is notably true for diagrams of points under a general quadratic distance and for power diagrams of spheres. In contrast, we have seen that for diagrams with additive weights, such an approach does not lead to optimal algorithms in dimension 2.

Section 19.1 describes an algorithm that computes the Voronoi diagram of a set of points in the plane. This algorithm is remarkable for several features: it uses the sweep method, it is simple and optimal, and it can be generalized in a number of ways. For instance, it can be adapted to compute the Voronoi diagram of a set of segments or to use other metrics such as the L_1 or L_∞ distances.

Voronoi diagrams of line segments have important applications, such as the motion planning of a disk (see subsection 19.2.5). In dimension 2, they are well understood and we present, in addition to the generalization of the sweep algorithm, a randomized algorithm and its accelerated version when the set of segments is connected, and the segments intersect only at common vertices.

Section 19.3 studies an instance of the problem when the points belong to two planes in $\mathbb{E}^3$. This is a particular instance of three-dimensional diagram for which an algorithm is presented that is output-sensitive and optimal.

19.1 A sweep algorithm

In this section, we present a sweep algorithm that computes the Voronoi diagram of a set $\mathcal{M} = \{M_1, \ldots, M_n\}$ of n points in $\mathbb{E}^2$. To simplify the presentation, the points are supposed to be in L_2-general position, meaning that no four points are co-circular.

As shown in subsection 18.3.1, if $\mathbb{E}^2$ is identified with the xy-plane $z = 0$ in

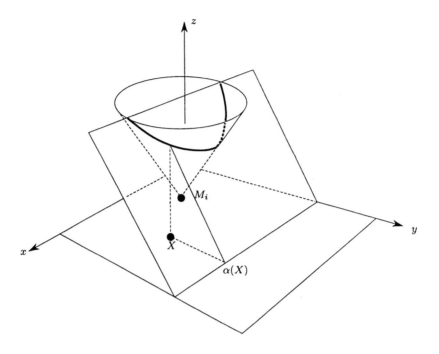

Figure 19.1. The projection used in the sweep algorithm.

$\mathbb{E}^3$, we may place cones $\mathcal{C}_i$ on top of each M_i: $\mathcal{C}_i$ is the upward vertical cone of revolution of angle $\frac{\pi}{4}$ that has apex M_i; its equation is

$$z = \|XM_i\|.$$

Then the Voronoi diagram of $\mathcal{M}$ is the projection onto the plane $z = 0$ of the lower envelope of the cones $\mathcal{C}_i$, $i = 1, \ldots, n$.

The algorithm computes this lower envelope. Rather than using a vertical projection, however, we project parallel to a line that generates the cones. More precisely, the plane is swept by a line parallel to the x-axis that moves along the increasing y-axis, and the direction of the projection onto this plane is given by the vector $(0, 1, -1)$.

In this way, we map a point $X = (x, y)$ to another point

$$\alpha(X) = (x, y + \min_{M_i \in \mathcal{M}} (\|XM_i\|)).$$

This point is obtained by first lifting X onto the lower envelope of the $\mathcal{C}_i$'s and then projecting the result onto the plane $z = 0$ parallel to the direction $(0, 1, -1)$. The map α is depicted in figure 19.1. As usual, $V(M_i)$ stands for the cell that corresponds to M_i in the Voronoi diagram. Because the lower envelope is continuous, the map α is also continuous. Moreover, its restriction to a line D that is parallel to the y-axis and that does not contain a point in $\mathcal{M}$ is injective. The images of the points on D are also on D, and if $X_1 = (x, y_1)$ and $X_2 = (x, y_2)$ are two points on D with $y_1 < y_2$, the triangle inequality shows that

$$\forall i, \quad \|X_1 M_i\| < \|X_1 X_2\| + \|X_2 M_i\|.$$

This shows that the n maps that return the value $y + \|X M_i\|$, given the ordinate y of a point X on D, are continuous functions that increase with y. Therefore the minimum of these functions is also an increasing function, and so α is injective on D. Now if D contains a point M_i, the inequality above still holds if the ordinate of X_2 is greater than that of M_i and so α is still injective. If both X_1 and X_2 have smaller ordinates than M_i, then

$$\|X_1 M_i\| = \|X_1 X_2\| + \|X_2 M_i\|.$$

Any point $X \in V(M_i)$ that has the same abscissa as, and a smaller ordinate than M_i is mapped by α onto M_i, so M_i is invariant. From the preceding discussion, it follows that the restriction of α to the edges of the Voronoi diagram of $\mathcal{M}$ is injective, since any ray cast from M_i intersects the boundary of the convex cell $V(M_i)$ at only one point.

The map α deforms the Voronoi diagram $Vor(\mathcal{M})$ into a new diagram $Vor'(\mathcal{M})$. Since α is injective on the edges of $Vor(\mathcal{M})$ and is continuous, the cell $V(M_i)$ is mapped by α onto a simply connected cell $V'(M_i)$ which we call a cell of $Vor'(\mathcal{M})$. The edge $V(M_i, M_j) = V(M_i) \cap V(M_j)$ in $Vor(\mathcal{M})$ that is supported by the perpendicular bisector of two points M_i and M_j is mapped by α onto a curved arc $V'(M_i, M_j)$ which we call an edge of $Vor'(\mathcal{M})$. Finally, α maps the vertices of $Vor(\mathcal{M})$ onto the endpoints of the edges of $Vor'(\mathcal{M})$: we call these points the vertices of $Vor'(\mathcal{M})$.

The curved arc $V'(M_i, M_j)$ is in fact contained in a branch of the hyperbola, that is, the projection of the intersection of the cones C_i and C_j. This branch passes through M_i or M_j, whichever has the greatest ordinate, and this point is the point of smallest ordinate on this branch. The hyperbola degenerates into a line parallel to the y-axis if M_i and M_j have identical ordinates. The cell $V'(M_i)$ is delimited by hyperbolic arcs and the point with smallest ordinate in $V'(M_i)$ is M_i (see figure 19.2). This property is crucial for the sweep algorithm. Observe that all the hyperbolic arcs that form the boundary of $V'(M_i)$ are y-monotone, except for the one that contains M_i. This one can be split into two y-monotone sub-arcs that have M_i as the endpoint of smallest ordinate.

The sweep line Δ is parallel to the x-axis and moves towards increasing ordinates. To simplify the presentation, we assume that no two points in $\mathcal{M}$ have identical ordinates. Like any sweep algorithm, this algorithm maintains two structures. The first structure is a dictionary $\mathcal{D}$ of the edges of $Vor'(\mathcal{M})$ that intersect Δ, in increasing order of their intersections along Δ, and the intervals induced

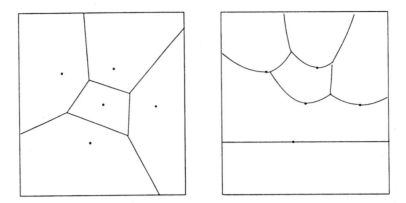

Figure 19.2. $\mathcal{V}or(\mathcal{M})$ and $\mathcal{V}or'(\mathcal{M})$.

by these points on Δ. Each interval is the intersection of Δ with a cell of the deformed Voronoi diagram. If this corresponds to the site M_i, the interval is labeled by i. As we observed above, the edges of $\mathcal{V}or'(\mathcal{M})$ that contain a site are not y-monotone, and they intersect Δ in two points. This slight problem is easily remedied by splitting the edge at M_i into two arcs as already mentioned.

The second data structure maintained by the algorithm is a priority queue $\mathcal{Q}$ that represents the events (sorted by increasing ordinate) for which Δ sweeps over a vertex or over a site, at which point the first data structure must be updated.

At the start of the algorithm, Δ is the line parallel to the x-axis that passes through the point of smallest ordinate. Only a single unbounded region exists and the dictionary $\mathcal{D}$ stores a single interval. As Δ moves towards increasing ordinates, the structure $\mathcal{D}$ must be updated for two kinds of events. In a *site event*, the sweep line sweeps over a site, say M_i. A new region and its two bounding hyperbolic arcs appear in the diagram, giving rise to a new interval with two new endpoints on Δ. More precisely, if M_i belongs to an interval labeled j, the hyperbola that supports the two arcs to be added to the diagram is the image under α of the perpendicular bisector of M_i and M_j. The interval labeled j is replaced in the structure $\mathcal{D}$ by three intervals labeled j, i, and j in that order. In the second kind of event, a *circle event*, Δ sweeps over the intersection point of two arcs. This situation corresponds to a vanishing interval on Δ whose two endpoints are the intersections of Δ and the two meeting hyperbolic arcs. Let j be the label of that interval, and i and k be the labels of the adjacent intervals in $\mathcal{D}$. These two arcs are the images under α of two incident edges that bound $V(M_iM_j)$ and $V(M_jM_k)$. Their meeting point is the image of a vertex of the Voronoi diagram, the center of the circle passing through M_i, M_j, and M_k (hence the name of this kind of event). The two arcs are replaced by the image of the edge of $V(M_iM_k)$ that is incident to this vertex.

As happens when computing the arrangement of a set of line segments using the sweep algorithm described in section 3.2, each time two arcs become consecutive in $\mathcal{D}$, we test whether they intersect beyond Δ and, if so, we insert this intersection point into $\mathcal{Q}$. When deleting an arc, we must also remove the two corresponding entries defined by this arc in $\mathcal{Q}$. In fact, for each arc $\mathcal{Q}$ must contain only an entry that stores the intersection point of smallest ordinate defined by this arc (beyond this point, the arc cannot exist), and for each arc in $\mathcal{D}$, we maintain a pointer to this entry. Each time an arc is considered, its pointer is updated, and the entry in $\mathcal{Q}$ is removed when the arc disappears.

We must point out that computing the hyperbolas is not needed. In fact, hyperbolas are the images under α of the perpendicular bisectors of two of the sites, and their intersections are the images of the intersections of the two corresponding perpendicular bisectors. Location in $\mathcal{D}$ can be performed by using the perpendicular bisectors rather than the hyperbolas. Location in $\mathcal{Q}$ can be carried out by computing the ordinates of the images of the intersection of two perpendicular bisectors. The Voronoi diagram is computed directly during the sweep.

The complexity of the algorithm can be estimated very simply. The sweep line stops over sites and vertices of $\mathcal{Vor}'(\mathcal{M})$ so the number of events processed by the sweep algorithm is no more than the size of $\mathcal{Vor}'(\mathcal{M})$, namely $O(n)$. During each event, only $O(1)$ locations, insertions, or deletions are performed in $\mathcal{D}$ and $O(1)$ events are added to or removed from $\mathcal{Q}$. The sizes of $\mathcal{D}$ and $\mathcal{Q}$ are thus $O(n)$ at any event throughout the algorithm. That the size of $\mathcal{D}$ is $O(n)$ can also be seen by noticing that Δ intersects any edge of $\mathcal{Vor}'(\mathcal{M})$ at most twice. Each update operation (insertion, deletion, or location) can therefore be carried out in time $O(\log n)$, for a grand total of $O(n \log n)$ operations. This is optimal as shown by corollary 17.3.2, and this proves the following theorem.

Theorem 19.1.1 *The Voronoi diagram of n points in the plane can be computed using a sweep algorithm in time $O(n \log n)$, using storage $O(n)$, and this is optimal.*

Application to diagrams with additive weights

Let $M_1, \ldots, M_n$ be n points in the plane, endowed with the weights $r_1, \ldots, r_n$. The distance with additive weights (additive distance for short) from a point X to M_i is defined as

$$\delta_+(X, M_i) = \|XM_i\| - r_i.$$

The additive diagram is defined as the Voronoi diagram for this additive distance.

As shown in subsection 18.3.1, the additive diagram of a set of n points in the plane is the projection onto this plane of vertical cones with aperture angle $\frac{\pi}{4}$.

Its complexity is $O(n)$. The aperture angles of the cones being identical, the sweep algorithm performs in a way that is strictly analogous to the Euclidean case. Only a few differences deserve to be pointed out.

First of all, the transformation α introduced above (see figure 19.1) becomes

$$\alpha(X) = (x, y) = (x, y + \min_{M_i \in \mathcal{M}} (\|X M_i\| - r_i))$$

and the M_i's are no longer invariant. Secondly, some sites may have an empty corresponding region. This can be detected during the sweep: when a new site M_i is encountered, it is contained in the region of a point M_{i^*} of weight r_{i^*} in the additive diagram, and its region is non-empty if and only if its weight r_i satisfies $\|M_i M_{i^*}\| - r_{i^*} \geq -r_i$.

The remaining details of the algorithm are strictly analogous to those in the Euclidean case, and hence:

Theorem 19.1.2 *The Voronoi diagram of n points in the plane for the additive distance can be computed in time $O(n \log n)$ by a sweep algorithm, and this is optimal.*

19.2   Voronoi diagram of a set of line segments

19.2.1   Definition and basic properties

Let $\mathcal{S} = \{S_1, \ldots, S_n\}$ be a set of n disjoint line segments in the plane. The *distance of a point A from a segment S_i*, denoted by $\delta(A, S_i)$, is defined as the smallest distance between A and any point in S_i. The segments are said to lie in L_2-general position if no point is equidistant from more than three segments in $\mathcal{S}$.

Let us consider the equivalence relation R, shared by two points if and only if they have identical subsets of nearest neighbors in $\mathcal{S}$. The equivalence classes of R subdivide $\mathbb{E}^2$ into (open) regions whose topological closures form the *faces* of the Voronoi diagram of $\mathcal{S}$. The interior of a cell in this diagram is formed by the points closer to some segment S_i in $\mathcal{S}$ than to the others, and this cell is denoted by $V(S_i)$ ($i = 1, \ldots, n$). The edges are the regions formed by the points equidistant from two segments in $\mathcal{S}$ and closer to these segments than to any other segment in $\mathcal{S}$. The vertices are the points that are equidistant from at least three segments in $\mathcal{S}$ and closer to these segments than to any other. The cells, edges, and vertices and their incidence relationships form the *Voronoi diagram of the set $\mathcal{S}$ of segments*, denoted by $\mathcal{V}or(\mathcal{S})$ (see figure 19.3).

The two following lemmas characterize the cells $V(S_i)$ in such a Voronoi diagram.

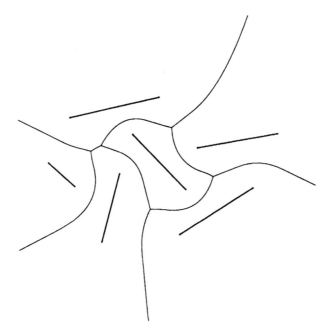

Figure 19.3. The Voronoi diagram of a set of line segments.

Lemma 19.2.1 *Let $V(S_i)$ be the cell of S_i in the diagram, X a point in $V(S_i)$, and X' its closest point in S_i. The ray originating at X' that contains X is either entirely contained in $V(S_i)$ or intersects $V(S_i)$ along a line segment whose endpoints are X' and the unique intersection point of the ray with the boundary of $V(S_i)$.*

Proof. With the notation of the lemma, any point Y on the segment XX' is closer to X' than to any other point in $\mathcal{S}$, so that Y belongs to $V(S_i)$. The segment XX' is therefore contained in $V(S_i)$, proving that $V(S_i)$ is star-shaped. The ray $X'X$ thus intersects $V(S_i)$ along a connected subset of the ray that contains X', which can only be the ray itself or a line segment whose endpoints are X' and a point Z on the boundary of $V(S_i)$. It remains to see that in the latter case, this point Z is uniquely defined as the intersection of the ray with the boundary of $V(S_i)$ (this intersection could contain several points, possibly infinitely many).

For this we show that the relative interior of $X'Z$ does not intersect the boundary of $V(S_i)$. Towards a contradiction, assume that there is a point Y on the relative interior of the segment $X'Z$ that is on the boundary of $V(S_i)$. This implies that Y is equidistant from S_i and another segment S_j in $\mathcal{S}$. Let us denote by Y' the point on S_j that is the closest to Y. The disk centered at Z of radius ZX' contains the disk centered at Y of radius YX', and only X' belongs to both their boundaries. So the first disk must contain Y' in its interior. Therefore Z is closer to Y' and hence S_j than to S_i, and it does not belong to $V(S_i)$, a contradiction.

It follows that if the ray $X'X$ intersects $V(S_i)$ along a line segment, only the endpoint other than X' of this segment belongs to the boundary of $V(S_i)$, whereas if the entire ray is contained in $V(S_i)$, then no point of this ray belongs to the boundary of $V(S_i)$. $\qquad\square$

Lemma 19.2.2 *The interior of any cell of the Voronoi diagram of S is simply connected.*

Proof. Let us first see that $V(S_i)$ is connected. Indeed, considering two points X and Y in $V(S_i)$, denote by X' and Y' their closest point in S_i. It follows from the previous lemma that the segments XX' and YY' are contained in $V(S_i)$. Obviously, S_i is contained in $V(S_i)$, and $X'Y'$ is also contained in $V(S_i)$ since it is a subset of S_i. Thus X and Y can be connected by the polygonal line $XX'Y'Y$ which is entirely contained in $V(S_i)$. Note that its interior is also connected since the segments are disjoint and thus S_i is contained in the interior of $V(S_i)$.

Let us now show that the interior of $V(S_i)$ is simply connected. Assuming the contrary, there exists a point X that does not belong to $V(S_i)$ but is contained in the interior of a closed Jordan curve Γ that is entirely contained in $V(S_i)$. If X' is the closest point to X on S_i, the ray originating at X' that contains X intersects Γ in at least two points (see exercise 11.1), one of which is further from X' than from X; call this point I. The interior of the disk centered at I and of radius $\|IX'\|$ does not intersect any segment in S, hence neither does the interior of the disk centered at X of radius $\|XX'\|$. So X is closer to S_i than to any other segment of S, hence belongs to the interior of $V(S_i)$, a contradiction. $\qquad\square$

The edges of the diagram are formed by the points that are equidistant from two segments and closer to these segments than to the others. An edge is thus contained in the *bisector of two segments*, which is the set of points at equal distance from two segments. In general, such a bisector can be split into several components: line segments, contained in the perpendicular bisector of two endpoints or in the bisector of the lines supporting the two segments, and parabolic arcs formed by the points at equal distance from an endpoint of one segment and the line supporting the other segment. The following lemmas explore the nature of these bisectors more precisely.

Lemma 19.2.3 *The bisector of two disjoint line segments is a simple curve that disconnects the plane into two connected components and that can be split into at most seven line segments and parabolic arcs.*

Proof. Consider two disjoint line segments S_1 and S_2 whose endpoints are A_1, B_1 and A_2, B_2 respectively. At least one of these segments (say S_1) is contained in one of the two half-spaces bounded by the line that supports the other (S_2). Each segment is identified with a flat rectangle oriented counter-clockwise, and its four elements are conceptually separated: its endpoints A_i, B_i, and its two oriented sides $E_i = (A_i, B_i)$ and $F_i = (B_i, A_i)$. Each arc of the bisector D_{12} is the locus of points at equal distance from two elements, one of S_1 and the other of S_2, and closer to these than to any other element. The arcs of the bisector D_{12} are labeled by the two corresponding elements.

When following the boundary of the region $V(S_i)$ counter-clockwise, the labels of the arcs of D_{12} are enumerated in counter-clockwise order for those that belong to S_i, and in clockwise order for those that belong to S_j ($i, j = 1, 2$ and $i \neq j$). Indeed, consider two points X and Y in D_{12} that are labeled by two distinct elements of S_i, say L_X and L_Y. Let X' be the closest point to X in L_X and let Y' be the closest point to Y in L_Y. It is easy to see that the relative interiors of XX' and YY' cannot intersect. Thus the labels of S_i are seen in counter-clockwise order along the boundary of $V(S_i)$ oriented counter-clockwise.

Let us construct the following oriented graph G (see figures 19.4 and 19.5). A node in the graph represents a pair formed by an element of S_1 and an element of S_2. There is an arc in G between the nodes $N = (L_1, L_2)$ and $N' = (L'_1, L_2)$ if and only if L'_1 follows L_1 in counter-clockwise order along S_1. Similarly, there is an arc in G between the nodes $N = (L_1, L_2)$ and $N' = (L_1, L'_2)$ if and only if L'_2 follows L_2 in clockwise order along S_2. The graph G can be drawn on a torus (the topological product of the boundaries of S_1 and S_2). The bisector of S_1 and S_2 is represented by an oriented path in this graph. We distinguish two cases according to whether the entire segment S_1 appears on the boundary of the convex hull of S_1 and S_2, which is the convex hull of A_1, A_2, B_1, and B_2. (Note that S_2 is always entirely contained in this boundary.)

Case 1. The boundary of the convex hull of S_1 and S_2 is the polygonal line A_1, B_1, B_2, A_2, in clockwise order (see figure 19.4). The bisector of S_1 and S_2 has two infinite branches supported by the perpendicular bisectors of A_1 and A_2, and of B_1 and B_2. From the preceding discussion, it is clear that the bisector of S_1 and S_2 corresponds to an oriented path in the graph G that connects the two nodes (A_1, A_2) and (B_1, B_2). The number of arcs on this bisector equals the number of vertices of such a path, namely five.

Case 2. Only one endpoint of S_1, say A_1, appears on the boundary of the convex hull $conv(S_1, S_2)$, which is the polygonal line A_1, B_2, A_2 in clockwise order (see figure 19.5). The bisector of S_1 and S_2 has two infinite branches supported by the perpendicular bisectors of A_1 and A_2, and of A_1 and B_2. The bisector of S_1 and S_2 corresponds to an oriented path in the graph G that connects the two nodes (A_1, A_2) and (A_1, B_2). The number of arcs on this bisector is thus seven.

□

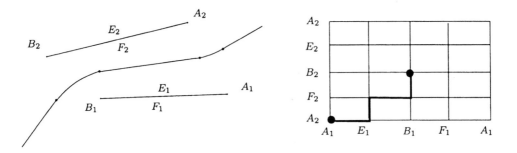

Figure 19.4. A bisector of segments with five arcs.

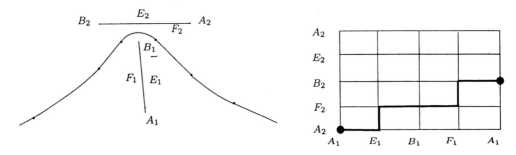

Figure 19.5. A bisector of segments with seven arcs.

Lemma 19.2.4 *Let S_1, S_2, and S_3 be three disjoint segments. Any two of the three bisectors of these segments intersect in at most two points.*

Proof. Consider three segments S_1, S_2, S_3, and denote by $V(S_1)$, $V(S_2)$, and $V(S_3)$ the three cells in the Voronoi diagram of the three segments. Let us assume that two of the three bisectors, say D_{12} and D_{13}, have three points in common. Then these points also belong to D_{23}: they are at the same distance from S_1 and S_2, and from S_1 and S_3, so they are also at the same distance from S_2 and S_3. This shows that the three bisectors have three points in common. Then $V(S_1)$ has at least three vertices I_1, I_2, and I_3, say, in clockwise order along its boundary, each belonging to the intersection of $V(S_1)$, $V(S_2)$, and $V(S_3)$. Two successive edges of $V(S_1)$ cannot belong to the same bisector, otherwise the cells $V(S_1)$, $V(S_2)$, and $V(S_3)$ would all contain the endpoint common to both edges, and since they cover the plane entirely, one would not be simply connected, violating lemma 19.2.2. The edge E_1 of $V(S_1)$ that precedes I_1, and the edge E_3 of $V(S_1)$ that connects I_2 to I_3, both belong to the same bisector, say D_{12}. Similarly, the edge E_2 of $V(S_1)$ that connects I_1 to I_2 and the edge E_4 of $V(S_1)$ that follows I_3 both belong to the same bisector D_{13}. The situation is shown in figure 19.6.

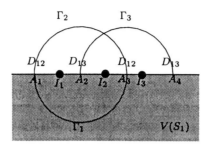

Figure 19.6. For the proof of lemma 19.2.4.

Now pick a point A_i on E_i $(i = 1, \ldots, 4)$. Since the interiors of $V(S_1)$, $V(S_2)$, and $V(S_3)$ are simply connected, there exists a path Γ_1 connecting A_1 to A_3 that is entirely contained in the interior of $V(S_1)$ (except for its two endpoints A_1 and A_3). Similarly, there exists a path Γ_2 connecting A_1 to A_3 that is entirely contained in the interior of $V(S_2)$ (except for A_1 and A_3), and there exists a path Γ_3 connecting A_2 to A_4 that is entirely contained in the interior of $V(S_3)$ (except for A_2 and A_4). Now the union of Γ_1 and Γ_2 is a closed simple curve Γ_{12} which splits the plane into two connected components (by Jordan's theorem), one containing A_2 and the other containing A_4. Therefore Γ_3 must intersect Γ_{12}, which contradicts the fact that the cells have disjoint interiors. $\quad\square$

Remark. The bisector of two segments may contain an arbitrarily high number of edges of $Vor(\mathcal{S})$.

Lemma 19.2.5 *The Voronoi diagram of n segments has complexity $O(n)$.*

Proof. Let $\mathcal{S}$ be a set of n segments. Lemma 19.2.2 assures us that the diagram has exactly n connected cells. Moreover, each vertex in the diagram has degree three when the segments are in L_2-general position. The Voronoi diagram of $\mathcal{S}$ is thus a planar map with n cells and vertices of degree three. Lemma 19.2.3 implies that each edge in the diagram has at most seven arcs, so Euler's relation can be used to show that the number of faces of this map is $O(n)$. The complexity of the diagram, defined equivalently (within constant factors) as the number of arcs, or the number of faces, is thus $O(n)$. It is easy to see that the L_2-general position assumption does not matter for the result. Indeed, the segments in $\mathcal{S}$ may be slightly perturbed so as to achieve L_2-general position, and perturbed back into their primary positions. During this second perturbation, no face is created; on the contrary, some vertices are merged and the zero-length edges that join them disappear. The complexity of the diagram only decreases from that of a diagram of segments in L_2-general position. $\quad\square$

19.2.2   A sweep algorithm

Consider n segments $S_1 = A_1 B_1, \ldots, S_n = A_n B_n$ that are disjoint and in L_2-general position. This means that no point is at equal distance from four segments. By convention, A_i is the endpoint of S_i that has the smallest abscissa. If we don't care which endpoint A_i or B_i of S_i we consider, we use the notation M_i. We define and use a mapping α_s that is analogous to the mapping α used in section 19.1:

$$\alpha_s(X) = \left(x, y + \min_{S_i \in \mathcal{S}} \delta(X, S_i) \right).$$

Its geometric interpretation is identical to that of α, except that C_i $(i = 1, \ldots, n)$ now denotes the union of the vertical cones $C_i(X)$ of angle $\frac{\pi}{4}$ whose vertices are the points X on segment S_i. An argument similar to that used in section 19.1 shows that the restriction of the application α_s to the edges of the diagram is continuous and injective.

This mapping deforms the Voronoi diagram $Vor(\mathcal{S})$ into a new diagram $Vor'(\mathcal{S})$ whose faces are in one-to-one correspondence with the faces of $Vor(\mathcal{S})$. The cell $V(S_i)$ is transformed into a cell $V'(S_i)$ whose points all have a greater ordinate than A_i. An arc of $Vor(\mathcal{S})$, contained in the parabolic locus of the points at equal distance from an endpoint M_i of S_i and the line L_j that supports S_j, is transformed by α_s into an arc contained in a parabola that is the projection of the intersection of C_i and of the dihedron D_j (the union of the cones $C_i(X)$ for $X \in L_j$). An arc of $Vor(\mathcal{S})$ contained in the bisector of the lines supporting two segments S_i and S_j is mapped by α_s onto a line segment contained in the projection of the intersection of the two dihedra D_i and D_j. Finally, as is the case for the Voronoi diagram of points, an edge of $Vor(\mathcal{S})$ contained in the perpendicular bisector of the endpoints M_i and M_j of two segments is mapped onto an arc contained in the branch of the hyperbola that is the projection of the intersection of the cones C_i and C_j. This branch contains the points M_i or M_j, whichever has the greatest ordinate, and this point has the smallest ordinate among all the points on this branch. The hyperbola degenerates into a line parallel to the y-axis if M_i and M_j have the same ordinate y.

The sweep line Δ is parallel to the x-axis and moves towards increasing ordinates. To simplify the presentation, we assume that all the ordinates of the endpoints are distinct. In particular, no segment is parallel to the x-axis. The sweep uses data structures similar to those used for computing the diagram of points. There are now three kinds of events. The *first kind* happens when Δ sweeps over an endpoint of smallest ordinate A_i, and is analogous to a site event for the sweep algorithm that computes the diagram of points. The only significant difference arises because the bisector of two segments consists of several arcs. These arcs are computed when the bisector is encountered for the first position of the sweep line, and their endpoints correspond to as many events of the *third kind* that must be inserted into the event priority queue. An event of the *second kind* is an intersection point of two arcs of the transformed diagram. It can be handled in very much the same way as a circle event for the sweep algorithm that computes the diagram of a set of points. Processing the events of the third kind is straightforward: we must simply change the description of the arc when Δ sweeps over the endpoint of greatest ordinate of that arc, and also update the event in the priority queue that corresponds to the disappearance of the edge that contains that arc.

Theorem 19.2.6 *The sweep algorithm described above computes the Voronoi diagram of a set of n line segments in the plane in time $O(n \log n)$, and this is optimal.*

19.2.3 An incremental algorithm

This section presents an incremental algorithm that computes the Voronoi diagram of a set of n line segments. This algorithm is an extension of the incremental algorithm for points that is given in exercise 17.10. Before we describe the algorithm, we must point out that any algorithm that computes the Voronoi diagram of n points (and *a fortiori* segments) incrementally may require time $\Omega(n^2)$. Indeed, as is shown in figure 19.7, the Delaunay triangulation of n points on the moment curve in $\mathbb{E}^2$ contains the triangles connecting the point of smallest abscissa to the edges of the convex hull. Adding these points in order of decreasing abscissae will create $i - 2$ new triangles at the i-th step, so $\Omega(n^2)$ triangles must be created overall in the worst case. If the points are inserted in random order, however, the expected complexity of an incremental algorithm is $O(n \log n)$ as was shown in chapter 5.[1]

The algorithm we present now is an on-line incremental algorithm that uses an influence graph (see section 5.3). The term "on-line" means that the algorithm maintains a representation of the Voronoi diagram under insertions of arbitrary segments. This algorithm is deterministic, and if the n segments are inserted in random order, a randomized analysis shows that its expected complexity is $O(n \log n)$. Again, we point out that the algorithm does not assume any geometric distribution of the segments: the expectation is taken for a given set of n segments and only the insertion order is random. Our only assumption is that the segments are in L_2-general position.

To fit the framework of chapter 4, we must redefine the problem in terms of objects, regions, and conflicts.

The *objects* are naturally the line segments. The *regions* are the edges in the Voronoi diagrams of any subset S' of S. A region E is *defined* by a subset S' of

---

[1]This is the case for the randomized algorithm that computes the convex hull in $\mathbb{E}^3$ of the points lifted on the paraboloid, but is also the case for the algorithm described in this section.

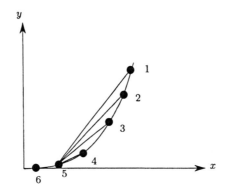

Figure 19.7. A difficult instance for the incremental algorithm.

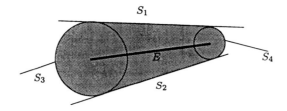

Figure 19.8. A region E, in bold, and its domain of influence D_E, shaded.

segments if and only if E is an edge in the Voronoi diagram of S' but not in the diagram of any subset of S'. A segment in S' is called a *determinant* of E. A region is determined by at most four segments: the two line segments S_1 and S_2 whose bisector contains E, and occasionally one or two segments S_3 and S_4 that determine the endpoints of E.

To a region E corresponds a domain of influence D_E that is the union of the open disks centered on E and not intersecting the segments defining E (see figure 19.8). An object S and a region E *conflict* if S and the interior of D_E have a non-empty intersection. The edges of the Voronoi diagram are precisely the regions defined and without conflict over S, that is, the regions determined by segments of S that do not conflict with any segment in S. The edges determined by a segment S that are without conflict over S are the edges of the Voronoi cell $V(S)$ and all the edges incident to a vertex of $V(S)$.

The algorithm proceeds by inserting the segments one by one. At an incremental step, the Voronoi diagram of the current subset S_c of already inserted segments is stored in the influence graph. Each edge of the current diagram is a region without conflict over the current subset of segments, and has pointers towards the segments that define it. Updating the diagram upon inserting a segment S involves removing all the edges or portions of edges in $\mathcal{V}or(\mathcal{S}_c)$ that are contained in the Voronoi cell $V(S)$ of the new diagram $\mathcal{V}or(\mathcal{S}_c \cup \{S\})$, and adding the new edges that are on the boundary of $V(S)$. We note that the edges of $\mathcal{V}or(\mathcal{S}_c)$ that intersect $V(S)$ are the regions that conflict with S. The following lemma will be very useful later on:

Lemma 19.2.7 *The set $\mathcal{A}$ of edges or portions of edges in $\mathcal{V}or(\mathcal{S}_c)$ that are contained in the cell $V(S)$ of the new diagram $\mathcal{V}or(\mathcal{S}_c \cup \{S\})$ is connected.*

Proof. Assume the contrary. Then there is a path Γ contained in $V(S)$ that connects two points X and Y on the boundary of $V(S)$, does not intersect the edges in $\mathcal{A}$, and subdivides $V(S)$ into two connected components that each contain a connected component in $\mathcal{A}$. Since Γ does not intersect the edges in $\mathcal{A}$, it is entirely contained in a single cell of $\mathcal{V}or(\mathcal{S}_c)$, say the cell $V_c(R)$ that corresponds to segment $R \in \mathcal{S}_c$. But then X and Y belong to the boundary of the cell $V(R)$ of R in the new diagram $\mathcal{V}or(\mathcal{S}_c \cup \{S\})$, so there must exist a path Γ' contained in $V(R)$ with endpoints X and Y. The union of Γ and Γ' is a simple closed curve in $V_c(R)$ whose interior contains a connected component of $\mathcal{A}$. Thus $V_c(R)$ is not simply connected, which contradicts lemma 19.2.2. $\qquad\square$

To find the conflicting edges rapidly, the algorithm also maintains an influence graph. We may recall that the influence graph is a structure used for detecting conflicts between the new segment and the regions defined and without conflict over the current subset of segments $\mathcal{S}_c$. The influence graph is an oriented acyclic graph that contains a node for each region that was defined and without conflict over the current subset at some previous incremental step. At each step of the algorithm, the regions defined and without conflict over $\mathcal{S}_c$ are stored in the leaves of the influence graph. The arcs in the influence graph connect two nodes in such a way that a segment that conflicts with the region stored at a node also conflicts with the region stored in one of this node's parents. Thus the influence domain D_E of a region E stored at a node is contained in the union of the influence domains D_{E_i} of the regions E_i stored at this node's parents.

The algorithm proceeds in two phases, first locating the edges or portions of edges to be removed, and then updating the current Voronoi diagram and the corresponding influence graph.

Locating. The location phase aims at retrieving all the leaves in the influence graph that conflict with the new segment S inserted during the current incremental step. This can be achieved by traversing the influence graph from the root only through the nodes that conflict with S.

Updating. The reader is referred to figure 19.9. All the regions that conflict with S, found during the location phase, correspond to edges in the Voronoi

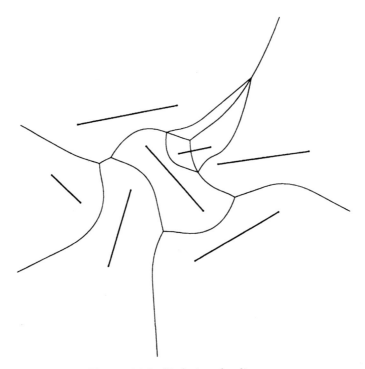

Figure 19.9. Updating the diagram.

diagram that are modified or disappear in the new diagram. Consider such an edge E, that belongs to the bisector of two segments S_1 and S_2. E intersects the region $V(S)$ in the Voronoi diagram of $\mathcal{S}_c \cup \{S\}$. If E does not intersect the boundary of $V(S)$, then E disappears from the new diagram. Otherwise, the intersection points between E and the boundary of $V(S)$ are new vertices in the Voronoi diagram: they are the vertices that are at equal distance from S_1, S_2, and S. Owing to lemma 19.2.4, there are at most two of them for each edge E, and they can be computed in constant time. The at most two portions of E that belong to $V(S)$ disappear from the new diagram while the at most two portions outside $V(S)$ become two new edges. Each of these portions is a new region that becomes a child of E in the influence graph.

It remains to compute the new edges that form the boundary of $V(S)$ in the diagram. These edges have their endpoints at the vertices that we have just computed and they are contained in the edges of $V(S, S_i)$ for some already inserted segments S_i. These edges are computed in counter-clockwise order along the boundary of $V(S)$ by the following operations. Let V be a new vertex that belongs to an edge that is equidistant from S and S_1. V is the endpoint of a new edge E to be created. From V, the next vertex V' of E is found by following the edges in $V(S_1)$ that conflict with S in the previous diagram. (Lemma 19.2.7

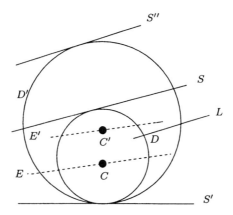

Figure 19.10. The algorithm is correct.

proves that these edges are connected.) V' is equidistant from S, S_1, and S_3. We then create the new edge E; the corresponding node in the influence graph has for parents all the edges visited between V and V'. Starting from V' and following the boundary of $V(S_3)$, we discover a new edge and we repeat this procedure until V is encountered again. Then all the edges on the boundary of $V(S)$ have been created, and the update phase is over.

It is easy to see that the update procedure described above creates all the edges in the Voronoi diagram. It remains to show that it correctly updates the influence graph. For this, we show that a segment that conflicts with a node of the influence graph conflicts with at least one of its parents. Consider for instance a new edge E and a segment L that conflicts with E (see figure 19.10). There exists a maximal disk D centered on E whose interior does not intersect any of the segments inserted, including S, but that intersects L. D is tangent[2] to S and to another segment S'. Let D' be the disk tangent to S' at the same point as D, maximal among those that cut S but no other segment in the current set of segments $\mathcal{S}_c$. Then D' contains D and its center C' belongs to an edge E' of the cell $V(S')$ of the Voronoi diagram of $\mathcal{S}_c$ that conflicts with S. Thus L conflicts with E'. Moreover E' intersects $V(S)$ and C' belongs to $V(S)$. It follows that the portion of $E' \cap V(S)$ that contains C' was traversed during the update phase, and so E received E' as a parent. This finishes the proof of the correctness of the update phase.

To use the results of the analysis of randomized incremental algorithms, we must verify the three clauses of the update condition 5.3.3: detecting a conflict is performed in constant time (condition 1), the number of children of a node must

---

[2]We say that a disk is tangent to a segment if their intersection consists of a single point. It could either be tangent to the line that supports the segment or contain only one of its endpoints.

be bounded by a constant (condition 2), and the update phase can be performed in time proportional to the number of conflicts between S and the edges of the Voronoi diagram of S_c (condition 3).

Condition 1 is clearly satisfied.

As we have seen, an edge of the current Voronoi diagram is split into at most three pieces by the new region $V(S)$: two pieces outside $V(S)$ and one piece inside, or two pieces inside and one piece outside, or one piece inside and one piece outside. To a portion of E outside corresponds a new node in the influence graph that is a child of E. A portion of E inside $V(S)$ belongs to two cells in the previous diagram. It is therefore traversed twice during the update phase and two children are attached to E. The maximum number of children of E is thus five, which shows that condition 2 is satisfied.

For condition 3, we note that the first part of the update phase requires only constant time per conflicting edge. For the second part of the update phase, during which the new edges are created, the incidence graph of the edges in the Voronoi diagram is traversed, and each edge is visited only four times (since each edge has at most two portions in $V(S)$, each visited at most twice). This shows that condition 3 is satisfied.

Finally, the results of theorem 5.3.4 apply, and we must estimate the maximum number $f_0(n, S)$ of edges defined and without conflict over a set of n segments. This number is simply the number of edges in the Voronoi diagram, which is $O(n)$. Theorem 5.3.4 shows that the expected cost of inserting the n-th segment is $O(\log n)$.

Theorem 19.2.8 *The Voronoi diagram of n segments in the plane can be computed by a randomized on-line algorithm in expected time $O(n \log n)$. The expected cost of inserting the n-th segment is $O(\log n)$.*

19.2.4   The case of connected segments

Up to now, we have required that the segments are disjoint. It is very natural to ask for the Voronoi diagram of a polygon, or of a collection of polygons, or more generally of segments that do not intersect except possibly at their endpoints. In such a situation, the set of points at equal distance from two segments may be a region of the plane (see figure 19.11a). We modify the definition of the Voronoi diagram in one of two ways, in order to ensure that the bisectors remain curves. One possibility is to consider the interiors of the segments and their common endpoint as three sites. This leads to diagrams of the kind depicted in figure 19.11b. The other possibility is to consider the interiors of the segments but not their endpoints: this leads to diagrams of the kind depicted in figure 19.11c. This latter definition allows us to consider only disjoint segments and to extend the

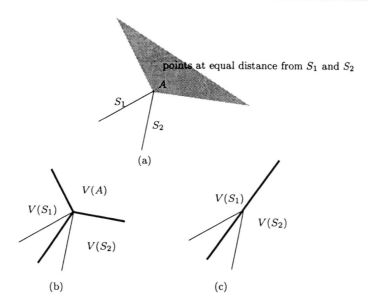

points at equal distance from S_1 and S_2

(a)

(b)

(c)

Figure 19.11. The two possible ways to define the Voronoi diagram of a set of connected segments.

previous results to the case of non-intersecting segments that may share their endpoints. The two diagrams differ, but one may be computed from the other in linear time.

In this section, we show that the general technique of accelerated randomized algorithms (see section 5.4) can be used to compute the Voronoi diagram of a connected set S of n segments with disjoint relative interiors in time $O(n \log^* n)$. The same algorithm can be used to compute the Voronoi diagram of a simple polygon $\mathcal{P}$, which is the portion of the Voronoi diagram of the edges of $\mathcal{P}$ contained in the interior of $\mathcal{P}$. Such a diagram is depicted in figure 19.12.

The algorithm is essentially the same as that presented in the previous subsection. The main difference is that, at certain incremental steps in the construction, the algorithm computes the conflict graph between the regions, i.e. edges of the current Voronoi diagram, and the remaining objects, i.e. segments yet to be inserted. (We recall that the conflict graph is a bipartite graph that stores an arc between a region and the remaining objects that conflict with it.) This graph is used to speed up the subsequent locations. Section 5.4 explains the operations for such algorithms in detail.

The general analysis presented in section 5.4 applies, if we can show how to compute the conflict graph at step r in time $O(n)$, meaning that we must compute all the conflicts between the regions defined and without conflict over the subset $\mathcal{R}$ of segments inserted before or during step r and the segments in $S \setminus \mathcal{R}$. This

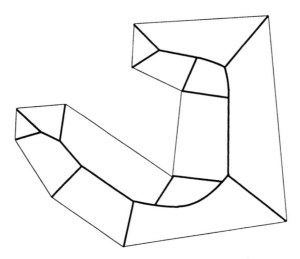

Figure 19.12. The Voronoi diagram of a simple polygon.

can be carried out by traversing the incidence graph of the segments in S. Pick any segment in the sample, say S_0, and find all the edges of $Vor(\mathcal{R})$ that are determined by S_0, in time $O(n)$. Then traverse the incidence graph starting at S_0, say, in a depth-first fashion, and for each segment that has not yet been inserted, enumerate the edges of $Vor(\mathcal{R})$ with which it conflicts.

Denote by S_c the current segment in this traversal of the graph, and let S_p be a segment incident to S_c that has been already visited. Either S_p belongs to $\mathcal{R}$, or we already know the edges of $Vor(\mathcal{R})$ that conflict with S_p. If S_c belongs to $\mathcal{R}$, the traversal proceeds to the next incident segment not yet visited (case 1). Otherwise, we seek an edge of $Vor(\mathcal{R})$ that conflicts with S_c. If S_p belongs to $\mathcal{R}$ (case 2a), we identify, among the edges of $Vor(\mathcal{R})$ determined by S_p (the edges on the boundary of $V(S_p)$), an edge R that conflicts with S_c. If S_p does not belong to $\mathcal{R}$ (case 2b), among all the edges of $Vor(\mathcal{R})$ that conflict with S_p, we look for an edge R that conflicts with S_c. In either case, it is easily checked that we find such an edge R. The other edges of $Vor(\mathcal{R})$ that conflict with S_c are connected by lemma 19.2.7, so they can be found by traversing the incidence graph of the edges of $Vor(\mathcal{R})$, starting at R. At the end of the traversal, all the conflicts between the segments in $S \setminus \mathcal{R}$ and the edges in $Vor(\mathcal{R})$ have been identified.

The cost of processing cases of type 1 is bounded by $r \leq n$. Each edge of $Vor(\mathcal{R})$ is determined by at most four segments in $\mathcal{R}$, so it is examined at most four times during the search for a first conflict in cases 2a. The cost of processing cases 2a is thus $O(r)$. The cost of finding the remaining conflicts or a first conflict in cases 2b is proportional to the number of conflicts detected. Indeed, each vertex of

$Vor(\mathcal{R})$ is incident to three edges, assuming L_2-general position, and we suppose that the incidence relationships are stored explicitly, or at least can be retrieved in constant time so that it takes constant time to find one conflict from another. Theorem 4.2.6 and corollary 4.2.7 show that the expected number of conflicts detected is $O\left(\frac{n-r}{r}f_0(r,\mathcal{S})\right)$, which is $O(n)$.

The hypotheses of theorem 5.4.2 are satisfied, and we may conclude that:

Theorem 19.2.9 *There is an incremental algorithm that computes the Voronoi diagram of a connected set of n segments that have disjoint relative interiors in expected time $O(n \log^* n)$.*

19.2.5 Application to the motion planning of a disk

Let $\mathcal{E}$ be a bounded polygonal region, possibly with polygonal holes. Let n denote the total number of edges in $\mathcal{E}$. By the *Voronoi diagram of $\mathcal{E}$*, we mean the portion of the Voronoi diagram of the edges of $\mathcal{E}$ that is contained in the interior of $\mathcal{E}$. If the Voronoi diagram of $\mathcal{E}$ is precomputed, then a greatest disk contained in $\mathcal{E}$ can be found in linear time. Indeed, such a disk is tangent to three edges in $\mathcal{E}$ (it has only one intersection point with each edge), and hence it is a vertex of the Voronoi diagram.

If our goal is to plan the motion of a disk $\mathcal{D}$ of radius r inside $\mathcal{E}$, then the Voronoi diagram can also be used efficiently. Intuitively, the edges of $\mathcal{E}$ are the obstacles, and the edges of the Voronoi diagram are the positions that are the farthest possible from any obstacle. We may restrict the motion of the disk to these positions rather than the entire domain $\mathcal{E}$: a geometric problem is thus cast into a graph-theoretic problem.

Let X be a point inside $\mathcal{E}$. We map it to a point $\nu(X)$ as follows. If X belongs to an edge of the Voronoi diagram of $\mathcal{E}$, then we put $\nu(X) = X$. Otherwise, X belongs to a single cell in this Voronoi diagram, say $V(S_X)$, attached to a segment S_X. Let X' be the closest point in S_X to X, and draw the ray from X' to X. It follows from lemma 19.2.1 that the segment $X'X$ is contained in the cell, and the ray intersects the boundary of $V(S_X)$ in at most one point. Since the cells are bounded, this point always exists and is our choice for $\nu(X)$. Thus $\nu(X)$ is well-defined.

Equivalently, $\nu(X)$ is the center of the greatest disk contained in $\mathcal{E}$ in the pencil of disks tangent to S_X at X'.

Any point X is mapped onto a single point $\nu(X)$, and the boundary of $V(S_X)$ is connected by lemma 19.2.2. As a consequence, the restriction of $\nu(X)$ to a single cell of the diagram is continuous. As the boundaries of the cells are invariant under ν, ν is continuous over its entire domain of definition.

To each point X inside $\mathcal{E}$ corresponds the distance from X to $\mathcal{E}$, which is the

Figure 19.13. Motion planning of a disk.

same as its distance to X' defined above. It also equals the radius $\rho(X)$ of the greatest disk centered at X that is contained in $\mathcal{E}$.

We say that an arc Γ that connects two points A and B is an *admissible path* if $\mathcal{D}$ moving along this path remains entirely inside $\mathcal{E}$.

Lemma 19.2.10 *If there exists an admissible path γ that connects A to B then there exists an admissible path γ' that consists of the segment $A\nu(A)$, a path from $\nu(A)$ to $\nu(B)$ contained in the edges of the Voronoi diagram, and the segment $\nu(B)B$.*

Proof. Let γ'' be the path obtained by applying ν to all the points in γ. Then γ'' connects $\nu(A)$ to $\nu(B)$ and is contained in the edges of the Voronoi diagram. Moreover, we have $\rho(X) \leq \rho(\nu(X))$, since ρ is non-decreasing on the segment from X to $\nu(X)$. Therefore $\mathcal{D}$ remains inside $\mathcal{E}$ when moving along γ'', showing that γ'' is admissible. If A and B are admissible positions for $\mathcal{D}$, then so are the paths $A\nu(A)$ and $\nu(B)B$. $\square$

The algorithm that computes an admissible path from a position A to another position B proceeds as follows (see figure 19.13). We first test in time $O(n)$ whether A and B are admissible positions, meaning that the disk $\mathcal{D}$ centered at A or B is entirely contained inside $\mathcal{E}$. If this is not the case, then there is no solution to the problem. Otherwise, we compute the Voronoi diagram of $\mathcal{E}$ in time $O(n \log n)$, and remove the portions of the edges for which $\rho(X) < r$ in time $O(n)$. Then we compute $\nu(A)$ and $\nu(B)$ and we find a path γ'' in the incidence graph of the remaining edges of the Voronoi diagram. This also takes time $O(n)$.

Theorem 19.2.11 *Let $\mathcal{D}$ be a disk and $\mathcal{E}$ be a polygonal region with a total of n edges. Knowing the Voronoi diagram of $\mathcal{E}$, one can in time $O(n)$ either find a path from a position A to a position B along which $\mathcal{D}$ remains entirely inside $\mathcal{E}$, if it exists, or otherwise correctly conclude that $\mathcal{D}$ cannot reach B from A while remaining entirely inside $\mathcal{E}$.*

19.3 The case of points distributed in two planes

Studying the Voronoi diagram of points distributed in two planes in $\mathbb{E}^3$ presents a level of difficulty intermediate between the planar case and the three-dimensional case. We will see that such a diagram may have quadratic size, but that it can be computed optimally in time that depends on the size of the input and of the output. This is very interesting in practice, since the quadratic bound is rarely achieved.

19.3.1 The two planes are parallel

Let P_1 and P_2 be two parallel planes in $\mathbb{E}^3$, $\mathcal{M}_1$ a set of n_1 points in P_1, and $\mathcal{M}_2$ a set of n_2 points in P_2. We put $\mathcal{M} = \mathcal{M}_1 \cup \mathcal{M}_2$. To simplify the explanations, we assume that the points are in L_2-general position, meaning that no five points in $\mathcal{M}$ are co-spherical. This also implies that no four points of $\mathcal{M}_1$ or $\mathcal{M}_2$ are co-circular.

The following lemma shows that the planar Delaunay triangulations $\mathcal{D}el(\mathcal{M}_1)$ of $\mathcal{M}_1$ and $\mathcal{D}el(\mathcal{M}_2)$ of $\mathcal{M}_2$ can be computed easily from the three-dimensional Delaunay triangulation $\mathcal{D}el(\mathcal{M})$ of $\mathcal{M}$. Conversely, we show later how to compute $\mathcal{D}el(\mathcal{M})$ knowing $\mathcal{D}el(\mathcal{M}_1)$ and $\mathcal{D}el(\mathcal{M}_2)$.

Lemma 19.3.1 *The intersection of $\mathcal{D}el(\mathcal{M})$ with P_i, $i = 1, 2$, yields exactly $\mathcal{D}el(\mathcal{M}_i)$.*

Proof. $\mathcal{D}el(\mathcal{M})$ is a simplicial 3-complex whose vertices are the points of $\mathcal{M}_1$ (which belong to P_1) and the points of $\mathcal{M}_2$ (which belong to P_2). Its intersection with P_i is thus a 2-complex whose vertices are the points of $\mathcal{M}_i$ ($i = 1, 2$). A face F of $\mathcal{D}el(\mathcal{M}) \cap P_i$ is a face of $\mathcal{D}el(\mathcal{M}_i)$. Indeed, it follows from theorem 17.3.4 that there exists a sphere Σ_F passing through the vertices of F whose interior contains no point in $\mathcal{M}$. The intersection of Σ_F with P_i therefore is a circle that contains the vertices of F and whose interior contains no point of $\mathcal{M}_i$. Theorem 17.3.4, this time used in the plane P_i, shows that F is a face of $\mathcal{D}el(\mathcal{M}_i)$. Conversely, if F is a face of $\mathcal{D}el(\mathcal{M}_i)$, then F can always be circumscribed by a sphere whose interior contains no point of $\mathcal{M}$: simply place the center of this sphere far away from the plane P_j, $j \neq i$. Theorem 17.3.4 then shows that F is a face of $\mathcal{D}el(\mathcal{M})$.

$\square$

Notice that this result does not depend on the distance between the two planes, but it is important that they are parallel for proving the second statement in the lemma.

We now show how to compute $\mathcal{D}el(\mathcal{M})$ knowing $\mathcal{D}el(\mathcal{M}_1)$ and $\mathcal{D}el(\mathcal{M}_2)$. The previous lemma shows that we already have the faces of $\mathcal{D}el(\mathcal{M})$ contained in P_1 and P_2. The following lemma characterizes the others.

Lemma 19.3.2 *The faces of $\mathcal{D}el(\mathcal{M})$ that are not faces of $\mathcal{D}el(\mathcal{M}_1)$ or $\mathcal{D}el(\mathcal{M}_2)$ are in one-to-one correspondence with the 2-complex obtained by projecting the edges of $\mathcal{V}or(\mathcal{M}_1)$ onto $\mathcal{V}or(\mathcal{M}_2)$ orthogonally to P_1 and P_2.*

Proof. Consider two points C_1 in P_1 and C_2 in P_2 that belong to a line orthogonal to P_1 and hence to P_2 (since P_1 and P_2 are parallel). We denote by Σ_i the greatest circle in P_i centered at C_i and whose interior does not contain any point in $\mathcal{M}_i$ ($i = 1, 2$). There exists a unique sphere Σ_{12} that intersects P_1 along Σ_1 and P_2 along Σ_2. By construction, the interior of this sphere does not contain any point in $\mathcal{M}$.

We must distinguish between several cases, according to whether C_1 (resp. C_2) belongs to a cell, to an edge, or is a vertex of $\mathcal{V}or(\mathcal{M}_1)$ (resp. $\mathcal{V}or(\mathcal{M}_2)$).

Case 1. C_1 belongs to the interior of a cell of $\mathcal{V}or(\mathcal{M}_1)$, say $V(A_1)$, and C_2 belongs to the interior of a cell of $\mathcal{V}or(\mathcal{M}_2)$, say $V(A_2)$. Then Σ_1 passes through A_1, Σ_2 passes through A_2, and hence Σ_{12} contains both A_1 and A_2. Theorem 17.3.4 shows that $A_1 A_2$ is an edge in $\mathcal{D}el(\mathcal{M})$.

Reciprocally, let $A_1 A_2$ be an edge in $\mathcal{D}el(\mathcal{M})$. By theorem 17.3.4, there is a sphere Σ_{12} that passes through A_1 and A_2 and whose interior does not contain any point in $\mathcal{M}$. This sphere intersects P_1 and P_2 along two circles Σ_1 and Σ_2 whose interiors do not contain points in $\mathcal{M}_1$ or $\mathcal{M}_2$. The center of Σ_1 thus belongs to $V(A_1)$ and that of Σ_2 belongs to $V(S_2)$.

As a consequence, the orthogonal projection of $V(A_1)$ onto P_2 and $V(A_2)$ have a non-empty intersection if and only if $A_1 A_2$ is an edge of $\mathcal{D}el(\mathcal{M})$. There exists a bijection between the edges of $\mathcal{D}el(\mathcal{M})$ that connect a point of $\mathcal{M}_1$ to a point of $\mathcal{M}_2$ and the cells in the planar subdivision obtained by projecting $\mathcal{V}or(\mathcal{M}_1)$ orthogonally onto $\mathcal{V}or(\mathcal{M}_2)$ in P_2.

Let us denote by $\mathcal{C}$ this planar 2-complex in P_2. Thus $\mathcal{C}$ is the *overlay* of the orthogonal projection of $\mathcal{V}or(\mathcal{M}_1)$ onto P_2 and of $\mathcal{V}or(\mathcal{M}_2)$. In the rest of this proof, the only projection used will be the orthogonal projection onto P_2.

Case 2. C_1 belongs to an edge of $Vor(\mathcal{M}_1)$, say $V(A_1B_1)$, and C_2 belongs to the interior of a cell of $Vor(\mathcal{M}_2)$, say $V(A_2)$. Σ_{12} thus passes through the two points A_1 and B_1 in $\mathcal{M}_1$, and through the point A_2 in $\mathcal{M}_2$. It does not contain any point of $\mathcal{M}$ in its interior by construction, so $A_1B_1A_2$ is a face of $Del(\mathcal{M})$.

The converse can be shown in a manner almost identical to the first case. As a result, the edges of $\mathcal{C}$ that are intersections of the projection of an edge of $Vor(\mathcal{M}_1)$ with a cell of $Vor(\mathcal{M}_2)$ correspond bijectively to the faces in $Del(\mathcal{M})$ that have two vertices in $\mathcal{M}_1$ and one in $\mathcal{M}_2$.

Case 3. C_1 belongs to the interior of a cell of $Vor(\mathcal{M}_1)$, and C_2 belongs to an edge of $Vor(\mathcal{M}_2)$. This case is entirely symmetric to the second case, and establishes a bijection between the edges of $\mathcal{C}$ obtained by intersecting an edge of $Vor(\mathcal{M}_2)$ with the projection of a cell of $Vor(\mathcal{M}_1)$, and the faces in $Del(\mathcal{M})$ that have two vertices in $\mathcal{M}_2$ and one in $\mathcal{M}_1$.

Case 4. C_1 belongs to an edge of $Vor(\mathcal{M}_1)$, say $V(A_1B_1)$, and C_2 belongs to an edge of $Vor(\mathcal{M}_2)$, say A_2B_2. Then Σ_{12} passes through the two points A_1 and B_1 in $\mathcal{M}_1$, and through A_2 and B_2 in $\mathcal{M}_2$. By construction, Σ_{12} does not contain points in $\mathcal{M}$, and thus $A_1B_1A_2B_2$ is a tetrahedron in $Del(\mathcal{M})$.

The converse is also true, showing a bijection between the vertices of $\mathcal{C}$ obtained by intersecting the projection of an edge of $Vor(\mathcal{M}_1)$ with an edge of $Vor(\mathcal{M}_2)$, and the tetrahedra in $Del(\mathcal{M})$ that have two vertices in $\mathcal{M}_1$ and two vertices in $\mathcal{M}_2$.

Case 5. C_1 is a vertex of $Vor(\mathcal{M}_1)$, say $V(A_1B_1C_1)$, and C_2 belongs to the interior of a cell of $Vor(\mathcal{M}_2)$, say $V(A_2)$. Then Σ_{12} passes through three points in $\mathcal{M}_1$ and one point in $\mathcal{M}_2$, and contains no point in $\mathcal{M}$ by construction. Thus $A_1B_1C_1A_2$ is a tetrahedron in $Del(\mathcal{M})$.

The converse is also true, showing a one-to-one correspondence between the vertices of $\mathcal{C}$ that are the projections of vertices of $Vor(\mathcal{M}_1)$ and the tetrahedra in $Del(\mathcal{M})$ that have a facet in P_1.

Case 6. C_1 belongs to the interior of a cell of $Vor(\mathcal{M}_1)$, and C_2 is a vertex of $Vor(\mathcal{M}_2)$. This case is entirely symmetrical to the fifth case, and so provides a bijection between the vertices of $\mathcal{C}$ that are vertices of $Vor(\mathcal{M}_2)$ and the tetrahedra in $Del(\mathcal{M})$ that have a facet in P_2. $\qquad\square$

Figure 19.14 shows a simple example. There are three points in P_1, A_1, B_1, and C_1, and three points in P_2, A_2, B_2, and C_2. $Vor(\mathcal{M}_1)$ and $Vor(\mathcal{M}_2)$ have each one vertex and three edges. $Vor(\mathcal{M})$ contains a vertex corresponding to case 5, namely the center of the sphere circumscribed to $A_1B_1C_1A_2$, a vertex corresponding to case 6, namely the center of the sphere circumscribed to $A_1A_2B_2C_2$, and three vertices corresponding to case 4, the centers of the spheres circumscribed to $A_1C_1A_2C_2$, $A_1C_1B_2C_2$, and $B_1C_1B_2C_2$.

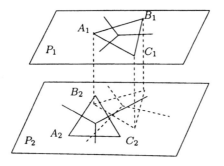

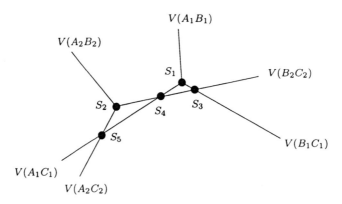

Figure 19.14. Construction of $\mathcal{V}or(\mathcal{M})$ knowing $\mathcal{V}or(\mathcal{M}_1)$ and $\mathcal{V}or(\mathcal{M}_2)$. The triangulation $\mathcal{D}el(\mathcal{M})$ consists of five tetrahedra $S_1 = A_1B_1C_1B_2$, $S_2 = A_1A_2B_2C_2$, $S_3 = B_1C_1B_2C_2$, $S_4 = A_1C_1B_2C_2$, and $S_5 = A_1C_1A_2C_2$.

Lemma 19.3.2 gives a method for computing $\mathcal{V}or(\mathcal{M})$ knowing $\mathcal{V}or(\mathcal{M}_1)$ and $\mathcal{V}or(\mathcal{M}_2)$. In fact, it reduces the problem to that of computing the 2-complex $\mathcal{C}$ that is the overlay of the orthogonal projection of $\mathcal{V}or(\mathcal{M}_1)$ onto P_2 and of $\mathcal{V}or(\mathcal{M}_2)$. We may compute this complex by using any of the randomized algorithms that compute the intersection of a set of line segments presented in subsections 5.2.2 and 5.3.2, or a more sophisticated deterministic algorithm (see the bibliographical notes of chapter 3). The complexity of these algorithms is $O(m \log m + t)$ if m is the number of segments and t the number of intersection points. Here, $m = O(n)$ is the total number of edges in the planar Voronoi diagrams $\mathcal{V}or(\mathcal{M}_1)$ and $\mathcal{V}or(\mathcal{M}_2)$, and t is the number of vertices of $\mathcal{V}or(\mathcal{M})$. This finishes the proof of the following theorem.

Theorem 19.3.3 *Given a set $\mathcal{M}$ of n points that belong to two parallel planes, we can compute its Voronoi diagram $\mathcal{V}or(\mathcal{M})$ in time $O(n \log n + t)$, where t is the number of tetrahedra in $\mathcal{V}or(\mathcal{M})$. This is optimal as a function of the input and output sizes.*

19.3.2 The two planes are not parallel

In this section, we are now interested in a set $\mathcal{M}$ of points in $\mathbb{E}^3$ that belong to two planes P_1 and P_2 that intersect along a line L. We write $\mathcal{M}_i = \mathcal{M} \cap P_i$, $i = 1, 2$, and for simplicity we assume that the points in $\mathcal{M}$ are in L_2-general position.

Let us first assume that the points in $\mathcal{M}_1$ are contained in a half-plane H_1 of P_1 bounded by L, and that the points in $\mathcal{M}_2$ are also contained in a half-space H_2 of P_2 bounded by L. Each plane has an orthonormal system of coordinates, by choosing a common origin O and a common unit vector contained in L, and the other unit vectors in P_1 and P_2 perpendicular to L contained in H_1 and H_2. Thus the equation of H_i is $y_i \geq 0$, $i = 1, 2$.

Lemma 19.3.1 still holds and its proof is unchanged. Lemma 19.3.2 can be generalized as shown in this section. As in section 17.2, we represent the circles in the planes P_1 and P_2 by the points in a three-dimensional Euclidean space, which we denote by $\mathcal{E}$ in order not to confuse it with the Euclidean space that contains $\mathcal{M}$. More precisely, the circle Σ_i in the plane P_i ($i = 1$ or 2) whose center has coordinates C_i in P_i and with respect to which the origin has a power of σ_i (in that plane), is mapped to the point $\phi(\Sigma_i) = (C_i, \sigma_i)$ in $\mathcal{E}$. We identify the x-axes of P_1 and P_2 and the first axis of $\mathcal{E}$, which we also call the x-axis of $\mathcal{E}$, so that there is no confusion between the notation x for the points in $\mathcal{E}$ and in $\mathbb{E}^3$. Also, we identify the y_1-axis of P_1, the y_2-axis of P_2, and the second axis of $\mathcal{E}$, which we call the y-axis. The third axis of $\mathcal{E}$ is naturally called the z-axis. Note that, with the identification of x and y in P_1, P_2, and $\mathcal{E}$, $\mathcal{M}_1$ and $\mathcal{M}_2$ induce two point sets in the xy-plane of $\mathcal{E}$, which we also denote by $\mathcal{M}_1$ and $\mathcal{M}_2$.

We denote by $\mathcal{V}(\mathcal{M}_i)$ ($i = 1, 2$) the Voronoi polytope of $\mathcal{M}_i$. Recall from subsection 17.2.7 that $\mathcal{V}(\mathcal{M}_i)$ is the polytope in $\mathcal{E}$ defined as the intersection of the half-spaces lying above the hyperplanes tangent to the paraboloid $\mathcal{P}$ in $\mathcal{E}$ at the points of $\mathcal{M}$ lifted onto $\mathcal{P}$. $\mathcal{V}(\mathcal{M}_i)$ projects vertically onto P_i along the Voronoi diagram of $\mathcal{M}_i$ in the plane P_i.

Remark. Under the assumption that the points in $\mathcal{M}_1$ are contained in the half-plane H_1 and those of $\mathcal{M}_2$ in H_2, the points of $\mathcal{M}_1$ and $\mathcal{M}_2$ are lifted onto the portion of the paraboloid contained in $y \geq 0$. Hence the projection parallel to the y-axis onto the plane $y = 0$ of the polytope $\mathcal{V}(\mathcal{M}_1)$ covers the entire plane, and so does the projection of $\mathcal{V}(\mathcal{M}_2)$. (This projection map plays an important role below.)

Lemma 19.3.4 *The faces of $\mathcal{D}el(\mathcal{M})$ that are not faces of $\mathcal{D}el(\mathcal{M}_1)$ or $\mathcal{D}el(\mathcal{M}_2)$ can be put in one-to-one correspondence with the faces of the 2-complex C obtained by projecting onto the plane $y = 0$, parallel to the y-axis, the edges of the polytopes $\mathcal{V}(\mathcal{M}_1)$ and $\mathcal{V}(\mathcal{M}_2)$ in $\mathcal{E}$.*

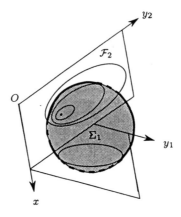

Figure 19.15. The pencil of spheres $\mathcal{F}_2$.

Proof. The proof is analogous to that of lemma 19.3.2, so we only mention how to construct the bijection between the 0-faces of $\mathcal{C}$ and the tetrahedra of $\mathcal{D}el(\mathcal{M})$.

A vertex of $\mathcal{C}$ is either the projection of a vertex of $\mathcal{V}(\mathcal{M}_1)$ or $\mathcal{V}(\mathcal{M}_2)$, or the intersection of the projections of two edges of $\mathcal{V}(\mathcal{M}_1)$ or $\mathcal{V}(\mathcal{M}_2)$.

Case 1. Consider a vertex S_1 of $\mathcal{V}(\mathcal{M}_1)$. S_1 is the image under ϕ of a circle Σ_1 in P_1 that passes through three points A_1, B_1, C_1 in $\mathcal{M}_1$, and the interior of Σ_1 contains no point in $\mathcal{M}_1$. Let $\mathcal{F}$ be the pencil of spheres in $\mathbb{E}^3$ that intersect along Σ_1 (see figure 19.15).

This pencil of spheres intersects P_2 along a pencil of circles $\mathcal{F}_2$ in P_2 whose image under ϕ is the line $\phi(\mathcal{F}_2)$ parallel to the y-axis that contains S_1. Indeed, the radical axis of the circles in $\mathcal{F}_2$ is just $L = P_1 \cap P_2$ since P_1 is the radical plane of $\mathcal{F}$. The line that supports the centers of the circles in $\mathcal{F}_2$ is thus orthogonal to L and so $\phi(\mathcal{F}_2)$ is contained in a plane perpendicular to the x-axis. Since O belongs to L, it has the same power with respect to all the circles in $\mathcal{F}_2$, so $\phi(\mathcal{F}_2)$ is also contained in a plane perpendicular to the z-axis. Moreover, $\phi(\mathcal{F}_2)$ contains S_1, since the center of Σ_1 has the same abscissa as the centers of the spheres in the pencil $\mathcal{F}_2$ and since the power of O with respect to any circle in $\mathcal{F}_2$ equals the power of O with respect to any sphere of $\mathcal{F}$ and hence with respect to Σ_1. This shows that $\phi(\mathcal{F}_2)$ is the line parallel to the y-axis that contains S_1.

The line $\phi(\mathcal{F}_2)$ intersects $\mathcal{V}(\mathcal{M}_2)$ in exactly one point. This point is the image under ϕ of the sphere Σ_2 in $\mathcal{F}_2$ that contains a point of $\mathcal{M}_2$, say A_2, but contains no point of $\mathcal{M}_2$ in its interior. There exists a unique sphere that intersects P_1 along Σ_1 and P_2 along Σ_2, and this sphere belongs to $\mathcal{F}$. This sphere passes through the four points A_1, B_1, C_1 in $\mathcal{M}_1$ and A_2 in $\mathcal{M}_2$. Its interior contains no point of $\mathcal{M}_1$ or $\mathcal{M}_2$. It follows from theorem 17.3.4 that $A_1 B_1 C_1 A_2$ is a tetrahedron in the Delaunay triangulation $\mathcal{D}el(\mathcal{M})$.

Case 2. An analogous discussion establishes a map from the projections of the vertices of $\mathcal{V}(\mathcal{M}_2)$ to the tetrahedra in $\mathcal{D}el(\mathcal{M})$ that have three vertices in P_2 and one in P_1.

Case 3. Consider now a vertex S of $\mathcal{C}$ that is the intersection of the projection of two edges, E_1 in $\mathcal{V}(\mathcal{M}_1)$ and E_2 in $\mathcal{V}(\mathcal{M}_2)$. The line Δ that contains S and is parallel to the y-axis intersects E_1 in exactly one point, say S_1, and intersects E_2 in another point S_2. S_i $(i = 1, 2)$ is the image under ϕ of a circle Σ_1 in P_i that passes through two points of $\mathcal{M}_1$, A_i and B_i, and whose interior in P_i contains no point in $\mathcal{M}_i$. The powers of O with respect to Σ_1 and Σ_2 are equal, so there exists a unique sphere Σ that intersects P_1 along Σ_1 and P_2 along Σ_2. This sphere passes through the four points A_1, B_1, A_2, B_2 and its interior contains no point in $\mathcal{M}_1$ or $\mathcal{M}_2$. Thus $A_1 B_1 A_2 B_2$ is a tetrahedron in the Delaunay triangulation $\mathcal{D}el(\mathcal{M})$.

Converse of cases 1, 2, 3. Let T be a tetrahedron in $\mathcal{D}el(\mathcal{M})$. It is circumscribed by a sphere Σ that intersects P_1 along a circle Σ_1 and P_2 along a circle Σ_2. Neither circle contains points of $\mathcal{M}_1$ or $\mathcal{M}_2$ in its interior. Both circles are mapped by ϕ onto points that belong to a line Δ parallel to the y-axis, since all the points on the x-axis have the same power with respect to Σ_1 and Σ_2. Note that Δ intersects $\mathcal{V}(\mathcal{M}_i)$ in only one point $(i = 1, 2)$.

The tetrahedron T has either three vertices in P_1 and one in P_2 (first case), three vertices in P_2 and one in P_1 (second case), or two vertices in each of P_1 and P_2 (third case). In the first case, Σ_1 passes through three points in $\mathcal{M}_1$ and Σ_2 passes through only one point in $\mathcal{M}_2$. So Δ contains a vertex of $\mathcal{V}(\mathcal{M}_1)$ and intersects the interior of a facet of $\mathcal{V}(\mathcal{M}_2)$. The second case is entirely symmetrical. In the third case, Σ_i contains exactly two points in $\mathcal{M}_i$ $(i = 1, 2)$ so that Δ intersects the relative interiors of two edges, one of $\mathcal{V}(\mathcal{M}_1)$ and the other of $\mathcal{V}(\mathcal{M}_2)$. $\square$

In the preceding discussion, we assumed that the points in $\mathcal{M}_1$ were contained in a half-plane H_1^+ of P_1 bounded by L, and that similarly the points in $\mathcal{M}_2$ were contained in a half-plane H_2^+ of P_2 bounded by L. If this is not true, lemma 19.3.1 collapses and the triangles in $\mathcal{D}el(\mathcal{M}_i)$ are not necessarily all faces in $\mathcal{D}el(\mathcal{M})$, since some triangles in $\mathcal{D}el(\mathcal{M}_i)$ may admit no circumscribed sphere whose interior contains no points in $\mathcal{M}_j$ $(j \neq i)$. Nevertheless, we can salvage our construction by considering successively the four dihedra defined by P_1 and P_2. More precisely, denote by $\mathcal{V}(\mathcal{M}_i)^+$ (resp. $\mathcal{V}(\mathcal{M}_i)^-$) the portion of $\mathcal{V}(\mathcal{M}_i)$ visible from $(0, +\infty, 0)$ (resp. $(0, -\infty, 0)$). Each pair of the kind $(\mathcal{V}(\mathcal{M}_1)^\star, \mathcal{V}(\mathcal{M}_2)^\star)$, where $\star, \star \in \{-, +\}$, corresponds to a dihedron. The previous construction for this dihedron applies to this pair with almost no modification. The only difference comes from the fact that the projection of $\mathcal{V}(\mathcal{M}_1)^\star$ parallel to the y-axis may not cover the plane $y = 0$ entirely, but only a region $R_1^\star$, and similarly the projection of $\mathcal{V}(\mathcal{M}_2)^\star$ parallel to the y-axis may cover only a region $R_2^\star$ of the

plane $y = 0$. (See also the remark preceding lemma 19.3.4). The previous lemma
still holds if the complex C is restricted to the region $R_1^* \cap R_2^*$. It is also worth
observing that a vertex of $\mathcal{V}(\mathcal{M}_i)^+$ or $\mathcal{V}(\mathcal{M}_i)^-$ that does not project onto R_j^+ or
R_j^- $(j \neq i)$ corresponds to a triangle of $\mathcal{D}el(\mathcal{M}_i)$ that is not a face of $\mathcal{D}el(\mathcal{M})$.
A vertex of the resulting internal subdivision is in bijection with a tetrahedron
of some dihedron, meaning that the center of its circumscribed sphere belongs
to that dihedron. Proceeding similarly for all four dihedra retrieves the entire
triangulation $\mathcal{D}el(\mathcal{M})$.

   This finishes the proof of the following theorem.

Theorem 19.3.5 *Given a set $\mathcal{M}$ of n points that belong to two planes, we can
compute their Voronoi diagram $Vor(\mathcal{M})$ in time $O(n \log n + t)$, if t is the size of
this diagram. This is optimal both in the size of the input and in the size of the
output.*

   The previous construction is the strict analogue of the construction for points
distributed in two parallel planes, when embedded in the space that represents
circles. The reader will also notice similarities with section 18.5 devoted to hyperbolic diagrams. Exercise 19.11 explains the reasons for these similarities.

19.4   Exercises

Exercise 19.1 (Divide-and-conquer) Adapt the divide-and-conquer method to compute the Voronoi diagram of n points in the plane in optimal time $O(n \log n)$.

Exercise 19.2 (Optimal L_1 diagrams) Devise an optimal deterministic algorithm
that computes the Voronoi diagram of n points in the plane for the L_1 distance in
time $O(n \log n)$. Assume that the points are in L_1-general position.

Hint: Use either the divide-and-conquer method, or the sweep method.

Exercise 19.3 (Polygonal distance) Let $\mathcal{P}$ be a polygon that contains the origin O.
We denote by $\lambda \mathcal{P}$ the image of this polygon under the homothety centered at O and
of ratio λ. The polygonal distance $\delta_{\mathcal{P}}(X, A)$ from point X to point A is defined as the
smallest real $\lambda \geq 0$ such that $X - A$ belongs to $\lambda \mathcal{P}$. Show that the Voronoi diagram of n
points under this polygonal distance has complexity $O(np)$ if p is the number of vertices
of $\mathcal{P}$ and that it may be computed in time $O(np \log(np))$.

Exercise 19.4 (Convex polygon) Show how to compute the Voronoi diagram of a
convex polygon in the plane with a randomized algorithm in expected time $O(n)$.

Hint: Maintain the convex hull of the polygon while pulling off the vertices one by one
and in a random order. Then insert the points back into the polygon in the reverse
order, while maintaining the Voronoi diagram. During each insertion, we already know

a region (half-plane) that conflicts with the vertex to be inserted. The other conflicts can be retrieved and the structure updated, without paying the logarithmic cost of the location phase in the usual incremental algorithm.

Exercise 19.5 (Dual of the Voronoi diagram of segments) Consider a set of segments in the plane, assumed to be in L_2-general position. A vertex of the Voronoi diagram is incident to two arcs if the arc is contained in the relative interior of an edge, or to three arcs if it is an endpoint of an edge. For each vertex contained in two arcs, the maximal circle centered on this vertex that does not properly intersect the segments is tangent to two segments, and we draw the edge that joins the two points of tangency. Show that these edges partition the convex hull of the segments into $O(n)$ regions which are either triangles or trapezoids.

Exercise 19.6 (Dynamic algorithm) Devise a dynamic randomized algorithm to compute the Voronoi diagram of a set of segments in the plane.

Exercise 19.7 (Constrained Delaunay triangulation) Consider a set S of segments in the plane with disjoint relative interiors (some may be reduced to a point). We denote by $\mathcal{E}$ the collection of their endpoints. A constrained triangulation of S is a triangulation of $\mathcal{E}$ that contains all segments in S as edges. This triangulation is a *constrained Delaunay triangulation* if the interior of a circle circumscribed to some triangle T contains no point that is visible from all three vertices of T. Two points are visible if the segment that connects them does not intersect the relative interior of any segment in S. Show that such a triangulation exists for any S, and devise an algorithm that computes it in time $O(n \log n)$.

Exercise 19.8 (Motion planning of a polygon) Let $\mathcal{P}$ be a convex polygon with m edges. Extend the notion of a Voronoi diagram for the polygonal metric defined by $\mathcal{P}$ (see exercises 19.3 and 19.2) to the metric defined by a bounded polygonal region $\mathcal{R}$. If $\mathcal{R}$ totals n edges, show how to compute the greatest homothet of $\mathcal{P}$ contained in $\mathcal{R}$ in time $O(mn \log(mn))$. If I and J are two points in $\mathcal{R}$, show how to determine whether there exists a path from I to J such that the polygon $\mathcal{P}$ moving on this path by translation remains entirely contained in $\mathcal{R}$. If so, show how to compute this path.

Exercise 19.9 (Segments in $\mathbb{E}^3$) Characterize the surfaces bisecting two segments in $\mathbb{E}^3$.

Exercise 19.10 (Segments in two planes) Show that the complexity of the Voronoi diagram of a set of n segments distributed in two planes is $t = O(n^2)$. Show how to compute such a diagram in time $O(n \log n + t)$. Extend these results to polygonal regions distributed in two planes.

Exercise 19.11 (Two non-parallel planes) Consider a set $\mathcal{M}$ of points in $\mathbb{E}^3$ distributed on two half-planes P_1 and P_2 bounded by a common line L. Denote by $\mathcal{M}_1$ and $\mathcal{M}_2$ the sets of points in either plane. Show that the tetrahedra of $\mathcal{D}el(\mathcal{M})$ circumscribed by a sphere that does not intersect L are in one-to-one correspondence with the vertices of the subdivision obtained by overlaying the hyperbolic diagram $\mathcal{V}or_h(\mathcal{M}_2)$ onto the hyperbolic diagram $\mathcal{V}or_h(\mathcal{M}_1)$, with a rotation about the axis L that maps P_2 onto P_1.

Hint: Adapt the discussion in section 19.3 (for two non-parallel planes) using arguments borrowed from section 18.5.

Exercise 19.12 (The case of several planes) Show how to compute the Delaunay triangulation of a set $\mathcal{M}$ of n points in $\mathbb{E}^3$ distributed in k given planes in time $O(kn \log n)$.

Hint: Compute the triangulation by a greedy method, finding the tetrahedra one by one. If F is a facet that belongs to a single computed tetrahedron T, we seek the tetrahedron T' that shares the facet F with T. Notice that the spheres circumscribed to T and T' are the two greatest spheres in the pencil of spheres that contain the circle circumscribed to F and whose interiors contain no points of $\mathcal{M}$. This pencil intersects each of the k planes along a pencil of circles. In each plane P_i, we seek the greatest circle in the corresponding pencil whose interior contains no point in $\mathcal{M}_i = \mathcal{M} \cap P_i$. This problem reduces to finding the intersection between a line and a polytope in $\mathbb{E}^3$, which can be solved in logarithmic time (see exercise 9.6). Among all k candidates, pick the one that gives the fourth vertex of T'.

Exercise 19.13 (Conformal triangulation) Consider a set $\mathcal{S}$ of n segments in the plane, and let $\mathcal{E}$ denote the set of their endpoints. Show that by adding a finite set $\mathcal{P}$ of points on the segments of $\mathcal{S}$, each segment becomes an edge of the Delaunay triangulation of $\mathcal{E} \cup \mathcal{P}$. Show that one may have to add $\Omega(n^2)$ points to $\mathcal{E}$ in the worst case.

19.5   Bibliographical notes

The sweep algorithm described in section 19.1 and its extension to the case of additive weights and line segments is due to Fortune [106]. Its generalization to L_1 and L_∞ distances is described by Shute, Deneen, and Thomborson in [209].

The incremental algorithm that computes the Voronoi diagram of a set of line segments in the plane is presented by Boissonnat, Devillers, Schott, Teillaud, and Yvinec [28], and is dynamized by Dobrindt and Yvinec [86] (see exercise 19.6). The analysis of the case when the segments are connected is due to Devillers [80]. The case when the segments are the edges of a convex polygon is treated by Aggarwal, Guibas, Saxe, and Shor who propose a deterministic algorithm [5]. The remarkably simple randomized solution presented in exercise 19.4 is due to Chew [59].

Optimal algorithms that compute the constrained Delaunay triangulation of a set of points (see exercise 19.7) are described by Chew [60] and by Wang and Schubert [217]. A randomized linear time algorithm by Klein and Lingas computes the constrained triangulation of a polygon [140].

Conformal triangulations were used first by Boissonnat [25]. Edelsbrunner and Tan [100] give a polynomial solution to exercise 19.13.

Ó'Dúnlaing and Yap were the first to use the Voronoi diagram for planning the motion of a disk. The case when the robot is polygonal rather than circular, and can move by translation and rotation in the plane (with three degrees of freedom), is studied by Ó'Dúnlaing, Sharir, and Yap in [180, 181] and by Chew and Kedem in [62].

Bounding the complexity of the Voronoi diagram of a set of line segments in $\mathbb{E}^3$ is a long-standing open question. Related results for the L_1 and L_∞ metrics were established by Chew, Kedem, Sharir, Tagansky, and Welzl [63].

The case of points distributed in two parallel planes has been studied by Boissonnat in the context of three-dimensional shape reconstruction from cross-sections in tomography [25] and generalized to two non-parallel planes and to a (small) number of planes (see exercise 19.12) by Boissonnat, Cérézo, Devillers, and Teillaud [26].

References

[1] P. K. Agarwal. *Intersection and Decomposition Algorithms for Planar Arrangements*. Cambridge University Press, New York, NY, 1991.

[2] P. K. Agarwal, M. de Berg, J. Matoušek, and O. Schwarzkopf. Constructing levels in arrangements and higher order Voronoi diagrams. In *Proc. 10th Ann. ACM Symp. Comp. Geom.*, 67–75, 1994.

[3] P. K. Agarwal, D. Eppstein, and J. Matoušek. Dynamic half-space reporting, geometric optimization, and minimum spanning trees. In *Proc. 33rd Ann. IEEE Symp. Found. Comp. Sci.*, 80–89, 1992.

[4] P. K. Agarwal and J. Matoušek. Ray shooting and parametric search. In *Proc. 24th Ann. ACM Symp. Theory Comp.*, 517–526, 1992.

[5] A. Aggarwal, L. J. Guibas, J. Saxe, and P. W. Shor. A linear-time algorithm for computing the Voronoi diagram of a convex polygon. *Discrete Comp. Geom.*, 4(6):591–604, 1989.

[6] A. V. Aho, J. E. Hopcroft, and J. D. Ullman. *The Design and Analysis of Computer Algorithms*. Addison-Wesley, Reading, MA, 1974.

[7] P. Alevizos, J.-D. Boissonnat, and F. P. Preparata. An optimal algorithm for the boundary of a cell in a union of rays. *Algorithmica*, 5:573–590, 1990.

[8] P. D. Alevizos, J.-D. Boissonnat, and M. Yvinec. Non-convex contour reconstruction. *J. Symbolic Comp.*, 10:225–252, 1990.

[9] B. Aronov and M. Sharir. Triangles in space or building (and analyzing) castles in the air. *Combinatorica*, 10(2):137–173, 1990.

[10] B. Aronov and M. Sharir. Castles in the air revisited. In *Proc. 8th Ann. ACM Symp. Comp. Geom.*, 146–156, 1992.

[11] B. Aronov, M. Pellegrini, and M. Sharir. On the zone of a surface in a hyperplane arrangement. *Discrete Comp. Geom.*, 9(2):177–186, 1993.

[12] P. Ash, E. Bolker, H. Crapo, and W. Whiteley. Convex polyhedra, Dirichlet tessellations, and spider webs. In Marjorie Senechal and George Fleck, editors, *Shaping Space: A Polyhedral Approach*, chapter 17, 231–250. Birkhäuser, Boston, MA, 1988.

[13] M. J. Atallah. Dynamic computational geometry. In *Proc. 24th Ann. IEEE Symp. Found. Comp. Sci.*, 92–99, 1983.

[14] F. Aurenhammer. Power diagrams: properties, algorithms and applications. *SIAM J. Comp.*, 16:78–96, 1987.

[15] F. Aurenhammer. A new duality result concerning Voronoi diagrams. *Discrete Comp. Geom.*, 5:243–254, 1990.

[16] F. Aurenhammer. Voronoi diagrams: A survey of a fundamental geometric data structure. *ACM Comp. Surv.*, 23:345–405, 1991.

[17] F. Aurenhammer and H. Imai. Geometric relations among Voronoi diagrams. *Geom. Dedicata*, 27:65–75, 1988.

[18] F. Aurenhammer and O. Schwarzkopf. A simple on-line randomized incremental algorithm for computing higher order Voronoi diagrams. *Internat. J. Comp. Geom. Appl.*, 2:363–381, 1992.

[19] D. Avis and H. El-Gindy. Triangulating point sets in space. *Discrete Comp. Geom.*, 2:99–111, 1987.

[20] C. Bajaj and T. K. Dey. Polygon nesting and robustness. *Inf. Proc. Lett.*, 35:23–32, 1990.

[21] L. J. Bass and S. R. Schubert. On finding the disc of minimum radius containing a given set of points. *Math. Comp.*, 12:712–714, 1967.

[22] A. F. Beardon. *The Geometry of Discrete Groups*. Graduate Texts in Mathematics. Springer-Verlag, 1983.

[23] J. L. Bentley and T. A. Ottmann. Algorithms for reporting and counting geometric intersections. *IEEE Trans. Comp.*, C-28:643–647, 1979.

[24] M. Berger. *Géométrie 3. Convexes et Polytopes, Polyèdres Réguliers, Aires et Volumes*. Fernand Nathan, Paris, 1974.

[25] J.-D. Boissonnat. Shape reconstruction from planar cross-sections. *Comp. Vision Graph. Image Proc.*, 44(1):1–29, October 1988.

[26] J.-D. Boissonnat, A. Cérézo, O. Devillers, and M. Teillaud. Output-sensitive construction of the Delaunay triangulation of points lying in two planes. *Internat. J. Comp. Geom. Appl.*, 6(1):1–14, 1996.

[27] J.-D. Boissonnat, O. Devillers, and F. Preparata. Computing the union of 3-colored triangles. *Internat. J. Comp. Geom. Appl.*, 1(2):187–196, 1991.

[28] J.-D. Boissonnat, O. Devillers, R. Schott, M. Teillaud, and M. Yvinec. Applications of random sampling to on-line algorithms in computational geometry. *Discrete Comp. Geom.*, 8:51–71, 1992.

[29] J.-D. Boissonnat, O. Devillers, and M. Teillaud. A semidynamic construction of higher-order Voronoi diagrams and its randomized analysis. *Algorithmica*, 9:329–356, 1993.

[30] J.-D. Boissonnat and K. Dobrindt. On-line construction of the upper envelope of triangles and surface patches in three dimensions. *Comp. Geom. Theory Appl.*, 5:303–320, 1996.

[31] J.-D. Boissonnat and M. Teillaud. A hierarchical representation of objects: The Delaunay tree. In *Proc. 2nd Ann. ACM Symp. Comp. Geom.*, 260–268, 1986.

[32] J.-D. Boissonnat and M. Teillaud. On the randomized construction of the Delaunay tree. *Theor. Comp. Sci.*, 112:339–354, 1993.

[33] J.-D. Boissonnat and M. Yvinec. Probing a scene of non-convex polyhedra. *Algorithmica*, 8:321–342, 1992.

[34] J.-D. Boissonnat, M. Sharir, B. Tagansky, and M. Yvinec. Voronoi diagrams in higher dimensions under certain polyhedral distance functions. In *Proc. 11th Ann. ACM Symp. Comp. Geom.*, 79–88, 1995.

[35] H. Brönnimann. *Derandomization of Geometric Algorithms*. Ph.D. thesis, Dept. Comp. Sci., Princeton University, Princeton, NJ, May 1995.

[36] H. Brönnimann, B. Chazelle, and J. Matoušek. Product range spaces, sensitive sampling, and derandomization. In *Proc. 34th Ann. IEEE Symp. Found. Comp. Sci.*, 400–409, 1993.

[37] A. Brönsted. *An Introduction to Convex Polytopes*. Springer-Verlag, New York, NY, 1983.

[38] K. Q. Brown. Voronoi diagrams from convex hulls. *Inf. Proc. Lett.*, 9:223–228, 1979.

[39] H. Bruggesser and P. Mani. Shellable decompositions of cells and spheres. *Math. Scand.*, 29:197–205, 1971.

[40] C. Burnikel, K. Mehlhorn, and S. Schirra. On degeneracy in geometric computations. In *Proc. 5th ACM-SIAM Symp. Discrete Algorithms*, 16–23, 1994.

[41] B. Chazelle. A theorem on polygon cutting with applications. In *Proc. 23rd Ann. IEEE Symp. Found. Comp. Sci.*, 339–349, 1982.

[42] B. Chazelle. Convex partitions of polyhedra: a lower bound and worst-case optimal algorithm. *SIAM J. Comp.*, 13:488–507, 1984.

[43] B. Chazelle. On the convex layers of a planar set. *IEEE Trans. Inform. Theory*, IT-31:509–517, 1985.

[44] B. Chazelle. Triangulating a simple polygon in linear time. *Discrete Comp. Geom.*, 6:485–524, 1991.

[45] B. Chazelle. An optimal algorithm for intersecting three-dimensional convex polyhedra. *SIAM J. Comp.*, 21(4):671–696, 1992.

[46] B. Chazelle. An optimal convex hull algorithm in any fixed dimension. *Discrete Comp. Geom.*, 10:377–409, 1993.

[47] B. Chazelle, R. L. Drysdale, III, and D. T. Lee. Computing the largest empty rectangle. *SIAM J. Comp.*, 15:300–315, 1986.

[48] B. Chazelle and H. Edelsbrunner. An optimal algorithm for intersecting line segments in the plane. In *Proc. 29th Ann. IEEE Symp. Found. Comp. Sci.*, 590–600, 1988.

[49] B. Chazelle and H. Edelsbrunner. An optimal algorithm for intersecting line segments in the plane. *JACM*, 39:1–54, 1992.

[50] B. Chazelle, H. Edelsbrunner, M. Grigni, L. Guibas, J. Hershberger, M. Sharir, and J. Snoeyink. Ray shooting in polygons using geodesic triangulations. *Algorithmica*, 12:54–68, 1994.

[51] B. Chazelle, H. Edelsbrunner, L. Guibas, and M. Sharir. A singly-exponential stratification scheme for real semi-algebraic varieties and its applications. *Theor. Comp. Sci.*, 84:77–105, 1991.

[52] B. Chazelle, H. Edelsbrunner, L. Guibas, M. Sharir, and J. Snoeyink. Computing a face in an arrangement of line segments and related problems. *SIAM J. Comp.*, 22:1286–1302, 1993.

[53] B. Chazelle and J. Friedman. A deterministic view of random sampling and its use in geometry. *Combinatorica*, 10(3):229–249, 1990.

[54] B. Chazelle and L. J. Guibas. Visibility and intersection problems in plane geometry. *Discrete Comp. Geom.*, 4:551–581, 1989.

[55] B. Chazelle, L. J. Guibas, and D. T. Lee. The power of geometric duality. *BIT*, 25:76–90, 1985.

[56] B. Chazelle and J. Incerpi. Triangulation and shape-complexity. *ACM Trans. Graph.*, 3(2):135–152, 1984.

[57] B. Chazelle and J. Matoušek. On linear-time deterministic algorithms for optimization problems in fixed dimension. In *Proc. 4th ACM-SIAM Symp. Discrete Algorithms*, 281–290, 1993.

[58] B. Chazelle and L. Palios. Triangulating a non-convex polytope. *Discrete Comp. Geom.*, 5:505–526, 1990.

[59] L. P. Chew. Building Voronoi diagrams for convex polygons in linear expected time. Technical Report PCS-TR90-147, Dept. Math. Comp. Sci., Dartmouth College, Hanover, NH, 1986.

[60] L. P. Chew. Constrained Delaunay triangulations. *Algorithmica*, 4:97–108, 1989.

[61] L. P. Chew and R. L. Drysdale, III. Voronoi diagrams based on convex distance functions. In *Proc. 1st Ann. ACM Symp. Comp. Geom.*, 235–244, 1985.

[62] L. P. Chew and K. Kedem. Placing the largest similar copy of a convex polygon among polygonal obstacles. In *Proc. 5th Ann. ACM Symp. Comp. Geom.*, 167–174, 1989.

[63] L. P. Chew, K. Kedem, M. Sharir, B. Tagansky, and E. Welzl. Voronoi diagrams of lines in 3-space under polyhedral convex distance functions. In *Proc. 6th ACM-SIAM Symp. Discrete Algorithms*, 197–204, 1995.

[64] K. Clarkson, H. Edelsbrunner, L. Guibas, M. Sharir, and E. Welzl. Combinatorial complexity bounds for arrangements of curves and spheres. *Discrete Comp. Geom.*, 5:99–160, 1990.

[65] K. Clarkson, R. E. Tarjan, and C. J. Van Wyk. A fast Las Vegas algorithm for triangulating a simple polygon. *Discrete Comp. Geom.*, 4:423–432, 1989.

[66] K. L. Clarkson. Linear programming in $O(n3^{d^2})$ time. *Inf. Proc. Lett.*, 22:21–24, 1986.

[67] K. L. Clarkson. New applications of random sampling in computational geometry. *Discrete Comp. Geom.*, 2:195–222, 1987.

[68] K. L. Clarkson. A Las Vegas algorithm for linear programming when the dimension is small. In *Proc. 29th Ann. IEEE Symp. Found. Comp. Sci.*, 452–456, 1988.

[69] K. L. Clarkson, R. Cole, and R. E. Tarjan. Randomized parallel algorithms for trapezoidal diagrams. *Internat. J. Comp. Geom. Appl.*, 2(2):117–133, 1992.

[70] K. L. Clarkson, K. Mehlhorn, and R. Seidel. Four results on randomized incremental constructions. *Comp. Geom. Theory Appl.*, 3(4):185–212, 1993.

[71] K. L. Clarkson and P. W. Shor. Applications of random sampling in computational geometry, II. *Discrete Comp. Geom.*, 4:387–421, 1989.

[72] T. H. Cormen, C. E. Leiserson, and R. L. Rivest. *Introduction to Algorithms*. The MIT Press, Cambridge, Mass., 1990.

[73] H. S. M. Coxeter. A classification of zonohedra by means of projective diagrams. *J. Math. Pure Appl.*, 41:137–156, 1962.

[74] H. Davenport and A. Schinzel. A combinatorial problem connected with differential equations. *Amer. J. Math.*, 87:684–689, 1965.

[75] M. de Berg. *Ray Shooting, Depth Orders and Hidden Surface Removal*, volume 703 of *Lecture Notes in Computer Science*. Springer-Verlag, Berlin, Germany, 1993.

[76] M. de Berg, K. Dobrindt, and O. Schwarzkopf. On lazy randomized incremental construction. In *Proc. 26th Ann. ACM Symp. Theory Comp.*, 105–114, 1994.

[77] M. de Berg, L. J. Guibas, and D. Halperin. Vertical decompositions for triangles in 3-space. *Discrete Comp. Geom.*, 15:35–61, 1996.

[78] B. Delaunay. Sur la sphère vide. A la memoire de Georges Voronoi. *Izv. Akad. Nauk SSSR, Otdelenie Matematicheskih i Estestvennyh Nauk*, 7:793–800, 1934.

[79] P. Desnoguès. *Triangulations et Quadriques*. Thèse de doctorat en sciences, Université de Nice-Sophia Antipolis, France, 1996.

[80] O. Devillers. Randomization yields simple $O(n \log^* n)$ algorithms for difficult $\Omega(n)$ problems. *Internat. J. Comp. Geom. Appl.*, 2(1):97–111, 1992.

[81] O. Devillers, S. Meiser, and M. Teillaud. Fully dynamic Delaunay triangulation in logarithmic expected time per operation. *Comp. Geom. Theory Appl.*, 2(2):55–80, 1992.

[82] O. Devillers, S. Meiser, and M. Teillaud. The space of spheres, a geometric tool to unify duality results on Voronoi diagrams. Report 1620, INRIA Sophia-Antipolis, Valbonne, France, 1992.

[83] M. T. Dickerson, S. A. McElfresh, and M. Montague. New algorithms and empirical findings on minimum weight triangulation heuristics. In *Proc. 11th Ann. ACM Symp. Comp. Geom.*, 238–247, 1995.

[84] M. Dietzfelbinger, A. Karlin, K. Mehlhorn, F. Meyer auf der Heide, H. Rohnert, and R. E. Tarjan. Dynamic perfect hashing – upper and lower bounds. In *Proc. 29th Ann. IEEE Symp. Found. Comp. Sci.*, 524–531, 1988.

[85] D. P. Dobkin and D. G. Kirkpatrick. A linear algorithm for determining the separation of convex polyhedra. *J. Algorithms*, 6:381–392, 1985.

[86] K. Dobrindt and M. Yvinec. Remembering conflicts in history yields dynamic algorithms. In *Proc. 4th Ann. Internat. Symp. Algorithms Comp. (ISAAC 93)*, volume 762 of *Lecture Notes in Computer Science*, 21–30. Springer-Verlag, 1993.

[87] M. E. Dyer. Linear time algorithms for two- and three-variable linear programs. *SIAM J. Comp.*, 13:31–45, 1984.

[88] M. E. Dyer. On a multidimensional search technique and its application to the Euclidean one-centre problem. *SIAM J. Comp.*, 15:725–738, 1986.

[89] H. Edelsbrunner. *Algorithms in Combinatorial Geometry*, volume 10 of *EATCS Monographs on Theoretical Computer Science*. Springer-Verlag, Heidelberg, West Germany, 1987.

[90] H. Edelsbrunner, L. Guibas, J. Pach, R. Pollack, R. Seidel, and M. Sharir. Arrangements of curves in the plane: Topology, combinatorics, and algorithms. *Theor. Comp. Sci.*, 92:319–336, 1992.

[91] H. Edelsbrunner, L. Guibas, and M. Sharir. The complexity of many cells in arrangements of planes and related problems. *Discrete Comp. Geom.*, 5:197–216, 1990.

[92] H. Edelsbrunner and L. J. Guibas. Topologically sweeping an arrangement. *J. Comp. Syst. Sci.*, 38:165–194, 1989. Corrigendum in 42 (1991), 249–251.

[93] H. Edelsbrunner, L. J. Guibas, and M. Sharir. The complexity and construction of many faces in arrangements of lines and of segments. *Discrete Comp. Geom.*, 5:161–196, 1990.

[94] H. Edelsbrunner and H. A. Maurer. Finding extreme points in three dimensions and solving the post-office problem in the plane. *Inf. Proc. Lett.*, 21:39–47, 1985.

[95] H. Edelsbrunner, F. P. Preparata, and D. B. West. Tetrahedrizing point sets in three dimensions. *J. Symb. Comp.*, 10(3–4):335–347, 1990.

[96] H. Edelsbrunner and R. Seidel. Voronoi diagrams and arrangements. *Discrete Comp. Geom.*, 1:25–44, 1986.

[97] H. Edelsbrunner, R. Seidel, and M. Sharir. On the zone theorem for hyperplane arrangements. *SIAM J. Comp.*, 22(2):418–429, 1993.

[98] H. Edelsbrunner and N. R. Shah. Incremental topological flipping works for regular triangulations. *Algorithmica*, 15:223–241, 1996.

[99] H. Edelsbrunner and W. Shi. An $O(n \log^2 h)$ time algorithm for the three-dimensional convex hull problem. *SIAM J. Comp.*, 20:259–277, 1991.

[100] H. Edelsbrunner and T. S. Tan. An upper bound for conforming Delaunay triangulations. In *Proc. 8th Ann. ACM Symp. Comp. Geom.*, 53–62, 1992.

[101] H. Edelsbrunner and T. S. Tan. A quadratic time algorithm for the minmax length triangulation. *SIAM J. Comp.*, 22:527–551, 1993.

[102] H. Edelsbrunner, T. S. Tan, and R. Waupotitsch. An $O(n^2 \log n)$ time algorithm for the minmax angle triangulation. *SIAM J. Sci. Statist. Comp.*, 13(4):994–1008, 1992.

[103] P. Erdős, L. Lovász, A. Simmons, and E. Straus. Dissection graphs of planar point sets. In J. N. Srivastava, editor, *A Survey of Combinatorial Theory*, 139–154. North-Holland, Amsterdam, Netherlands, 1973.

[104] H. Everett, J.-M. Robert, and M. van Kreveld. An optimal algorithm for the $(\leq k)$-levels, with applications to separation and transversal problems. *Internat. J. Comp. Geom. Appl.*, 6:247–261, 1996.

[105] S. Fortune. Voronoi diagrams and Delaunay triangulations. In D.-Z. Du and F. K. Hwang, editors, *Computing in Euclidean Geometry*, volume 1 of *Lecture Notes Series on Computing*, 193–233. World Scientific, Singapore, 1992.

[106] S. J. Fortune. A sweepline algorithm for Voronoi diagrams. *Algorithmica*, 2:153–174, 1987.

[107] A. Fournier and D. Y. Montuno. Triangulating simple polygons and equivalent problems. *ACM Trans. Graph.*, 3(2):153–174, 1984.

[108] C. Froidevaux, M. C. Gaudel, and M. Soria. *Type de Données et Algorithmes*. McGraw-Hill, Paris, 1990.

[109] M. R. Garey, D. S. Johnson, F. P. Preparata, and R. E. Tarjan. Triangulating a simple polygon. *Inf. Proc. Lett.*, 7:175–179, 1978.

[110] P. J. Giblin. *Graphs, Surfaces and Homology*. Chapman and Hall, London, 1977.

[111] R. L. Graham. An efficient algorithm for determining the convex hull of a finite planar set. *Inf. Proc. Lett.*, 1:132–133, 1972.

[112] R. L. Graham and F. F. Yao. Finding the convex hull of a simple polygon. *J. Algorithms*, 4:324–331, 1983.

[113] P. J. Green and R. R. Sibson. Computing Dirichlet tessellations in the plane. *Comp. J.*, 21:168–173, 1978.

[114] B. Grünbaum. *Convex Polytopes*. Wiley, New York, NY, 1967.

[115] L. J. Guibas, D. Halperin, J. Matoušek, and M. Sharir. On vertical decomposition of arrangements of hyperplanes in four dimensions. In *Proc. 5th Canad. Conf. Comp. Geom.*, 127–132, Waterloo, Canada, 1993.

[116] L. J. Guibas, J. Hershberger, D. Leven, M. Sharir, and R. E. Tarjan. Linear-time algorithms for visibility and shortest path problems inside triangulated simple polygons. *Algorithmica*, 2:209–233, 1987.

[117] L. J. Guibas, D. E. Knuth, and M. Sharir. Randomized incremental construction of Delaunay and Voronoi diagrams. *Algorithmica*, 7:381–413, 1992.

[118] L. J. Guibas, M. Sharir, and S. Sifrony. On the general motion planning problem with two degrees of freedom. *Discrete Comp. Geom.*, 4:491–521, 1989.

[119] D. Halperin and M. Sharir. Near-quadratic bounds for the motion planning problem for a polygon in a polygonal environment. In *Proc. 34th Ann. IEEE Symp. Found. Comp. Sci. (FOCS 93)*, 382–391, 1993.

[120] D. Halperin and M. Sharir. New bounds for lower envelopes in three dimensions, with applications to visibility in terrains. *Discrete Comp. Geom.*, 12:313–326, 1994.

[121] D. Halperin and M. Sharir. Almost tight upper bounds for the single cell and zone problems in three dimensions. In *Proc. 10th Ann. ACM Symp. Comp. Geom.*, 11–20, 1994.

[122] S. Hart and M. Sharir. Nonlinearity of Davenport–Schinzel sequences and of generalized path compression schemes. *Combinatorica*, 6:151–177, 1986.

[123] D. Haussler and E. Welzl. Epsilon-nets and simplex range queries. *Discrete Comp. Geom.*, 2:127–151, 1987.

[124] J. Hershberger. Finding the upper envelope of n line segments in $O(n \log n)$ time. *Inf. Proc. Lett.*, 33:169–174, 1989.

[125] J. Hershberger and S. Suri. Applications of a semi-dynamic convex hull algorithm. In *Proc. 2nd Scand. Workshop Algorithm Theory*, volume 447 of *Lecture Notes in Computer Science*, 380–392. Springer-Verlag, 1990.

[126] J. Hershberger and S. Suri. Offline maintenance of planar configurations. In *Proc. 2nd ACM-SIAM Symp. Discrete Algorithms*, 32–41, 1991.

[127] J. Hershberger and S. Suri. Efficient computation of Euclidean shortest paths in the plane. In *Proc. 34th Ann. IEEE Symp. Found. Comp. Sci. (FOCS 93)*, 508–517, 1993.

[128] S. Hertel and K. Mehlhorn. Fast triangulation of simple polygons. In *Proc. 4th Internat. Conf. Found. Comp. Theory*, volume 158 of *Lecture Notes in Computer Science*, 207–218. Springer-Verlag, 1983.

[129] H. Imai, M. Iri, and K. Murota. Voronoi diagrams in the Laguerre geometry and its applications. *SIAM J. Comp.*, 14:93–105, 1985.

[130] R. A. Jarvis. On the identification of the convex hull of a finite set of points in the plane. *Inf. Proc. Lett.*, 2:18–21, 1973.

[131] B. Joe. Construction of three-dimensional Delaunay triangulations using local transformations. *Comp. Aided Geom. Design*, 8(2):123–142, May 1991.

[132] B. Joe. Delaunay versus max-min solid angle triangulations for three-dimensional mesh generation. *Internat. J. Num. Methods Eng.*, 31(5):987–997, April 1991.

[133] G. Kalai. A subexponential randomized simplex algorithm. In *Proc. 24th Ann. ACM Symp. Theory Comp.*, 475–482, 1992.

[134] N. Karmarkar. A new polynomial-time algorithm for linear programming. *Combinatorica*, 4:373–395, 1984.

[135] L. G. Khachiyan. Polynomial algorithm in linear programming. *USSR Comp. Math. and Math. Phys.*, 20:53–72, 1980.

[136] D. G. Kirkpatrick. Optimal search in planar subdivisions. *SIAM J. Comp.*, 12:28–35, 1983.

[137] D. G. Kirkpatrick and R. Seidel. The ultimate planar convex hull algorithm? *SIAM J. Comp.*, 15:287–299, 1986.

[138] V. Klee and G. J. Minty. How good is the simplex algorithm? In O. Shisha, editor, *Inequalities III*, 159–175. Academic Press, 1972.

[139] R. Klein. *Concrete and Abstract Voronoi Diagrams*, volume 400 of *Lecture Notes in Computer Science*. Springer-Verlag, 1989.

[140] R. Klein and A. Lingas. A linear-time randomized algorithm for the bounded Voronoi diagram of a simple polygon. *Int. J. Comp. Geom. Appl.*, 6:263–278, 1996.

[141] R. Klein, K. Mehlhorn, and S. Meiser. Randomized incremental construction of abstract Voronoi diagrams. *Comp. Geom. Theory Appl.*, 3(3):157–184, 1993.

[142] D. E. Knuth. *Fundamental Algorithms*, volume 1 of *The Art of Computer Programming*. Addison-Wesley, Reading, MA, 1968.

[143] B. Lacolle, N. Szafran, and P. Valentin. Geometric modelling and algorithms for binary mixtures. *Internat. J. Comp. Geom. Appl.*, 4:243–260, 1994.

[144] C. L. Lawson. Software for C^1 surface interpolation. In J. R. Rice, editor, *Mathematical Software III*, 161–194, New York, NY, 1977. Academic Press.

[145] D. T. Lee. On k-nearest neighbor Voronoi diagrams in the plane. *IEEE Trans. Comp.*, C-31:478–487, 1982.

[146] D. T. Lee. On finding the convex hull of a simple polygon. *Int. J. Comp. Inform. Sci.*, 12:87–98, 1983.

[147] D. T. Lee and R. L. Drysdale, III. Generalization of Voronoi diagrams in the plane. *SIAM J. Comp.*, 10:73–87, 1981.

[148] D. T. Lee and F. P. Preparata. Euclidean shortest paths in the presence of rectilinear barriers. *Networks*, 14:393–410, 1984.

[149] D. T. Lee and C. K. Wong. Voronoi diagrams in L_1 (L_∞) metrics with 2-dimensional storage applications. *SIAM J. Comp.*, 9:200–211, 1980.

[150] J. Matoušek. Approximations and optimal geometric divide-and-conquer. In *Proc. 23rd Ann. ACM Symp. Theory Comp.*, 505–511, 1991. Also to appear in *J. Comp. Syst. Sci.*

[151] J. Matoušek. Cutting hyperplane arrangements. *Discrete Comp. Geom.*, 6:385–406, 1991.

[152] J. Matoušek. Efficient partition trees. *Discrete Comp. Geom.*, 8:315–334, 1992.

[153] J. Matoušek. Linear optimization queries. *J. Algorithms*, 14:432–448, 1993.

[154] J. Matoušek. Randomized optimal algorithm for slope selection. *Inf. Proc. Lett.*, 39:183–187, 1991.

[155] J. Matoušek. Reporting points in halfspaces. *Comp. Geom. Theory Appl.*, 2(3):169–186, 1992.

[156] J. Matoušek and O. Schwarzkopf. Linear optimization queries. In *Proc. 8th Ann. ACM Symp. Comp. Geom.*, 16–25, 1992.

[157] J. Matoušek, M. Sharir, and E. Welzl. A subexponential bound for linear programming. In *Proc. 8th Ann. ACM Symp. Comp. Geom.*, 1–8, 1992.

[158] P. McMullen. On zonotopes. *Trans. Amer. Math. Soc.*, 159:91–109, 1971.

[159] P. McMullen and G. C. Shephard. *Convex Polytopes and the Upper Bound Conjecture*. Cambridge University Press, Cambridge, England, 1971.

[160] N. Megiddo. Linear-time algorithms for linear programming in R^3 and related problems. *SIAM J. Comp.*, 12:759–776, 1983.

[161] N. Megiddo. Linear programming in linear time when the dimension is fixed. *JACM*, 31:114–127, 1984.

[162] K. Mehlhorn. *Multi-dimensional Searching and Computational Geometry*, volume 3 of *Data Structures and Algorithms*. Springer-Verlag, Heidelberg, West Germany, 1984.

[163] K. Mehlhorn. *Sorting and Searching*, volume 1 of *Data Structures and Algorithms*. Springer-Verlag, Heidelberg, West Germany, 1984.

[164] K. Mehlhorn, S. Meiser, and C. Ó'Dúnlaing. On the construction of abstract Voronoi diagrams. *Discrete Comp. Geom.*, 6:211–224, 1991.

[165] K. Mehlhorn and S. Näher. Bounded ordered dictionaries in $O(\log \log n)$ time and $O(n)$ space. *Inf. Proc. Lett.*, 35:183–189, 1990.

[166] K. Mehlhorn and S. Näher. Dynamic fractional cascading. *Algorithmica*, 5:215–241, 1990.

[167] K. Mehlhorn, M. Sharir, and E. Welzl. Tail estimates for the space complexity of randomized incremental algorithms. In *Proc. 3rd ACM-SIAM Symp. Discrete Algorithms*, 89–93, 1992.

[168] K. Mehlhorn, M. Sharir, and E. Welzl. Tail estimates for the efficiency of randomized incremental algorithms for line segment intersection. *Comp. Geom. Theory Appl.*, 3:235–246, 1993.

[169] A. Melkman. On-line construction of the convex hull of a simple polyline. *Inf. Proc. Lett.*, 25:11–12, 1987.

[170] D. M. Mount and A. Saalfeld. Globally-equiangular triangulations of cocircular points in $O(n \log n)$ time. In *Proc. 4th Ann. ACM Symp. Comp. Geom.*, 143–152, 1988.

[171] K. Mulmuley. A fast planar partition algorithm, I. In *Proc. 29th Ann. IEEE Symp. Found. Comp. Sci.*, 580–589, 1988.

[172] K. Mulmuley. An efficient algorithm for hidden surface removal. *Comp. Graph.*, 23(3):379–388, 1989.

[173] K. Mulmuley. A fast planar partition algorithm, II. *JACM*, 38:74–103, 1991.

[174] K. Mulmuley. On levels in arrangements and Voronoi diagrams. *Discrete Comp. Geom.*, 6:307–338, 1991.

[175] K. Mulmuley. Randomized multidimensional search trees: Dynamic sampling. In *Proc. 7th Ann. ACM Symp. Comp. Geom.*, 121–131, 1991.

[176] K. Mulmuley. Randomized multidimensional search trees: Lazy balancing and dynamic shuffling. In *Proc. 32nd Ann. IEEE Symp. Found. Comp. Sci.*, 180–196, 1991.

[177] K. Mulmuley. *Computational Geometry: An Introduction Through Randomized Algorithms*. Prentice Hall, Englewood Cliffs, NJ, 1994.

[178] K. Mulmuley and S. Sen. Dynamic point location in arrangements of hyperplanes. In *Proc. 7th Ann. ACM Symp. Comp. Geom.*, 132–141, 1991.

[179] G. M. Nielson and R. Franke. Surface construction based upon triangulations. In R. Barnhill and W. Boehm, editors, *Surfaces in Computer-Aided Geometric Design*, 163–177. North Holland, 1983.

[180] C. Ó'Dúnlaing, M. Sharir, and C. K. Yap. Generalized Voronoi diagrams for a ladder: I. Topological analysis. *Commun. Pure Appl. Math.*, 39:423–483, 1986.

[181] C. Ó'Dúnlaing, M. Sharir, and C. K. Yap. Generalized Voronoi diagrams for a ladder: II. Efficient construction of the diagram. *Algorithmica*, 2:27–59, 1987.

[182] A. Okabe, B. Boots, and K. Sugihara. *Spatial Tessellations: Concepts and Applications of Voronoi Diagrams*. John Wiley & Sons, Chichester, England, 1992.

[183] J. O'Rourke. *Art Gallery Theorems and Algorithms*. Oxford University Press, New York, NY, 1987.

[184] M. H. Overmars and J. van Leeuwen. Maintenance of configurations in the plane. *J. Comp. Syst. Sci.*, 23:166–204, 1981.

[185] J. Pach and M. Sharir. The upper envelope of piecewise linear functions and the boundary of a region enclosed by convex plates: combinatorial analysis. *Discrete Comp. Geom.*, 4:291–309, 1989.

[186] J. Pach, W. Steiger, and E. Szemerédi. An upper bound on the number of planar k-sets. *Discrete Comp. Geom.*, 7:109–123, 1992.

[187] C. H. Papadimitriou and K. Steiglitz. *Combinatorial Optimization: Algorithms and Complexity*. Prentice Hall, Englewood Cliffs, NJ, 1982.

[188] M. Pocchiola and G. Vegter. Computing the visibility graph via pseudo-triangulations. In *Proc. 11th Ann. ACM Symp. Comp. Geom.*, 248–257, 1995.

[189] R. Pollack, M. Sharir, and S. Sifrony. Separating two simple polygons by a sequence of translations. *Discrete Comp. Geom.*, 3:123–136, 1987.

[190] F. P. Preparata. An optimal real-time algorithm for planar convex hulls. *Commun. ACM*, 22:402–405, 1979.

[191] F. P. Preparata and S. J. Hong. Convex hulls of finite sets of points in two and three dimensions. *Commun. ACM*, 20:87–93, 1977.

[192] F. P. Preparata and M. I. Shamos. *Computational Geometry: An Introduction*. Springer-Verlag, New York, NY, 1985.

[193] V. T. Rajan. Optimality of the Delaunay triangulation in R^d. In *Proc. 7th Ann. ACM Symp. Comp. Geom.*, 357–363, 1991.

[194] S. Rippa and B. Schiff. Minimum energy triangulations for elliptic problems. *Computer Methods in Applied Mechanics and Engineering*, 84:257–274, 1990.

[195] S. Rippa. Minimal roughness property of the Delaunay triangulation. *Comp. Aided Geom. Design*, 7:489–497, 1990.

[196] N. Sarnak and R. E. Tarjan. Planar point location using persistent search trees. *Commun. ACM*, 29:669–679, 1986.

[197] J. T. Schwartz and M. Sharir. Algorithmic motion planning in robotics. In J. van Leeuwen, editor, *Algorithms and Complexity*, volume A of *Handbook of Theoretical Computer Science*, 391–430. Elsevier, Amsterdam, 1990.

[198] O. Schwarzkopf. Dynamic maintenance of geometric structures made easy. In *Proc. 32nd Ann. IEEE Symp. Found. Comp. Sci.*, 197–206, 1991.

[199] O. Schwarzkopf. *Dynamic Maintenance of Convex Polytopes and Related Structures*. Ph.D. thesis, Fachbereich Mathematik, Freie Universität Berlin, Berlin, Germany, June 1992.

[200] R. Sedgewick. *Algorithms*. Addison-Wesley, Reading, MA, 1983.

[201] R. Seidel. A convex hull algorithm optimal for point sets in even dimensions. M.Sc. thesis, Dept. Comp. Sci., Univ. British Columbia, Vancouver, BC, 1981. Report 81/14.

[202] R. Seidel. Constructing higher-dimensional convex hulls at logarithmic cost per face. In *Proc. 18th Ann. ACM Symp. Theory Comp.*, 404–413, 1986.

[203] R. Seidel. Backwards analysis of randomized geometric algorithms. Manuscript, ALCOM Summerschool on efficient algorithms design, Århus, Denmark, 1991.

[204] R. Seidel. A simple and fast incremental randomized algorithm for computing trapezoidal decompositions and for triangulating polygons. *Comp. Geom. Theory Appl.*, 1:51–64, 1991.

[205] R. Seidel. Small-dimensional linear programming and convex hulls made easy. *Discrete Comp. Geom.*, 6:423–434, 1991.

[206] M. Sharir. Improved lower bounds on the length of Davenport–Schinzel sequences. *Combinatorica*, 8:117–124, 1988.

[207] M. Sharir and P. K. Agarwal. *Davenport–Schinzel Sequences and their Geometric Applications*. Cambridge University Press, New York, 1995.

[208] M. Sharir and E. Welzl. A combinatorial bound for linear programming and related problems. In *Proc. 9th Symp. Theoret. Aspects Comp. Sci.*, volume 577 of *Lecture Notes in Computer Science*, 569–579. Springer-Verlag, 1992.

[209] G. M. Shute, L. L. Deneen, and C. D. Thomborson. An $O(n \log n)$ plane-sweep algorithm for L_1 and L_∞ Delaunay triangulations. *Algorithmica*, 6:207–221, 1991.

[210] J. Stolfi. Oriented Projective Geometry. In *Proc. 3rd Ann. ACM Symp. Comp. Geom.*, 76–85, 1987.

[211] J. Stolfi. *Oriented Projective Geometry: A Framework for Geometric Computations*. Academic Press, New York, NY, 1991.

[212] B. Tagansky. A new technique for analyzing substructures in arrangements. Unpublished manuscript, Tel-Aviv University, 1994.

[213] R. E. Tarjan and C. J. Van Wyk. An $O(n \log \log n)$-time algorithm for triangulating a simple polygon. *SIAM J. Comp.*, 17:143–178, 1988. Erratum in 17 (1988), 106.

[214] C. Thomassen. The Jordan–Schönflies theorem and the classification of surfaces. *Amer. Math. Monthly*, 99:116–130, 1992.

[215] G. T. Toussaint. A historical note on convex hull finding algorithms. *Pattern Recog. Lett.*, 3:21–28, 1985.

[216] P. van Emde Boas, R. Kaas, and E. Zijlstra. Design and implementation of an efficient priority queue. *Math. Syst. Theory*, 10:99–127, 1977.

[217] C. A. Wang and L. Schubert. An optimal algorithm for constructing the Delaunay triangulation of a set of line segments. In *Proc. 3rd Ann. ACM Symp. Comp. Geom.*, 223–232, 1987.

[218] E. Welzl. Smallest enclosing disks (balls and ellipsoids). In H. Maurer, editor, *New Results and New Trends in Computer Science*, volume 555 of *Lecture Notes in Computer Science*, 359–370. Springer-Verlag, 1991.

[219] A. Wiernik and M. Sharir. Planar realizations of nonlinear Davenport–Schinzel sequences by segments. *Discrete Comp. Geom.*, 3:15–47, 1988.

[220] A. C. Yao. A lower bound to finding convex hulls. *JACM*, 28:780–787, 1981.

Notation

This appendix offers a collection of the notational conventions used in the book. Apart from a few exceptions, we have tried to abide by the following rules: Lower case italic letters represent integers and lower case Greek letters represent real numbers. Upper case letters (whether italic or Greek) represent elementary geometric objects (points, lines, etc.), and upper case script letters represent sets thereof, geometric structures, or data structures. Bold upper case letters represent objects, sets, and structures in projective spaces.

Mathematical symbols

$\mathbb{E}^d$	Euclidean space of dimension d
$\mathbb{H}^d$	hyperbolic space
$\mathbb{I}$	identity matrix
$\mathbb{I}^d$	identity matrix of order d
$\mathbb{P}^d$	projective space of dimension d
$\mathbb{P}^d_o$	oriented projective space
$aff()$	affine hull
$conv()$	convex hull
$int()$	open interior
$\overline{int()}$	closed interior
$ext()$	exterior
$\cdot$	dot product
$|\ |$	cardinality
$\|\ \|$	Euclidean norm
$\lceil\ \rceil$	upper integer part, also called ceiling
$\lfloor\ \rfloor$	lower integer part, also called floor
	Note that for any integer d, $\lfloor d/2 \rfloor + \lceil d/2 \rceil = d$ and that $\lfloor (d+1)/2 \rfloor = \lceil d/2 \rceil$.
$\log^*$	number of iterations of the log function necessary and sufficient to reach a value smaller than 1
$\binom{n}{k}$	number of distinct subsets of size k in a set of size n

Lower case italic

a	number of intersecting pairs of a set of segments
b	maximum number of objects that define a region
	number of distinct intersection points of a set of segments
c	integer constant
d	dimension of the space
$e(\mathcal{C})$	Euler characteristic of the complex $\mathcal{C}$
f	number of faces or facets of a polytope, triangulation, or arrangement
$f_0(r, \mathcal{S})$	expected number of regions defined and without conflict over a sample of size r of a set $\mathcal{S}$
$f_j^i(r, \mathcal{S})$	expected number of regions determined by i objects, defined and with j conflicts over a sample of size r of a set $\mathcal{S}$
g	degree of a vertex
h	height of a tree
i	index variable, number of objects that define a region
j	index variable, number of objects that conflict with a region
k	index variable, rank of item in a list, level in an arrangement
l	index variable, level in an arrangement
m	number of edges of a polytope
$m_k(\mathcal{R}, \mathcal{S})$	moment of order k of a subset $\mathcal{R}$ of $\mathcal{S}$
$m_k(r, \mathcal{S})$	expected moment of order k of a random sample of size r of $\mathcal{S}$
n	number of objects, cardinality of a set of objects
$n_k(\mathcal{C})$	number of k-faces of a complex $\mathcal{C}$
p	probability
$p_{j,k}^i(r)$	probability that a region in $\mathcal{F}_j^i(\mathcal{S})$ be a region defined and with k conflicts over a random sample of size r of $\mathcal{S}$
$p_j^i(r)$	probability that a region in $\mathcal{F}_j^i(\mathcal{S})$ be a region defined and without conflict over a random sample of size r of $\mathcal{S}$
p'^i_j	probability that a region F in $\mathcal{F}_j^i(\mathcal{S})$ be created by a randomized incremental algorithm
$p'^i_j(r)$	probability that a region F in $\mathcal{F}_j^i(\mathcal{S})$ be created at step r of a randomized incremental algorithm
r	size of a sample, radius of a sphere
s	maximum number of alternations of two symbols in a Davenport–Schinzel sequence
t	number of triangles, or of simplices of maximal dimension in a triangulation
x, y, z	index variables for the coordinates in $\mathbb{E}^2$ or $\mathbb{E}^3$
$(x_1, x_2, \ldots, x_d)$	real coordinates of a point in $\mathbb{E}^d$

Upper case italic

A	vertex, point, or vector
AB	segment, vector
$A_1 A_2 \ldots A_{d+1}$	d-simplex
$[A_1 A_2 \ldots A_{d+1}]$	determinant
C	vertex, point, vector, center of a sphere
	cylindrical cell, cell in an arrangement
D	line
E	edge
F	region, face (generally of dimension $d-1$)
G	region, face (generally of dimension $d-2$)
H	hyperplane
H^+, H^-	open half-spaces bounded by H
$\overline{H^+}, \overline{H^-}$	closed half-spaces bounded by H
H^*	pole of a hyperplane H
K	hyperplane, face of dimension $d-3$
L	line
M, N	vertex, point, vector
O	object, origin
OA	point or vector
$O(\ )$	asymptotic upper bound
P, Q, R	vertex, point, vector
P^*	hyperplane polar to a point P
$P^\#$	half-space bounded by the hyperplane polar to a point P
S	object, segment, simplex
T	triangle, simplex of maximal dimension in a complex
U, V, W	vertex, point, vector
$V(M)$	Voronoi cell of a point M.
$V_k(\mathcal{M}_k)$	Voronoi cell of order k of a subset $\mathcal{M}_k$
X, Y, Z	vertex, point, vector
$X(x_1, x_2, \ldots, x_d)$	point or vector in $\mathbb{E}^d$
$\underline{X}(x_1, x_2, \ldots, x_{d+1})$	point or vector in $\mathbb{E}^{d+1}$

Upper case script

$\mathcal{A}$	arrangement
$\mathcal{A}(\mathcal{H})$	arrangement of a set of hyperplanes $\mathcal{H}$
$\mathcal{A}_k$	sub-complex of $\mathcal{A}$ consisting of all cells at level k and of their subfaces
$\mathcal{A}_{\leq k}$	sub-complex of $\mathcal{A}$ consisting of all the faces at level at most k

$\mathcal{B}$	quadric
$\mathcal{C}$	polytope, complex
$\mathcal{D}$	dictionary
$\mathcal{D}(\mathcal{M})$	the Delaunay polytope, dual to $\mathcal{V}(\mathcal{M})$
$\mathcal{D}el(\mathcal{M})$	the Delaunay complex
$\mathcal{D}ec(\mathcal{S})$	vertical decomposition of $\mathcal{S}$
$\mathcal{D}ec_s(\mathcal{S})$	simplified vertical decomposition
$\mathcal{E}$	set of edges, envelope
$\mathcal{F}$	pencil of spheres
$\mathcal{F}(\mathcal{S})$	set of regions defined over a set $\mathcal{S}$ of objects
$\mathcal{F}_0(\mathcal{R})$	set of regions defined and without conflict over a set $\mathcal{S}$ of objects
$\mathcal{F}_j^i(\mathcal{S})$	set of regions defined by i objects and with j conflicts over a set $\mathcal{S}$ of objects
$\mathcal{F}_{\leq k}(\mathcal{S})$	set of regions defined and with at most k conflicts over a set $\mathcal{S}$ of objects
$\mathcal{G}$	graph
$\mathcal{GH}$	horizon graph
$\mathcal{H}$	set of hyperplanes
$\mathcal{I}$	set of indices
$\mathcal{L}$	list, lower envelope
$\mathcal{M}$	set of points
$\mathcal{M}_d$	moment curve in $\mathbb{E}^d$
$\mathcal{P}$	polytope, polyhedron, paraboloid
$\mathcal{P}\#$	polytope dual to a polytope $\mathcal{P}$
$\mathcal{P}ow(\mathcal{S})$	power diagram
$\mathcal{Q}$	priority queue, quadric
$\mathcal{R}$	sample of size r
$\mathcal{S}$	set of objects, of segments, of sites, of spheres
$\mathcal{T}$	triangulation
$\mathcal{T}(\mathcal{M})$	triangulation of a set $\mathcal{M}$ of points
$\mathcal{U}$	universe
$\mathcal{V}(\mathcal{M})$	Voronoi polytope of a set $\mathcal{M}$
$\mathcal{V}or(\mathcal{M})$	Voronoi diagram of a set $\mathcal{M}$
$\mathcal{V}or_k(\mathcal{M})$	Voronoi diagram of order k of a set $\mathcal{M}$
$\mathcal{V}or_+(\mathcal{M})$	Voronoi diagram with additive weights
$\mathcal{V}or_*(\mathcal{M})$	Voronoi diagram with multiplicative weights
$\mathcal{V}or_{L_1}(\mathcal{M})$	Voronoi diagram for the L_1 norm
$\mathcal{V}or_{L_\infty}(\mathcal{M})$	Voronoi diagram for the L_∞ norm

Lower case Greek

$\alpha(n)$	Ackermann function
$\delta_+(X, M_i)$	distance with additive weights
$\delta_*(X, M_i)$	distance with multiplicative weights
$\delta_1(X, M_i)$	L_1 distance
$\delta_\infty(X, M_i)$	L_∞ distance
$\delta_\mathcal{Q}(X, M_i)$	generalized quadratic distance with respect to $\mathcal{Q}$
ε	real constant that can be made arbitrarily small
$\lambda_s(n)$	maximal length of a Davenport–Schinzel sequence
ρ	radius of a sphere
σ	power of the origin with respect to the sphere Σ
ϕ	bijective mapping from spheres in $\mathbb{E}^d$ onto $\mathbb{E}^{d+1}$

Upper case Greek

Γ	curved arc
Δ	sweep line
$\Delta_\mathcal{Q}$	matrix of the quadric $\mathcal{Q}$ in homogeneous coordinates
$\Theta(\ )$	asymptotic equivalent
Σ	chronological sequence, sphere
$\Sigma(C, r)$	sphere of center C and radius r
$\Sigma(X)$	power of a point X with respect to a sphere Σ
$\Omega(\ )$	asymptotic lower bound

Bold upper case

$\boldsymbol{H}$	projective hyperplane
$\boldsymbol{P}$	projective point
$\neg \boldsymbol{P}$	point antipodal to $\boldsymbol{P}$ in an oriented projective space
$\mathcal{P}$	projective polytope
$\mathcal{P}\#$	polar transform of a projective polytope $\mathcal{P}$

Index

Printed in the United States
64624LVS00003B/15-52